Classical and Quantum Mechanics with Lie Algebras

Second Edition

Classical and Quantum Mechanics with Lie Algebras

Second Edition

Yair Shapira

Technion - Israel Institute of Technology, Israel

World Scientific

NEW JERSEY • LONDON • SINGAPORE • BEIJING • SHANGHAI • HONG KONG • TAIPEI • CHENNAI • TOKYO

Published by

World Scientific Publishing Co. Pte. Ltd.
5 Toh Tuck Link, Singapore 596224
USA office: 27 Warren Street, Suite 401-402, Hackensack, NJ 07601
UK office: 57 Shelton Street, Covent Garden, London WC2H 9HE

Library of Congress Control Number: 2026025155

British Library Cataloguing-in-Publication Data
A catalogue record for this book is available from the British Library.

CLASSICAL AND QUANTUM MECHANICS WITH LIE ALGEBRAS
Second Edition

ISBN 978-981-98-3138-8 (hardcover)
ISBN 978-981-98-3247-7 (paperback)
ISBN 978-981-98-3139-5 (ebook for institutions)
ISBN 978-981-98-3140-1 (ebook for individuals)

For any available supplementary material, please visit
https://www.worldscientific.com/worldscibooks/10.1142/14826#t=suppl

Preface

This edition has massive changes (in Parts II, IV-V, VIII, and more), and two new chapters:

- Feynman diagrams and the Killing form.
- Spin and polarization.

How to study physics? Can this be done with no math at all? No! You can't escape: you need a little bit of math. After all, this is the language of nature. Still, keep math to its easy face: only algebra, and only little calculus. With this background, physics gets as easy as ever. Math will translate into a vivid physical picture.

How to see physics in its full glory? This book offers a new approach: use simple tools from algebra and geometry, and avoid heavy analysis. This will let physics flow easy: from Newtonian to quantum mechanics and relativity.

Indeed, Part I uses elementary calculus to introduce Newtonian mechanics. On top of this, Part II introduces a key concept in classical mechanics: stability. This marks the start of modern physics: you can never predict the future! This principle will lead to quantum mechanics and relativity.

Next, Part III gives a little more background in algebra: the binomial formula, useful in statistical mechanics. Part IV introduces special relativity from an algebraic-geometrical point of view.

Next, Parts V-VI use elementary linear algebra to introduce quantum physics and chemistry, including energy levels and spin. Finally, Parts VII-VIII use just mathematical induction to introduce Lie algebras, and even decompose them in terms of simple ideals, useful in quantum and Hamiltonian mechanics.

The book could be used as a textbook in three undergraduate courses (two in physics, and one in math):

- Introduction to Newtonian mechanics and special relativity (Parts I and IV).
- Introduction to quantum physics and chemistry (Parts V-VI and VIII).
- Introduction to Lie algebras and their applications (Parts VII-VIII).

Each chapter ends with a lot of relevant exercises and their solutions. The exercises are not boring at all: they teach new material, step by step, exercise by exercise, in an active pedagogical way.

The only prerequisites are linear algebra and calculus. To make the book self-contained, this background is available in the appendix.

Yair Shapira
Haifa, Israel

Contents

Part I

Introduction to Newtonian Physics

Part I: Introduction to Newtonian Physics

To learn Newtonian mechanics, we only need elementary calculus (available in the appendix). But this is not the end of it: later on, we'll go into modern physics too. This will give us a yet better understanding of nature.

Thanks to geometry and calculus, we'll be able to study motion and dynamics. For this, we need an important physical quantity: speed (or velocity).

Speed is not so innocent as it looks. On the contrary: it is quite tricky and mysterious. After all, it is not constant, but changes all the time. To see how, we must differentiate it, and obtain its derivative: acceleration. This will lead us to Newton's law: force is proportional to both mass and acceleration.

Moreover, we can also integrate force along a path, and obtain a new important quantity: work. Better yet, work has a yet more fundamental face: energy, written as a sum: kinetic plus potential energy.

Energy is quite useful: it tells us a lot about the physical process, even without knowing its explicit form. This helps solve practical problems, especially in gravity.

To describe motion more completely, we actually need three fundamental concepts: energy, momentum, and angular momentum. Energy tells us the capacity to do work. Momentum, on the other hand, tells us how difficult it is to stop the motion (or change its direction). Angular momentum, on the other hand, tells us how difficult it is to stop a rotation. All three are conserved: in a closed system, they never change or get lost. This may tell us a lot about the dynamics, even without knowing it explicitly.

Later on, we'll put this in a deeper context: Hamiltonian mechanics. This will give physics not only geometrical but also an algebraic face: Lie algebras.

1

Introduction to Newtonian Mechanics

To see motion, we need a basic tool: speed (or velocity). By differentiating it, we'll get another important quantity: acceleration. Force is proportional to both mass and acceleration. This will help solve for the dynamics in its explicit form. These are the basic ingredients in the laws of nature.

Better yet, thanks to integration, we can even define a yet more fundamental quantity: energy. This is Newtonian mechanics: a comprehensive theory to explain nature [14, 33, 44].

But this theory has its own limits: in astronomic macroscale, it is inferior to Einstein's relativity. In subatomic microscale, on the other hand, it is inferior to quantum mechanics. We'll come to this later. Fortunately, it is still good enough in intermediate scales, used in practical engineering problems. Moreover, it offers new conceptual tools, and opens horizons towards a better understanding of nature.

1.1 Motion: Zeno's Paradox

1.1.1 Speed – Velocity

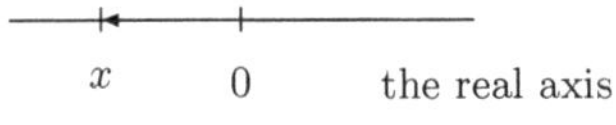

Fig. 1.1. The infinite real axis: $\mathbb{R}$. The arrow leading from 0 to $x < 0$ makes angle π (180°) with the positive part of the real axis.

To understand motion, we need a new quantity: speed (velocity). For this purpose, assume that a particle moves along the real axis $\mathbb{R}$ (Figure 1.1). At time $t \geq 0$, the particle is at $x \equiv x(t)$. Thus, x depends on t, whereas t depends on nothing. This way, t is the independent variable, and $x \equiv x(t)$ is a function of it.

At the initial time $t = 0$, assume that the particle lies at the origin:

$$x(0) = 0.$$

(After all, in theory, we could place the origin wherever we like.) Then, the particle starts to move. At time $t > 0$, it reaches the point $x(t)$ on the real axis.

The velocity is a ratio: the distance that the particle makes per second. In general, the velocity is nonconstant: it may change too. In other words, the velocity is a function of t too.

For instance, at the beginning, the particle may be at rest: zero velocity. Then, it may accelerate, and pick more and more speed. Still, there is a paradox here: since it had zero velocity, how could it start moving in the first place? This is Zeno's paradox.

1.1.2 Average vs. Momentary Velocity

We must therefore be careful. First, define the *average* velocity. For this purpose, pick some parameter $h > 0$. Then, define the average velocity from time t to time $t + h$:

$$\frac{x(t+h) - x(t)}{h}.$$

Still, this is not good enough: we'd like to have not only the average velocity in the entire time interval $[t, t+h]$ but also the *momentary* velocity at the precise time t. Fortunately, for this purpose, we could use a limit process.

How to define momentary velocity? At first glance, this might seem impossible. After all, at the precise time t, there is no motion at all, so how could velocity be relevant? This is Zeno's paradox again.

How to "solve" Zeno's paradox? In the context of quantum mechanics and general relativity, use Zorn's lemma (Chapters 11-12 in [48]). Here, on the other hand, we'll look at it both geometrically and analytically.

1.1.3 Zeno's Paradox in Geometrical Terms

Zeno's paradox also has a geometrical version: how could the real axis be covered by dimensionless points? After all, each individual point has no size at all, so even infinitely many are not enough!

The ancient Greeks were puzzled with Zeno's paradox. Why? Because they avoided the infinity. Indeed, in Euclidean geometry, a line is not *made* of points. On the contrary: it is a fundamental object on its own right, defined implicitly by a few plausible axioms.

1.1.4 Line and Point in Euclidean Geometry

The line may contain many points: every two distinct points on it must have a third point in between. Still, these points never *make* the original line. They have a different nature, and remain isolated and discrete. As such, they could never produce the continuous line in the first place. This is why Zeno's paradox is so relevant: there is no way to "jump" from point to point to model a physical motion. With no analytic terms at hand, the original language of Euclidean geometry seems too poor.

Moreover, in Euclidean geometry, there is no notion of length or dimension at all ([15] and Chapter 6 in [46]). An interval is not "one-dimensional" in any way, and has no length. In fact, intervals are never measured by any length unit. Length is just a relative property, useful to compare intervals: one interval could be longer than another, or shorter.

1.1.5 Line and Point in Analytical Geometry

In analytical geometry, on the other hand, we can already use infinite sets. The line is no longer interpreted as a pure abstract object, but as a set of those points that satisfy some linear equation. Thus, the line is a high-level object, made of lower-level objects: points. This is how the line indeed takes its one-dimensional nature, ready to model a one-dimensional motion. Later on, in Chapter 5, the concept of dimension will be defined in some more detail.

1.1.6 Line and Point in Set Theory

How many points are needed to make a line? Thanks to set theory, we already know the answer: $\aleph$. This kind of infinity is bigger than the discrete infinity, used to count the natural numbers:

$$\aleph > \aleph_0.$$

$\aleph$ is big enough: it helps introduce a new dimension, with a substantial size or length. This is how points of zero measure could join and make a new continuous interval of a positive measure.

Why isn't $\aleph_0$ big enough? Because an infinite sequence of points, placed one by one in a row, could never make a continuous line. Fortunately, we can still use "more" points: a "greater" infinity: $\aleph$.

More precisely, why is $\aleph_0$ too small? Because the natural numbers make an ordinal: every subset must start somewhere. $\aleph$, on the other hand, is not an ordinal. Indeed, in the real axis, there are many subsets that start nowhere. For example, let t be fixed, and look at an open semiaxis, containing those points greater than t:

$$(t, \infty) \equiv \left\{ \tilde{t} \mid t < \tilde{t} < \infty \right\} \subset \mathbb{R}.$$

Since t itself is excluded, this is an open subset of $\mathbb{R}$: it has no start at all! Thanks to this, we can now use a one-sided limit process, and approach t from the right.

1.1.7 Limit Process

How to do this? Easy: let $h > 0$ be a positive parameter. This way, $t + h$ already belongs to our semiaxis, as required. Now, make h smaller and smaller. This way, $t + h$ is still in our semiaxis, and gets ever so close to t. This way, $t + h$ indeed approaches t from the right. This is denoted by

$$t + h \to_{h \to 0^+} t^+.$$

Likewise, t could also be approached from the left. This will be useful to define the velocity at t. Before diving into this, let's talk a little more about words and their meaning.

1.2 Words: What Are They?

1.2.1 Concept and Word

The ancient Greeks were not yet ready to use infinite sets or analytical geometry. Why? Because, in their philosophy, a concept or idea has a life of its own: it mustn't depend on any other object of any other kind. For example, the line mustn't depend on a point.

Still, the ancient Greeks managed to "solve" Zeno's paradox in their own way: the line has a "soul" or a nature of its own. This is what makes it more than just a collection of points. It has a new meaning, planted into it by the "god" of linearity. This is what makes it not just the sum of its own ingredients, but more.

How would a dog view a line? Well, a dog can understand only a concrete line: a long straight path to run freely on. We humans, on the other hand, can speak. For this purpose, we must use words, which describe not only concrete but also abstract ideas. After all, the word "line" doesn't necessarily stand for a straight path: it could stand for any other straight thing.

1.2.2 Plato's Philosophy

Thus, the word "line" is an atom word. Where does it come from? There are two approaches. In Plato's theory, every term must have its own private "god" (or source) to define it uniquely, and make it what it is: not only concrete instances but also a general idea, with its own characteristics and features.

This is the idea of being linear. In Plato's theory, this must be represented by the "god" of linearity, who gives each particular line its own linear nature. This is how the line can indeed model a one-dimensional physical motion. Zeno's paradox gets solved!

1.2.3 Aristotle's Philosophy

In Aristotle's theory, on the other hand, there is no such god. Every term has its own private nature, built in it. For example, the nature of the line is to be straight and long, to allow one-dimensional motion along it.

We can now see the mathematical roots behind Aristotle's theory. Indeed, analytical geometry is more free and relative: everyone is free to draw his/her own coordinate system. No need to wait for anyone to tell us where the origin is, or what a line is!

Plato and Aristotle answered Zeno's paradox philosophically. Newton, on the other hand, answered it analytically.

1.2.4 The Language of Analysis

To answer Zeno's paradox, Newton used a limit process. This enriched our language, giving us a new word: derivative (Chapter 26).

1.3 Velocity and Acceleration

1.3.1 Average Velocity and Its Limit

Thanks to the derivative, the momentary velocity can now be defined as the limit of the average velocity in shorter and shorter intervals:

$$v(t) \equiv \lim_{h \to 0} \frac{x(t+h) - x(t)}{h} = x'(t) = \frac{dx}{dt}(t).$$

Thus, velocity is the derivative of position. This answers Zeno's paradox once and for all.

In reality, however, the particle may move not only in one but also in three dimensions. This way, $x(t)$ could be not only a scalar but also a vector with three components:

$$x(t) = \begin{pmatrix} x_1(t) \\ x_2(t) \\ x_3(t) \end{pmatrix}.$$

In this case, the particle may move in every spatial direction. This is where the term velocity is better than speed. Indeed, speed is just a scalar, whereas velocity could also be a vector:

$$v(t) \equiv \begin{pmatrix} v_1(t) \\ v_2(t) \\ v_3(t) \end{pmatrix} \equiv \begin{pmatrix} x_1'(t) \\ x_2'(t) \\ x_3'(t) \end{pmatrix}.$$

For simplicity, however, we stick here to our original particle, confined to one dimension only. (After all, in theory, we could always pick our axis system to align with the motion.) This way, for now, velocity is the same as speed: a scalar.

1.3.2 Laws of Nature

To govern motion, nature uses three laws. The first law says that, with no force, the particle would move undisturbed, at a constant velocity:

$$v(t) \equiv v(0) \quad (t \geq 0)$$

(Figure 1.2). Why is this law first? Because this is an elementary case.

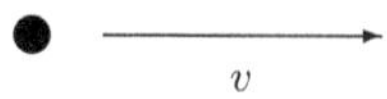

Fig. 1.2. Newton's first law: if there is no force, then the velocity v remains constant.

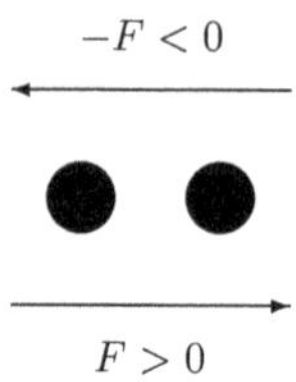

Fig. 1.3. Newton's third law: the left bullet applies force F upon the right bullet. As a reaction, the right bullet applies force $-F$ back upon the left bullet.

Later on, we'll define momentum as a product: mass times velocity. How could an outer force affect the particle? In one way only: changing its momentum. For a particle of constant mass, this means changing the velocity. This is indeed acceleration. This is the second law.

Finally, the third law talks about action and reaction: if a force is applied to our particle by some other particle that hits it, then our particle reacts symmetrically: it applies the same amount of force back upon the other particle, in the opposite direction (Figure 1.3). This way, the entire system remains in balance: the forces cancel out, and sum to zero. After all, no outer force is applied!

1.3.3 Acceleration

What is acceleration? It tells us how the velocity changes per second. In other words, it is the derivative of the velocity with respect to time:

$$a \equiv a(t) \equiv v'(t) \equiv \frac{dv}{dt}(t).$$

This way, the acceleration is momentary too: it is nonconstant, and may change in time. In the beginning, for example, you might want to put your foot on the gas, and accelerate a lot. Once you gained a good velocity, you might want to relax, and stop accelerating, keeping velocity the same. Finally, you need to break, have a negative acceleration, and make a complete stop.

This is why the acceleration $a \equiv a(t)$ is a function of time in its own right. Now, recall that velocity is the derivative of position. Thus, acceleration is actually the second derivative of position (with respect to time):

$$a(t) \equiv v'(t) \equiv x''(t) \equiv \frac{d^2x}{dt^2}(t).$$

1.3.4 Towards Motion in 3-D

In the real 3-D world, x, v, and a have three components:

$$a(t) \equiv \begin{pmatrix} a_1(t) \\ a_2(t) \\ a_3(t) \end{pmatrix} \equiv \begin{pmatrix} v_1'(t) \\ v_2'(t) \\ v_3'(t) \end{pmatrix} \equiv \begin{pmatrix} x_1''(t) \\ x_2''(t) \\ x_3''(t) \end{pmatrix}.$$

This could be quite complicated: the velocity vector can now change not only magnitude but also direction, leading the particle along a complicated curve (trajectory) in 3-D.

Here, however, we stick to our original 1-D model, in which the velocity (and the acceleration) is just a scalar, not a vector. This way, the velocity may change in two ways only: either increase (positive acceleration), or decrease (negative acceleration).

1.4 Force: Derivative of Momentum

1.4.1 Mass

Here, we assume that our particle has a constant mass:

$$m \equiv \text{const.}$$

Still, this is not a must. In general, the mass may depend on time: $m \equiv m(t)$. For example, think of a spaceship, traveling in space. During its journey, it burns fuel, and loses mass. This way, its mass has a negative time derivative: $m' < 0$. Fortunately, we don't look at such a complicated case here, but only at a constant mass that never changes: $m' \equiv 0$.

1.4.2 Inertial Mass

The mass m is also called inertial mass: the amount by which the particle resists any change to its original velocity. Thanks to this resistance, a force is needed: with no force, the velocity would never change.

Still, the full resistance comes not from mass but from a more fundamental property: momentum. This really tells us how the particle resists any change. This is why, with no external force, momentum could never change. This is indeed a conservation law. Later on, we'll see this from a geometrical point of view.

1.4.3 Momentum

The momentum tells us how difficult it is to change the course of the system. In our case, it tells us how difficult it is to make the particle change its original motion.

To do this, the momentum should be proportional to both mass and velocity. For this, it should be their product:

$$p \equiv mv.$$

Again, in general, this is a function of time:

$$p(t) \equiv m(t)v(t).$$

In this form, the momentum is nonconstant, and may change in time. How fast? This is determined by the force.

1.4.4 Force: Derivative of Momentum

The force tells us how fast the momentum changes. In other words, the force is the derivative of the momentum. In fact, at each individual time t, the force is

$$F \equiv p' = (mv)' = m'v + mv' = m'v + ma.$$

1.4.5 The Second Law of Nature

Could this be simplified yet more? Recall that our particle has a constant mass:

$$m \equiv \text{const.} \quad \text{or } m' \equiv 0.$$

Thus, the force takes the simpler form of

$$F = m'v + ma = ma.$$

This is Newton's second law (Figure 1.4): at each individual time t,

$$F = ma = mv' = mx''$$

(where m is the constant mass of the particle, and F is the force applied to it). What is the physical meaning of this? Well, the force is linearly proportional to both mass and acceleration. For example, assume that the force is applied from the left: $F > 0$. (Say, our particle is hit by some other particle from the left.) In this case, the velocity of our particle increases, and the acceleration is positive ($a > 0$). If, on the other hand, the force comes from the right ($F < 0$), then the velocity decreases, and the acceleration is negative ($a < 0$).

Assume now that there is no force at all. In this case, momentum never changes: it is conserved. Let's see this geometrically.

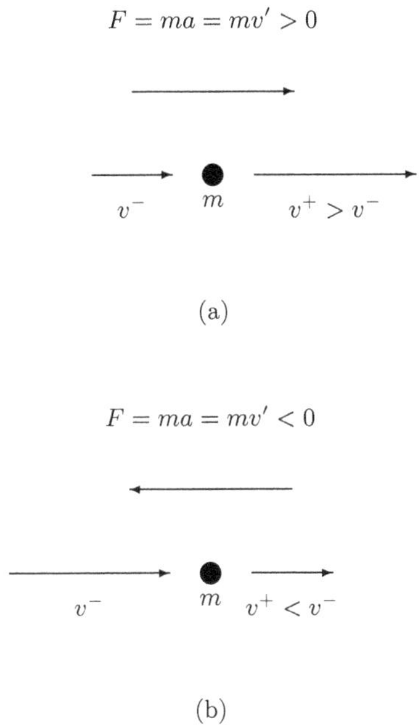

Fig. 1.4. Newton's second law: the force $F(t)$ is linearly proportional to both the mass m and the acceleration $a(t) \equiv v'(t) \equiv x''(t)$. This could be either positive, increasing the velocity from v^- to v^+ (a), or negative, decreasing the velocity from v^- to v^+ (b).

1.5 Momentum and Its Conservation

1.5.1 Elastic Particles

Consider a system of M particles, with velocities $v_i(t)$ and masses m_i $(1 \le i \le M)$. They may hit each other, and apply forces upon one another. Still, they are completely elastic: a particle gets hit, changes shape only temporarily, and soon gets to its original shape back again. In the process, it applies the same amount of force (in the opposite direction) back to the particle that hit it.

This way, the forces never make any permanent (plastic) change in any particle. On the contrary: the forces reflect (or return) fully, as in Newton's third law (Figure 1.3). No friction or heat is generated. This could model balls on a billiards (or pool, or snooker) table.

1.5.2 Closed System

The system is closed: isolated from the outer world: no external force is applied from the outside. Thus, the total force in the entire system must be zero:

$$0 = F = p' = \left(\sum_{i=1}^{M} m_i v_i\right)' = \sum_{i=1}^{M} m_i v_i' = \sum_{i=1}^{M} m_i a_i.$$

Now, let's look at two discrete times: the beginning, and the end. Let the ith particle have velocity v_i^- at the initial time, and v_i^+ at the end. Thanks to the above equation, the total momentum remains the same:

$$\sum_{i=1}^{M} m_i v_i^- = \sum_{i=1}^{M} m_i v_i^+.$$

1.5.3 Conservation of Momentum

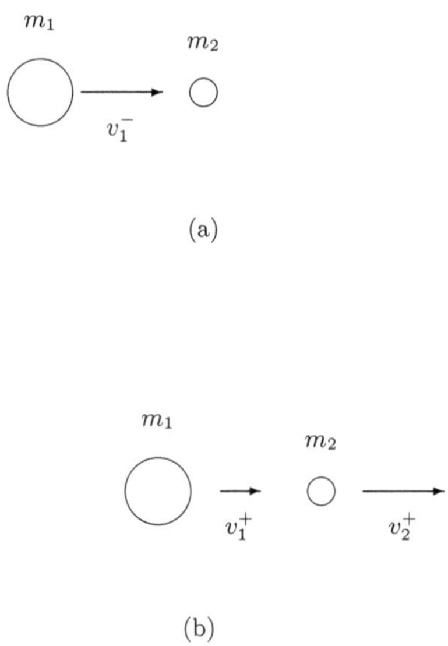

Fig. 1.5. Conservation of momentum: before the collision, the little ball on the right is at rest (no momentum at all). This way, the momentum $m_1 v_1^-$ is concentrated in the big ball only (a). After the collision, on the other hand, the original momentum splits into two parts: $m_1 v_1^+$ in the big ball, and $m_2 v_2^+$ in the little ball (b). This way, the velocity of the big ball decreases from v_1^- to v_1^+. Still, momentum is never lost, but only transferred to the little ball, to give it the new velocity v_2^+.

This is indeed conservation of momentum. In Figure 1.5, this is illustrated for two particles ($M = 2$):

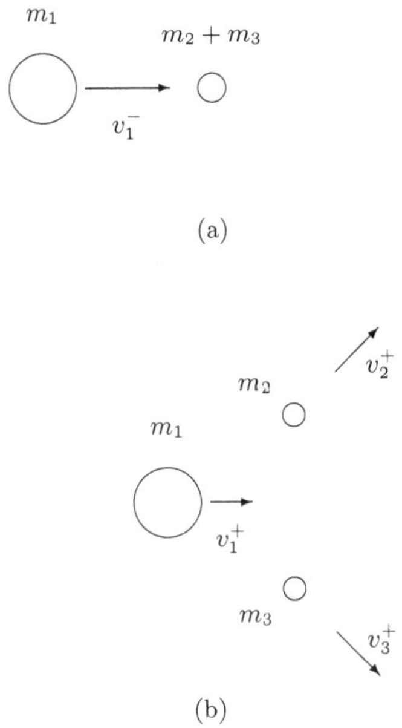

Fig. 1.6. At the collision, the little ball on the right may even break into two smaller balls. This way, the original momentum $m_1 v_1^-$ in (a) splits into three parts: $m_1 v_1^+$ is the momentum left in the big ball, whereas $m_2 v_2^+$ and $m_3 v_3^+$ are the new momenta introduced in the new little balls (b). The horizontal momentum remains positive, and the vertical momentum remains zero.

$$m_1 v_1^- + m_2 v_2^- = m_1 v_1^+ + m_2 v_2^+ .$$

Furthermore, some particles may collide too hard, and even break and split, so the total number of particles may increase from M to $N > M$:

$$\sum_{i=1}^{M} m_i v_i^- = \sum_{i=1}^{N} m_i v_i^+$$

(Figure 1.6).

1.5.4 Equilibrium: Zero Total Momentum

Consider now a static football that lies at rest. Assume that the temperature is above the absolute zero (−273°C). This means that the molecules in the football are never at a complete rest: they move all the time, and have nonzero momenta. Furthermore, they often hit each other, and change direction. How come the football stands still? Is there no momentum?

Well, on average, there is no momentum at all. After all, the molecules move at random, towards no specific direction. Later on, we'll call this Brownian motion (Sections 6.4.1–6.4.2). Therefore, their momenta cancel each other, and sum to zero in each and every direction.

Even when molecules hit each other, their momenta are never lost, but only transferred to one another, as in Figure 1.5 or 1.6. For this reason, the momentum of each individual molecule remains random, with no favorite direction, and the total momentum remains zero. This is why the football remains at rest, at least until someone comes and kicks it. This opens the system up: it introduces a new force from the outside, and gives all molecules a new (uniform, nonrandom) velocity and momentum, in the same spatial direction.

1.6 Harmonic Oscillator

1.6.1 A Spring

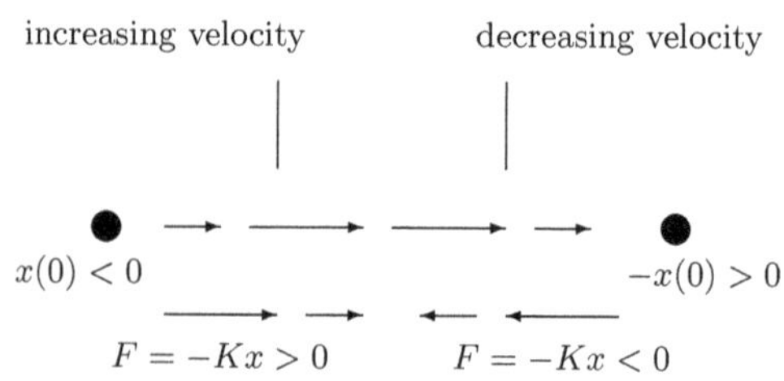

Fig. 1.7. Hooke's law: in the left part of the x-axis, the spring pulls the particle rightwards at force $F = -Kx > 0$. Once the particle passes the origin on its way rightwards, the spring pulls it back leftwards, at force $F = -Kx < 0$.

Consider now a more specific problem. Assume that the particle is connected to the origin by a spring (or an elastic wire, made of rubber). This is a harmonic oscillator.

At the initial time $t = 0$, the particle is placed at some negative point:

$$x(0) < 0$$

(to the left of the origin on the real axis). At $t = 0$, the particle is still held firmly, so it has no initial velocity:

$$v(0) = 0.$$

Suddenly, we let the particle go free (Figure 1.7). This way, the spring applies a positive force to it, to make it move rightwards.

For simplicity, assume no friction: the force generate no heat, and only increases the velocity. This is a fair assumption: the friction is often negligible.

At the initial time $t = 0$, the spring is well drawn, so the force is strong. As time goes by, though, the spring gets loose, so the force gets weaker and weaker. At some unknown time, the particle reaches the origin. At this point, the spring gets so loose that it applies no force at all any more. Still, the velocity accumulated so far makes the particle keep going rightwards, pass the origin, into the right part of the real axis.

There, the spring applies a negative force, in an attempt to pull the particle back to the origin. This also introduces a negative acceleration that gradually reduces the velocity. Thanks to symmetry, the rightmost point that the particle can reach is $-x(0) > 0$.

At this point, the velocity decreased to zero. Still, the negative force doesn't stop: it keeps reducing the velocity yet more, and make it even negative, pulling the particle leftwards, back to the origin.

Once the particle passes the origin on its way leftwards, the force gets positive again, and the velocity increases again. By the time the particle gets to $x(0) < 0$ back again, the velocity gets to its initial value back again: zero. This completes one cycle (period). The process can now start all over again.

1.6.2 Hooke's Law

This long story is summarized in Hooke's law:

$$F(t) = -Kx(t),$$

where K is a positive constant, dependent only on the material the spring is made of. At the initial time $t = 0$, F is positive, so it forces the particle to start moving rightwards:

$$F(0) = -Kx(0) > 0$$

(Figure 1.7). As time goes by, $x(t)$ decreases (in magnitude), and $F(t)$ decreases too. Once the particle reaches the origin, both x and F vanish. Still, the velocity accumulated so far moves the particle ahead, into the right part of the real axis. There, $x(t) > 0$, so F gets negative, and pulls the particle back leftwards.

Still, it takes some time to get the velocity to vanish. This happens only at the rightmost point: $-x(0) > 0$. At this point, the velocity gets negative. Why? Because the negative force wins: it reverses the motion, forcing the particle to move leftwards rather than rightwards.

Once the particle passes the origin on its way leftwards, $x(t)$ gets negative again, and $F(t)$ gets positive again. This way, the force produces a positive acceleration again, slowing the particle down (on its way leftwards), to arrive safely back home: to the initial position $x(0) < 0$. The entire period is complete, ready to start all over again.

All this story is now summarized in just one formula: Hooke's law. This will give us our solution: $x(t)$ (at each and every time t). How? Combine Hooke's law with Newton's second law. For this, we must first get our physical units right.

1.6.3 Physical Units

In math, we often use pure numbers. In physics, on the other hand, numbers are not good enough. To measure a physical quantity, they must have a physical unit.

One popular system of units is cgs (centimeters, grams, seconds). These are small units: length is measured in centimeters, weight in grams, time in seconds, velocity in centimeters per second (or cm/s), acceleration in (cm/s)/second (or cm/s^2), etc.

In an equation, both sides must have the same physical unit. In Newton's second law, for instance, force must be measured in weight times acceleration, or g·cm/s^2. This unit must also be used in Hooke's law. For this purpose, K must have a new unit: g/s^2. This way, both sides have the same unit: g·cm/s^2.

The cgs system is often used in science. A more practical system, on the other hand, is MKS: meter, kilogram, second. These are bigger, hence more practical.

1.6.4 The Differential Equation

We are now ready to combine Hooke's law with Newton's second law:

$$-Kx(t) = F(t) = mx''(t) = m\frac{d^2x}{dt^2}(t),$$

or

$$x''(t) = -\frac{K}{m}x(t).$$

This is a differential equation. The unknown solution is $x(t)$: the position of the particle on the real axis (at time t).

Does this make sense? Let's compare the physical units. On the left, $x(t)$ is differentiated twice (with respect to time). So, the unit is cm/s^2. On the right, the unit is cm/s^2 as well. After all, K has unit g/s^2 (Section 1.6.3).

1.6.5 Frequency

Both K and m are positive. So, we can now introduce a new quantity: frequency:

$$\omega \equiv \sqrt{\frac{K}{m}}.$$

(Why is it called frequency? In a moment, we'll see why.) Thanks to it, our differential equation simplifies to read

$$x''(t) = -\omega^2 x(t).$$

Let's solve it for $x(t)$.

1.6.6 How to Solve The Differential Equation?

Look at our differential equation. What is the solution? Well, we already know how it should look like:

$$x(t) = a\sin(\omega t) + b\cos(\omega t),$$

where a and b are some parameters (to be specified later). Indeed, the first derivative is

$$v(t) = x'(t) = \frac{dx}{dt}(t) = \omega\left(a\cos(\omega t) - b\sin(\omega t)\right).$$

Furthermore, the second derivative is

$$a(t) = v'(t) = \frac{dv}{dt}(t) = \omega^2\left(-a\sin(\omega t) - b\cos(\omega t)\right) = -\omega^2 x(t),$$

as required. Actually, this is a family of solutions. Indeed, for any a and b, this is a legitimate solution. How to specify a and b uniquely? For this purpose, we must use more physical information.

1.6.7 Initial-Value Problem

How to specify a and b? Use the initial state. Initially, the particle is at $-L$ (for some $L > 0$):

$$-L = x(0) = a\sin(0) + b\cos(0) = b.$$

This uncovers b. How to uncover a? Well, the initial velocity vanishes:

$$0 = v(0) = \omega\left(a\cos(0) - b\sin(0)\right) = \omega a.$$

Thus,

$$a = 0 \quad \text{and} \quad b = -L.$$

So, our unique solution is

$$x(t) = -L\cos(\omega t).$$

Thus, thanks to our initial data, our differential equation makes a new initial-value problem, with the unique solution: $x(t)$. Let's use it.

1.6.8 The Period Time

Our period time is:

$$T \equiv \frac{2\pi}{\omega}.$$

Indeed, after time T, the particle is back at:

$$x(T) = -L\cos(\omega T) = -L\cos(2\pi) = -L = x(0).$$

Furthermore, after time T, the velocity is again

$$v(T) = x'(T) = \omega L\sin(\omega T) = \omega L\sin(2\pi) = 0 = v(0).$$

So, the particle is ready to restart the next cycle all over again, and so on.

1.6.9 Muzzle Velocity

What happens after a quarter of a period, at time

$$\frac{T}{4} = \frac{\pi}{2\omega}?$$

At this time, the particle arrives at the origin:

$$x\left(\frac{T}{4}\right) = -L\cos\left(\omega\frac{T}{4}\right) = -L\cos\left(\frac{\pi}{2}\right) = 0.$$

Then, its velocity is at its maximum:

$$v\left(\frac{T}{4}\right) = x'\left(\frac{T}{4}\right) = L\omega\sin\left(\omega\frac{T}{4}\right) = L\omega\sin\left(\frac{\pi}{2}\right) = L\omega.$$

Assume now that, at time $T/4$, the spring is suddenly removed. This models a slingshot.

In this case, our solution is valid until time $T/4$ only. Afterwards, there is no force any more. As in Newton's first law (Figure 1.2), the particle continues to fly freely rightwards, at the same speed: the muzzle speed.

As a matter of fact, to calculate the muzzle speed, there is no need to solve the differential equation at all! To see this, we must first introduce a new fundamental concept: energy.

1.7 Work and Energy

1.7.1 Work

How to calculate the period? So far, we did this the hard way: we solved our differential equation explicitly. Still, is there a better way? After all, in many physical phenomena, the differential equation is too difficult to solve!

Fortunately, even before being solved, our mathematical model may still put us in the right direction. No need to solve: just read between the lines! For this, we need two new fundamental concepts: work and energy.

How does F depend on t? Not directly, but indirectly: through $x(t)$:

$$F \equiv F(x) \equiv F(x(t)).$$

On the real axis, consider a path: an interval $[a, b] \subset \mathbb{R}$. From a to b, how much work does the force do? This is the work: force times length (proportional to both force and length). Still, from a to b, the force may change. Thus, the work should better be the integral:

$$\int_a^b F(x)dx$$

(Section 26.13.1). For instance, to get the particle from $-L$ to 0, the spring (in Sections 1.6.1–1.6.2) does work

$$\int_{-L}^{0} F(x)dx = -K\int_{-L}^{0} xdx = -K\frac{x^2}{2}\bigg|_{-L}^{0} = -\frac{K}{2}(0^2 - L^2) = \frac{KL^2}{2}.$$

1.7.2 Energy

What is energy? It mirrors work. More precisely, what is potential energy? The ability to do work.

At time $t = 0$, the particle can still travel to 0 (the reference point). Thus, it has potential energy $KL^2/2$ (required to go to 0). Upon arriving at 0, its potential energy was transformed into kinetic energy (incorporated in its new velocity and momentum).

1.8 First Law of Thermodynamics

1.8.1 Energy in a Closed System

In a closed system, energy is never lost: it is only transformed from one form to another. This is the first law of thermodynamics: conservation of energy. Later on, we'll illustrate this geometrically.

1.8.2 From Potential to Kinetic Energy

How to use potential energy? To see this, differentiate the composite function v^2 (with respect to t):

$$\frac{d\left(v^2\right)}{dt} = 2v(t)v'(t)$$

(Section 26.7.2). Better yet, divide this by 2:

$$\left(\frac{v^2(t)}{2}\right)' = v(t)v'(t).$$

Thus, $v^2/2$ is the primitive function (indefinite integral) of $v'v$ (with respect to t). Let's use this to get rid of t altogether, and have work (and energy) in terms of v alone. After all, v is more physical. In fact, v tells us what happens. t, on the other hand, only tells us when. But who cares?

Consider a general force F, acting from time $t = 0$ to time $s > 0$. Under it, the particle travels from $x(0)$ to a new position: $x(s)$. Thanks to Newton's second law (Figure 1.4), the work along this path is:

$$\begin{aligned}
\int_{x(0)}^{x(s)} F(x)dx &= \int_0^s F(x(t))dx(t) \\
&= \int_0^s F(x(t))\frac{dx(t)}{dt}dt \\
&= \int_0^s F(x(t))v(t)dt \\
&= m\int_0^s a(t)v(t)dt \\
&= m\int_0^s v'(t)v(t)dt
\end{aligned}$$

$$
\begin{aligned}
&= \frac{m}{2} v^2(t) \Big|_0^s \\
&= \frac{m}{2} \left(v^2(s) - v^2(0) \right) \\
&= \frac{m}{2} v^2(s)
\end{aligned}
$$

(assuming that the initial velocity was zero). What does this mean physically? On the left-hand side, we have the work that the force invested in the particle. What happened to this work? It wasn't lost, but transformed into kinetic energy (on the right-hand side).

1.8.3 Kinetic Energy at The Muzzle

This is true for any force. For instance, look at our slingshot. Assume that, at time s, the particle arrives at the muzzle. Actually, we already know that

$$
s = \frac{T}{4}.
$$

Still, let's pretend we don't. (After all, we want to forget about time, and focus on energy.) Thanks to Section 1.7.1,

$$
\frac{KL^2}{2} = \frac{mv^2(s)}{2},
$$

or

$$
v^2(s) = \frac{KL^2}{m} = \omega^2 L^2.
$$

This gives us the same muzzle velocity as in Section 1.6.9:

$$
v(s) = L\omega.
$$

But here, we deduced this from energy alone, without solving the differential equation at all!

1.9 Hamiltonian and Phase Plane

1.9.1 The Phase Plane

Here, we only ask "what," not "when." We only ask what happened, but don't care when. This is why we don't care what s is. We are only interested in energy (transformed from potential to kinetic).

Thus, we better drop t altogether, and focus on the physical state (or phase) of the particle: its location x, and its velocity v. Both may depend on time, but we don't care how: leave this hidden and implicit.

This way, t is gone. x and v take its place as the new independent variables. Together, they span a new plane: the (x, v)-phase plane (Figure 1.8).

In the phase plane, each point (x, v) tells us the whole physical state: position x, and velocity v. Virtually, the particle "travels" along a curve (or trajectory): from

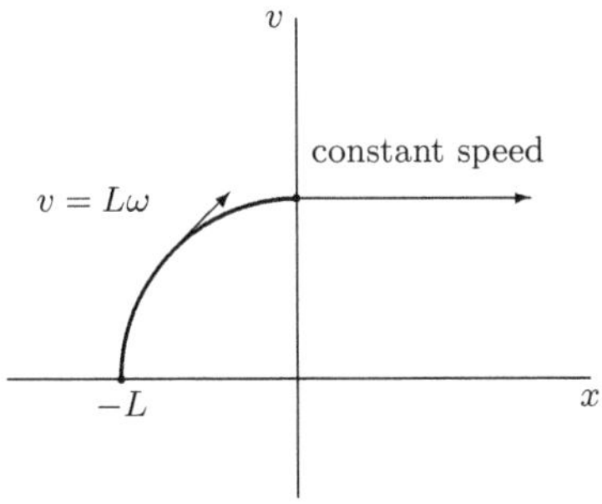

Fig. 1.8. The (x, v) (position-velocity) phase plane. The particle goes from $(-L, 0)$ to $(0, L\omega)$ (the muzzle). (This is nonphysical: physically, the particle remains on the horizontal x-axis.) Behind the muzzle, the force stops, and the particle flies freely rightwards.

$$(x(0), x'(0)) = (-L, 0)$$

to the muzzle:

$$(x(s), x'(s)) = (0, L\omega)\,.$$

But don't forget: this is just virtual and nonphysical. Physically, the particle is confined to the horizontal axis. To find it, project downwards, to obtain the "shadow" on the x-axis. This is where our 1-D motion takes place.

Before the projection, the (x, v)-point tells us more: not only position but also velocity: v. How to filter v out? Start from (x, v), and project rightwards (rather than downwards), onto the (vertical) v-axis.

1.9.2 Virtual Curve

In the phase plane, the particle draws a virtual curve: from the initial state, to the muzzle state. In between, there are infinitely many intermediate states.

How to draw the entire curve? For this, no need to solve the differential equation any more. Instead, use energy alone. How to do this? Consider an intermediate time $0 \le t \le s$. By this time, the force invested a lot of work in the particle. This was already converted into kinetic energy. How much? To calculate this, work as in Section 1.8.2, but the other way around: from energy back to work (with t playing the role of s):

$$\begin{aligned} m\frac{v^2(t)}{2} &= m\left(\frac{v^2(t)}{2} - \frac{v^2(0)}{2}\right) \\ &= \int_{-L}^{x(t)} F(x)dx \\ &= -K\int_{-L}^{x(t)} xdx \\ &= -K\frac{x^2}{2}\Big|_{-L}^{x(t)} \end{aligned}$$

$$= -\frac{K}{2}\left(x^2(t) - L^2\right)$$
$$= \frac{K}{2}\left(L^2 - x^2(t)\right).$$

How does this look like? Better drop t altogether, and use x and v only:

$$\frac{m}{2}v^2 + \frac{K}{2}x^2 = \frac{K}{2}L^2.$$

In our (x, v)-phase plane, this draws a quarter of an ellipse (Figure 1.8). Thanks to energy, we got this for free.

1.9.3 Conservation of Energy

Physically, what is this? Conservation of energy! On this virtual curve, at each individual point, the potential and kinetic energies sum to the same constant: total energy. This is the first law of thermodynamics: energy just changes form: from potential (at the beginning) to kinetic (at the muzzle).

What happens behind the muzzle? The force stops, and the particle keeps flying freely rightwards, at the same speed (as in Newton's first law). In the phase plane, this draws a horizontal line, from the muzzle state rightwards (constant v).

1.9.4 Hamiltonian

This drops t altogether! After all, in Figure 1.8, t is just an implicit parameter, to help "push" along the curve. This helps make a movie, to show the (virtual) dynamics along the curve.

To see the picture more clearly, define a new function – the Hamiltonian:

$$H(x, v) \equiv \frac{m}{2}\left(v^2 + \omega^2 x^2\right).$$

This defines H in the phase plane: completely independent of t. Better yet, write H in terms of momentum mv (rather than velocity v):

$$H(x, mv) = \frac{1}{2m}(mv)^2 + \frac{m}{2}\omega^2 x^2 = \frac{m\omega^2}{2}\left(\frac{1}{m^2\omega^2}(mv)^2 + x^2\right).$$

This will be useful in classical and quantum mechanics alike. Why? Because momentum is more fundamental than velocity. This is why it is conserved (at least when there is no force). In either style, our elliptic curve (Figure 1.8) makes a level set, on which H is constant:

$$H = \text{const.}$$

This is conservation of energy (in its geometrical face): on the ellipse, the total energy H (potential plus kinetic) never changes. This principle governs not only motion but also many other physical phenomena.

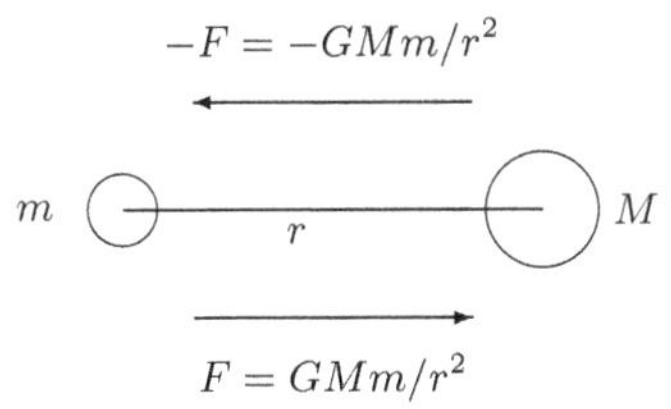

Fig. 1.9. Gravity: the balls attract each other symmetrically, at a force of $F = GMm/r^2$ (as in Newton's third law).

1.10 Gravity

1.10.1 Gravity

In outer space, look at two planets (or any other objects) of mass M and m, and distance r (between their centers of mass). Thanks to Kepler-Newton's model (to be introduced in classical mechanics below), they attract one another at force

$$F = \frac{GMm}{r^2},$$

where G is a positive universal constant (of suitable units). For instance, look at Figure 1.9. Here, there are two balls. The big ball (of mass M) attracts the small ball (of mass $m < M$) at a positive force: $F = GMm/r^2 > 0$ (where r is the distance between their centers). As a reaction, the small ball attracts the big ball at a negative force: $-F = -GMm/r^2 < 0$. (This is as in Newton's third law.) Let's see a practical example.

1.10.2 Gravity on The Earth

How does an object fall to the ground? To see this, let M be the mass of the entire Earth. Let R be the radius of the Earth. Consider an object of mass $m \ll M$ that lies on the face of the Earth. Since the object is so small, its distance from the center of the Earth is R (or so). Thanks to our formula, the Earth pulls it at force

$$F = -\frac{GMm}{R^2}.$$

Here, the minus sign tells us that the force is downwards, towards the ground. Thanks to Newton's second law, the object falls to the ground at a constant acceleration (denoted by $-g$):

$$-g = -\frac{GM}{R^2}.$$

1.10.3 Galileo's Experiment

Again, the minus sign tells us that the acceleration is downwards, towards the ground (so $g > 0$). This way, as the object approaches the ground, it picks more and more (negative) velocity, and falls faster and faster. It hits the ground at maximum velocity and momentum (in magnitude). This is independent of m (as in Galileo's experiment). Why? Because F is linear in m:

$$F = ma = -mg.$$

(Here, both F and a are negative, as required.) As a reaction, the object pulls the entire Earth upwards, at force

$$-F = mg.$$

This is as in Newton's third law (Figure 1.9).

1.10.4 An Airplane

The constant acceleration $-g$ is inaccurate, but still quite useful: not only on the face of the Earth but also higher. For instance, look at an airplane, flying at height $H \ll R$. It feels force $F = -mg$ too. (See exercises below.)

This is good enough for an airplane, but not for a rocket, flying in outer space. As it leaves the atmosphere (on its way out), it feels a decreasing force, as in our original formula (Figure 1.9).

1.10.5 Rocket and Its Energy

Consider a rocket (of mass m), lunged towards outer space. Assume that it feels force from the Earth only. What is the escape velocity? No need to solve any differential equation. Instead, just use energy: the initial velocity $v(0)$ should supply just enough kinetic energy to overcome the Earth's gravity, and fly to infinity. During this journey, the initial kinetic energy won't get lost or wasted, but fully invested. It will do valuable work: to fly against the Earth's gravity. In the end, at infinity, the entire kinetic energy will get transformed into new potential energy:

$$\begin{aligned}
\frac{mv^2(0)}{2} &= -\frac{m}{2}\left(v^2(\infty) - 0^2\right) \\
&= -\int_R^\infty F(r)dr \\
&= \int_R^\infty \frac{GMm}{r^2}dr \\
&= GMm\int_R^\infty \frac{1}{r^2}dr \\
&= -GMm\frac{1}{r}\bigg|_R^\infty \\
&= -GMm\left(0 - \frac{1}{R}\right) \\
&= \frac{GMm}{R}.
\end{aligned}$$

Thus, the escape velocity is:

$$v(0) = \sqrt{\frac{2GM}{R}}.$$

1.10.6 Laws of Thermodynamics

This process is reversible (symmetrically). From infinity, let the rocket fall all the way back to the Earth. This will pay you your entire investment back again. The entire potential energy (accumulated in the rocket) will transform into kinetic energy back again. Thanks to this, the rocket will eventually hit the ground at speed $-v(0)$. This energy will then split into three new kinds of energy: heat, sound, and some momentum (to help push the entire Earth a little downwards). After all, energy is never lost, but only changes. This is the first law of thermodynamics.

Still, what about heat and sound? These have higher entropy. Thus, although the total energy remains the same, it gets less useful: it has now a higher entropy. This is the second law of thermodynamics. We'll come back to this later.

1.11 Exercises: Galileo's Odd Numbers

1.11.1 Speed and Its Derivative

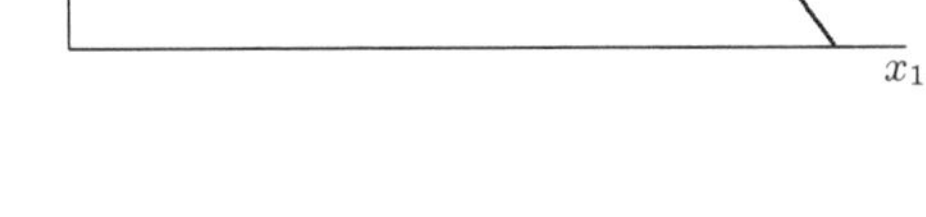

Fig. 1.10. The entire path of the bomb in the (x_1, x_2)-plane, from drop time to hit time.

1. Let R be the radius of the Earth.
2. For simplicity, assume that the Earth has a flat face.
3. On the face of the Earth, draw the x_1-axis: a straight horizontal line.
4. Above the x_1-axis, imagine an aircraft, flying ahead (parallel to the x_1-axis) at height $H \ll R$ and speed V.
5. Once the aircraft is above the origin $x_1 = 0$, it suddenly drops a bomb (of mass m).

6. At time $t \geq 0$, where is the bomb? Hint: in the (x_1, x_2)-plane, the bomb is at
$$x(t) \equiv (x_1(t), x_2(t).$$
7. Physically, where is the bomb? Hint: it is at height $x_2(t)$ above $x_1(t)$ (Figure 1.10).
8. What force does it feel? Hint: thanks to Newton's second law (Figure 1.4), the gravity force is:
$$F \equiv (F_1, F_2) \equiv (0, -mg)$$
(Section 1.10.2).
9. What is the acceleration? Hint:
$$a \equiv (a_1, a_2) \equiv (0, -g).$$
10. For the horizontal component, what is the differential equation? Hint: thanks to Newton's second law,
$$x_1''(t) = a_1 = 0.$$
11. What is its general solution? Hint:
$$x_1(t) = b + ct,$$
for some unspecified parameters b and c (Sections 26.5.2–26.5.3).).
12. Use the initial condition
$$x_1(0) = 0$$
to uncover b:
$$0 = x_1(0) = b + c \cdot 0 = b.$$
13. Furthermore, use the second initial condition
$$v_1(0) = x_1'(0) = V$$
to uncover c:
$$V = v_1(0) = x_1'(0) = c.$$
14. Conclude that, at time t, the bomb is above
$$x_1(t) = Vt.$$
15. Conclude that the horizontal component of the velocity is constant:
$$v_1(t) = x_1'(t) \equiv V.$$
16. Is this as expected from Newton's first law? Hint: yes! See Figure 1.2.
17. Obtain the horizontal velocity $v_1(t)$ in yet another way, using its own differential equation. Hint: thanks to Newton's second law,
$$v_1'(t) = a_1 = 0.$$
18. What is its general solution? Hint:
$$v_1(t) \equiv c,$$
for some unspecified parameter c (Figure 26.9).

19. To specify c, use the initial condition

$$V = v_1(0) = c.$$

20. Does this give the same horizontal velocity as before?
21. What is the kinetic energy thanks to the horizontal velocity? Hint: it is constant as well:

$$\frac{mv_1^2(t)}{2} \equiv \frac{mV^2}{2}$$

(never transformed to any other kind of energy).
22. Did you know this even before solving for $v_1(t)$ explicitly? Hint: horizontally, there is no force or work:

$$\int F_1 dx_1 = \int 0 \cdot dx_1 = 0.$$

23. Next, work vertically (rather than horizontally). What is the differential equation? Hint: thanks to Newton's second law,

$$x_2''(t) = a_2(t) = -g.$$

24. What is its general solution? Hint:

$$x_2(t) = b + ct - g\frac{t^2}{2},$$

for some unspecified parameters b and c (Sections 26.5.1–26.5.3).
25. Use the initial condition

$$x_2(0) = H$$

to uncover b:

$$H = x_2(0) = b.$$

26. Furthermore, use the second initial condition

$$x_2'(0) = v_2(0) = 0$$

to uncover c:

$$0 = x_2'(0) = c.$$

27. Conclude that, at time t, the bomb is at height

$$x_2(t) = H - g\frac{t^2}{2}.$$

28. In particular, look at a few specific times (measured in integer seconds):

$$t = l = 1, 2, 3, 4, \ldots.$$

29. During the lth second, how much closer did the bomb get to the ground? Hint:

$$x_2(l-1) - x_2(l) = \frac{g}{2}l^2 - \frac{g}{2}(l-1)^2 = \frac{g}{2}(2l-1).$$

30. This is proportional to the odd number

$$2l-1=1,3,5,7,\ldots.$$

31. This is Galileo's law of odd numbers.
32. Use your results (vertical and horizontal). At time t, where is our bomb (in 2-D)? Hint:

$$x(t)\equiv(x_1(t),x_2(t))=\left(Vt,H-g\frac{t^2}{2}\right).$$

33. In the x_1-x_2 plane, how does the curve $x_1(t)$-$x_2(t)$ look like? Hint: this is a parabola (Figure 1.10).
34. Physically, what is it? Hint: the position of the bomb in 2-D.
35. In the velocity, look at the vertical component on its own. Hint:

$$v_2(t)=x_2'(t)=-gt.$$

36. How to solve for $v_2(t)$ alone, from its own differential equation? Hint: thanks to Newton's second law,

$$v_2'(t)=a_2(t)=-g.$$

37. What is its general solution? Hint:

$$v_2(t)=c-gt$$

(for some unspecified parameter c).
38. How to specify c? Hint: use the initial condition

$$0=v_2(0)=c.$$

39. Could you now obtain $x_2(t)$ in yet another way? Hint: it solves the new differential equation

$$x_2'(t)=v_2(t)=-gt.$$

40. What is its general solution? Hint:

$$x_2(t)=b-g\frac{t^2}{2}$$

(for some unspecified parameter b).
41. How to specify b? Hint: use the initial condition

$$H=x_2(0)=b.$$

42. Does this agree with the above?
43. What is the hit time? In other words, when does the bomb hit the ground? Hint: the hit time is

$$s\equiv\sqrt{\frac{2H}{g}}.$$

Indeed, s satisfies

$$x_2(s)=0.$$

44. What is the hit velocity? Hint: at time s, the vertical velocity is

$$v_2(s) = -gs = -g\sqrt{\frac{2H}{g}} = -\sqrt{2Hg}.$$

45. What is the hit kinetic energy? Hint: the kinetic energy thanks to the vertical velocity is

$$m\frac{v_2^2(s)}{2} = m\frac{2Hg}{2} = mgH.$$

46. What is the total work made by the Earth's gravity on the bomb? Hint:

$$\int_H^0 F_2 dy = \int_H^0 (-mg)dy = mg\int_0^H dy = mgy|_0^H = mg(H-0) = mgH.$$

47. What is the physical meaning of this? Hint: the initial potential energy of the bomb (thanks to its initial height H) was converted fully into new (vertical) kinetic energy at hit time.
48. Did you know this even before solving for $v_2(t)$ explicitly?
49. At hit time, what will happen to this (vertical) kinetic energy? Hint: it will split into three new kinds of energy: heat, sound, and some momentum (to help push the entire Earth a little downwards).
50. Is the same true not only at hit time but also at any intermediate time $0 \le t \le s$? Hint: yes! The work made by the Earth's gravity (accumulated so far) will convert into new (vertical) kinetic energy (gained so far).
51. Prove this mathematically. Hint:

$$\begin{aligned}
m\frac{v_2^2(t)}{2} &= \frac{m}{2}\left(v_2^2(t) - v_2^2(0)\right) \\
&= \int_H^{x_2(t)} F_2 dy \\
&= \int_H^{x_2(t)} (-mg)dy \\
&= -mg\int_H^{x_2(t)} dy \\
&= -mgy\,|_H^{x_2(t)} \\
&= mgy\,|_{x_2(t)}^H \\
&= mg\left(H - x_2(t)\right).
\end{aligned}$$

52. To simplify this, get rid of t, and use x_2 and v_2 as new independent variables:

$$m\frac{v_2^2}{2} + mgx_2 = mgH.$$

53. Did you know this already?
54. What is the physical meaning of this? Hint: conservation of energy (vertically): at any (x_1, x_2)-position, sum the kinetic and potential energy (vertically), and obtain the initial potential energy.

55. For this, was there any need to know $x_2(t)$ or $v_2(t)$ explicitly?
56. Conclude that there was actually no need to solve for $x(t)$ or $v(t)$ at all! Energy considerations are good enough. After all, energy is more fundamental.
57. Still, substitute our explicit $x_2(t)$ and $v_2(t)$, and make sure that no energy was lost. Hint: potential energy was only converted to new (vertical) kinetic energy:

$$\begin{aligned} mg\left(H - x_2(t)\right) &= mg\left(H - \left(H - g\frac{t^2}{2}\right)\right) \\ &= m\frac{g^2t^2}{2} \\ &= m\frac{v_2^2(t)}{2}. \end{aligned}$$

1.11.2 System: Open or Closed?

1. Look at the bomb on its own. Is it a closed system? Hint: no! It is attracted to the Earth.
2. Why? Hint: there is an outer force: gravity.
3. So, the bomb is an open system.
4. Why then is its energy conserved? Hint: gravity acts from the outside, but has an effect inside: its potential. With this potential, the system is closed (even without the Earth itself): energy is conserved, but momentum is not (as in a harmonic oscillator). At hit time, however, the system opens up: new force gets in, and momentum-energy leak out.
5. Next, look at the bomb and the Earth together. Is this a closed system? Hint: yes: they attract one another.
6. In this system, what is the total momentum? Hint: zero. Indeed, while the bomb accelerates down, the entire Earth accelerates up a little. Although this is tiny, the Earth has a huge mass. Thus, it offsets the bomb, keeping the total momentum at zero, all the time.
7. Why? Hint: at each given time, the bomb is attracted down, and its momentum increases downwards. Thanks to Newton's third law, the Earth reacts: it is attracted up, and its momentum increases upwards.
8. Write this in a more general (differential) form. Hint: let the momentum p and the force F be vector functions, defined in a closed system $V \subset \mathbb{R}^3$. Since the total force vanishes, the total momentum is conserved:

$$\frac{d}{dt}\int\int\int_V p\,dxdydz = \int\int\int_V \frac{dp}{dt}dxdydz = \int\int\int_V F\,dxdydz = \mathbf{0} \in \mathbb{R}^3.$$

9. In Figure 1.8, where is the time variable? Why is it missing? Hint: this is a phase plane. In it, time remains hidden: an implicit parameter that "pushes" along the virtual curve.
10. To see this more vividly, make a movie that follows the dynamics along the virtual curve.

2

Angular Momentum and Inertia

So far, we studied motion in 1-D. How about 3-D? This opens a new field: geometrical mechanics (with angular momentum and velocity). For this, we need new (rotating) coordinates. These are related to the self-system (in special relativity), and even to Riemann's normal coordinates (in general relativity).

In the rotating coordinates, we feel new (fictitious) forces: Euler force, Coriolis force, and more. But, on our Earth, they are not fictitious at all, but real. After all, we live in the coordinates of the Earth, spinning on its own axis, all the time. Thus, for us, these forces are as real as ever: we actually feel them. For this reason, a spaceship flying to outer space must take them into account. Later on, we'll put this in a yet deeper context: Hamiltonian mechanics, and Lie algebras.

2.1 Background: Inner Product and Orthogonality

2.1.1 Inner (Dot) Product

To move on to 3-D, we need some (good old) background. Let u and v be column vectors in the 3-D:

$$u \equiv \begin{pmatrix} u_1 \\ u_2 \\ u_3 \end{pmatrix} \in \mathbb{R}^3$$
$$v \equiv \begin{pmatrix} v_1 \\ v_2 \\ v_3 \end{pmatrix} \in \mathbb{R}^3.$$

The transpose vector, on the other hand, is a row vector:

$$u^t = (u_1, u_2, u_3)$$
$$v^t = (v_1, v_2, v_3).$$

The inner product of u and v is a new number, defined by

$$(u, v) = u^t v = u_1 v_1 + u_2 v_2 + u_3 v_3.$$

2.1.2 Vector Norm

For example, look at the inner product of u with itself:

$$(u, u) \equiv u^t u = u_1^2 + u_2^2 + u_3^2 \geq 0.$$

Could this vanish? Only if u is the zero vector (the origin). We can now define the norm (or length, or magnitude) of u:

$$\|u\| \equiv \sqrt{(u, u)} \geq 0.$$

In particular, every vector of norm 1 is called a unit vector. For example, there are three standard unit vectors:

$$(1, 0, 0)^t, \quad (0, 1, 0)^t, \quad \text{and} \quad (0, 0, 1)^t.$$

2.1.3 Orthogonal Vectors and Matrix

Assume that u and v have zero inner product:

$$(u, v) = 0.$$

Then they are orthogonal. If they also have norm 1, then they are even orthonormal. A matrix with orthonormal columns is called orthogonal.

2.1.4 Orthogonal Projection

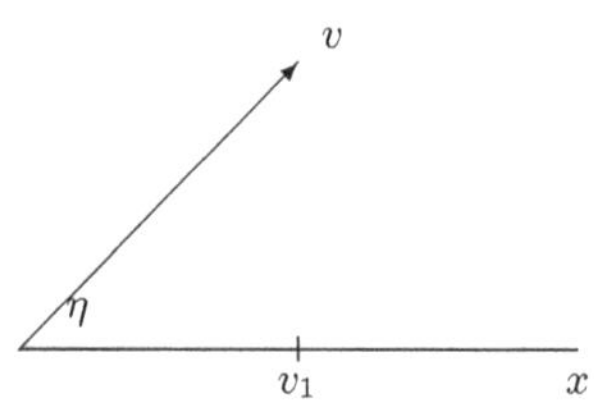

Fig. 2.1. The vector v makes angle η with the positive part of the x-axis: $\cos(\eta) = v_1/\|v\|$.

For simplicity, assume that u is horizontal. Let η be the angle between u and v. The inner product takes now its geometrical meaning:

$$(u, v) = \|u\| \cdot \|v\| \cos(\eta).$$

For example, orthogonality means perpendicularity: angle 90° in between. For yet another example, assume that u is the standard unit vector that points in the x-direction:

$$u \equiv \begin{pmatrix} 1 \\ 0 \\ 0 \end{pmatrix}.$$

In this case, the inner product filters out the x-coordinate in v:

$$(u, v) = v_1 = \|v\| \cos(\eta)$$

(Figure 2.1). In other words, the inner product with u is just the orthogonal projection on u.

And what if u wasn't in the x-direction? Easy: rotate, until u matches the x-axis. This would change nothing: neither (u, v) nor η. (See exercises below, and Sections 28.3.2 and 24.6, where rotation will be written as an orthogonal exponent.)

2.2 Vector Product and the Right-Hand Rule

2.2.1 Vector (Cross) Product

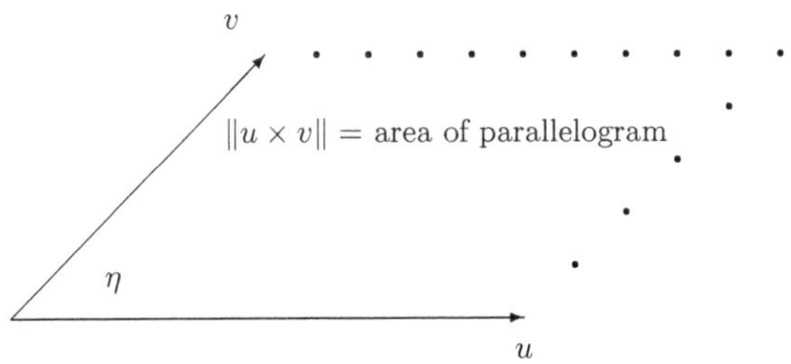

Fig. 2.2. $\|u \times v\|$ is the area of the parallelogram made by u and v.

Your clock has two hands. Do they match? If they do, then their inner product is 1 (as at 12 o'clock). If, on the other hand, they are perpendicular, then their inner product is 0 (as at 3 o'clock). And if they are opposite, then their inner product is -1 (as at 6 o'clock).

Vector product, on the other hand, works the other way around: if they are perpendicular, then it is ± 1 (as at 3 o'clock). If, on the other hand, they are (anti-)parallel, then it is 0 (as at 12 or 6 o'clock).

In a sense, vector product measures how much it takes to switch from one vector to another, and change direction. This is why it is so useful to measure angular momentum. For this purpose, however, it can't be just a number: it must be a new vector, with a direction of its own.

Consider our vectors, u and v. Their vector product is a new vector: $u \times v$. What is its direction? Well, it is perpendicular (or orthogonal) to both u and v. In other words, it has zero inner product with either one:

$$(u \times v, u) = (u \times v, v) = 0.$$

Thanks to linearity, $u \times v$ is perpendicular not only to u and v themselves but also to every linear combination of them, and even to the entire plane that they span. Thus, it could serve as an axis of rotation to turn from u to v. Later on, we'll see that this must be done with your right hand. (Otherwise, pick a minus sign.)

By now, we already have an idea about $u \times v$, and its direction. But what is its length (or norm)? This is the area of the parallelogram made by u and v (Figure 2.2):

$$\|u \times v\| = \|u\| \cdot \|v\| \sin(\eta)$$

(where η is the angle in between). For this reason, if u and v coincide, then we get

$$u \times u = \mathbf{0}$$

(the zero vector). This is called alternativity.

2.2.2 The Right-Hand Rule

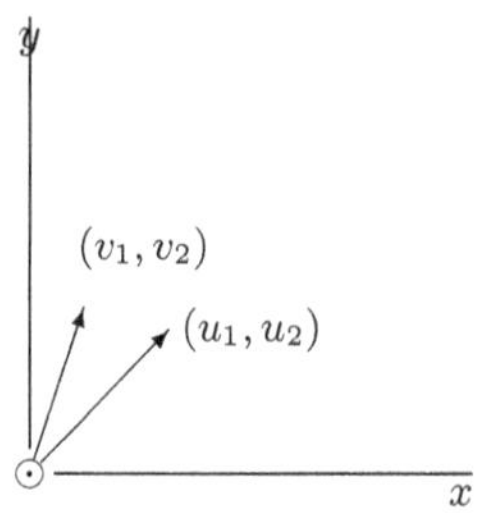

Fig. 2.3. The right-hand rule: take your right hand, and match your thumb to (u_1, u_2), and your index finger to (v_1, v_2). Then, your middle finger will point upwards, towards your own eyes, as indicated by the '$\odot$' at the origin.

By now, $u \times v$ is specified up to sign. Relative to the u-v plane, where should it point? Up or down? To find out, use the right-hand rule: take your right hand, and match your thumb to u, and your index finger to v. Then your middle finger will point in the correct direction of $u \times v$ (Figure 2.3).

To turn from u to v, turn your thumb towards your index finger. While doing this, keep your middle finger pointing towards your own eyes. The turn will then look counterclockwise: a positive mathematical turn.

Thanks to the right-hand rule, we also have antisymmetry:

$$u \times v = -v \times u.$$

Indeed, interchanging u and v only reverses the direction of your middle finger.

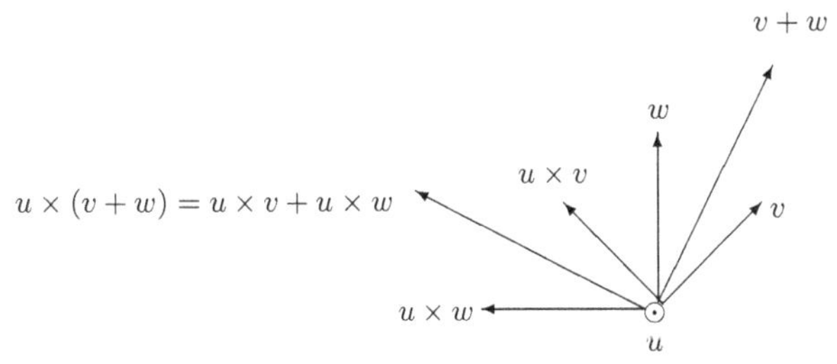

Fig. 2.4. The distributive law. u points upwards, towards your eyes, as indicated by the '$\odot$' at the origin. Thus, $u\times$ is a linear operator: it just rotates by 90°, and multiplies by $\|u\|$.

2.2.3 Linearity: The Distributive Law

Is the vector product linear? To check on this, let α be a real number. Then, α could be pulled out:

$$(\alpha u) \times v = u \times (\alpha v) = \alpha(u \times v).$$

Indeed, the parallelogram in Figure 2.2 is just multiplied by factor α.

Next, let w be yet another vector. Then, the distributive law holds. In other words, you can open parentheses:

$$u \times (v + w) = u \times v + u \times w.$$

To see this, look at three vectors: v, w, and $v + w$. How to apply $u\times$ to them (from the left)? First, project them on the plane perpendicular to u. After all, this has no effect on their vector product with u: the parallelogram in Figure 2.2 just transforms into a rectangle of the same area. The rest follows from Figure 2.4.

Thanks to these properties, our vector product has a new algebraic meaning: Lie brackets, in a Lie algebra. We'll come back to this later, with many more examples. But first, let's use it in physics.

2.3 Linear and Angular Momentum

2.3.1 Linear Momentum

Consider a particle of constant mass m, traveling in 3-D. At time t, it is at position

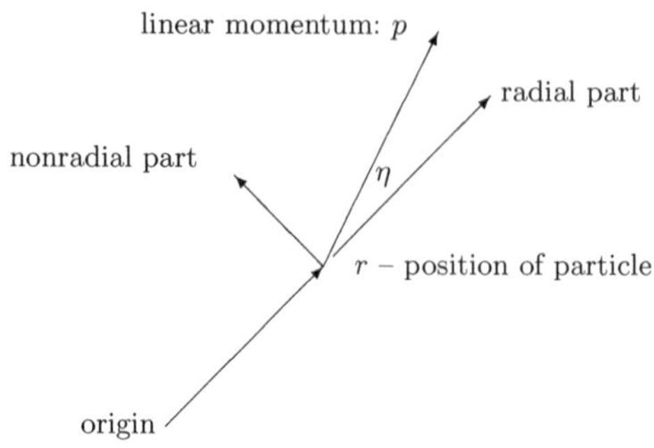

Fig. 2.5. At time t, the particle is at $r \equiv r(t) \in \mathbb{R}^3$, with linear momentum $p \equiv p(t) \in \mathbb{R}^3$. Split it into two parts: the radial part (proportional to r), and the nonradial part (perpendicular to r).

$$r \equiv r(t) \equiv \begin{pmatrix} x(t) \\ y(t) \\ z(t) \end{pmatrix} \in \mathbb{R}^3.$$

Later on, in quantum mechanics, we'll see that this is not so simple. Still, for now, let's accept this.

How to obtain the velocity? Differentiate (with respect to time):

$$r' \equiv r'(t) \equiv \begin{pmatrix} x'(t) \\ y'(t) \\ z'(t) \end{pmatrix} \in \mathbb{R}^3.$$

How to obtain the linear momentum? Multiply by m (the mass of the particle):

$$p \equiv p(t) \equiv mr'(t) \in \mathbb{R}^3.$$

Later on, in special relativity, we'll do this better. Still, for now, let's accept this.

Finally, differentiate p too. This will give you the force: $p'(t)$. Thus, at each particular time t, p describes the entire motion, in three spatial directions in 3-D. Still, we'll soon see that only two directions matter: the third one vanishes.

Indeed, p will soon split into two (orthogonal) components. The first will tell us how fast the particle gets farther and farther away from the origin, or how fast $\|r\|$ grows. This component must be radial: proportional to r, pointing in the same direction. In fact, it is just the orthogonal projection of p on the unit vector $r/\|r\|$:

$$\frac{(r,p)}{\|r\|^2}r = \left(\frac{r}{\|r\|}, p\right)\frac{r}{\|r\|} = \cos(\eta)\|p\|\frac{r}{\|r\|},$$

where η is the angle between p and r (Figures 2.1 and 2.5). This part indeed tells us the radial change: how fast the particle gets away from the origin (and how fast $\|r\|$ grows in time), as required.

2.3.2 Angular Momentum

Still, this is not the whole story. After all, as time goes by, r may not only change magnitude but also change direction. To understand this part of the motion, let's look at the other component of p, which will tell us how fast the particle rotates around the origin.

Where does this rotation take place? Well, infinitesimally, it takes place in the r-p plane: the plane spanned by r and p. This is the plane perpendicular to $r \times p$.

Besides, the second component of p must be nonradial: perpendicular (or orthogonal) to r. This way, it will indeed tell us how fast the particle rotates around a new vector: the angular momentum $r \times p$.

What is the norm of $r \times p$? We already know what it is:

$$\|r \times p\| = \|r\| \cdot \|p\| \sin(\eta)$$

(Section 2.2.1). To rotate around $r \times p$, the particle must make a small (infinitesimal) arc in the r-p plane. For this purpose, the second component of p must be tangent to this arc, or orthogonal not only to $r \times p$ but also to r (Figure 2.5). Thus, it must be proportional to

$$(r \times p) \times r.$$

Fortunately, we already know the norm of this vector:

$$\|(r \times p) \times r\| = \|r \times p\| \cdot \|r\| \sin\left(\frac{\pi}{2}\right) = \|r \times p\| \cdot \|r\| = \|p\| \sin(\eta) \|r\|^2.$$

The second (nonradial) component of p should be proportional to this, but could have a different norm. To have the correct norm, just divide by $\|r\|^2$:

$$\frac{r \times p}{\|r\|^2} \times r = \frac{r \times p}{\|r\|} \times \frac{r}{\|r\|}.$$

Indeed, the norm of this vector is

$$\left\| \frac{r \times p}{\|r\|^2} \times r \right\| = \sin(\eta)\|p\|,$$

as required in Pythagoras' theorem. In summary, we now have the complete orthogonal decomposition of the original linear momentum:

$$p = \frac{(r,p)}{\|r\|^2} r + \frac{r \times p}{\|r\|^2} \times r = \frac{(r,p)}{\|r\|} \cdot \frac{r}{\|r\|} + \frac{r \times p}{\|r\|} \times \frac{r}{\|r\|}.$$

What is so nice about this decomposition? Well, it is uniform: both terms mirror one another. The only difference is that the former uses inner product, and the latter vector product. Better yet, they are orthogonal (perpendicular) to one another. (After all, this is how they were designed in the first place.) This is why they obey Pythagoras' theorem:

$$\begin{aligned}
\left\| \frac{(r,p)}{\|r\|} \cdot \frac{r}{\|r\|} \right\|^2 + \left\| \frac{r \times p}{\|r\|} \times \frac{r}{\|r\|} \right\|^2 &= \left(\frac{(r,p)}{\|r\|} \right)^2 + \left\| \frac{r \times p}{\|r\|} \right\|^2 \\
&= \cos^2(\eta)\|p\|^2 + \sin^2(\eta)\|p\|^2 \\
&= \left(\cos^2(\eta) + \sin^2(\eta)\right) \|p\|^2 \\
&= \|p\|^2,
\end{aligned}$$

as required.

2.4 Angular Velocity

2.4.1 Angular Velocity

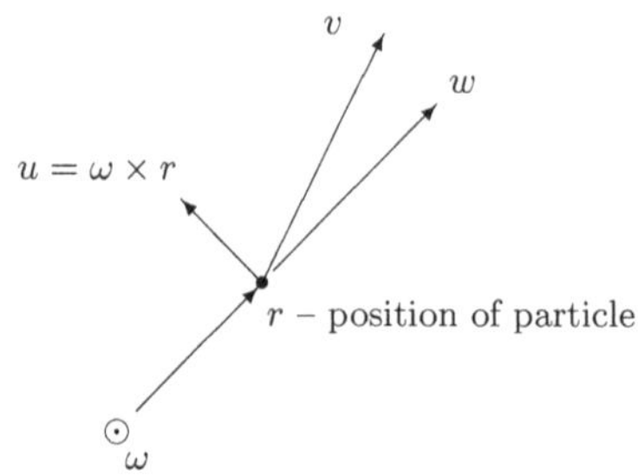

Fig. 2.6. Here, the angular velocity ω is assumed to be perpendicular to r. This way, it points from the page towards your eyes, as indicated by the '$\odot$' at the origin.

What is momentum? It is mass times velocity. This is true in 3-D too. So, to obtain the velocity vector, just divide by the mass of the particle:

$$v = \frac{p}{m} = \frac{r \times p}{m\|r\|^2} \times r + \frac{(r, p)}{m\|r\|^2} r.$$

This is just a special case of a more general (not necessarily orthogonal) decomposition:

$$v = u + w$$

(Figure 2.6). Here,

$$u \equiv \omega \times r,$$

and

$$\omega \equiv \omega(t) \equiv \begin{pmatrix} \omega_1(t) \\ \omega_2(t) \\ \omega_3(t) \end{pmatrix} \in \mathbb{R}^3$$

is a new vector: the angular velocity. This way, the particle rotates around ω. The norm $\|\omega\|$ tells us how fast: by what angle per second. By definition, u must be perpendicular to both ω and r.

The angular velocity could change in time, not only in magnitude but also in direction. For simplicity, however, we often look at some fixed time. Thus, we often drop the '(t)' for short.

2.4.2 The Rotating Axis System

In general, ω may point in just any direction, not necessarily perpendicular to r or v. (See exercises below.) For simplicity, however, we often assume that ω is perpendicular to r:

$$(\omega, r) = 0.$$

Otherwise, just redefine the origin, and shift it along the ω-axis, until obtaining new orthogonal ω and r. This way, ω-r-u make a new right-hand system, rotating around the ω-axis.

As a matter of fact, the rotating coordinates are more fundamental and physical. They are closely related to the physical phenomenon, much more than the original Cartesian coordinates, which are rather arbitrary and artificial. In terms of these new coordinates, the particle may "feel" a few new forces.

2.4.3 Velocity and Its Decomposition

At time t, the particle rotates (infinitesimally) around the ω-axis (at a rate of angle $\|\omega\|$ per second), making an infinitesimal arc. In our velocity

$$v = u + w,$$

the former term

$$u \equiv \omega \times r$$

is tangent to this arc. The remainder

$$w \equiv v - u,$$

on the other hand, may be nontangential, and even perpendicular to the arc.

For the sake of better visualization, we often assume that ω is perpendicular not only to r but also to w (Figure 2.6). Still, this is not a must: in our final example below, w is nonradial, and contains a component parallel to ω. In most of our discussion, however, ω is orthogonal to w (and v) too. In this case, it makes sense to define

$$\omega \equiv \frac{r \times p}{m\|r\|^2},$$

in agreement with the formula at the beginning of Section 2.4.1. This way, w is radial, so ω-w-u make the same right-hand system as ω-r-u: the rotating coordinate system (Section 2.4.2).

2.5 Fictitious Forces

2.5.1 The Centrifugal Force

What is the centrifugal force? This is the force that the particle "feels" in its own ideal "world:" the rotating coordinate system.

In general (even if ω is not perpendicular to r), the centrifugal force is

$$-m\omega \times (\omega \times r).$$

To illustrate, assume that ω and r are perpendicular:

$$(\omega, r) = 0.$$

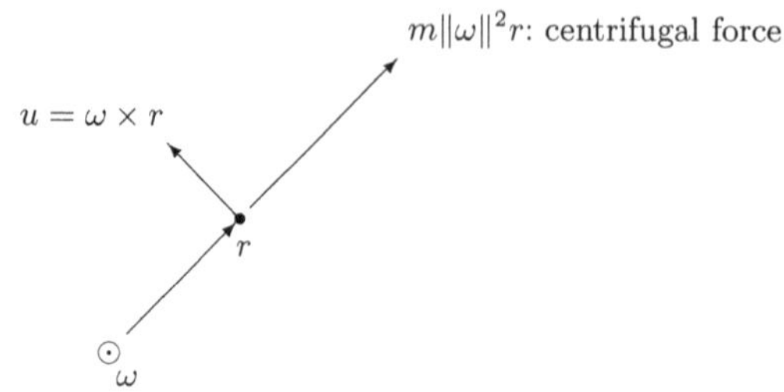

Fig. 2.7. The fictitious centrifugal force: $-m\omega \times (\omega \times r)$. If ω and r are perpendicular, then this is radial: $m\|\omega\|^2 r$.

In this case,

$$\|\omega \times r\| = \|\omega\| \cdot \|r\|,$$

and the centrifugal force is radial:

$$-m\omega \times (\omega \times r) = m\|\omega\|^2 r$$

(Figure 2.7). Still, this is "felt" in the rotating system only. In reality, on the other hand, there is no centrifugal force at all. Indeed, in the static system used in Figures 2.5–2.7, the real force is just $p' = mv'$. So long as $v' = \mathbf{0}$, there is no force at all: Newton's first law holds, and the linear momentum is conserved. As a result, v can never change physically: what could change is just its writing style in rotating coordinates. This nonphysical "change" is due to the fictitious centrifugal "force."

Still, the rotating coordinates are legitimate too, and we might want to work in them. In fact, if the particle kept the same rotating coordinates $(0, \|r\|, 0)$ all the time, then it would rotate physically round and round forever. For this purpose, however, a new counter-force must be applied, to cancel the centrifugal force out.

2.5.2 The Centripetal Force

How to balance (or cancel) the centrifugal force? For this purpose, "connect" the particle to the origin with a wire. In rotating coordinates, this reacts to the centrifugal force. Indeed, thanks to Newton's third law, this supplies the required counter-force: the centripetal force:

$$m\omega \times (\omega \times r).$$

If ω and r are perpendicular, then this force is radial as well:

$$m\omega \times (\omega \times r) = -m\|\omega\|^2 r$$

(Figure 2.8). This is how the wire pulls the particle back towards the origin. After all, Newton's third law works in any coordinate system, rotating or not.

In the rotating system, the centripetal force cancels the original centrifugal force, leaving the particle at the same (rotating) coordinates $(0, \|r\|, 0)$ forever. In the

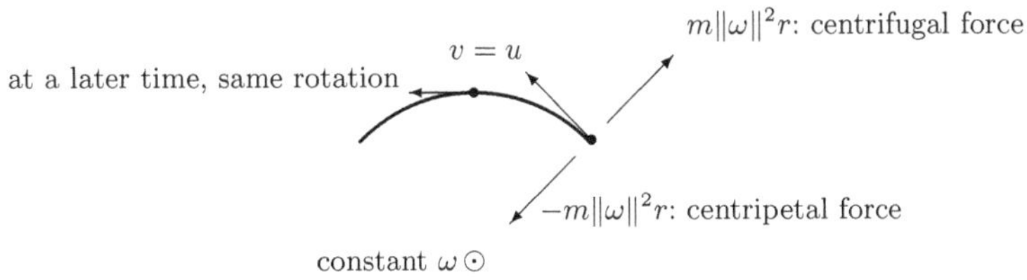

Fig. 2.8. Here, the particle is connected to the origin by a wire. This supplies the centripetal force required to cancel the original centrifugal force, and keep the particle at the constant distance $\|r\|$ from the origin, rotating at the constant angular velocity ω.

static coordinates, on the other hand, the centripetal force has a more physical job: to make u turn.

In fact, the centripetal force pulls the particle towards the origin in just the correct amount: it keeps the particle at a constant distance $\|r\|$ from the origin, rotating at a constant angular velocity ω. This way, the particle makes not only infinitesimal but also global arc around the ω-axis (Figure 2.8). In fact, this arc is as big as a complete circle of radius $\|r\|$. As a result, the velocity vector is always tangent to the circle, with no nontangential part any more:

$$u = v \quad \text{and} \quad w = \mathbf{0}.$$

2.5.3 Euler Force

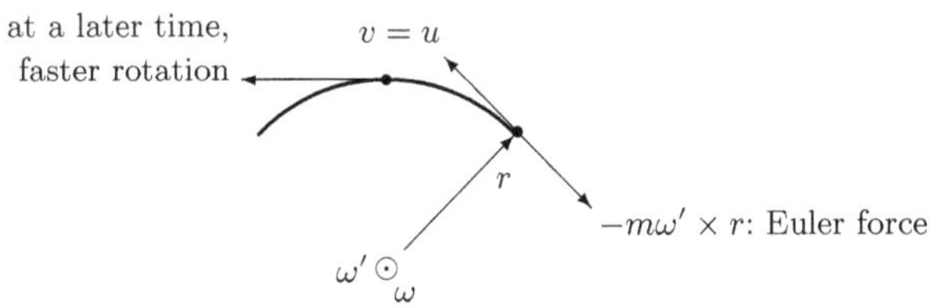

Fig. 2.9. Here, the particle rotates counterclockwise, faster and faster, so ω' points in the same direction as ω. In this case, the Euler force "pulls" the particle clockwise, so it "feels" no force at all: in the rotating system, it is static.

In the above example, the particle rotates at a constant angular velocity: ω. But what if ω changed in time? In this case, ω might have a nonzero time derivative:

$$\omega' \equiv \omega'(t) \equiv \begin{pmatrix} \omega_1'(t) \\ \omega_2'(t) \\ \omega_3'(t) \end{pmatrix} \neq \mathbf{0}.$$

In the rotating coordinate system, this introduces a new force: the Euler force:

$$-m\omega' \times r.$$

What is the direction of this force? In general, we can't tell. After all, ω' could point anywhere. As a result, ω may change not only magnitude but also direction, making the particle rotate in all sorts of new r-u planes.

The rotating coordinates may now rotate in all sorts of directions, changing in time. In them, the particle "feels" nothing, and remains at $(0, \|r\|, 0)$ (where only $\|r\|$ could change, due to a radial force that never changes ω). Why? Because all this drama gets canceled by the Euler force.

In standard (static) coordinates, on the other hand, the Euler force is missing, but is still quite useful: want to know the outer (nonradial) force? Take the Euler force, and change sign: $m\omega' \times r$.

Is this familiar? We already saw something like this before. In rotating coordinates, the centrifugal force cancels the centripetal force out. This is why the particle never feels it. Still, in static coordinates, it is quite useful: want to know the centripetal force? Take the centrifugal force, and change sign.

For example, assume that ω' keeps pointing upwards (Figure 2.9). This way, ω keeps pointing upwards too, and only gets longer and longer:

$$\|\omega\|' \equiv \frac{d\|\omega\|}{dt} > 0.$$

What happens physically? The particle rotates counterclockwise, faster and faster. What could possibly supply the required energy? Assume that there is an angular accelerator that keeps increasing the frequency $\|\omega\|$. This way, the particle keeps rotating counterclockwise (in the same r-u plane), at a bigger and bigger angle $\|\omega(t)\|$ per second.

In the rotating coordinates, the Euler force pulls the particle back clockwise, in an attempt to oppose this motion, slow it down, and even cancel it out. This balances the accelerator, which keeps applying force

$$m\|\omega' \times r\| = m\|\omega'\| \cdot \|r\|$$

counterclockwise.

In static coordinates, the Euler force is missing, so the accelerator has an effect. In the rotating coordinates, on the other hand, the particle feels nothing, and remains at rest at $(0, \|r\|, 0)$, all the time. In its own subjective (moving) "world," it keeps floating effortless, allowing the rotating axes to carry it round and round, faster and faster.

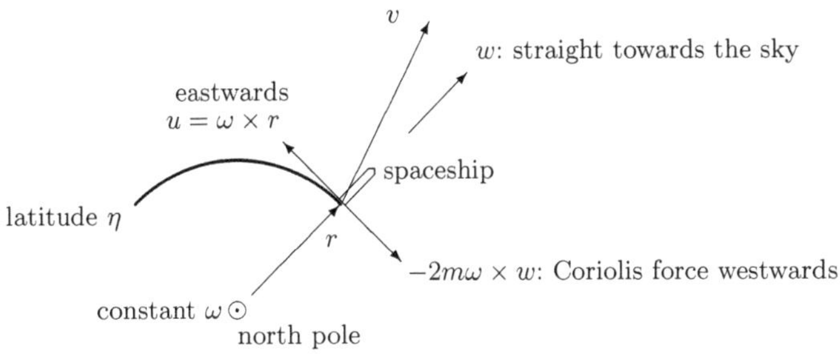

Fig. 2.10. The Earth: a view from above. Cut the northern hemisphere (at latitude $0 \le \eta < \pi/2$) by a horizontal cross section. Place the origin $(0,0,0)$ at the center of this cross section (above the center of the Earth). This way, r is horizontal too: it lies in the cross section. But w (the direction of the spaceship) is not: it makes angle η with the horizontal r-axis. The Coriolis force pulls the entire spaceship westwards.

2.5.4 The Earth and Its Rotation

The rotating coordinates are not just theoretical. They may also be quite real, and easy to work with. Let's use them in practice.

In Figures 2.8–2.9, the particle makes a complete circle around the ω-axis. In fact, the velocity is tangent to this circle, with no nontangential part at all:

$$u = v \quad \text{and} \quad w = \mathbf{0}.$$

Consider now a more complicated case, in which v does contain a nontangential part:

$$v = u + w, \quad \text{where} \quad w \ne \mathbf{0}.$$

As a matter of fact, we already saw such a case. In Figure 2.6, however, w may seem radial. Here, on the other hand, w makes angle η with the horizontal r-axis, and angle $\pi/2 - \eta$ with the vertical ω-axis:

$$(\omega, w) \ge 0$$

(Figure 2.10), Consider, for example, a spaceship standing somewhere in the northern hemisphere, at latitude $0 \le \eta < \pi/2$. Its nose points upwards: straight towards the sky.

The entire Earth rotates eastwards: this is why we see the sun rising from the east. For this reason, in Figure 2.10, ω points northwards: from the page towards your eyes.

Actually, ω is not quite constant: it changes direction (but very slowly). In fact, as the Earth rotates, ω rotates too. Why? Because the Earth is not a perfect sphere. For this reason, the north pole is not quite fixed: it loops clockwise. This is very slow: the loop takes $27,000$ year to close. This is called precession.

Besides, the north pole makes yet another (small) loop clockwise. This loop is quicker: it takes 1.3 year to close. Still, it is very small: just ten meters in radius. So, both loops can be ignored.

In Figure 2.10, we look at the Earth from above: say, from the North star. In the northern hemisphere, at latitude $0 \leq \eta < \pi/2$, we cut a horizontal cross section. Recall that the origin $(0,0,0)$ can be picked arbitrarily. Place it at the middle of the cross section (above the center of the Earth). This way, r is horizontal: it lies in the cross section. Thus, $\|r\|$ is not the radius of the Earth but the radius of the cross section.

Initially, at $t = 0$, the spaceship stands still at r (on the boundary of the cross section). Its body points towards the sky, making angle η with the horizontal r-axis.

At $t > 0$, on the other hand, the spaceship will fly away. Extend the cross section into an infinite (horizontal) plane. On it, r will mark the shadow (orthogonal projection) of the spaceship.

If $\eta = 0$, then this is easy. The cross section cuts through the equator, so its center coincides with the center of the Earth. r lies in it, and coincides with the radius of the Earth. This way, r is perpendicular to the face of the Earth, and points straight into the sky. Fortunately, the centrifugal force in this direction is balanced by gravity, which supplies the required centripetal force.

By the way, this answers an interesting question: why isn't the Earth a perfect sphere? Well, look at the equator. There, some gravity is wasted to balance the centrifugal force. So, at the equator, gravity is a little weaker. This is how the Earth got a little "far:" not quite a sphere, but an oblate spheroid.

If, on the other hand, $\eta > 0$, then things are more complicated. The cross section is now above the center of the Earth. r is still horizontal, but shorter:

$$\|r\| = \cos(\eta) \text{ times the radius of the Earth.}$$

Furthermore, r is no longer perpendicular to the face of the Earth: it also has a new component that points southwards.

What is the norm of this new component? Clearly, it is $\sin(\eta)\|r\|$. This produces a new centrifugal force in the amount of $m\|\omega\|^2 \sin(\eta)\|r\|$ southwards, which is not balanced by gravity any more. After all, gravity pulls downwards, towards the ground, not northwards.

Recall that here we work in the rotating axis system: we stand on the face of the Earth, unaware of any rotation. Therefore, to us, the above force is real, and is truly felt. Likewise, the spaceship feels it as well, and its route could be affected, including its shadow r on the (extended) cross section.

Fortunately, the above force can never affect ω, which produced it in the first place. Can it affect the shadow r? Not much: it needs time to act. Thus, for a small t, its effect on r is as small as t^2. (See exercises at the end of Chapter 1.) For this reason, it hardly affects the original motion, illustrated in Figure 2.10 at the initial time of $t = 0$. Still, after a while, the effect may accumulate and grow, and should be taken into account. To balance this, the nose of the spaceship should better point a little obliquely northwards, from the start.

2.5.5 Coriolis Force

With the entire Earth, the spaceship also rotates eastwards (at a rate of angle $\|\omega\|$ per second). This supplements its velocity with a tangential part:

$$u \equiv \omega \times r.$$

Still, in our ω-r-u system, there is no tangential motion at all. This is why, here on the Earth, we feel no rotation at all.

Now, at time $t = 0$, the spaceship suddenly gets an initial velocity w: upwards, straight towards the sky. This is liftoff: the spaceship really feels it. Still, at the same time, it also feels a new force:

$$-2m\omega \times w.$$

This is the Coriolis force (Figure 2.10). What is its direction? Well, it is perpendicular to the entire ω-w plane. Thanks to the right-hand rule, it is westwards. And what is its norm? Well, this depends on the angle $\pi/2 - \eta$ between ω and w:

$$2m \|\omega \times w\| = 2m \|\omega\| \cdot \| w\| \sin\left(\frac{\pi}{2} - \eta\right) = 2m \|\omega\| \cdot \| w\| \cos(\eta).$$

Fortunately, this force doesn't have much time to act: for a small t, its effect on w is as small as t. (See exercises at the end of Chapter 1.) Thus, it hardly affects the motion, illustrated in Figure 2.10 at the initial time of $t = 0$. Still, as time goes by, it may accumulate and grow, and mustn't be ignored. To balance it, the spaceship should better point a little obliquely eastwards, from the start.

2.6 Exercises: Inertia and Principal Axes

2.6.1 Determinant of a 2 × 2 Matrix

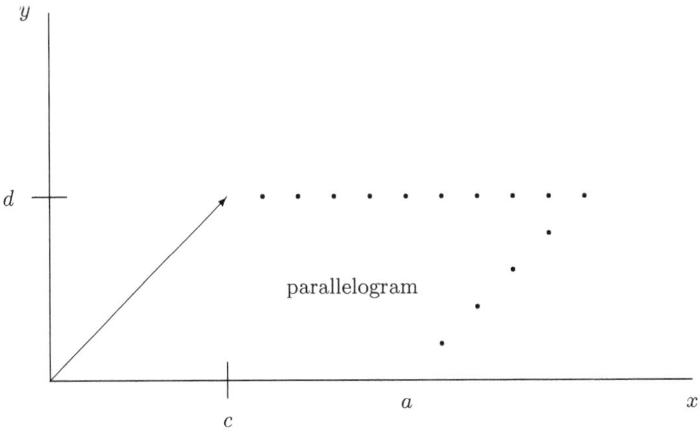

Fig. 2.11. Our matrix contains two columns: r$(a, 0)^t$ (horizontal), followed by $(c, d)^t$ (oblique). Together, they make a parallelogram of area $ad - bc$ (the determinant).

1. What is a determinant? Hint: later on, we'll define it in general (Chapter 13). But here, we only look at a 2×2 matrix (containing two column vectors):

$$\det\left(\begin{pmatrix} a & c \\ b & d \end{pmatrix}\right) \equiv \det\left(\left(\begin{pmatrix} a \\ b \end{pmatrix} \begin{pmatrix} c \\ d \end{pmatrix}\right)\right) \equiv ad - bc.$$

2. What is this geometrically? Hint: for simplicity, assume that a, b, c, and d are positive (or zero). Start with $b = 0$. This way, the first column vector is horizontal: $(a, b)^t = (a, 0)^t$ (Figure 2.11). This way, $(a, 0)^t$ and $(c, d)^t$ make a parallelogram: base a, height d, and area ad (the determinant).
3. Sum the areas of two parallelograms of the same base. Hint: in Figure 2.11, draw a new oblique vector: $(\tilde{c}, \tilde{d})^t$. This makes a new parallelogram: base a, height $\tilde{d}$, and area $a\tilde{d}$. The total area is, thus,

$$a(d + \tilde{d}) = ad + a\tilde{d}.$$

4. What is this geometrically? Hint: the parallelograms are united into one big parallelogram.
5. What is this algebraically? Hint: the distributive law: linearity in d.
6. Work the other way around: replace the horizontal column vector $(a, 0)^t$ by a vertical one: $(0, b)^t$. This way, your parallelogram is now now upside-down: base b (vertical), height c (horizontal), and area $-bc$. Why the minus sign? Hint: in our matrix, $(c, d)^t$ follows $(0, b)^t$, violating the right-hand rule. This makes a wrong parallelogram, which lies under its own base, with negative height and area.
7. Use the same (geometrical) trick to show linearity in c too.
8. Work the other way around, and show linearity in a and b too.
9. Add these two, and obtain the determinant: $ad-bc$: the area of the parallelogram made by $(a, b)^t$ and $(c, d)^t$.
10. Look at two 2×2 matrices: A and B. Show that

$$\det(AB) = \det(A)\det(B).$$

Hint: in Section 13.4.3, we'll prove this in general, and use it in quantum chemistry.
11. How to define the determinant of a bigger 3×3 matrix, keeping the same formula? Hint: see Chapter 13.
12. How to define the determinant of a yet bigger 4×4 matrix, keeping the same formula? Hint: see Chapter 13.

2.6.2 Rotation Matrix

1. Let Q be an $n \times n$ orthogonal matrix. Show that it preserves inner product: for every two n-dimensional vectors u and v,

$$(Qu, Qv) = (u, Q^t Qv) = (u, Iv) = (u, v)$$

(where I is the $n \times n$ identity matrix). Hint: see Section 28.3.4.
2. Conclude that Q also preserves norm:

$$\|Qv\|^2 = (Qv, Qv) = (v, v) = \|v\|^2.$$

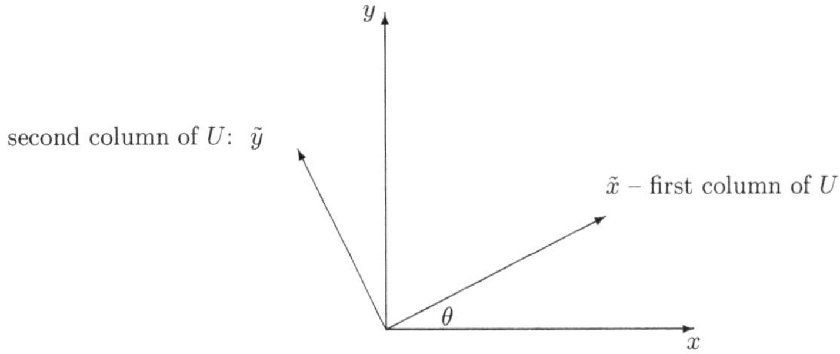

Fig. 2.12. The orthogonal matrix U rotates the entire x-y plane by angle θ counterclockwise, and maps it to the new $\tilde{x}$-$\tilde{y}$ plane.

3. Let O and Q be two orthogonal matrices of the same order. Show that their product OQ is an orthogonal matrix as well. Hint: thanks to associativity,
$$(OQ)^t(OQ) = \left(Q^tO^t\right)(OQ) = Q^t\left(O^tO\right)Q = Q^tIQ = Q^tQ = I.$$
4. Let $0 \le \theta < 2\pi$ be some (fixed) angle. Let U be the following 2×2 matrix:
$$U \equiv U(\theta) \equiv \begin{pmatrix} \cos(\theta) & -\sin(\theta) \\ \sin(\theta) & \cos(\theta) \end{pmatrix}.$$
5. Show that U rotates the x-axis by angle θ counterclockwise. Hint: apply U to the standard unit vector that points rightwards:
$$U\begin{pmatrix} 1 \\ 0 \end{pmatrix} = \begin{pmatrix} \cos(\theta) \\ \sin(\theta) \end{pmatrix}.$$
The result is the $\tilde{x}$-unit vector in Figure 2.12.
6. Show that U rotates the y-axis by angle θ (counterclockwise) too. Hint: apply U to the standard unit vector $(0, 1)^t$.
7. Conclude that U rotates the entire x-y plane by angle θ (counterclockwise). Hint: extend the above linearly. For this purpose, write a general two-dimensional vector as a linear combination of the standard unit vectors $(1, 0)^t$ and $(0, 1)^t$.
8. Show that the columns of U are orthogonal to one another.
9. Show that the columns of U are unit vectors of norm 1.
10. Conclude that the columns of U are orthonormal.
11. Conclude that the columns of U span new axes: the $\tilde{x}$- and $\tilde{y}$-axes in Figure 2.12.
12. Conclude also that U is an orthogonal matrix.
13. Verify that U indeed satisfies
$$U^tU = UU^t = I$$
(the 2×2 identity matrix).

14. Conclude that U^t is an orthogonal matrix as well.
15. Verify that the columns of U^t (the rows of U) are orthogonal as well.
16. Verify that the columns of U^t (the rows of U) are indeed unit vectors of norm 1.
17. Recall that U has a geometrical meaning: once applied to a two-dimensional vector, it rotates it by angle θ counterclockwise. Hint: check this for the standard unit vectors $(1, 0)^t$ and $(0, 1)^t$, and extend this linearly.
18. Interpret U^t geometrically as the inverse rotation: once applied to a vector, it rotates it by angle θ clockwise.
19. Look at the equation $U^tU = I$. What is its geometrical meaning? Hint: this is U (counterclockwise rotation), followed by U^t (clockwise rotation, back again). The result is the identity mapping that changes nothing.
20. Work the other way around: look at the equation $UU^t = I$. What is its geometrical meaning?
21. Show that U has determinant 1:

$$\det(U) = \det\left(U^t\right) = 1.$$

2.6.3 Euler Angles

1. Introduce a third spatial dimension: the z-axis. This makes the new x-y-z axis system in $\mathbb{R}^3$.
2. Consider yet another (right-hand) axis system: the $\tilde{x}$-$\tilde{y}$-$\tilde{z}$ axis system. (Assume that it shares the same origin.) How to map the original x-y-z system to the new $\tilde{x}$-$\tilde{y}$-$\tilde{z}$ system?
3. To do this, use three stages:
   - Rotate the entire $\tilde{x}$-$\tilde{y}$ plane by a suitable angle ψ clockwise, until the $\tilde{x}$-axis hits the x-y plane.
   - Then, rotate the entire x-y plane by a suitable angle ϕ counterclockwise, until the x-axis matches the up-to-date $\tilde{x}$-axis.
   - By now, the up-to-date x- and $\tilde{x}$-axes match. So, all that is left to do is to rotate the up-to-date y-z plane by a suitable angle θ counterclockwise, until the y- and z-axes match the up-to-date $\tilde{y}$- and $\tilde{z}$-axes, respectively.
4. Conclude that, to map the original x-y-z axis system to the original $\tilde{x}$-$\tilde{y}$-$\tilde{z}$ axis system, one could use three stages:
   - Rotate the entire x-y plane by angle ϕ counterclockwise.
   - Then, rotate the up-to-date y-z plane by angle θ counterclockwise.
   - Finally, rotate the up-to-date x-y plane by angle ψ counterclockwise.
5. The angles ϕ, θ, and ψ are called Euler angles.
6. Show that this is a triple product of three orthogonal matrices.
7. Conclude that this triple product is orthogonal too.
8. What is its determinant? Hint: 1.
9. What matrix represents the first stage? Hint:

$$\begin{pmatrix} U(\phi) & \\ & 1 \end{pmatrix}.$$

10. What matrix represents the second stage? Hint: note that this stage rotates the *up-to-date* y-z plane. Therefore, its matrix is

$$\begin{pmatrix} U(\phi) & \\ & 1 \end{pmatrix} \begin{pmatrix} 1 & \\ & U(\theta) \end{pmatrix} \begin{pmatrix} U(\phi) & \\ & 1 \end{pmatrix}^{-1}.$$

11. What is the product of these two matrices? Hint:

$$\begin{pmatrix} U(\phi) & \\ & 1 \end{pmatrix} \begin{pmatrix} 1 & \\ & U(\theta) \end{pmatrix}.$$

12. What matrix represents the third stage? Hint: note that this stage rotates the *up-to-date* x-y plane. Therefore, its matrix is

$$\begin{pmatrix} U(\phi) & \\ & 1 \end{pmatrix} \begin{pmatrix} 1 & \\ & U(\theta) \end{pmatrix} \begin{pmatrix} U(\psi) & \\ & 1 \end{pmatrix} \left(\begin{pmatrix} U(\phi) & \\ & 1 \end{pmatrix} \begin{pmatrix} 1 & \\ & U(\theta) \end{pmatrix} \right)^{-1}.$$

13. What matrix represents the entire mapping? Hint: multiply the former product (representing the first and second stages) by the latter (representing the third stage):

$$\begin{pmatrix} U(\phi) & \\ & 1 \end{pmatrix} \begin{pmatrix} 1 & \\ & U(\theta) \end{pmatrix} \begin{pmatrix} U(\psi) & \\ & 1 \end{pmatrix}.$$

14. What is its transpose? Hint:

$$\begin{pmatrix} U^t(\psi) & \\ & 1 \end{pmatrix} \begin{pmatrix} 1 & \\ & U^t(\theta) \end{pmatrix} \begin{pmatrix} U^t(\phi) & \\ & 1 \end{pmatrix}.$$

15. How does this matrix map the original x-y-z system onto the new $\tilde{x}$-$\tilde{y}$-$\tilde{z}$ system? Hint: symbolically:

$$\begin{pmatrix} \tilde{x}\text{-axis} \\ \tilde{y}\text{-axis} \\ \tilde{z}\text{-axis} \end{pmatrix} = \begin{pmatrix} U^t(\psi) & \\ & 1 \end{pmatrix} \begin{pmatrix} 1 & \\ & U^t(\theta) \end{pmatrix} \begin{pmatrix} U^t(\phi) & \\ & 1 \end{pmatrix} \begin{pmatrix} x\text{-axis} \\ y\text{-axis} \\ z\text{-axis} \end{pmatrix}.$$

16. Indeed, take the transpose:

$$(x\text{-axis}, y\text{-axis}, z\text{-axis}) \begin{pmatrix} U(\phi) & \\ & 1 \end{pmatrix} \begin{pmatrix} 1 & \\ & U(\theta) \end{pmatrix} \begin{pmatrix} U(\psi) & \\ & 1 \end{pmatrix} = (\tilde{x}\text{-axis}, \tilde{y}\text{-axis}, \tilde{z}\text{-axis}).$$

17. Multiply this formula by $(1, 0, 0)^t$ from the right:

$$(x\text{-axis}, y\text{-axis}, z\text{-axis}) \begin{pmatrix} U(\phi) & \\ & 1 \end{pmatrix} \begin{pmatrix} 1 & \\ & U(\theta) \end{pmatrix} \begin{pmatrix} U(\psi) & \\ & 1 \end{pmatrix} \begin{pmatrix} 1 \\ 0 \\ 0 \end{pmatrix}$$
$$= (\tilde{x}\text{-axis}, \tilde{y}\text{-axis}, \tilde{z}\text{-axis}) \begin{pmatrix} 1 \\ 0 \\ 0 \end{pmatrix} = \tilde{x}\text{-axis}.$$

18. How does the new $\tilde{x}$-axis look like? Hint: thanks to the above formula, it is spanned by the first column of the above matrix:

$$\begin{pmatrix} U(\phi) & \\ & 1 \end{pmatrix} \begin{pmatrix} 1 & \\ & U(\theta) \end{pmatrix} \begin{pmatrix} U(\psi) & \\ & 1 \end{pmatrix} \begin{pmatrix} 1 \\ 0 \\ 0 \end{pmatrix}.$$

19. How does the new $\tilde{y}$-axis look like? Hint: likewise, it is spanned by the second column:

$$\begin{pmatrix} U(\phi) & \\ & 1 \end{pmatrix} \begin{pmatrix} 1 & \\ & U(\theta) \end{pmatrix} \begin{pmatrix} U(\psi) & \\ & 1 \end{pmatrix} \begin{pmatrix} 0 \\ 1 \\ 0 \end{pmatrix}.$$

20. How does the new $\tilde{z}$-axis look like? Hint: likewise, it is spanned by the third column:

$$\begin{pmatrix} U(\phi) & \\ & 1 \end{pmatrix} \begin{pmatrix} 1 & \\ & U(\theta) \end{pmatrix} \begin{pmatrix} U(\psi) & \\ & 1 \end{pmatrix} \begin{pmatrix} 0 \\ 0 \\ 1 \end{pmatrix}.$$

21. Does the above matrix have determinant 1? Hint: it is the product of three rotation matrices of determinant 1.
22. In terms of group theory, what does the above discussion tell us? Hint: the group of 3-D rotations is *generated* by 2-D rotations of individual planes.

2.6.4 Principal Axes

1. Recall that

$$r \equiv \begin{pmatrix} x \\ y \\ z \end{pmatrix} \in \mathbb{R}^3$$

is the position of the particle in 3-D. Assume that r is fixed. Show that rr^t is a 3×3 matrix.
2. Write it explicitly. Hint:

$$rr^t \equiv \begin{pmatrix} x \\ y \\ z \end{pmatrix} (x, y, z) = \begin{pmatrix} x^2 & xy & xz \\ yx & y^2 & yz \\ zx & zy & z^2 \end{pmatrix}.$$

3. Use this to show that rr^t is symmetric.
4. Show this in yet another way. Hint:

$$\left(rr^t\right)^t = \left(r^t\right)^t r^t = rr^t.$$

5. What are the eigenvalues and eigenvectors of rr^t? Hint: see below.
6. Show that r is an eigenvector, with eigenvalue $\|r\|^2$. Hint: thanks to associativity,

$$\left(rr^t\right) r = r \left(r^t r\right) = (r, r)r = \|r\|^2 r.$$

7. Let q be a vector that is orthogonal to r. Show that q is an eigenvector in its own right, with eigenvalue 0. Hint: thanks to associativity,

$$\left(rr^t\right) q = r \left(r^t q\right) = (r, q)r = 0r = \mathbf{0}.$$

8. Design two different q's, both orthogonal to r, and also orthogonal to one another. Hint: pick two linearly independent q's, both orthogonal to r. Then, apply the Gram-Schmidt process to them (Section 9.11.1).
9. Show that this could be done in many different ways. Still, pick arbitrarily one particular pair of orthogonal q's.

10. Conclude that rr^t is positive semidefinite: its eigenvalues are greater than or equal to zero.
11. Conclude that rr^t has the following diagonal form:

$$rr^t = O \begin{pmatrix} \|r\|^2 & & \\ & 0 & \\ & & 0 \end{pmatrix} O^t,$$

where O is an orthogonal matrix, whose columns are proportional to r and the above q's. These columns are called principal axes. They span the entire Cartesian space, using new principal coordinates.
12. In terms of principal coordinates, where is the particle? Hint: its principal coordinates are: $(\|r\|, 0, 0)$.
13. Only one principal axis is unique: the two others could be picked in many ways. Still, pick them in one particular way.

2.6.5 The Inertia Matrix

1. Let I be the 3×3 identity matrix. Define the new 3×3 matrix

$$A \equiv A(r) \equiv \|r\|^2 I - rr^t.$$

2. Once A is multiplied by mass m, we obtain mA: the inertia matrix of the particle. It will be useful below.
3. Show that A is symmetric.
4. Show that r is an eigenvector of A, with eigenvalue 0.
5. Show that the above q's are eigenvectors of A, with eigenvalue $\|r\|^2$.
6. Conclude that A is positive semidefinite: its eigenvalues are positive or zero.
7. These eigenvalues (times m) are called moments of inertia. They will be useful below.
8. Show that A has the diagonal form

$$A = O \begin{pmatrix} 0 & & \\ & \|r\|^2 & \\ & & \|r\|^2 \end{pmatrix} O^t,$$

where O is as above.
9. Recall that

$$\omega \equiv \begin{pmatrix} \omega_1 \\ \omega_2 \\ \omega_3 \end{pmatrix} \in \mathbb{R}^3$$

is the angular velocity. So far, we assumed that ω was perpendicular to r. But this is not a must. From now on, let's drop this assumption. Show that, even if ω is no longer perpendicular to r, $A\omega$ still is:

$$(A\omega, r) = 0.$$

Hint: in $A\omega$, there is no radial component any more.

10. Show that
$$\|\omega \times r\|^2 = \|\omega\|^2 \|r\|^2 - (\omega, r)^2.$$
Hint: let τ be the angle between ω and r. Then
$$\begin{aligned}\|\omega \times r\|^2 &= \|\omega\|^2 \|r\|^2 \sin^2(\tau)\\ &= \|\omega\|^2 \|r\|^2 \left(1 - \cos^2(\tau)\right)\\ &= \|\omega\|^2 \|r\|^2 - \|\omega\|^2 \|r\|^2 \cos^2(\tau)\\ &= \|\omega\|^2 \|r\|^2 - (\omega, r)^2\end{aligned}$$
(Sections 2.1.1 and 2.2.1).
11. Conclude that
$$\|\omega \times r\|^2 = \omega^t A \omega.$$
Hint: thanks to associativity,
$$\begin{aligned}\|\omega \times r\|^2 &= \|\omega\|^2 \|r\|^2 - (\omega, r)^2\\ &= \|r\|^2 \omega^t \omega - \left(\omega^t r\right)\left(r^t \omega\right)\\ &= \|r\|^2 \omega^t I \omega - \omega^t \left(r r^t\right) \omega\\ &= \omega^t A \omega.\end{aligned}$$
12. Show that the entire principal axis system could rotate around the ω-axis, carrying the "passive" particle at angle $\|\omega\|$ per second. Hint: in this case, the particle keeps the same principal coordinates: $(\|r\|, 0, 0)$, all the time.
13. What is the necessary condition for this? Hint: to have such a rotation, we must have no nontangential velocity: $w = \mathbf{0}$.
14. Give an example where this condition holds, and another example where it doesn't. Hint: in Figure 2.10, for example, the entire Earth rotates around ω. The spaceship, on the other hand, has $w \neq \mathbf{0}$, to help it not only rotate (with the entire atmosphere) but also spiral towards outer space (until escaping from the Earth's gravity).

2.6.6 Triple Vector Product

1. Let $\mathbf{v}$ and $\mathbf{w}$ be linearly independent vectors in $\mathbb{R}^3$. This means that $\mathbf{v}$ is not a scalar multiple of $\mathbf{w}$, but points in a different direction (Figure 2.13). Let η be the angle between $\mathbf{v}$ and $\mathbf{w}$ $(0 < \eta < \pi)$. Let $\mathbf{p}$ be yet another vector, perpendicular to $\mathbf{v}$ in the $\mathbf{v}$-$\mathbf{w}$ plane. Show that
$$\mathbf{p} \times (\mathbf{v} \times \mathbf{w}) = (\mathbf{p}, \mathbf{w})\mathbf{v}.$$
Hint: use the fact that
$$\cos\left(\frac{\pi}{2} + \eta\right) = -\sin(\eta).$$
2. Conclude that
$$\mathbf{p} \times (\mathbf{w} \times \mathbf{v}) = -(\mathbf{p}, \mathbf{w})\mathbf{v}.$$
3. In this formula, interchange the roles of $\mathbf{v}$ and $\mathbf{w}$.

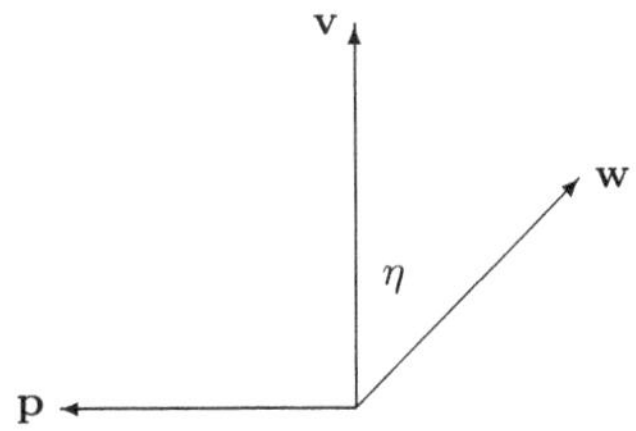

Fig. 2.13. In the **v**-**w** plane, **p** is perpendicular to **v**, but not to **w**.

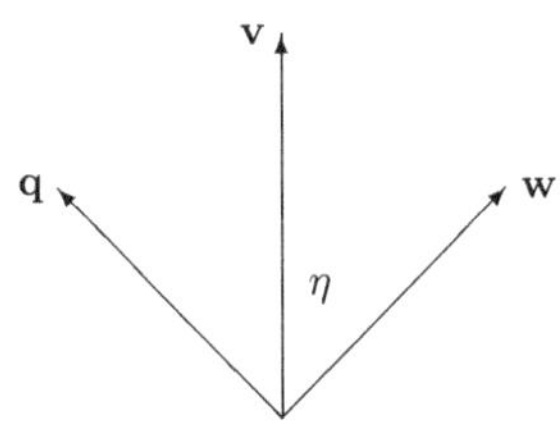

Fig. 2.14. In the **v**-**w** plane, **q** is perpendicular to **w**, but not to **v**.

4. Conclude that
$$\mathbf{q} \times (\mathbf{v} \times \mathbf{w}) = -(\mathbf{q}, \mathbf{v})\mathbf{w},$$
where **q** is a new vector, perpendicular to **w** in the **v**-**w** plane (Figure 2.14).
5. Add these formulas:
$$(\mathbf{p} + \mathbf{q}) \times (\mathbf{v} \times \mathbf{w}) = (\mathbf{p}, \mathbf{w})\mathbf{v} - (\mathbf{q}, \mathbf{v})\mathbf{w} = (\mathbf{p} + \mathbf{q}, \mathbf{w})\mathbf{v} - (\mathbf{p} + \mathbf{q}, \mathbf{v})\mathbf{w},$$
which is the same, since **p** is perpendicular to **v** (and **q** to **w**) in the **v**-**w** plane.
6. Show that every vector **u** in the **v**-**w** plane could be written as $\mathbf{u} = \mathbf{p} + \mathbf{q}$, where **p** is perpendicular to **v** (and **q** to **w**) in the **v**-**w** plane. Hint: **v** and **w** are linearly independent, and so are also **p** and **q**.
7. What does this mean geometrically? Hint: this is just the parallelogram rule.
8. Conclude that
$$\mathbf{u} \times (\mathbf{v} \times \mathbf{w}) = (\mathbf{u}, \mathbf{w})\mathbf{v} - (\mathbf{u}, \mathbf{v})\mathbf{w},$$
where **u** is just any vector in the **v**-**w** plane.

9. Extend this supplement $\mathbf{u}$ with a new component, proportional to $\mathbf{v} \times \mathbf{w}$. Hint: this new component is perpendicular to both $\mathbf{v}$ and $\mathbf{w}$, and contributes nothing to either side of the above formula.
10. Thus, our formula is true in 3-D as well. Let's use it in practice.

2.6.7 Linear Momentum and Its Decomposition

1. How to use this formula in practice? Give an example.
2. For example, set $\mathbf{u} \equiv \mathbf{v} \equiv r$ (the position of the particle), and $\mathbf{w} \equiv p$ (the linear momentum). What do you get? Hint:

$$r \times (r \times p) = (r, p)r - (r, r)p,$$

or

$$(r, r)p = -r \times (r \times p) + (r, p)r = (r \times p) \times r + (r, p)r.$$

3. Divide both sides by $\|r\|^2$.
4. Is this familiar? Hint: this is the orthogonal decomposition of the linear momentum p (end of Section 2.3.2).

2.6.8 Centrifugal and Centripetal Forces

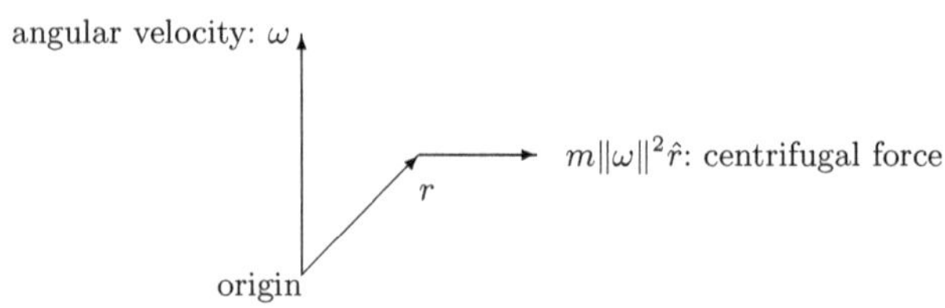

Fig. 2.15. A more general case, in which r is no longer perpendicular to ω (a view from the side). Let $\hat{r}$ be the part of r that is perpendicular to ω. Then, the centrifugal force is $m\|\omega\|^2\hat{r}$ rightwards.

1. Look again at our triple-vector-product formula. Can you give yet another practical example?
2. Use it to write the centrifugal force (Figure 2.7) in a more general case, in which ω and r are not necessarily perpendicular, so the rotating axes not necessarily align with them any more. Hint: set $\mathbf{u} = \mathbf{v} \equiv \omega$, and $\mathbf{w} \equiv r$:

$$\omega \times (\omega \times r) = (\omega, r)\omega - (\omega, \omega)r.$$

This way, the centrifugal force (divided by m) takes the form

$$\begin{aligned}-\omega \times (\omega \times r) &= -(\omega, r)\omega + (\omega, \omega)r \\ &= \|\omega\|^2 \left(r - \frac{(\omega, r)}{\|\omega\|^2}\omega \right) \\ &= \|\omega\|^2 \hat{r},\end{aligned}$$

where $\hat{r}$ is the part of r that is perpendicular to ω (Figure 2.15).

3. Prove this in a more geometrical way. Hint: note that $\omega \times r = \omega \times \hat{r}$. Thus,

$$-\omega \times (\omega \times r) = -\omega \times (\omega \times \hat{r}) = \|\omega\|^2 \hat{r}.$$

4. Does the centrifugal force really exist? Hint: only in the rotating axis system! In the standard (static) system, on the other hand, it has no business to exist. Indeed, in Figures 2.5–2.6, the velocity is often constant, meaning equilibrium: no force at all.
5. Does the centripetal force exist? Hint: only if supplied by some source, such as gravity.
6. What is the role of the centripetal force? Hint: in the rotating axis system, it balances the centrifugal force, and cancels it out. This way, there is no force at all, so the particle always has the same (rotating) coordinates, and is carried effortlessly by the rotating axes, round and round forever. In the static system, on the other hand, the centripetal force has a more "active" role: to make u turn. This indeed makes the particle go round and round, as required.

2.6.9 Inertia Matrix times Angular Velocity

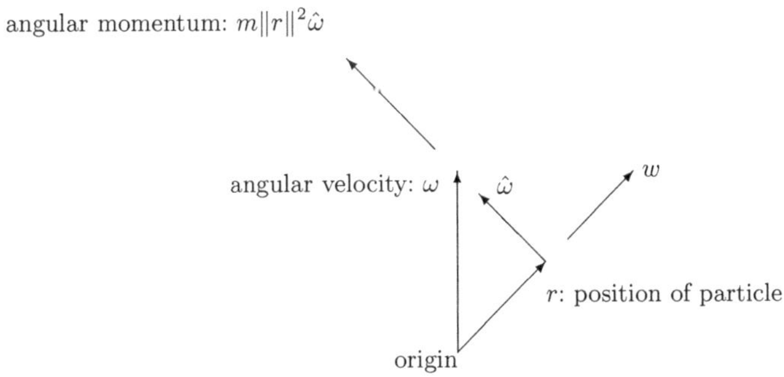

Fig. 2.16. Let $\hat{\omega}$ be the part of ω that is perpendicular to r. If w is radial, then the angular momentum is $m\|r\|^2\hat{\omega}$. This is nonconstant: it must change direction, to point not only upwards but also obliquely inwards. Why isn't the angular momentum conserved? Because the system is open: a horizontal centripetal force must be supplied from the outside.

1. How to define the angular velocity to align with the angular momentum? Hint:
$$\omega \equiv \frac{r \times p}{m\|r\|^2}$$
(end of Section 2.4.3).
2. This makes ω orthogonal to r.
3. Here, on the other hand, we don't need this kind of orthogonality, and don't assume it any more.
4. Work the other way around: assume that the angular velocity is given, and use it to uncover the angular momentum (assuming that w in Figure 2.6 is radial):
$$\begin{aligned} r \times p &= m \cdot r \times v \\ &= m \cdot r \times (u + w) \\ &= m \cdot r \times u \\ &= m \cdot r \times (\omega \times r) \\ &= m\left((r, r)\omega - (r, \omega)r\right) \\ &= m\|r\|^2 \left(\omega - \frac{(r, \omega)}{\|r\|^2} r\right) \\ &= m\|r\|^2 \hat{\omega}, \end{aligned}$$
where $\hat{\omega}$ is the part of ω that is perpendicular to r (Figure 2.16). Hint: in our triple-vector-product formula, set $\mathbf{u} \equiv \mathbf{w} \equiv r$ and $\mathbf{v} \equiv \omega$.
5. Prove this in a more geometrical way. Hint: note that $\omega \times r = \hat{\omega} \times r$. Thus,
$$r \times (\omega \times r) = r \times (\hat{\omega} \times r) = \|r\|^2 \hat{\omega}.$$
6. Conclude that, if w in Figure 2.6 is radial, then the angular momentum has a yet simpler form:
$$r \times p = mA\omega,$$
where $mA \equiv mA(r)$ is the inertia matrix of the particle.
7. What is the physical meaning of the inertia matrix? Hint: the inertia matrix is like mass. Indeed, linear momentum is mass times velocity. Likewise, angular momentum is inertia matrix times angular velocity (if w is radial).
8. In Figure 2.10, how to make w radial? Hint: shift the origin downwards, to hit the center of the Earth. This way, the new r is no longer horizontal, but oblique (parallel to w): it leads from the center of the Earth to the spaceship. As a result, w gets radial, and the angular momentum gets oblique: proportional to $\hat{\omega}$ (as in Figure 2.16).
9. Look at things the other way around. Assume now that the angular momentum is already available. How to define the angular velocity? Hint: design ω in two stages: first, let it have some radial component. Then, define its other component by
$$\hat{\omega} \equiv \frac{r \times p}{m\|r\|^2}.$$
10. This way, in Figure 2.6, is w radial? Hint: thanks to the above definition of $\hat{\omega}$,

$$\begin{aligned}r \times p &= m \cdot r \times v \\ &= m \cdot r \times (u + w) \\ &= m \cdot r \times u + m \cdot r \times w \\ &= m \cdot r \times (\omega \times r) + m \cdot r \times w \\ &= m\|r\|^2 \hat{\omega} + m \cdot r \times w \\ &= r \times p + m \cdot r \times w.\end{aligned}$$

Thus,

$$0 = m \cdot r \times w,$$

so

$$r \times w = 0.$$

This means that w must indeed be radial.

2.6.10 Angular Momentum and Its Conservation

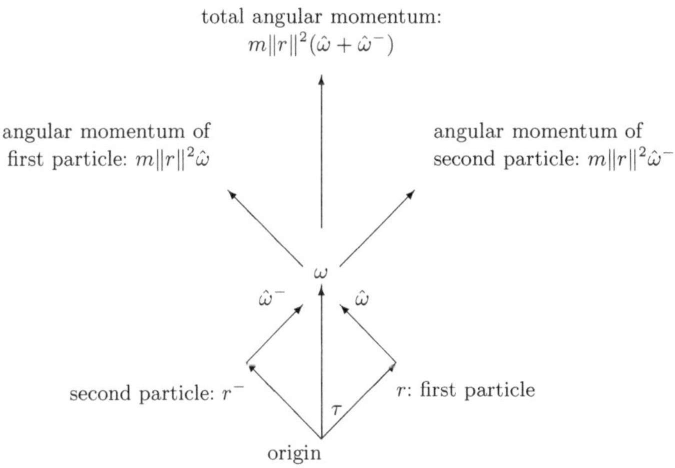

Fig. 2.17. The particles at r and r^- attract each other just enough to supply the horizontal centripetal force required to keep them rotating together around the vertical ω-axis. This is a closed system: no external force is needed. Angular momentum is conserved: $\hat{\omega} + \hat{\omega}^-$ points upwards, all the time.

1. In Figure 2.16, the angular momentum is not conserved. Indeed, it is proportional to $\hat{\omega}$, which must change direction to keep pointing obliquely: upwards-inwards. How come? Isn't angular momentum supposed to be constant? Hint: the system is open: it welcomes a horizontal centripetal force to keep the particle rotating around the vertical ω-axis, all the time.

2. To supply this centripetal force, introduce a second particle at position

$$r^- \equiv \begin{pmatrix} -x \\ -y \\ z \end{pmatrix},$$

on the other side of the vertical ω-axis (Figure 2.17).
3. What is its inertia matrix? Hint: $mA(r^-)$.
4. What are the eigenvectors of $A(r^-)$? Hint: one eigenvector is r^- itself. The two others are perpendicular to r^- (and to one another).
5. Could the eigenvectors of $A(r^-)$ be the same as those of $A(r)$? Hint: only one.
6. What are the principal axes of the second particle? Are they the same as those of the first particle? Hint: only one.
7. In terms of its own principal coordinates, where is the second particle? Hint: it is always at the same principal coordinates:

$$(\|r^-\|, 0, 0) = (\|r\|, 0, 0).$$

8. Let $\hat{\omega}^-$ be the part of ω that is perpendicular to r^-. Show that the angular momentum of the second particle is

$$m\|r^-\|^2\hat{\omega}^- = mA(r^-)\omega.$$

9. Together, these particles make a closed system: for a suitable $\|r\|$, they attract each other just enough to supply the horizontal centripetal force required to keep them rotating together around the vertical ω-axis, all the time.
10. This is the two-body system.
11. Around what point does it rotate? Hint: around its center of mass: the midpoint in between the particles.
12. What is the total angular momentum of the two-body system? Hint: see Figure 2.17.
13. What is the direction of this vector?
14. Must it be vertical?
15. Show that the total angular momentum is now conserved. Hint: the sum $\hat{\omega}+\hat{\omega}^-$ points now straight upwards, with no inward component any more.
16. Define the inertia matrix of the two-body system. Hint:

$$B \equiv B(r) \equiv mA(r) + mA(r^-).$$

17. Write the total angular momentum as

$$m\|r\|^2\left(\hat{\omega}+\hat{\omega}^-\right) = mA(r)\omega + mA(r^-)\omega = B\omega.$$

18. What does this mean algebraically? Hint: ω is an eigenvector of B.
19. Without calculating B explicitly, show that it is symmetric.
20. Conclude that B has three orthogonal eigenvectors. Hint: see Section 28.4.5.
21. What are they physically? Hint: the principal axes of the two-body system.
22. Could these new eigenvectors be the same as those of $A(r)$ and $A(r^-)$? Hint: only one.

23. Show that B is positive semidefinite: its eigenvalues are positive or zero. Hint: every three-dimensional vector q could be written as a linear combination of eigenvectors of $A(r)$ [or $A(r^-)$]. Thus,

$$(q, Bq) = (q, mA(r)q) + (q, mA(r^-)q) = m(q, A(r)q) + m(q, A(r^-)q) \geq 0.$$

24. When is this strictly positive (for all q's)? Hint: if $z \neq 0$, then $r^- \neq -r$, so either $(q, A(r)q) > 0$ or $(q, A(r^-)q) > 0$ (for all q's).
25. Conclude that, if $r^- \neq -r$, then $(q, Bq) > 0$ (for all q's).
26. Conclude that, in this case, B is positive definite, with positive eigenvalues.
27. What are they physically? Hint: the (positive) moments of inertia of the two-body system.
28. Design an eigenvector of B that is perpendicular to both r and r^-. Hint: if $r^- \neq -r$, then take their vector product: $r \times r^-$.
29. What is its eigenvalue? Hint: $2m\|r\|^2$.
30. Still, this eigenvector depends on r. Design yet another eigenvector that doesn't depend on r, and remains the same all the time. Hint: take ω.
31. What is its eigenvalue? Hint: let τ be the angle between r and ω (Figure 2.17). Now, in $B\omega$, what contributes is not ω but $\hat{\omega}$, which is $\sin(\tau)$ times as short. Once projected back onto ω, it picks yet another $\sin(\tau)$:

$$2m\|r\|^2 \sin^2(\tau) = 2m\|r\|^2 \cos^2\left(\frac{\pi}{2} - \tau\right) = 2m\left(\|r\|^2 - z^2\right) = 2m\left(x^2 + y^2\right),$$

where $r = (x, y, z)^t$.
32. Could this be negative or zero? Hint: no! If it were zero, then $r^- = r$, which is nonphysical.
33. Conclude that our vertical ω remains an eigenvector of B, all the time.
34. Conclude that the total angular momentum $B\omega$ remains vertical, all the time.
35. Conclude once again that the total angular momentum is conserved.

2.6.11 Rigid Body

1. This is just an example of a more general (closed) system of many particles, rotating together around a constant eigenvector ω of the inertia matrix of the system. This way, the total angular momentum remains constant, as required.
2. How to use this in practice? Hint: this holds not only for two but also for (infinitely) many particles, making a new rigid body.
3. In a rigid body, how to define B? Hint: use an integral rather than a sum:

$$B \equiv \int\int\int A(r)\rho(r)dxdydz.$$

Here, at $r = (x, y, z)^t$, ρ is the density: mass per unit volume. This way, $\rho dxdydz$ is the local mass, which multiplies the local inertia matrix $A(r)$, integrated upon.
4. Is B symmetric?
5. Conclude that B has three orthogonal eigenvectors: the new principal axes.
6. Conclude that B has three real eigenvalues: the new moments of inertia.
7. What is their sum? Hint: this is the trace of B.

8. How to calculate it? Hint: use linearity: inside the integral, take the trace of each individual $A(r)$:

$$\text{trace}(B) = \int\int\int 2\|r\|^2 \rho dxdydz.$$

9. Is B positive semidefinite? Hint: the proof is as in the two-body system.
10. Conclude that B has positive eigenvalues (moments of inertia) or zero (which is very rare – see below).
11. If our rigid body rotates around a constant ω (an eigenvector of B that doesn't change during the rotation), what happens to the angular momentum? Hint: $B\omega$ is constant as well: the conserved angular momentum.
12. Even if the rigid body travels wildly, it doesn't matter: ω remains constant relative to it.
13. Is the system closed? Hint: yes: no force acts on the rigid body from the outside.
14. Geometrically, where is the principal axis? Hint: it is spanned by an eigenvector of B.
15. Physically, what is the principal axis? Hint: it marks the direction of constant angular velocity and momentum, around which the entire rigid body could rotate, round and round forever.
16. Could B be singular (have eigenvalue 0)? Hint: only in a trivial case, in which all the $A(r)$'s commute, and share a common eigenbasis. For example, look at a straight wire, in which all atoms lie on the same line, one by one. This way, B has a singular eigenvector, parallel to the wire. Around it, no rotation could take place: the wire is just too thin! A real (nontrivial) rotation could take place only around a nonsingular eigenvector, perpendicular to the wire.
17. In the wire, where to place the origin? Hint: at the middle.
18. Does the wire have a more physical principal axis? Hint: from the origin, issue a vertical, perpendicular to the wire. This could serve as a better principal axis, around which the wire could really rotate.
19. Consider a more interesting example: a thin disk (of radius R and mass m).
20. Where to place the origin? Hint: at its center.
21. What principal axis must the disk have? Hint: from the origin, issue a vertical, perpendicular to the disk. Around it, the disk could rotate, round and round forever. The disk can even travel, and keep rotating around its own perpendicular, in the same way.
22. What is the moment of inertia? Hint: here, ω points upwards. This will be our eigenvector, common to all $A(r)$'s. What is its eigenvalue (with respect to B)? Since $dxdy = rdrd\theta$,

$$\int\int_{x^2+y^2 \le R^2} \left(x^2+y^2\right) \rho dxdy = \frac{m}{\pi R^2} \int_0^{2\pi} \int_0^R r^3 drd\theta = \frac{m}{\pi R^2} \cdot \frac{2\pi R^4}{4} = \frac{mR^2}{2}.$$

23. On top of this, rotate in yet another way. On the disk, draw a horizontal diameter. Rotate around it: upside-down and up again, and so on. (After all, w could be nonradial.)
24. What is the moment of inertia? Hint: here, ω points rightwards. This will be our new eigenvector (with respect to B). What is the eigenvalue? Look at some r on the disk (at angle τ from ω). It contributes $\|r\|^2$ not to ω but to $\hat{\omega}$, which

is $\sin(\tau)$ times as short (Figure 2.17). Once projected back onto ω, it picks yet another $\sin(\tau)$:

$$\frac{m}{\pi R^2}\int_0^{2\pi}\int_0^R r^3\sin^2(\tau)drd\tau = \frac{m}{\pi R^2}\cdot\frac{\pi R^4}{4} = \frac{mR^2}{4}.$$

25. On top of this, rotate in yet another way. On the disk, draw a vertical diameter. Rotate around it too: left-to-right and left again, and so on.
26. What is the moment of inertia? Hint: same as before.
27. How to verify this? Hint: see below.
28. What is their sum? Hint: mR^2.
29. Is this the same as the trace of B? Hint: look at the previous integral (for the first moment of inertia). Multiply by 2.
30. These three motions could take place together, at the same time. After all, each rotation uses its own principal axis, which remains constant (relative to the disk), regardless of the other two.
31. Later on, these will help model polarization.
32. Later on, these 2-D rotations will help generate a Lie group of 3-D rotations.
33. What other principal axes could the disk have? Hint: every diameter could serve as a new principal axis, around which the entire disk could rotate, upside-down and up again, and so on.
34. In Figure 2.7, even if ω is not perpendicular to r, how to write the centrifugal force easily? Hint:
$$-m\omega\times(\omega\times r) = mA(\omega)r.$$

2.6.12 Percussion Point

1. Hold your baseball bat vertically, ready to hit.
2. Assume that the bat has mass m and length d.
3. For simplicity, assume that the bat is uniform. This way, its center of mass is at the midpoint: at height $d/2$ (Figure 2.18). Furthermore, its density is m/d kilogram per meter.
4. Let r be the vertical vector leading from the bottom to the center of mass.
5. What is the norm of r? Hint: $\|r\| = d/2$.
6. What happens if you hit the ball at the midpoint r? Hint: this is no good: the hit would push the entire bat horizontally into your hands, and hurt you.
7. On the other hand, what happens if you hit the ball at the tip at $2r$? Hint: this is no good either: the hit would rotate the bat around its center of mass, pulling it out of your hands.
8. Where is the best place to hit? Hint: somewhere in between, where no angular momentum is wasted on rotating the bat needlessly. This is called the percussion point.
9. Where exactly is the percussion point? Hint: see below.
10. Let a ball come from the right, at a (horizontal) linear momentum p_1. Upon meeting the bat at height y, the ball returns rightwards, at a smaller linear momentum: $p_2 < p_1$. How much linear momentum did it lose? Hint: $p_1 - p_2$.

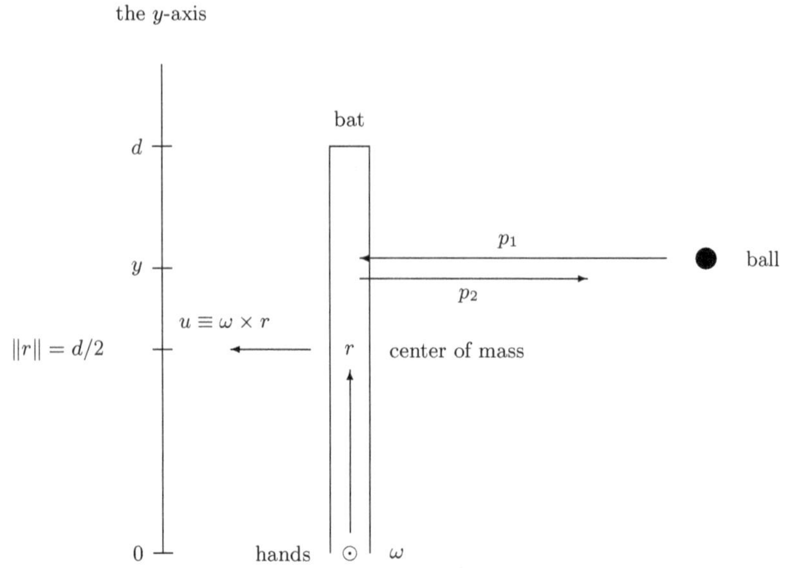

Fig. 2.18. A vertical baseball bat of height d. The ball meets it at height y, and loses momentum $p_1 - p_2$, which is transferred fully to the center of mass of the bat (at height $\|r\| = d/2$), producing a new velocity of $u \equiv \omega \times r$.

11. Where did this go? Hint: thanks to conservation of linear momentum, it went fully to the center of mass of the bat, to give it an initial linear momentum of
$$p_1 - p_2,$$
and an initial velocity of
$$\frac{p_1 - p_2}{m} = u = \omega \times r$$
leftwards (Figure 2.18).
12. What is ω? Hint: this is the angular-velocity vector, issuing from the page towards your eyes, as indicated by '$\odot$' at the origin (the bottom of the bat). Around ω, the entire bat rotates counterclockwise.
13. Conclude that ω-r-u is a right-hand system.
14. Use this in the above formula, and take the norm of both sides. Hint:
$$\frac{1}{m}\|p_1 - p_2\| = \|u\| = \|\omega\| \cdot \|r\| = \|\omega\|\frac{d}{2}.$$
15. Multiply both sides by m:
$$m\|\omega\|\frac{d}{2} = \|p_1 - p_2\|.$$

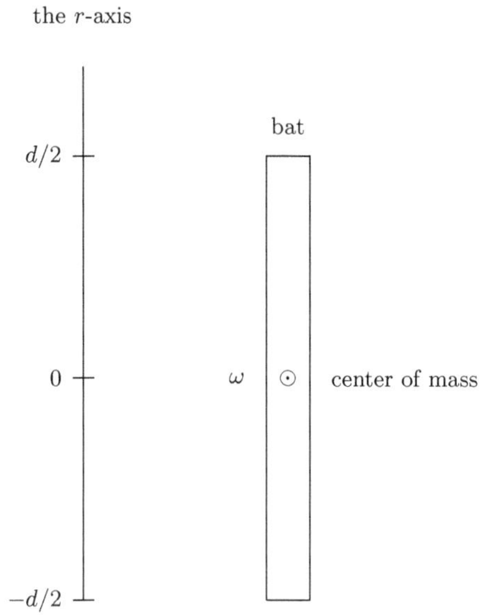

Fig. 2.19. Look at things the other way around. As the center rotates around the bottom, the bottom also rotates around the center (at the same angular velocity ω, issuing now from the center, towards your eyes). From the perspective of the center, the entire bat rotates around it, at a new moment of inertia.

This will be useful later.

16. Now, look at things the other way around. As the midpoint rotates around the bottom, the bottom also rotates (counterclockwise) around the midpoint, at the same angular velocity ω (issuing now from the midpoint, towards your eyes, as in Figure 2.19).
17. What is the moment of inertia around the midpoint? Hint:

$$\begin{aligned} \frac{m}{d}\int_{-d/2}^{d/2} \tilde{y}^2 d\tilde{y} &= 2\frac{m}{d}\int_{0}^{d/2} \tilde{y}^2 d\tilde{y} \\ &= 2\frac{m}{d} \cdot \frac{1}{3}\left(\frac{d}{2}\right)^3 \\ &= 2\frac{m}{d} \cdot \frac{d^3}{24} \\ &= \frac{md^2}{12}. \end{aligned}$$

18. From the perspective of the midpoint. How to write conservation of angular momentum? Hint: at hit time, the original angular momentum (of the ball around the midpoint) is used fully to introduce a new angular momentum in the bat: the above moment of inertia times ω:

$$\left(y - \frac{d}{2}\right) \frac{r}{\|r\|} \times (p_1 - p_2) = \frac{md^2}{12}\omega.$$

19. Take the norm of both sides:

$$\left(y - \frac{d}{2}\right) \|p_1 - p_2\| = \frac{md^2}{12}\|\omega\|.$$

20. Recall that

$$\|p_1 - p_2\| = m\|\omega\|\frac{d}{2}.$$

21. Plug this in:

$$\left(y - \frac{d}{2}\right) m\|\omega\|\frac{d}{2} = \frac{md^2}{12}\|\omega\|.$$

22. Divide both sides by $m\|\omega\|d/2$:

$$y - \frac{d}{2} = \frac{d}{6}.$$

23. Conclude that

$$y = \frac{d}{2} + \frac{d}{6} = \frac{2}{3}d.$$

24. This is the percussion point: the best point to hit the ball.
25. What is so good about this point? Hint: the midpoint is not fixed, but free to rotate around the bottom (with the entire bat). The angular momentum around it was used fully to rotate around the bottom, and hit the ball back. None was left to hurt your hands, or pull the bat out of your hands.

2.6.13 Gravity in The Solar System

1. Look at two physical phenomena: gravity and electrostatics. What do they have in common? Hint: both make a force of $\|r\|^{-2}$, and a potential of $\|r\|^{-1}$ (times a constant, which we disregard).
2. Look at the solar system. The sun is at the origin: $(0, 0, 0)$. Around it, there is a planet at r, orbiting the sun.
3. What is the centrifugal force? Hint: proportional to $\|\omega\|^2\|r\|$. (Forget about constants.)
4. What is its direction? Hint: radial, outwards.
5. What keeps the planet in orbit? What prevents it from escaping to outer space? Hint: gravity supplies the centripetal force, required to balance the centrifugal force, and cancel it out.
6. What is the direction of the centripetal force? Hint: radial, inwards.
7. What is its magnitude? Hint: the same as that of the centrifugal force.
8. What supplies the centripetal force? Hint: gravity.

9. What is the angular velocity? Hint: proportional to $\|r\|^{-3/2}$. This way, the centripetal force agrees with gravity:

$$\|\omega\|^2\|r\| \sim \left(\|r\|^{-3/2}\right)^2 \|r\| = \|r\|^{-2}.$$

10. What is the angular momentum? Hint:

$$\|r \times p\| \sim \|A(r)\omega\| \sim \|r\|^2\|\omega\| \sim \|r\|^2\|r\|^{-3/2} = \|r\|^{1/2}.$$

11. Is it conserved? Hint: yes: $\|r\|$ is constant in the orbit. Furthermore, $r \times p$ keeps pointing in the same direction as ω: perpendicular to the plane containing the orbit (Figure 2.8).
12. What is the kinetic energy? Hint: in general, it is half momentum times velocity. Here, in a circular motion, it is half angular momentum times angular velocity:

$$\frac{1}{2}\|r \times p\| \cdot \|\omega\| \sim \|r\|^{1/2}\|r\|^{-3/2} = \|r\|^{-1}.$$

13. Is it conserved? Hint: yes: $\|r\|$ is constant in the orbit.
14. Where did it come from? Hint: from the gravitational potential, which is proportional to $\|r\|^{-1}$ too (see below).
15. Why doesn't the Earth spiral or fall into the sun? Hint: because it has just enough kinetic energy to remain in orbit all the time.
16. Why is a Neptune-year much longer than an Earth-year? Hint: because the angular velocity falls sharply with $\|r\|$:

$$\|\omega\| \sim \|r\|^{-3/2}.$$

17. Let's use this in the microscale too.

2.6.14 Bohr's Atom and Its Stability

1. Let's move to the microscale.
2. Look at the atom. In Bohr's model, the nucleus is at the origin, and the electrons "live" in their orbitals, around it.
3. First, let's use the same model as in the solar system. Hint: forget about constants, as above.
4. Later on, in quantum mechanics, we'll see that angular momentum comes in discrete quanta: say, n quanta, where n is an integer number.
5. Assume that the electron is at r, orbiting the nucleus. What could the radius $\|r\|$ be? Hint: not every radius is allowed. Indeed, the angular momentum must be n (times a constant):

$$\|r \times p\| \sim \|r\|^{1/2} \sim n.$$

Thus, the radius must be n^2 (times a constant):

$$\|r\| \sim n^2.$$

6. What energy level is allowed? Hint: the energy level must be n^{-2} (times a constant):

$$-\|r\|^{-1} \sim -n^{-2}.$$

7. Where did the minus sign come from? Hint: the electron has a negative charge. This is why it is attracted to the nucleus (which has a positive charge in the proton). In terms of energy, it wants to approach the nucleus, where the potential is minimal:
$$-\|r\|^{-1} \to_{r\to 0} -\infty.$$
8. Where did the electron get its kinetic energy from? Hint: assume that the electron was initially at infinity, where the potential is zero. Then, it started to spiral and fall towards the nucleus, gaining negative potential and positive kinetic energy, until reaching its current position: r. During its journey, it lost potential energy of
$$0 - \left(-\|r\|^{-1}\right) = \|r\|^{-1},$$
which was converted to kinetic energy.
9. While orbiting the nucleus, the electron should radiate an electromagnetic wave, and lose its kinetic energy very quickly (Maxwell's theory). Why doesn't it fall to the nucleus? Hint: in quantum mechanics, its position in its orbit is nondeterministic: a (periodic) wave function, with an integer wave number. This is a standing wave that travels nowhere, and radiates nothing. This is how Debroglie proved stability in Bohr's atom: the electron is not quite a particle, but more of a wave. We'll come back to this later.
10. By now, the electron is at r, at radius $\|r\|$ from the nucleus. Could it get yet closer to the nucleus? In other words, could it fall inwards, to an inner radius? Hint: only if the inner radius is allowed as well.
11. In the inner radius, how much potential energy would it have? Hint: less than before. Indeed, $-\|r\|^{-1}$ grows with $\|r\|$.
12. What happened to the potential energy that was lost? Hint: it transforms into light: the electron emits a photon.
13. On the other hand, could the electron "jump" outwards, to an outer radius? Hint: only if the outer radius is allowed as well.
14. In the outer radius, how much potential energy would it have? Hint: more than before. Indeed, $-\|r\|^{-1}$ grows with $\|r\|$.
15. What could possibly supply the required energy? Hint: a photon (light ray) comes from the sun, and hits the electron hard enough.
16. Could the photon hit the electron ever so hard, and knock it off the atom, all the way to infinity? Hint: only if the light is blue: high frequency, with energy as high as n^{-2} (the old energy level). Only this could pay to kick the electron to $\|r\| = \infty$, where the potential is as high as zero.
17. This is the photo-electric effect, discovered by Einstein. We'll come back to this later.
18. Bohr's model focuses on one electron. Later on, in quantum chemistry, we'll even look at many.

3

Geometrical Optics and Its Stability

To model motion, we used the real axis: $\mathbb{R}$. This is infinite: it has two endpoints: $\pm\infty$. How to "tie" them, and unite them into one point? For this, introduce a new point: the infinity point. This is not really a point, but an abstract concept. Still, it has a geometrical value: it helps tie the real axis, and make it look like a circle. Topologically, both have the same structure: you can now "walk" on the real axis, reach ∞, "jump" (continuously) to $-\infty$, and continue walking rightwards again. This is useful in geometrical optics.

On the 2-D plane, do the same: tie its "edges" into one new point: the infinity point. This would make a new (closed, compact) sphere.

This leads us to complex numbers, useful in quantum mechanics and electromagnetics. The same trick works in the complex plane too: "tie" it at infinity, and obtain the new Riemann sphere.

3.1 The Infinity Point

3.1.1 The Infinity Paradox

The adjective "infinite" contains two words: not finite. Thus, it is actually negative: it assumes that we already know what finite is, and tells us that it is not.

This is why the property of being infinite can't be checked. Assume that you're given a set of items. Is it finite or not? Let's try a naive experiment: count the items, one by one. If you finish, then the set is finite. Otherwise, it is not.

Is this any good? For a finite set, it works: you'll finish, and find out that it is finite. But what about an infinite set? You never finish, and never know the answer! This is the infinity paradox.

Thus, to comprehend infinity, we must use our own imagination. This is why one of the elementary axioms in set theory is the existence of an infinite set.

3.1.2 Mathematical Induction

Not convinced? Use another axiom: mathematical induction. This could help define the natural numbers from scratch: start from the first natural number: 1. (Some people start from 0.) Then, assume that you already have the natural number n. Use it to advance to the next one: $n+1$.

Could this stop? Never! Indeed, if you get tired, then just rest for a while, and start again: add 1 again, and define the next successor.

As before, this is negative: we know what stoppable means, so we can also believe (or accept) that some process is not. Likewise, we know what an end is, so we can trust that something is endless. This is why our "proof" is by contradiction: if the process had stopped, then a contradiction would emerge.

3.1.3 Natural vs. Real Numbers

The set of natural numbers, although infinite, is still discrete (countable). This is the smallest kind of infinity: $\aleph_0$. The real axis $\mathbb{R}$, on the other hand, is "bigger" (Figure 1.1). This is denoted by

$$|\mathbb{R}| = \aleph > \aleph_0.$$

Actually, even the unit interval is of type $\aleph$: it contains a continuum of real numbers (points). This is denoted by

$$|[0,1]| = \aleph > \aleph_0.$$

3.1.4 The Infinity Point: An Abstract Object

The real axis has no left or right end. Let's introduce an "endpoint" from scratch: ∞. This is a new abstract object, standing for both ∞ and $-\infty$ at the same time. This is not a number: it is as big as any number! So, it mustn't lie on the real axis, or it would be smaller than those numbers on its right. (If you haven't noticed, this is a proof by contradiction.) Thus, it must "lie" outside of $\mathbb{R}$. It is an auxiliary point that "exists" in our mind only.

3.1.5 The Extended Real Axis

What is this? Start from the real axis:

$$\mathbb{R} \equiv \{x \mid x \text{ is a real number}\}.$$

Then, extend it by one more "point:" the new infinity "point" ∞' (standing for $\pm\infty$). How to do this? First, place our new infinity point in a little set:

$$\{\infty\}.$$

This set contains just one item: our new infinity point ∞. This could be denoted by

$$\infty \in \{\infty\}$$

(∞ belongs to $\{\infty\}$). We are now ready to extend $\mathbb{R}$ by one item, and form

$$\mathbb{R} \cup \{\infty\}$$

(where '$\cup$' stands for the union of two sets). This is the extended real axis: $\mathbb{R}$ plus the infinity point. In a sense, this is like a circle.

3.2 The Lamp Mapping

3.2.1 The Lamp Mapping

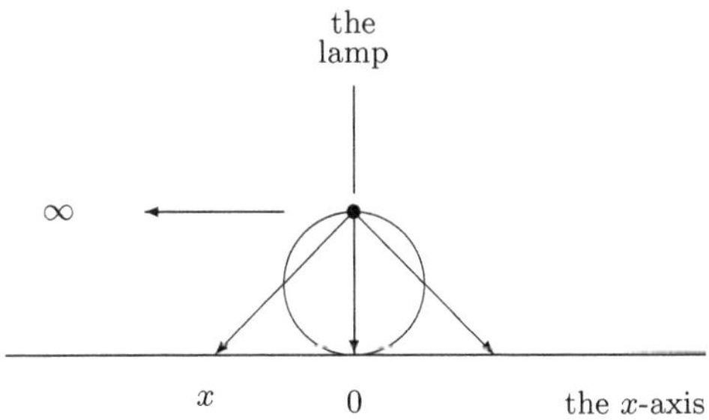

Fig. 3.1. The lamp mapping: each point on the circle projects a shadow on the x-axis. The lamp itself is mapped to ∞.

By now, we extended the real axis by one more point: the infinity point. What is the result? In a sense, it is the same as a circle.

Indeed, look at the Cartesian x-y plane (Figure 3.1). In it, look at the unit circle of radius 1, centered at the origin: $(0,0)$. Now, raise it, so its center is now at $(0,1)$ rather than $(0,0)$. This way, it "sits" on the x-axis, with its bottom at $(0,0)$, and its top at $(0,2)$.

Now, at $(0,2)$, place a little lamp. It shines in all directions. Ignore light going up, and focus on light going down only. Each light ray will pass through one point from the circle, and map it to its shadow on the x-axis. Together, all light rays map the entire circle onto its shadow: the entire x-axis, denoted by

$$\mathbb{R} \equiv \{(x,0) \mid \quad -\infty < x < \infty\}.$$

This is the lamp mapping: each point on the circle projects its own shadow on the x-axis below. There is just one special point (mapped to itself): the origin $(0,0)$. After all, it coincides with its own shadow. Still, there is a problem: what about the lamp itself? After all, it has no shadow at all!

3.2.2 Map The Lamp to Infinity

Look at the lamp at $(0,2)$. Where to map it? To our new point: the infinity point. For this, use the horizontal ray, tangent to the circle at the top.

Why does this make sense? In Figure 3.1, look at the lower-left oblique ray that makes angle 45° with the x-axis. This ray hits the x-axis at some $x < 0$. This is where the shadow is. Now, let x slide leftwards, slowly and smoothly:

$$x \to -\infty.$$

In this limit process, the ray changes as well: the angle it makes with the x-axis gets more and more acute, decreasing gradually, and approaching 0°. At this limit, the ray becomes horizontal: it maps the lamp to $\pm\infty$.

This completes the definition of the lamp mapping. The entire circle was projected onto the extended real axis. This is one-to-one: each individual point on the circle was mapped to a distinct point on the extended real axis.

Furthermore, the extended real axis is completely covered: the mapping is not only *into* but also *onto* it. This is why the lamp mapping is also invertible.

3.2.3 Topological Homeomorphism

In our limit process, x slides leftwards:

$$x \to -\infty$$

(Figure 3.1). But this is not the only option: it could also slide rightwards:

$$x \to \infty.$$

This is a little different: the ray points now down-right rather than down-left. Still, in the limit, it becomes horizontal again, tangent to the circle at the top again, but pointing rightwards rather than leftwards. Either way, it points to our unique infinity point: $\pm\infty$. After all, in the extended real axis, there is no difference:

$$+\infty = -\infty = \infty.$$

What is so good about this? Well, it makes the lamp mapping smooth and continuous: it never breaks, not even at the lamp itself! To see this, consider a particle that "lives" on the circle (Figure 3.1). The lamp maps it to its own shadow. Where is this? At x, on the x-axis below.

Assume that the particle starts from the origin: $(0, 0)$. Now, let it slide slowly, clockwise on the circle. At the same time, its shadow x will slide leftwards.

Once the particle approaches the lamp from the left, its shadow will approach $-\infty$. Furthermore, once the particle hits the lamp, its shadow will hit the infinity point itself.

Moreover, once the particle passes the lamp on its way clockwise, its shadow will proceed smoothly, "through" the infinity point, to the far right part of the x-axis. Finally, once the particle slides along the right part of the circle (back to the origin), its shadow will slide leftwards (back to the origin too).

Thus, the lamp mapping makes a complete topological homeomorphism between the circle and the extended real axis: every continuous move on one is mirrored by a continuous move on the other. Topologically, they are the same: homeomorphic, interchangable, and indistinguishable, with the same continuity properties. This will be useful in geometrical optics.

3.2.4 Reaching Infinity

Reaching infinity? Sounds fantastic. Is this possible at all? After all, this is too far! Still, we just did! Well, not quite directly, but only indirectly. Indeed, the lamp itself mirrors the infinity, and projects onto it. This makes the infinity much more friendly and close to heart.

Reaching the infinity takes too much time, which we don't have. Instead, use a trick: just walk on the circle, and let your shadow do the hard work. Still, this is a bit implicit. After all, we don't have the shadow, or its explicit location. Is there another way? Yes: mirror the infinity by the origin itself.

3.3 The Hyperbolic Mapping

3.3.1 The Hyperbolic Mapping on an Interval

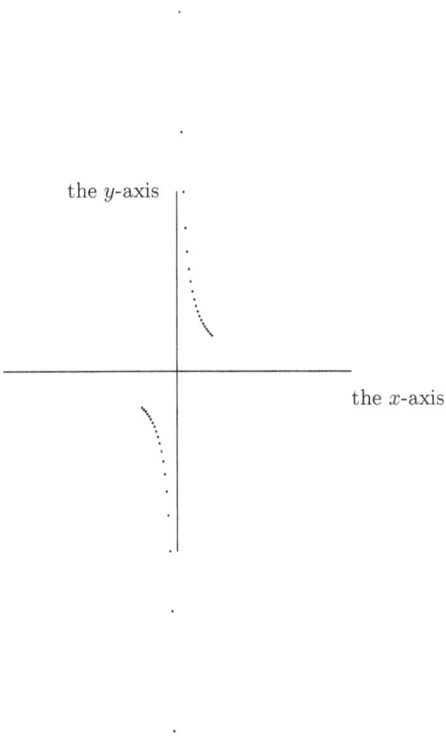

Fig. 3.2. The hyperbolic mapping maps the interval $[-1, 1]$ onto the complementary set $\{y \mid |y| \geq 1\} \cup \{\infty\}$.

On the extended real axis, define another useful mapping: the hyperbolic mapping. For this purpose, start from the closed interval

$$[-1,1] \equiv \{x \mid |x| \leq 1\}$$

(including the endpoints). On this interval, define the hyperbolic mapping:

$$x \to \frac{1}{x}$$

(Figure 3.2).

Let's start from $x = 1$. At this point, $1/x = 1$ as well. Now, let x decrease, gradually and continuously. At the same time, $1/x$ increases continuously. Once x approaches 0 from the right, $1/x$ approaches $+\infty$:

$$\frac{1}{x} \to_{x \to 0+} +\infty.$$

Fortunately, on the extended real axis, the infinity point is unique:

$$-\infty = +\infty = \infty.$$

Thus, once x hits 0, $1/x$ hits ∞.

3.3.2 Topological Homeomorphism

Next, let x decrease yet more: leftwards, pass 0, into the negative part of the original interval. At the same time, $1/x$ passes "through" the infinity point, and emerges from the far negative part of the real axis.

Finally, let x decrease yet more, and reach -1 from the right. At the same time, $1/x$ will increase smoothly, and reach -1 from the negative side. This shows us that $[-1,1]$ is topologically homeomorphic to the complementary set

$$\{y \mid |y| \geq 1\} \cup \{\infty\}.$$

3.3.3 The Inverse Mapping

Indeed, this also works the other way around. Thus, the original mapping has an inverse, which looks just the same:

$$x \to \frac{1}{x}.$$

In fact, the inverse mapping maps the set

$$\{y \mid |y| \geq 1\} \cup \{\infty\}$$

back onto $[-1,1]$ (Figure 3.3).

3.3.4 The Extended Hyperbolic Mapping

Better yet, from the start, define our hyperbolic mapping not only in $[-1,1]$ but also in the extended real axis, mapping it (continuously) onto itself (Figure 3.4). This will be useful in geometrical optics.

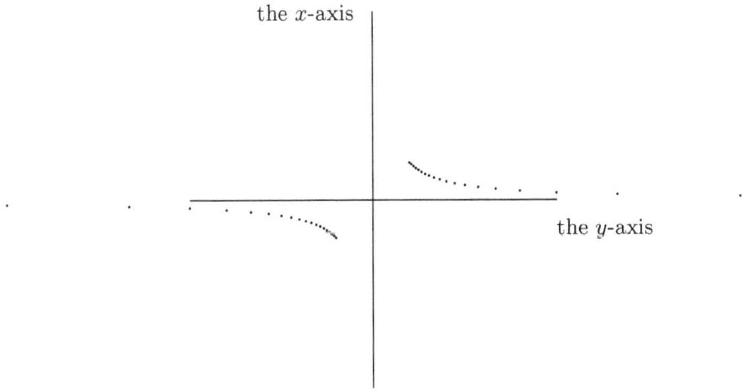

Fig. 3.3. The inverse mapping maps the set $\{y \mid |y| \geq 1\} \cup \{\infty\}$ back onto $[-1, 1]$.

3.4 Optical Focus

3.4.1 Plano Convex Lens

The hyperbolic mapping will be quite useful in geometrical optics [13, 21, 45]. Let's start from a plano convex lens (Figure 3.5). Why is it called so? Because its right face is flat (straight and vertical). Its left face, on the other hand, is round and convex. This way, the left face could be viewed as part of a circle of radius $r > 0$, centered at the bullet in the middle (Figure 3.5).

Assume now that two parallel (horizontal) light rays hit the lens from the left. Upon entering from the left, they break, and change direction towards the focus (marked by $f > 0$ in Figure 3.5).

This is just a geometrical model. Still, it models the true physical phenomenon quite well. To test this, place the lens against the sun. At f, on the other side, place a piece of paper. Will it burn?

The focus is also called the focal point. The positive number f is also called the focal length: the distance between the lens and the focus. In a lens of glass, for example, f satisfies

$$\frac{1}{f} = \frac{1}{2r},$$

or

$$f = 2r.$$

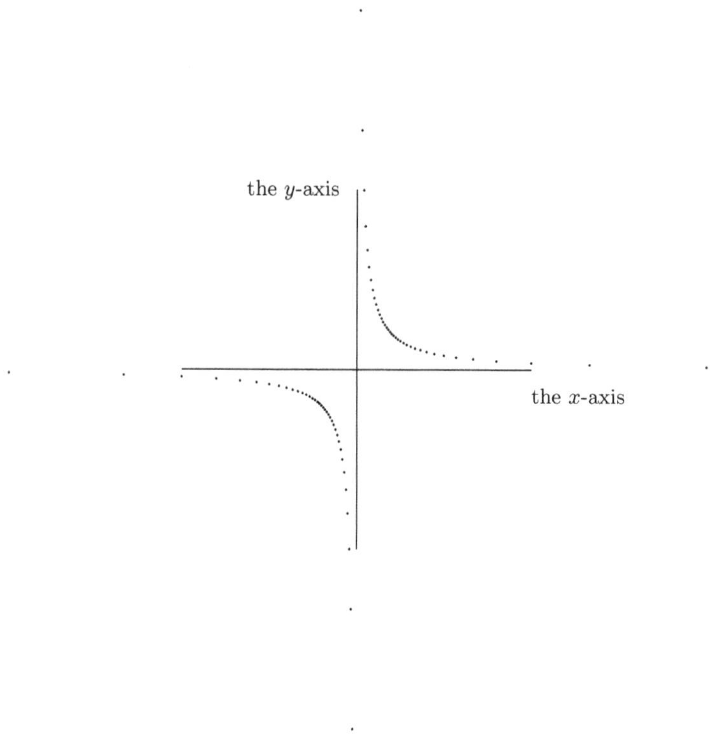

Fig. 3.4. The extended hyperbolic mapping maps the extended real axis onto itself.

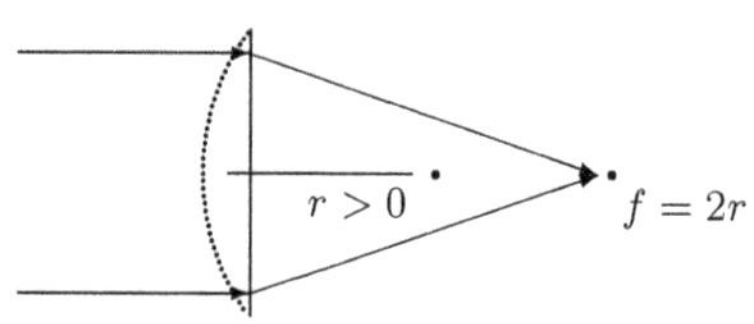

Fig. 3.5. A plano convex lens, made of glass of refractive index 1.5. Upon entering from the left, the parallel horizontal rays break, and meet at the focus $f = 2r$.

More precisely, the formula should read

$$\frac{1}{f} = \frac{\text{refractive index} \; - 1}{r}.$$

But here, for simplicity, we consider a glass of refractive index 1.5. Let's use our formula in a special case.

3.4.2 Flat Window of Glass

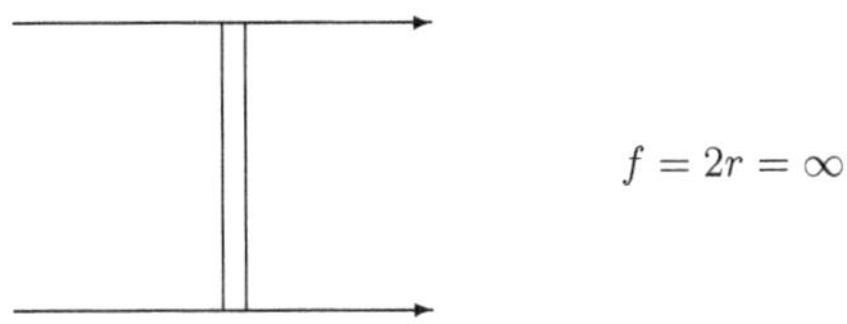

Fig. 3.6. A transparent window of glass. Parallel rays come from the left. They never break, but remain horizontal, and "meet" at $f = \infty$.

Now, let the radius r increase, gradually and continuously. At the same time, the focus travels rightwards smoothly as well, approaching the infinity point ∞:

$$f = 2r \to_{r \to \infty} \infty.$$

What happens when r hits ∞? Well, at the same time, f hits ∞ too. This means that the focus is at infinity.

What happens physically? Both faces are now parallel: flat, straight, and vertical. Thus, the original horizontal rays never break any more, but keep going rightwards: from $-\infty$, through the "lens," to the focus at ∞ (Figure 3.6). This is just a transparent window of glass.

3.4.3 Virtual Focus

So far, our r (and $f = 2r$) hit the infinity point. Next, let them pass through it, and emerge from the far left part of the real axis. What does this mean physically?

Physically, the left face gets now slightly concave: it is now part of a circle of radius $-r$, centered at the bullet marked by $r < 0$ (middle of Figure 3.7). Why is r negative? Because the circle is now on the left, not on the right.

In this case, what happens to the horizontal rays coming from the left? They break away from each other, and never meet any more. Still, to a viewer who looks from the right, they seem to come from one point: the virtual focus at $f = 2r < 0$ (Figure 3.7, on the left). For this reason, every object placed on the left may seem to the viewer smaller than it really is.

Thus, f still satisfies

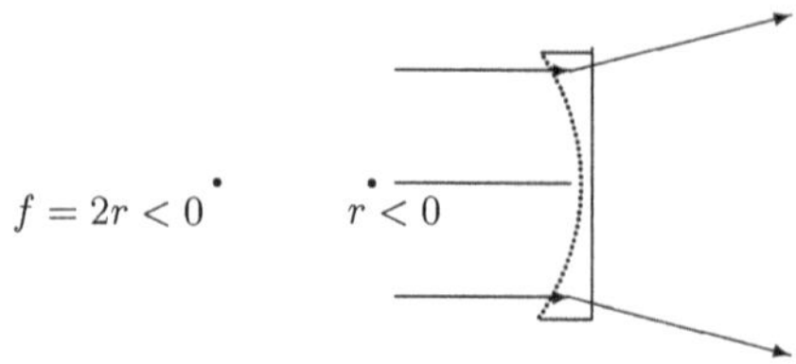

Fig. 3.7. A plano concave lens. Upon entering from the left, the parallel horizontal rays break away from each other, and never meet any more. To a viewer who looks from the right, they seem to come from the virtual focus at $f = 2r < 0$.

$$\frac{1}{f} = \frac{1}{2r},$$

or

$$f = 2r,$$

only this time both f and r are negative, not positive. After all, they are now on the left, not on the right! This way, the focus completed its journey: from $f > 0$, rightwards, through the infinity point, emerging at $f < 0$ on the far left. On the extended real axis, this is quite smooth and continuous.

3.4.4 Convex Lens

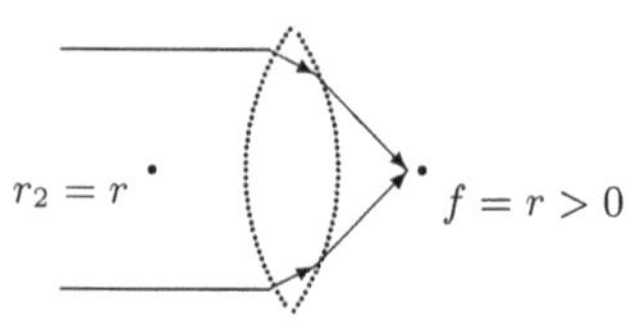

Fig. 3.8. A convex lens. Upon entering from the left, the parallel horizontal rays break. Upon exiting from the right, they break once again, and meet at the focus $f = r = r_2 > 0$.

Let's go back to our plano convex lens (Figure 3.5). So far, it was convex from the left only. Next, make it convex from the right too, by sticking yet another plano convex lens (symmetrically from the right). This makes a complete convex lens (Figure 3.8).

The convex lens has two round faces: the left face is part of a circle of radius $r > 0$ (as before), and the right face is part of a circle of radius $r_2 > 0$. Unlike r and f, however, r_2 is measured the other way around: from the lens and leftwards, not rightwards. This is why r_2 is positive as well.

What happens to parallel horizontal rays coming from the left? They break twice. Upon entering from the left, they break as in Figure 3.5. Upon exiting from the right, they break yet more, and meet at a new (closer) focus, satisfying

$$\frac{1}{f} = \frac{1}{2}\left(\frac{1}{r} + \frac{1}{r_2}\right).$$

Is this familiar? To obtain our plano convex lens back again, just pick $r_2 = \infty$. But it is more interesting to pick $r_2 < \infty$. This makes $1/f$ larger, and f smaller. In Figure 3.8, for example, we pick $r_2 = r$. This makes a symmetric convex lens, with $f = r_2 = r$.

3.4.5 Nonflat Window of Glass

To obtain our convex lens, r_2 decreased: from ∞ to $r_2 < \infty$. Now, go the other way around: increase r_2 back again, gradually and continuously. To see what's going on, we have now an algebraic tool: our formula.

What happens physically? The right face gets less and less convex. Thanks to our formula, this makes f increase as well.

Once r_2 hits ∞, the right face gets completely flat and vertical. Thanks to our formula, f gets as large as $2r$. This is no surprise: we already saw this (Figure 3.5).

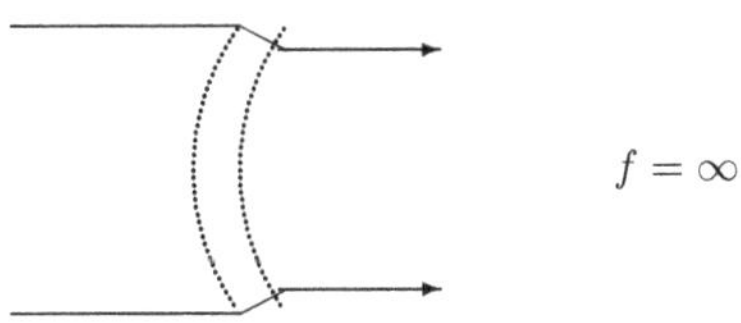

Fig. 3.9. A nonflat window of glass. The parallel rays remain horizontal, and "meet" at the focus at ∞.

But this is not the end of it. After all, in the extended real axis,

$$-\infty = +\infty = \infty.$$

Thus, r_2 may proceed smoothly: through the infinity point, to the far left (negative) part of the real axis. What happens physically? The right face gets now slightly concave, not convex. Thanks to our formula, this makes f increase yet more.

Next, increase r_2 even more: from $-\infty$ rightwards. At the same time, f will increase too. Once r_2 reaches $-r$ from the left, we have

$$\frac{1}{f} = \frac{1}{2}\left(\frac{1}{r} + \frac{1}{r_2}\right) = \frac{1}{2}\left(\frac{1}{r} - \frac{1}{r}\right) = 0,$$

so

$$f = \pm\infty.$$

This is a transparent (but not flat) window of glass. Indeed, our rays remain horizontal, and "meet" at infinity (Figure 3.9).

By now, the focus lies at ∞, which could be interpreted as $+\infty$ (real focus) or $-\infty$ (virtual focus). After all, to a viewer who looks from the right, the rays seem to come from $-\infty$.

3.4.6 Virtual Focus: Beyond Infinity

Finally, let r_2 increase a little more, so $-r < r_2 < 0$. Thanks to our formula, f must now be negative: it went through the infinity point, and emerged from the far left back again. What happens physically? The right face gets so concave that it wins: it overpowers the left face, and overrides its effect. This way, upon entering from the left, the rays still break towards one another. But upon exiting from the right, they break away even more sharply. The total effect is, thus, a divergence: to a viewer looking from the right, the rays seem to come from a virtual focus, to the left of the lens.

In summary, like r_2, f completed its journey rightwards: from r, through $2r$, and then through the infinity point, to emerge from the far left back again.

3.4.7 Symmetric Lens

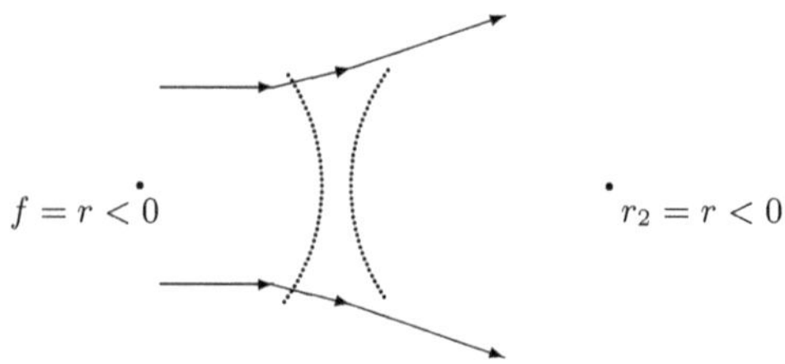

Fig. 3.10. A symmetric concave lens. Upon entering from the left, the rays break away. Upon exiting from the right, they break away even more, and never meet any more. To a viewer who looks from the right, they seem to come from the virtual focus at $f = r < 0$.

The same process also works from the other side. Instead of increasing r_2, increase r. This has the same effect on f.

What happens physically? Instead of the right face, the left face gets now less and less convex. This has the same effect as before: the focus travels rightwards again, pass the infinity point, to become virtual at the far left part of the real axis, as before.

Now, why not carry out both processes at the same time? This keeps the lens symmetric, all the time: start from convex, with $r_2 = r$ (Figure 3.8). Then, increase both $r_2 = r$ at the same time. This way, $f = r_2 = r$ increases too.

3.4.8 From Convex to Flat to Concave

What happens physically? Our symmetric lens gets less and less convex: its focus goes rightwards. At the infinity point $f = r_2 = r = \infty$, the lens gets completely flat, like a window of glass (Figure 3.6).

This is not the end of it. Carry on: make the (symmetric) lens concave, with two concave faces (Figure 3.10). This way, it has now a new virtual focus, at the left part of the real axis.

3.4.9 Convex and Concave: New Definition

How to tell whether a (symmetric or nonsymmetric) lens is convex or concave? Just look at the sign of f. If $f > 0$, then it is convex. (This actually defines a convex lens from scratch.) If, on the other hand, $f < 0$, then it is concave (which actually defines a concave lens too.) Finally, if $f = \infty$, then it is transparent: a window of glass.

3.5 Magnification

3.5.1 A Convex Lens

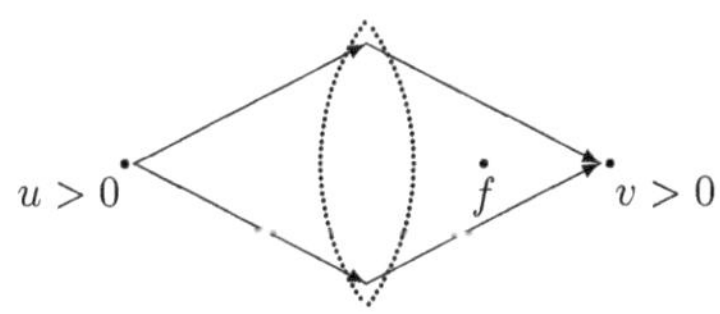

Fig. 3.11. No magnification at all: the object is at $u = 2f$, so its image at $v = 2f = u$ has the same size. Indeed, thanks to symmetry, the image is the same as the original object, only upside-down.

Let's use our convex lens in practice. For this purpose, place an object on the left, at distance u from the lens. Here, u is positive: it is measured from the lens leftwards, not rightwards (Figure 3.11).

Consider now a few (nonparallel) light rays, issuing from the object, and passing through the lens, to produce an (upside-down) image on the right, at distance v. Unlike u, v is measured from the lens rightwards, as usual. This way, $v > 0$ as well.

Fortunately, v could be obtained algebraically, from the following formula:

$$\frac{1}{u} + \frac{1}{v} = \frac{1}{f}.$$

This will be useful below. Moreover, the magnification is

$$\frac{|v|}{u}.$$

This means that a viewer whose eye is at v (looking leftwards, towards the lens) may think that the object is $|v|/u$ times as big as it really is.

3.5.2 No Magnification At All

Let's "play" with u. After all, this is our degree of freedom, which we can control and change. To calculate the magnification, we can now use our formula. Assume first that

$$u = 2f$$

(Figure 3.11). Thanks to our formula, we have

$$\frac{1}{v} = \frac{1}{f} - \frac{1}{u} = \frac{2}{u} - \frac{1}{u} = \frac{1}{u},$$

so we have complete symmetry:

$$v = u.$$

It is therefore no surprise that the magnification is

$$\frac{|v|}{u} = 1.$$

So, there is no gain at all: the image has the same size.

3.5.3 A Little Magnification

Don't worry: the lens is good. The problem is that the object is too far. Decrease u a little:

$$f < u < 2f,$$

to get the object a little closer. Could this help? Since $1/u$ increases,

$$\frac{1}{v} = \frac{1}{f} - \frac{1}{u}$$

decreases, so v must increase. As a result, the magnification is now larger than before:

$$\frac{|v|}{u} > 1.$$

3.5.4 Maximal Magnification

Now, let u decrease yet more, approaching f^+ from above. At the same time, v will increase, improving magnification yet more. Finally, once u hits f, we have

$$\frac{1}{v} = \frac{1}{f} - \frac{1}{u} = 0,$$

so

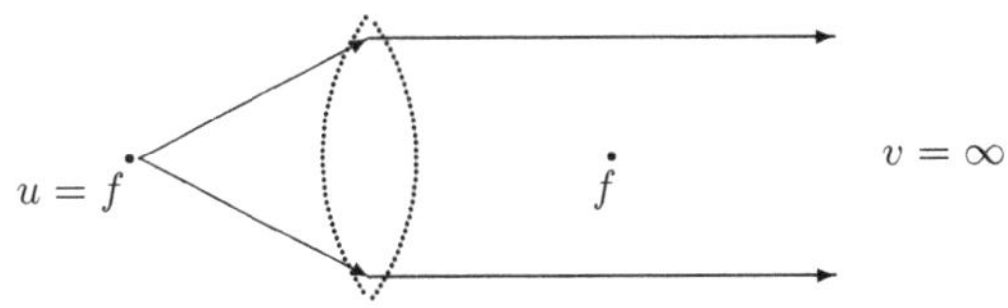

Fig. 3.12. Maximal magnification: the object is at $u = f$, so its image at $v = \infty$ has an "infinite" size.

$$v = \infty.$$

This is maximal magnification:

$$\frac{|v|}{u} = \infty.$$

Indeed, the infinitesimal object at $u = f$ "grows" into a sizable image at $v = \infty$ (Figure 3.12).

Clearly, this is just a theoretical model. A real object can't be so small. Still, our model tells us two plausible things:

- The magnification is pretty good, as evident from experiment.
- Physics is quite stable: the magnification remains good even "beyond" infinity!

Let's see this geometrically.

3.6 Stability

3.6.1 Virtual Image

To prove stability, let the object get even closer:

$$\frac{f}{2} < u < f.$$

This way,

$$\frac{1}{v} = \frac{1}{f} - \frac{1}{u} < 0.$$

What happened to v? It passed through the infinity point, and emerged from the far left back again. Looking from the right, the viewer sees now a virtual image: as if the rays came from the left. Although v is now negative, $|v|$ is still pretty large, so the magnification is still good.

3.6.2 Looking Through a Window of Glass

So far, we used a convex lens. What about a transparent window of glass (Figure 3.6 or 3.9)? In this case,

$$f = \infty.$$

After all, this is what transparency means: horizontal rays remain horizontal, and only "meet" at infinity. Thus,

$$\frac{1}{v} = \frac{1}{f} - \frac{1}{u} = -\frac{1}{u},$$

so

$$v = -u.$$

This is good: the viewer views the rays as coming from the object itself, as they really do. There is no magnification at all, as required.

3.6.3 Looking Through a Concave Lens

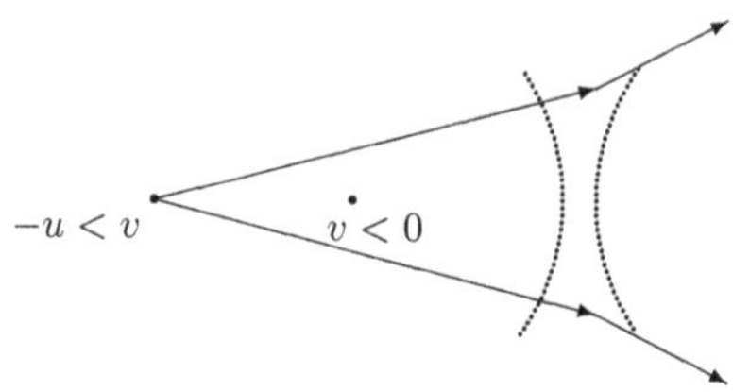

Fig. 3.13. A concave lens. To a viewer looking from the right, the virtual image at $v < 0$ seems smaller than the original object.

Finally, consider a concave lens, with a virtual focus: $f < 0$ (Figure 3.13). (After all, in Section 3.4.9, this is a legitimate definition.) In this case,

$$\frac{1}{v} = \frac{1}{f} - \frac{1}{u} < -\frac{1}{u},$$

so v is negative, but not as negative as $-u$:

$$-u < v < 0.$$

For this reason,

$$\frac{|v|}{u} < 1.$$

Thus, a viewer looking from the right will see a virtual image smaller than the original object (Figure 3.13). This is indeed as evident from experiment.

3.7 The Projective Line

3.7.1 From an Interval to a Circle

In the lamp mapping, the lamp is placed at the top of the circle. This maps the entire circle: each point projects its own shadow. Still, this is not the only way. Why not place the lamp at the center, and project the lower semicircle only? This makes a familiar function: the cotangent function. But, before doing this, we must "tie" the semicircle, and produce a complete circle.

For this purpose, consider the closed interval

$$[0, 2\pi] \equiv \{x \mid 0 \leq x \leq 2\pi\}$$

(including the endpoints). Now, identify both endpoints with one another, and consider them as one and the same:

$$0 = 2\pi.$$

In other words, map this interval to a circle:

$$x \to \eta,$$

where η is an angle in the circle. Since $\eta = 0$ is the same as $\eta = 2\pi$, we can also identify $x = 0$ with $x = 2\pi$, and treat them as one and the same.

3.7.2 The Composite Cotangent Function

Next, map to the subinterval

$$[0, \pi] \equiv \{x \mid 0 \leq x \leq \pi\}.$$

This could be done linearly:

$$x \to \frac{x}{2}.$$

On top of this, apply the cotangent function:

$$\operatorname{cotan}(x/2) \equiv \frac{\cos(x/2)}{\sin(x/2)}.$$

This is applied to the original interval $[0, 2\pi]$ (or the circle), and maps it onto the extended real axis. In particular, $0 = 2\pi$ is mapped to $\infty = \infty$, as required.

3.7.3 Project Onto The Cotangent Line

What is the geometrical meaning of this? To see this, look at the two-dimensional x-y plane. In it, draw a horizontal line, parallel to the x-axis:

$$\{(x, 1) \mid -\infty < x < \infty\}.$$

This is the cotangent line.

How to define the cotangent function geometrically? To do this, pick some angle $0 < \theta < \pi$. Now, in the x-y plane, draw the oblique line l that passes through the origin $(0, 0)$, and makes angle θ with the positive part of the x-axis (Figure 3.14). You could span l by a new running parameter $-\infty < \alpha < \infty$:

$$l \equiv \{(\cos(\theta), \sin(\theta))\alpha \mid -\infty < \alpha < \infty\}.$$

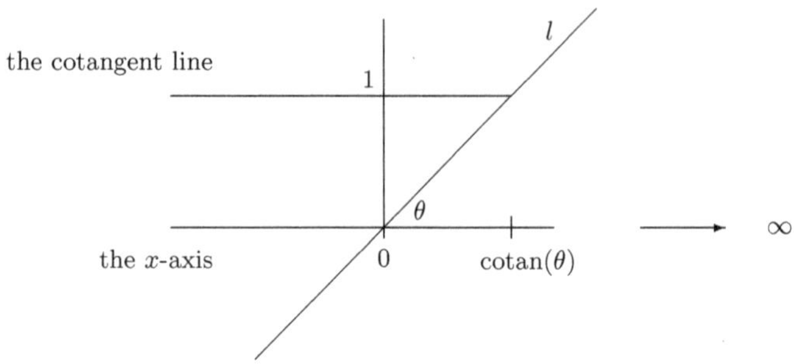

Fig. 3.14. Project onto the horizontal cotangent line. Draw the oblique line l that makes angle θ with the positive part of the x-axis. Look at the point where l meets the cotangent line. Look at its x-coordinate: this is $\text{cotan}(\theta)$. For $\theta = 0$, however, l "meets" the cotangent line at ∞.

We can now redefine the cotangent function geometrically: just project on the cotangent line. In other words, find the unique point where l meets the cotangent line. For this purpose, pick α cleverly:

$$\alpha \equiv \frac{1}{\sin(\theta)}.$$

This way, the y-coordinate is 1, so we are indeed on the cotangent line:

$$(\cos(\theta), \sin(\theta))\alpha = \frac{1}{\sin(\theta)}(\cos(\theta), \sin(\theta)) = (\text{cotan}(\theta), 1).$$

Now, look at the x-coordinate. This is what we wanted: $\text{cotan}(\theta)$. Finally, to have our composite function, substitute $x/2$ for θ:

$$\text{cotan}(x/2).$$

3.7.4 Project to Infinity

All this is very nice, but there is still one problem: what about

$$x = 0 = 2\pi,$$

or

$$\theta = 0 \quad \text{or} \quad \pi?$$

At this point, l coincides with the x-axis:

$$l \equiv \{(1,0)\alpha \mid -\infty < \alpha < \infty\} = \{(-1,0)\alpha \mid -\infty < \alpha < \infty\}.$$

How to define the cotangent? Easy: let l approach the x axis. For this, let θ approach 0 or π (which is just the same, thanks to our infinity point):

$$\lim_{\theta \to 0^+} \cot\text{an}(\theta) = \infty = -\infty = \lim_{\theta \to \pi^-} \cot\text{an}(\theta).$$

This is projective geometry: $x = 0 = 2\pi$ indeed mirrors the infinity point, as required.

3.8 The Extended Plane

3.8.1 The Extended Cartesian Plane

So far, we worked in the one-dimensional real axis. To it, we added one more point: the infinity point. Next, let's move on to the two-dimensional Cartesian plane:

$$\mathbb{R}^2 \equiv \{(x,y) \mid -\infty < x, y < \infty\}.$$

To it, add one more point: a new infinity point. This is the extended Cartesian plane:

$$\mathbb{R}^2 \cup \{\infty\}.$$

This "ties" the plane, and produces a complete sphere. To see this, we need to extend the lamp mapping.

3.8.2 The Lamp Mapping in The Sphere

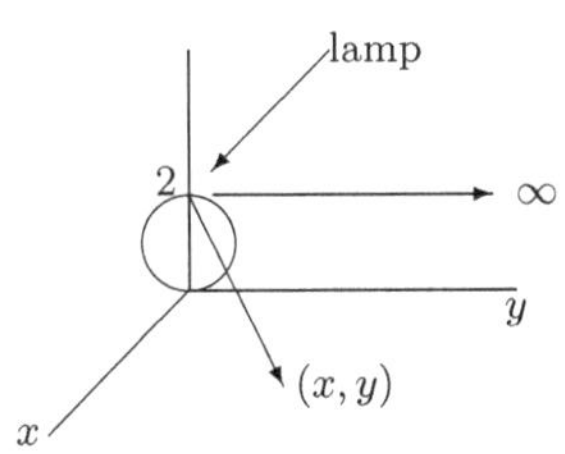

Fig. 3.15. Place the lamp at the top of the sphere: $(0,0,2)$. Each point on the sphere projects to its own shadow on the x-y plane. The lamp itself is mapped to ∞.

Consider the three-dimensional x-y-z Cartesian space. In it, consider the unit sphere of radius 1, centered at the origin $(0,0,0)$. Now, raise it a little, so its center is at $(0,0,1)$. This way, it "sits" on the x-y plane, with its bottom at $(0,0,0)$, and its top at $(0,0,2)$ (Figure 3.15).

Now, at the top, place a little lamp. Ignore light going up, and focus on light going down. Each light ray passes through one point from the sphere, and maps it to its shadow on the x-y plane. This maps the entire sphere onto the entire x-y plane:

$$\{(x, y, 0) \mid \quad -\infty < x, y < \infty\} .$$

For example, the origin $(0, 0, 0)$ is mapped to itself. After all, it coincides with its own shadow. But what about the lamp itself? Where to map it?

For this, consider a particle, placed at the origin $(0, 0, 0)$. Let it travel along some longitude, slowly and continuously, towards the lamp at $(0, 0, 2)$. At the same time, its shadow travels as well: from the origin, straight towards infinity. To preserve continuity, the lamp itself must be mapped to the new infinity point: ∞. This projects the entire sphere onto the extended Cartesian plane, as required. Furthermore, this is one-to-one: distinct points from the sphere are mapped to distinct points on the extended Cartesian plane.

Moreover, the extended Cartesian plane is completely covered, so the mapping is not only *into* but also *onto* it. As a result, this is invertible: a topological homeomorphism, useful in the complex plane.

3.9 Background: Complex Numbers

3.9.1 The Imaginary Axis

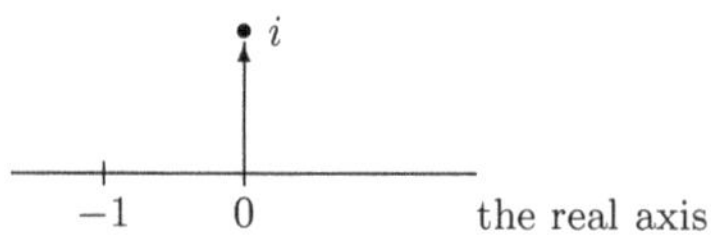

Fig. 3.16. The imaginary number i. The arrow leading from the origin to i makes a right angle of 90° with the real axis. In $i^2 = -1$, on the other hand, this angle doubles, to make a flat angle of 180° with the positive part of the real axis.

The complex plane is made of complex numbers, useful in quantum mechanics and electromagnetics. To design them, start from the negative number -1. It has no square root: there is no real number whose square is -1. Fortunately, we already know how to design a new auxiliary (not real) number. Let's use this trick once again, and introduce a new imaginary number (denoted by i), satisfying

$$i^2 = -1,$$

or

$$i \equiv \sqrt{-1}.$$

Because it lies outside the real axis, i may now span a new vertical axis, perpendicular to the real axis (Figure 3.16).

For this purpose, place i at $(0,1)$, above the origin. This way, i spans the entire imaginary axis (using a running parameter $-\infty < b < \infty$):

$$\{bi = (0,b) \mid \quad -\infty < b < \infty\}.$$

Here, b could be any real number. In the complex plane, bi looks like $(0,b)$: a new point on the vertical imaginary axis. For instance, $b = 1$ gives us i back again: $i = (0,1)$.

3.9.2 The Complex Plane

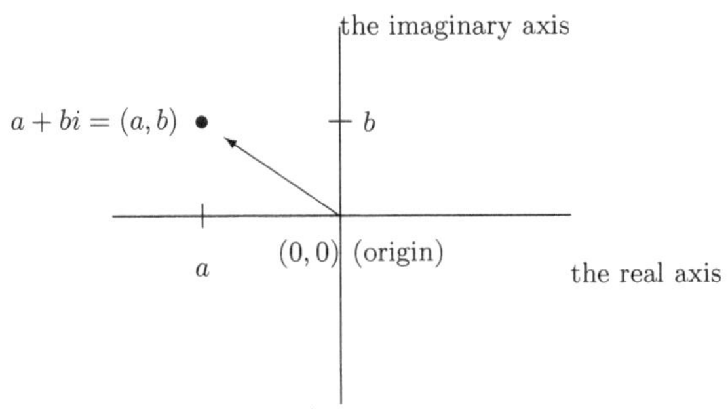

Fig. 3.17. The complex plane $\mathbb{C}$. The imaginary number $i \equiv \sqrt{-1}$ is at $(0,1)$. A complex number $a + bi$ is at (a,b).

Together, the real and imaginary axes span the entire complex plane, denoted by $\mathbb{C}$:

$$\mathbb{C} \equiv \{a + bi \equiv (a,b) \mid \quad -\infty < a,b < \infty\}.$$

Here, a and b are any real numbers. The new complex number $a + bi$ occupies the geometrical point (a,b) in $\mathbb{C}$ (Figure 3.17).

3.9.3 Arithmetic Operations

By now, our complex plane is just geometrical. It is time to get algebraic, and define arithmetic operations. Addition is easy: use the parallelogram rule: add coordinate by coordinate. What about multiplication? Well, we already know how to multiply i by itself:

$$i^2 = -1.$$

After all, this is how i was defined in the first plane.

3.9.4 Multiplication

This equation is our algebraic key. What is its geometrical face? Well, i points vertically upwards (Figure 3.16). This way, i makes a right angle of 90° with the real axis. Now, taking the square means doubling the angle. This is why i^2 makes a flat angle of 180° with the positive part of the real axis, and points leftwards: $i^2 = -1$.

Now, extend this linearly: let a, b, c, and d be some real numbers. Use the distributive and commutative laws to multiply $a + bi$ times $c + di$:

$$\begin{aligned}(a+bi)(c+di) &= a(c+di) + bi(c+di) \\ &= ac + adi + bci + bdi^2 \\ &= ac - bd + (ad+bc)i.\end{aligned}$$

3.9.5 The Complex Conjugate

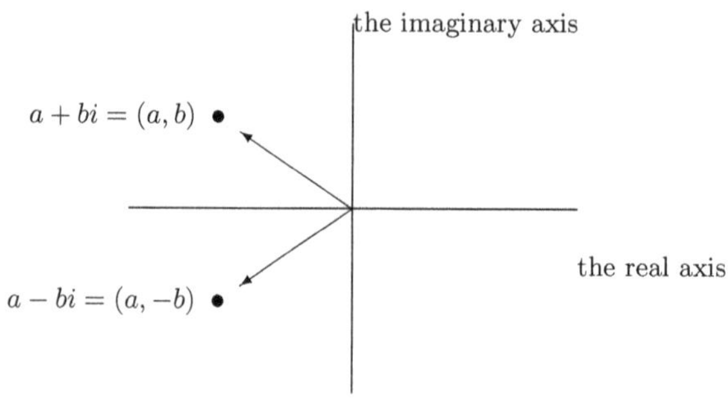

Fig. 3.18. $a + bi$ is mirrored by its complex conjugate: $a - bi$.

How to make the latter term vanish? Set

$$c = a \quad \text{and} \quad d = -b,$$

giving us

$$c + di = a - bi.$$

this is nearly the same as $a + bi$. There is just one change: the imaginary part picks a minus sign. This mirrors the original complex number: the real axis acts like a mirror (Figure 3.18).

This is the complex conjugate. It is denoted by $a + bi$, but with a small bar on top:

$$a - bi = (\overline{a + bi}).$$

Look at their product: it is completely real, with no imaginary part at all:

$$(a+bi)(a-bi) = a^2 + b^2.$$

3.9.6 Absolute Value

What is the square root of this? Thanks to Pythagoras' theorem, this is the length (or magnitude, or norm) of (a, b). This will also be the absolute value of $a + bi$, denoted by $|a + bi|$:

$$|a + bi|^2 \equiv (a + bi)(a - bi) = a^2 + b^2.$$

3.9.7 Reciprocal or Inverse

If $a + bi \neq 0$, then this is nonzero, and could be placed in the denominator:

$$(a + bi)\frac{a - bi}{a^2 + b^2} = 1.$$

this gives us the reciprocal (inverse):

$$(a + bi)^{-1} = \frac{a - bi}{a^2 + b^2}.$$

3.10 Riemann's Sphere

3.10.1 The Extended Complex Plane

To the complex plane, add now the infinity point. This forms the extended complex plane:

$$\mathbb{C} \cup \{\infty\}.$$

Thanks to the lamp mapping, this is topologically homeomorphic to the sphere. This is why it is also called Riemann's sphere.

3.11 Towards Differential Geometry

3.11.1 Tangent Plane: Chart

How to use the lamp mapping in practice? Place the lamp at the north pole. This will map the face of the Earth (excluding the north pole itself) onto the tangent plane on the south. On it, each latitude will have a circular shadow, and each longitude a radial shadow. Together, these will make new polar coordinates on the tangent plane, ready to serve as a new chart.

3.11.2 Atlas of Charts

But what about the north pole itself? It is still excluded! To cover it too, work also the other way around: place the lamp at the south, not the north. This time, the south pole will be excluded, but we don't care: it was already covered before. So, in two charts, we managed to cover the entire Earth.

This is our new atlas: two charts. Clearly, they have a big overlap. Still, they are compatible there: smoothly transferable from one to another.

3.11.3 Differentiable Manifold

This is a rather small atlas: two charts only. But you can also design many more atlases, each with (compatible) charts that cover the Earth fully (perhaps with less overlap). Is there a maximal atlas that includes all possible subatlases? Thanks to Zorn's lemma, there is (Chapter 12 in [48]). Thanks to this, the sphere makes a new differentiable manifold, useful in differential geometry. Likewise, in general relativity, spacetime makes a 4-D differentiable manifold, to help model gravity.

3.12 Exercises

1. Show that the lamp mapping (Figure 3.1) is one-to-one: distinct points are mapped to distinct points.
2. Show that it is not only *into* but also *onto* the extended real axis: it covers it completely.
3. Conclude that it is invertible.
4. Conclude that it is a topological homeomorphism.
5. Show that the inverse mapping is one-to-one as well.
6. Show that it is not only *into* but also *onto* the original circle.
7. Prove the same for the lamp mapping in the sphere (Figure 3.15).

Part II

Stability in Classical Mechanics

Part II: Stability in Classical Mechanics

By now, we already saw stability in two fields: geometrical mechanics and optics. This leads us to chaos theory. What is it? It studies dynamic systems, and their stability: sensitivity to initial conditions. It warns us: initial data are not always good enough to predict the future! Indeed, however accurate, they may still lead to a substantial error in the long run. This teaches us to be more humble, and not always trust our own eyes. Better have a little bit of doubt, and never take things for granted. This is a first step towards quantum mechanics and relativity, where doubt is the name of the game.

4

Poincar'e Stability

Consider a bounded domain in 3-D. In it, a particle could start from any initial point, and follow a complicated trajectory. This is its dynamics. How will it behave? Will it converge asymptotically to some fixed point? We don't know! Still, Poincar'e tells us one thing for sure: you can't place the trajectory in a pipe! There is just not enough room for this! In the end, the trajectory must (almost) return to the same point. More precisely, it has an accumulation point, near which it must pass, as close as you like. It doesn't have to converge to it: it could get close, then get away, then get close back again, and so on. This could be quite chaotic, with no pattern at all. Later on, we'll see more about dynamics: Hamiltonian mechanics, and its conservation laws.

4.1 Poincar'e Recurrence Theorem

4.1.1 The Domain

Consider a domain in 3-D:

$$V \subset \mathbb{R}^3.$$

(Actually, our discussion applies even to a higher dimension.) What is this? Well, V has a boundary: ∂V. (For instance, if V is a box, then ∂V contains the faces.) It also has an interior: $V \setminus \partial V$ (meaning V but the boundary). This is an open set: every point in it could be surrounded by a little ball, completely in it. The boundary, on the other hand, is not. Indeed, look at a point on the boundary. Every ball around it, however small, would remain half outside! For instance, if V is a box, then $V \setminus \partial V$ is the interior only, without the faces.

4.1.2 The Curved Trajectory

Let $v_0 \in V$ be our starting point. From it, issue an infinite trajectory. It starts from v_0, and wanders around in V. Where does it go? Well, it could approach one fixed point asymptotically, but this is quite rare and special. It could fool us: it may seem to approach some point, but then change its mind and start going towards another point, and then change its mind again and approach yet another point, and so on, with no pattern at all. This is called a soliton. Worse, it could be even more crazy, and approach no point at all! Can we say anything about its behavior?

Well, Poincar'e can. He says that, in a compact domain, there is an accumulation point, by which the trajectory passes as close as you like (but not necessarily converges).

4.1.3 Accumulation Point

To see this, assume that V is bounded, and can be placed in a box. Now, split the box into eight equal subboxes. In at least one subbox, the trajectory passes (and exits, end reenters) infinitely many times (or even stays there forever, which is even better). Pick one such subbox (using, say, the axiom of choice). In it, pick some arbitrary point (using the axiom of choice again). Now, in this subbox, repeat the same process recursively: split it into eight equal subsubboxes, and so on. In the end, the arbitrary points that you picked will form a Cauchy sequence, converging to a unique limit point. This is the accumulation point. Near it, the trajectory must pass, as close as you like.

4.1.4 Compact Domain

Where is our accumulation point? It is not necessarily in V: it could be in ∂V. Assume now that V is compact: not only bounded but also closed (contains its own boundary):

$$\partial V \subset V.$$

In this case, the accumulation point must be in V, as required.

4.2 Poincar'e Stability

4.2.1 The Mapping

Our trajectory illustrates how a particle propagates from the initial point $v_0 \in V$. After one second, it arrives at the new point $T(v_0)$. After two seconds, it arrives at $T^2(v_0) = T(T(v_0))$, and so on. Here, T maps the entire domain, telling us the whole dynamics. T^2 is not a square, but a composite function.

4.2.2 Mapping and Its Origin

In theory, two different trajectories could meet (or cross one another), and two particles could even occupy the same point. Thus, in general, T is not invertible: you can't go back in time! Thus, T^{-1} is *not* an inverse. What is T^{-1}? To see this, let

$$A \subset V$$

be a subset of our domain. Then, $T^{-1}(A)$ contains all those points that are mapped into A:

$$T^{-1}(A) \equiv \{v \in V \mid T(v) \in A\}.$$

We then say that $T^{-1}(A)$ is the origin of A under T.

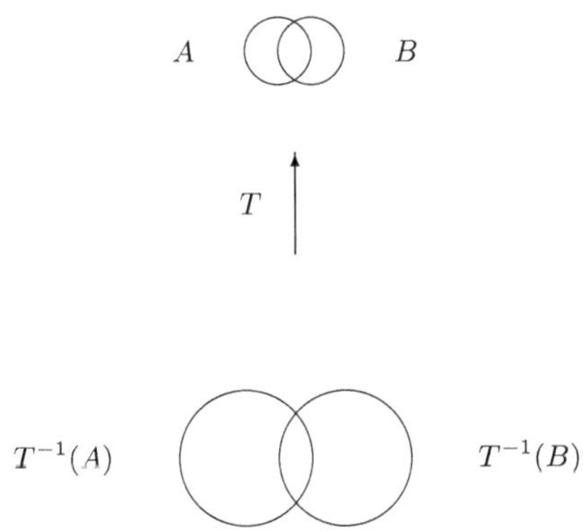

Fig. 4.1. T maps upwards: the lower-left circle goes to the upper-left circle, and the lower-right circle to the upper-right circle. Thus, the lower overlap goes to the upper overlap.

4.2.3 Intersection and Its Origin

What can we say about our (noninvertible) mapping T, and its origin? Consider two subsets of our domain:

$$A, B \subset V.$$

Their intersection (or overlap) contains their joint points:

$$A \cap B \equiv \{v \in V \mid v \in A \text{ and } v \in B\}.$$

In other words, the following statements are the same:

$$v \in A \cap B \Leftrightarrow v \in A \text{ and } v \in B.$$

In other words, $A \cap B$ is the maximal set included in both A and B. For example, if A and B are disjoint, then they have no overlap at all:

$$A \cap B = \emptyset$$

(the empty set). Still, this is quite rare. Usually, A and B may overlap, and have a few common points. Anyway, what is the origin of $A \cap B$ under T? Look at Figure 4.1:

$$\begin{aligned}
T^{-1}(A \cap B) &\equiv \{v \in V \mid T(v) \in A \cap B\} \\
&= \{v \in V \mid T(v) \in A \text{ and } T(v) \in B\} \\
&= \{v \in V \mid T(v) \in A\} \cap \{u \in V \mid T(u) \in B\} \\
&= T^{-1}(A) \cap T^{-1}(B).
\end{aligned}$$

Thus, we have a sort of "distributive" law, telling us how to open parentheses. Order doesn't matter: either intersect and then apply T^{-1}, or work the other way around:

apply T^{-1} and then intersect. In other words, the origin of the overlap is the overlap of the origins. This works not only for T but also for the composite mapping T^n ($n \geq 0$):

$$T^{-n}(A \cap B) = T^{-n}(A) \cap T^{-n}(B).$$

We're now ready to prove Poincar‘e stability.

4.2.4 Volume

What is volume? This is a real function (denoted by $|\cdot|$): to each subset of V, it assigns a real value (positive or zero). Furthermore, it is also additive, in the following sense: for every two subsets $A, B \subset V$,

$$|A \cap B| = 0 \Rightarrow |A \cup B| = |A| + |B|.$$

This is quite useful in classical mechanics.

4.2.5 Stationarity: Volume Preservation

Assume that our domain has a finite volume:

$$|V| < \infty.$$

Furthermore, assume that our mapping is stationary: it preserves volume. In other words, the origin has the same volume:

$$\left|T^{-1}(A)\right| = |A|$$

(for any subset $A \subset V$). In theory, A could even be as thin as air, and have no volume at all:

$$|A| = 0.$$

But we're more interested in a more substantial A, with a positive volume:

$$|A| > 0.$$

Now, use A to form an infinite sequence of sets: the origin of A, the origin of the origin, and so on:

$$A,\ T^{-1}(A),\ T^{-2}(A),\ \ldots,\ T^{-n}(A),\ \ldots.$$

Thanks to stationarity, all have the same volume: $|A| > 0$. Thus, they can't be completely disjoint: there is just not enough room! There must be two sets that really overlap:

$$\left|T^{-i}(A) \cap T^{-j}(A)\right| > 0$$

(for some indices $j > i \geq 0$). Thanks to our "distributive" law (Section 4.2.3), we can now factor T^{-i} out:

$$T^{-i}(A) \cap T^{-j}(A) = T^{-i}\left(A \cap T^{-(j-i)}(A)\right).$$

Thanks to volume preservation, we now have

$$\begin{aligned}\left|A \cap T^{-(j-i)}(A)\right| &= \left|T^{-i}\left(A \cap T^{-(j-i)}(A)\right)\right| \\ &= \left|T^{-i}(A) \cap T^{-j}(A)\right| \\ &> 0.\end{aligned}$$

Is this familiar? This mirrors the process used before to find the accumulation point. To make things clearer, define

$$N \equiv j - i \geq 1.$$

Thus, for every A of positive volume, there is an $N \geq 1$ for which $T^{-N}(A)$ overlaps with A substantially:

$$\left|A \cap T^{-N}(A)\right| > 0.$$

As a matter of fact, each A may have its own private N: different A's may have different N's. Thus, N is not quite uniform, but actually depends on A:

$$N \equiv N(A).$$

(But here we know our A, so we can safely drop this.) We're now ready to show stability.

4.2.6 Poincar'e Stability

So far, infinity was never mentioned. Besides, points were never mentioned: we used (sub)sets only. This is a good sign: sets are more fundamental. Still, it is time to go infinite. For this, let $A \subset V$ be a substantial set (of positive volume):

$$|A| > 0$$

(however small). Let $B \subset A$ contain "bad" points that never return to A:

$$B \equiv \{v \in A \mid T^n(v) \notin A,\ n = 1, 2, 3, \ldots\}.$$

The rest of the points in A, on the other hand, are "good:" eventually, they return to A. Our job is to show that B is as thin as air. Let's prove this by contradiction. Assume momentarily that B was substantial:

$$|B| > 0.$$

In this case, thanks to Section 4.2.5, there must be an

$$N \equiv N(B) \geq 1$$

for which

$$\left|B \cap T^{-N}(B)\right| > 0.$$

But these are disjoint:

$$B \cap T^{-N}(B) = \emptyset.$$

After all, $T^{-N}(B)$ contains "good" points that return to B after N applications of T, and can never be in B (which contains "bad" points that never return to A).

Thus, our momentary assumption was false: B could never be substantial! On the contrary:

$$|B| = 0.$$

This is Poincar'e stability: almost all points in A must eventually return to A.

Where was the infinity used? Only implicitly: to design B in the first place. Indeed, look at a given point in A. Is it good or bad? If good, then you'll get to know this: just apply T time and again, until getting back into A. But what if it was bad? You'd never get to know! This is the infinity paradox again. This is why our proof was by contradiction (Section 3.1.1).

4.3 Exercises

1. Define the real axis:
$$\mathbb{R} \equiv \{x \mid -\infty < x < \infty\}.$$
2. Define the three-dimensional Cartesian space:
$$\mathbb{R}^3 \equiv \{(x, y, z) \mid -\infty < x, y, z < \infty\}.$$
3. To it, add a fourth dimension: the w-coordinate:
$$\mathbb{R}^4 \equiv \{(x, y, z, w) \mid -\infty < x, y, z, w < \infty\}.$$
4. Prove Poincar'e recurrence theorem in 4-D too.
5. Consider a three-dimensional compact domain $V \subset \mathbb{R}^3$. Does it have a finite volume? Hint: since V is bounded, its volume can be defined as
$$|V| \equiv \int\int\int_V dxdydz < \infty.$$
6. Use this to establish Poincar'e stability.
7. Extend this to 4-D as well.
8. To define B in Section 4.2.6, how to "solve" the infinity paradox?

5

Chaos and Stability

Chaos theory tells us how difficult it is to predict the future. Given the initial conditions and the laws of nature, you might want to use them to draw the entire trajectory of the entire process. Unfortunately, there is too much instability and sensitivity to initial conditions. In fact, even a tiny error could grow exponentially, and produce a huge relative error in the long run. This makes the results useless. Still, chaos theory finds a nice pattern even in chaos. In a sense, there is some order, even in chaos. Let's see this in a simple model.

5.1 Process and Its Error

5.1.1 Relative vs. Absolute Error

To illustrate chaos, let's discuss a simple algebraic process. It starts from a given initial number: $v_0 \neq 0$. How does it continue? Step by step. At each step, it gets multiplied by a fixed parameter: $\alpha \neq 0$. For $n = 0, 1, 2, \ldots$, we get

$$v_{n+1} \equiv \alpha v_n.$$

(As a matter of fact, this is mathematical induction.) This is a linear process: in each step, v_n gets multiplied by a constant number: α. (Later on, we'll also see a nonlinear process, in which v_n gets multiplied by itself as well.) Thanks to mathematical induction, v_n also has a more explicit form:

$$v_n = \alpha v_{n-1} = \alpha^2 v_{n-2} = \cdots = \alpha^n v_0.$$

Assume now that v_n is some physical quantity, measured to some finite precision only. This way, the v_n's can never be known precisely, but only within an error. As a matter of fact, even the initial data could be inaccurate. To account for this, assume that v_0 is available only approximately, as

$$v_0 + \delta_0,$$

where the error δ_0 is small in magnitude relative to v_0:

$$|\delta_0| \ll |v_0|.$$

Under these circumstances, we can no longer compute v_n exactly, but only approximately: with an error δ_n. Indeed, mathematical induction can still be used to define the approximations

$$\begin{aligned} v_n + \delta_n &\equiv \alpha(v_{n-1} + \delta_{n-1}) \\ &= \alpha^2(v_{n-2} + \delta_{n-2}) \\ = \cdots &= \alpha^n(v_0 + \delta_0) \\ &= \alpha^n v_0 + \alpha^n \delta_0 \\ &= v_n + \alpha^n \delta_0. \end{aligned}$$

Thus, the error δ_n undergoes the same linear process as well:

$$\delta_n = \alpha^n \delta_0.$$

If $|\alpha| > 1$, then the absolute error $|\delta_n|$ may grow exponentially with n. Fortunately, what matters is not the absolute but the relative error, which doesn't grow at all:

$$\left|\frac{\delta_n}{v_n}\right| = \left|\frac{\alpha^n \delta_0}{\alpha^n v_0}\right| = \left|\frac{\delta_0}{v_0}\right|.$$

This is a stable process: the relative error remains small, so $v_n + \delta_n$ approximates v_n quite well. For this reason, $v_n + \delta_n$ is good enough in practice. In a nonlinear process, on the other hand, this is not always the case.

5.2 Stability vs. Instability

5.2.1 Instability in a Quadratic Process

To design a nonlinear process, we need yet another parameter: $\beta \neq 0$. Thanks to it, we can now define a quadratic process:

$$v_{n+1} \equiv \alpha v_n + \beta v_n^2.$$

Let's write this in terms of a new function y:

$$v_{n+1} - v_n = y(v_n),$$

where the quadratic function

$$y(x) \equiv (\alpha - 1)x + \beta x^2$$

tells us how the process progresses. For simplicity, assume that

$$\beta > 0.$$

(Otherwise, just work with $-v_n$ and $-\beta$.) This way, the graph of $y(x)$ is a straight parabola, as in Figure 5.1.

Do the v_n's converge to any limit? Well, let v_∞ denote such a (hypothetical) limit. It must satisfy

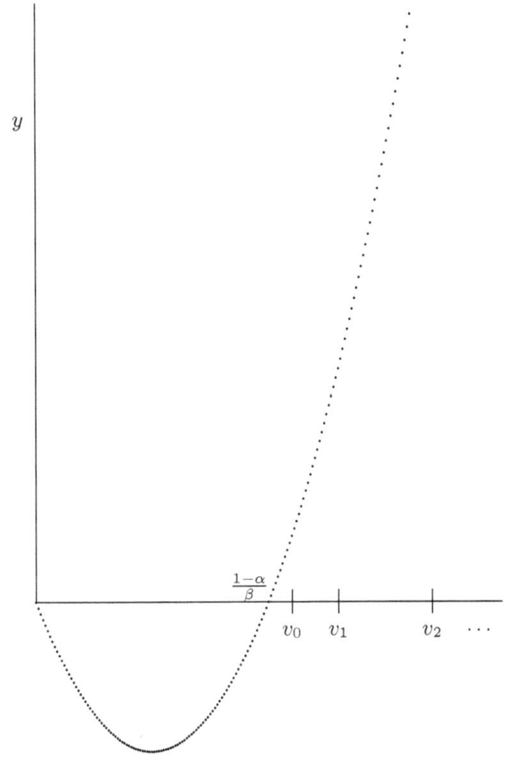

Fig. 5.1. An unstable quadratic process. The v_n's increase monotonically towards infinity, away from $(1-\alpha)/\beta$. The calculation is unstable and meaningless.

$$\begin{aligned}
0 &= v_\infty - v_\infty \\
&= \lim_{n\to\infty} v_n - \lim_{n\to\infty} v_n \\
&= \lim_{n\to\infty} v_{n+1} - \lim_{n\to\infty} v_n \\
&= \lim_{n\to\infty} (v_{n+1} - v_n) \\
&= \lim_{n\to\infty} y(v_n) \\
&= y(v_\infty).
\end{aligned}$$

Thus, v_∞ must solve the quadratic equation

$$y(x) = 0.$$

We already have one solution: $x = 0$. Still, there is a more interesting solution:

$$v_\infty \equiv \frac{1-\alpha}{\beta}.$$

In what follows, we'll distinguish between two major cases. In the first one,

$$\alpha \le 1,$$

so our solution is positive or zero:

$$v_\infty \ge 0.$$

But is this a good limit? Well, assume that our initial value was bigger:

$$v_0 > v_\infty$$

(Figure 5.1). By mathematical induction, the process will go rightwards, and diverge to infinity:

$$v_{n+1} - v_n = y(v_n) > 0.$$

How rapid is this? Well, as a bonus, we also have the ratio

$$1 \le \frac{v_{n+1}}{v_n} = \alpha + \beta v_n,$$

leading to a rapid divergence:

$$v_\infty < v_0 < v_1 < v_2 < \cdots < v_n \to_{n\to\infty} \infty.$$

Worse, the entire calculation is often misleading. After all, in practice, the calculation is inexact: it is only up to an error δ:

$$v_{n+1} + \delta_{n+1} \equiv \alpha\,(v_n + \delta_n) + \beta(v_n + \delta_n)^2.$$

For simplicity, neglect quadratic errors that are as small as δ^2:

$$\begin{aligned} v_{n+1} + \delta_{n+1} &= \alpha(v_n + \delta_n) + \beta(v_n + \delta_n)^2 \\ &= \alpha(v_n + \delta_n) + \beta\left(v_n^2 + 2v_n\delta_n + \delta_n^2\right) \\ &= v_{n+1} + \alpha\delta_n + \beta\left(2v_n\delta_n + \delta_n^2\right) \\ &\doteq v_{n+1} + \alpha\delta_n + 2\beta v_n\delta_n \\ &= v_{n+1} + (\alpha + 2\beta v_n)\,\delta_n. \end{aligned}$$

Thus, the new error is approximately

$$\delta_{n+1} \doteq (\alpha + 2\beta v_n)\,\delta_n.$$

Usually, $\delta_n \ne 0$, so we can also divide:

$$\frac{\delta_{n+1}}{\delta_n} \doteq \alpha + 2\beta v_n.$$

Is this familiar? This is just the derivative of v_{n+1} at v_n, denoted by $v'_{n+1}(v_n)$. This tells us how fast the error grows. Let's use this to estimate the relative error:

$$\frac{\delta_{n+1}}{v_{n+1}} \doteq \frac{(\alpha + 2\beta v_n)\,\delta_n}{v_{n+1}} = \frac{(\alpha + 2\beta v_n)\,\delta_n}{\alpha v_n + \beta v_n^2} = \frac{\alpha + 2\beta v_n}{\alpha + \beta v_n} \cdot \frac{\delta_n}{v_n}.$$

Thus, upon advancing from n to $n+1$, the relative error gets multiplied by the approximate factor

$$\frac{\alpha + 2\beta v_n}{\alpha + \beta v_n} = 2 - \frac{\alpha}{\alpha + \beta v_n} \geq \begin{cases} 2 & \text{if } \alpha \leq 0 \\ 2 - \frac{\alpha}{\alpha + \beta v_0} & \text{if } \alpha > 0. \end{cases}$$

Either way, this is bad news: the relative error grows exponentially:

$$\left| \frac{\delta_n}{v_n} \right| \geq \left(\frac{|\alpha| + 2\beta v_0}{|\alpha| + \beta v_0} \right)^n \left| \frac{\delta_0}{v_0} \right|.$$

Thus, the computed value $v_n + \delta_n$ is no good: it tells us nothing about v_n itself! Let's see a better case.

5.2.2 The Logistic Equation

So far, we assumed that $\alpha \leq 1$. Unfortunately, this produced an unstable process that can never be calculated in practice. Let's move on to a fundamentally different case [35]. For this purpose, assume now that

$$\alpha > 1.$$

For simplicity, assume now that

$$\beta < 0.$$

(Otherwise, just work with $-v_n$ and $-\beta$.) This way, the parabola $y(x)$ is now upside-down (Figure 5.2). As before, it still meets the x-axis at the hypothetic limit

$$v_\infty \equiv \frac{1 - \alpha}{\beta} \geq 0,$$

which solves the logistic equation:

$$y(v_\infty) = (\alpha - 1)v_\infty + \beta v_\infty^2 = 0.$$

In terms of stability and convergence, the situation is now much better than before. Indeed, if we start from some $0 < v_0 < v_\infty$, then we have

$$v_1 - v_0 = y(v_0) > 0.$$

If v_1 is still smaller than v_∞, then we still have

$$v_2 - v_1 = y(v_1) > 0.$$

But v_2 could be too large, and even skip v_∞ (Figure 5.2). In this case, we have

$$v_3 - v_2 = y(v_2) < 0,$$

so v_3 may approach v_∞ again (from the right). This is good. But this is not the end of it: v_4 may skip v_∞ again, and be on its left again. Fortunately, v_5 may still approach v_∞ (from the left again), and so on. This is good: the process may approach v_∞ from alternating sides.

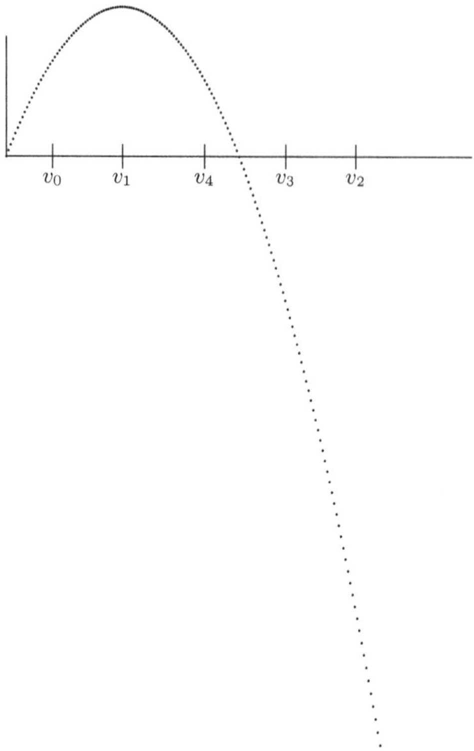

Fig. 5.2. A stable quadratic process. The v_n's approach v_∞. The calculation is now stable and meaningful.

5.2.3 Control Parameter

By now, it seems that our v_n's may approach their limit: v_∞. Do they indeed? To check on this, assume that they are already pretty close to their limit. How close? To measure this, redefine our δ's as the distance from the limit:

$$\delta_n \equiv v_n - v_\infty.$$

Thanks to this new definition,

$$\begin{aligned} v_\infty + \delta_{n+1} &= v_{n+1} \\ &= \alpha v_n + \beta v_n^2 \\ &= \alpha(v_\infty + \delta_n) + \beta(v_\infty + \delta_n)^2 \\ &= \alpha(v_\infty + \delta_n) + \beta\left(v_\infty^2 + 2v_\infty\delta_n + \delta_n^2\right) \end{aligned}$$

$$\begin{aligned}
&= v_\infty + \alpha\delta_n + \beta\left(2v_\infty\delta_n + \delta_n^2\right) \\
&\doteq v_\infty + \alpha\delta_n + 2\beta v_\infty\delta_n \\
&= v_\infty + \alpha\delta_n + 2(1-\alpha)\delta_n \\
&= v_\infty + (2-\alpha)\delta_n.
\end{aligned}$$

Thus,

$$\delta_{n+1} \doteq (2-\alpha)\delta_n.$$

Assuming that $\delta_n \neq 0$, we can also divide:

$$\frac{\delta_{n+1}}{\delta_n} \doteq 2-\alpha.$$

Is this familiar? This is just the derivative of $v_{n+1}(v_n)$ at v_∞, denoted by $v'_{n+1}(v_\infty)$. So, is the process stable? This depends on α: if it is not too large,

$$1 < \alpha \leq 3,$$

then the error (which now means distance from the limit) remains small:

$$|\delta_{n+1}| \doteq |(2-\alpha)\delta_n| = |2-\alpha| \cdot |\delta_n| \leq |\delta_n|.$$

In this case, the relative error remains small as well:

$$\left|\frac{\delta_{n+1}}{v_\infty}\right| \leq \left|\frac{\delta_n}{v_\infty}\right|,$$

so the calculation is indeed meaningful. This is independent of β: only α tells us about stability! This is why α is called the control parameter. Still, β is important too: its sign tells us how the parabola looks like.

5.3 Chaos and Hidden Order

5.3.1 Bifurcation Points

We can now look at different kinds of α's. In particular, the special point

$$\alpha_0 \equiv 3$$

distinguishes "good" from "bad" α's. This is why α_0 is called a bifurcation point. For α's larger than α_0, there is no convergence any more. Still, even for these "bad" α's, there is still some hope. In fact, there is a bigger bifurcation point:

$$\alpha_1 \doteq 3.45.$$

For those α's in between

$$\alpha_0 < \alpha < \alpha_1,$$

although the original process no longer converges, it still splits into two converging subprocesses: the even-numbered items converge to one limit,

$$v_0, v_2, v_4, v_6, \ldots, v_{2n} \rightarrow_{n\to\infty} v_{2\infty},$$

and the odd-numbered items converge to another limit:

$$v_1, v_3, v_5, v_7, \ldots, v_{2n+1} \rightarrow_{n\to\infty} v_{2\infty+1}$$

(just a notation). And what happens beyond α_1? The next bifurcation point is

$$\alpha_2 \doteq 3.55.$$

For those α's in between

$$\alpha_1 < \alpha < \alpha_2,$$

there are four disjoint subprocesses, converging to four different limits:

$$\begin{aligned} v_{4n} &\rightarrow_{n\to\infty} v_{4\infty} \\ v_{4n+1} &\rightarrow_{n\to\infty} v_{4\infty+1} \\ v_{4n+2} &\rightarrow_{n\to\infty} v_{4\infty+2} \\ v_{4n+3} &\rightarrow_{n\to\infty} v_{4\infty+3}, \end{aligned}$$

and so on.

5.3.2 Bifurcation Tree

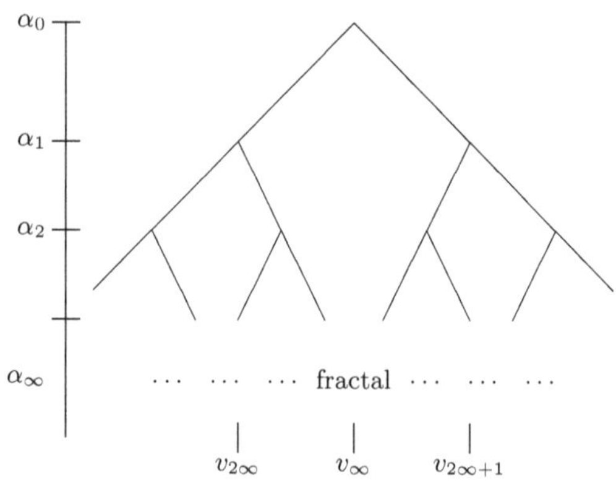

Fig. 5.3. The bifurcation tree.

This goes on: one could go on defining infinitely many bifurcation points:

$$\alpha_0 < \alpha_1 < \alpha_2 < \alpha_3 < \alpha_4 < \cdots < \alpha_m \to_{m \to \infty} \alpha_\infty \doteq 3.57.$$

What happens at $\alpha = \alpha_\infty$? Chaos! No subsequence converges any more. In other words, the original process has no accumulation point any more (Section 4.1.3).

Fortunately, even in chaos, there is still some hope. Indeed, from deep chaos, a beautiful geometrical pattern emerges: a new bifurcation tree (Figure 5.3). At its "bottom," it is quite transparent: it approaches a fractal of dimension< 1.

5.4 Fractal and Its Dimension

5.4.1 Binary Tree and Its Dimension

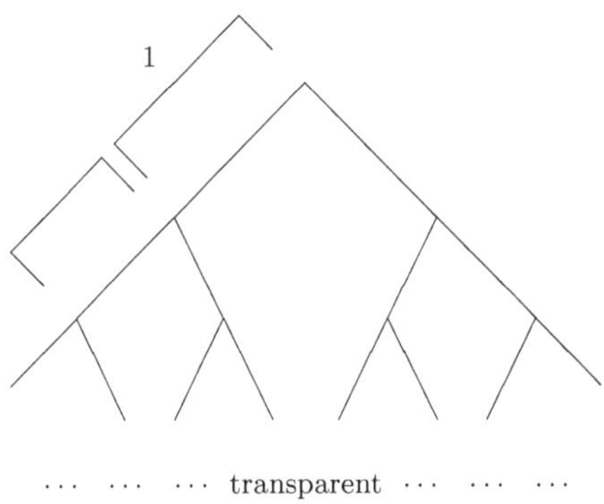

Fig. 5.4. The binary tree is a fractal of 1 <dimension< 2. Why? Because, at its bottom, it is quite transparent: it approaches a set of dimension< 1.

Look at an infinite binary tree, with infinitely many levels (Figure 5.4). What is its dimension? Well, to draw it, we must use 2-D. Still, is it really 2-D? After all, it is quite sparse, and contains more air than matter!

To answer this, let's try and cover it with thin strips. From the head, two branches issue. Cover each with a $1 \times \varepsilon/2^2$ strip (where $\varepsilon > 0$ is a small parameter). This is the first level. The next lower level contains four branches. Cover each with a thinner $1 \times \varepsilon/2^4$ strip. The next lower level contains twice as many: eight branches. Cover each with a yet thinner $1 \times \varepsilon/2^6$ strip, and so on. In the end, you'll cover infinitely many levels with strips of total area ε. But ε could be as small as you like. Thus, our binary tree can't be 2-D. It must be a fractal, of 1 <dimension< 2. Why? Because, at its bottom, it is quite transparent: it is covered from below by a set of dimension< 1.

5.4.2 Feigenbaum's Constant

Our binary tree is infinitely tall: all branches are of the same length: 1. This is why our strips were of length 1 too. Our bifurcation tree (Figure 5.3), on the other hand, is not so tall: upon going from some level to the next lower level, the branches "shrink" by a nearly constant factor. Indeed, if you go up the tree, then the ratio approaches

$$\frac{\alpha_m - \alpha_{m-1}}{\alpha_{m+1} - \alpha_m} \to_{m \to \infty} \delta \doteq 4.669.$$

Here, δ is a new constant: Feigenbaum's constant, which governs many phenomena in science.

5.5 Exercises: Picard Iteration

5.5.1 Fixed-Point Iteration

1. Let F be a real (differentiable) function:

$$F : \mathbb{R} \to \mathbb{R}.$$

2. Define our process in a more general way:

$$v_{n+1} \equiv F(v_n).$$

3. Use F to define y in general. Hint:

$$y(x) \equiv F(x) - x.$$

4. Differentiate this. Hint:

$$F' = y' + 1.$$

5. To be a legitimate limit, what should v_∞ satisfy? Hint:

$$y(v_\infty) = 0$$
$$|F'(v_\infty)| < 1.$$

6. Write this inequality in terms of y. Hint:

$$|F'(v_\infty)| = |y'(v_\infty) + 1| < 1.$$

7. Our quadratic process is just an example. How? Hint: see below.
8. In our case, what is F? Hint:

$$F(x) = \alpha x + \beta x^2.$$

9. And what is v_∞? Hint: to be a root of y (and make y vanish), v_∞ must be either 0 (which is not so good) or $(1-\alpha)/\beta$ (which is better).
10. At the latter root, what else must v_∞ satisfy? Hint:

$$|F'(v_\infty)| = \left| F'\left(\frac{1-\alpha}{\beta}\right) \right| = \left| \alpha + 2\beta \frac{1-\alpha}{\beta} \right| = |2 - \alpha| < 1.$$

11. This depends on α only. This is why α is called the control parameter.
12. In Section 5.2.1, why did we get instability? Hint: because $\alpha \leq 1$ was too small.
13. At $\alpha = 3$, could there still be stability in some sense? Hint: a subprocess could still converge. This is why $\alpha_0 \equiv 3$ is a bifurcation point.
14. In general, look at the composite function:

$$F^2(x) \equiv F(F(x)).$$

(Don't get confused: this is a composition, not a square.)
15. Use this to define two subprocesses.
16. Differentiate F^2. Hint: this is a composite function:

$$(F^2(x))' = (F(F(x)))' = F'(F(x))F'(x).$$

17. In the subprocess, what should the limit satisfy?
18. Look at a limit of F (if exists). What about F^2? How would it behave at this limit? Hint: at the limit, we already know two things: $F(v_\infty) = v_\infty$, and $|F'(v_\infty)| < 1$. Now, look at F^2. It is easy to see that, at this limit, it satisfies the same conditions! This is why the subprocesses converge (even faster) to the same limit.
19. And what happens at an x that is not a limit? Hint: there, $F(x) \neq x$, but a subprocess may still converge, as at a bifurcation point.
20. Extend this to a yet higher dimension as well. Hint: let F be a vector function, and F' its Jacobian. Replace absolute value by norm.

Part III

Towards Quantum Statistical Mechanics

Part III: Towards Quantum Statistical Mechanics

So far, we looked at instability: the error grows too fast: exponentially. To model this, look at a binary tree. It grows exponentially too: the next level contains twice as many branches. This is quite fast, common in chaos theory.

Pascal's triangle, on the other hand, grows more modestly – linearly: the next level contains just one more cell. But underneath, you can still find an exponential growth too. To see this, draw geometrical paths, from cell to cell, level to level. Together, they make a familiar structure: a new binary tree. This is the basis for the binomial formula, useful in statistical mechanics.

In this part, we introduce the binomial formula from two points of view: combinatorial and geometrical. Thanks to mathematical induction, we'll get to learn a lot about the binomial coefficients: how they are mirrored by geometrical paths, and how they help model an interesting phenomenon in statistical physics: Brownian motion.

Furthermore, in its extended form, the binomial formula is useful in quantum mechanics as well. Indeed, thanks to conservation of mass and energy, we can estimate the total number of particles in each energy level. This is related to three types of distributions: molecules are distributed in Maxwell-Boltzmann statistics, bosons in Bose-Einstein statistics, and electrons in Fermi-Dirac statistics.

6

Newton's Binomial and Brownian Motion

So far, we used a binary tree to model chaos from a geometrical point of view. Here, on the other hand, we use it for yet another purpose: to draw geometrical paths in Pascal's triangle, and count them. This leads to two important formulas: the binomial and trinomial formulas. Later on, we'll extend them even more, and use them in physics.

What is a binary tree? It is a good example of multilevel hierarchy, with an exponential growth. Indeed, the next level contains twice as many branches.

In terms of cells, on the other hand, Pascal's triangle grows quite modestly: only linearly: the next (lower) level contains just one more cell. Still, we'll soon get our good old binary tree back again: in Pascal's triangle, draw geometrical paths, leading from level to level. Together, they form a binary tree back again, with its familiar exponential growth. We can now study these paths, and count them. This leads to the binomial coefficients, and to Newton's formula. Thanks to mathematical induction, we'll even extend it to more advanced formulas, useful in statistical physics.

6.1 Geometrical Point of View

6.1.1 Pascal's Triangle

The binomial formula is most useful in (quantum) statistical mechanics. To introduce it from scratch, let's use Pascal's triangle. This is made of oblique cells, listed row by row (Figure 6.1). Later on, each cell will contain a number (entry).

How are the cells indexed? By two indices: n and k. The rows (or levels) are indexed by $n = 0, 1, 2, 3, \ldots$, top to bottom. For instance, the zeroth level (at the top) contains just one cell (and one entry in it). The next lower level contains two cells, indexed by $k = 0, 1$. The next lower level contains three cells, indexed by $k = 0, 1, 2$, and so on.

How to define the entries in the cells? By mathematical induction, of course! In the top cell, place the entry 1 (Figure 6.2). This fills the zeroth level, as required. Now, for $n = 0, 1, 2, 3, \ldots$, assume that the nth level was already filled (the induction hypothesis). How to fill the $(n + 1)$st level below?

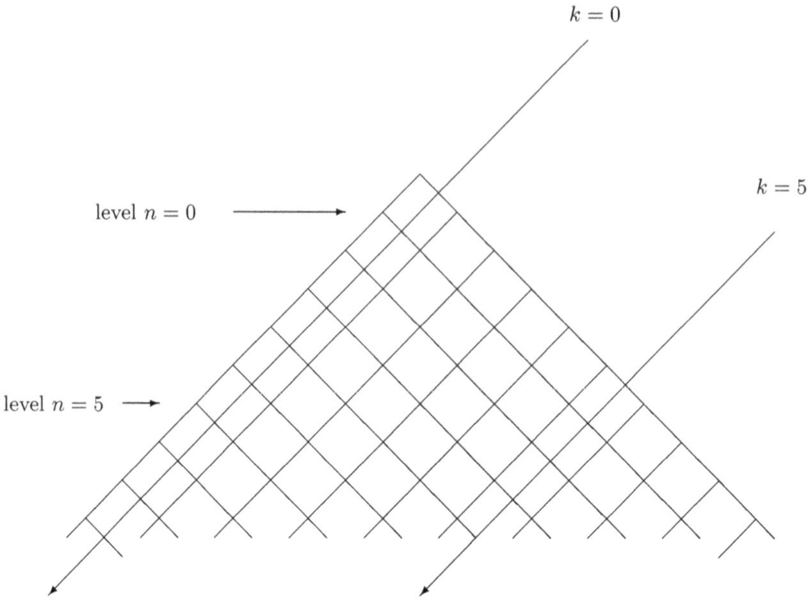

Fig. 6.1. Pascal's triangle: to index the levels (or rows), use $n = 0, 1, 2, \ldots$. To index the individual cells in the nth level, use $k = 0, 1, 2, \ldots, n$.

Like this: in each cell in the $(n + 1)$th level, place the sum of two well-defined entries: from the upper-left and upper-right cell. Is this always possible? Well, there are two exceptions: the leftmost cell has no upper-left neighbor, so the entry in it is the upper-right entry: 1. Likewise, the rightmost cell has no upper-right neighbor, so the entry in it is the upper-left entry: 1. This goes on at all levels: the leftmost and rightmost entries are 1. The intermediate entries, on the other hand, are greater than 1. To prove this, use mathematical induction as well.

This process is constructive: it offers a practical algorithm to fill Pascal's triangle (recursively, level by level). Still, what if we only need one particular entry? How to calculate it on its own, without filling the entire triangle?

6.1.2 The Binomial Coefficient

How to calculate an individual entry, without calculating all others? For this purpose, recall the factorial function:

$$n! \equiv \begin{cases} 1 & \text{if } n = 0 \\ (n-1)! \cdot n & \text{if } n > 0. \end{cases}$$

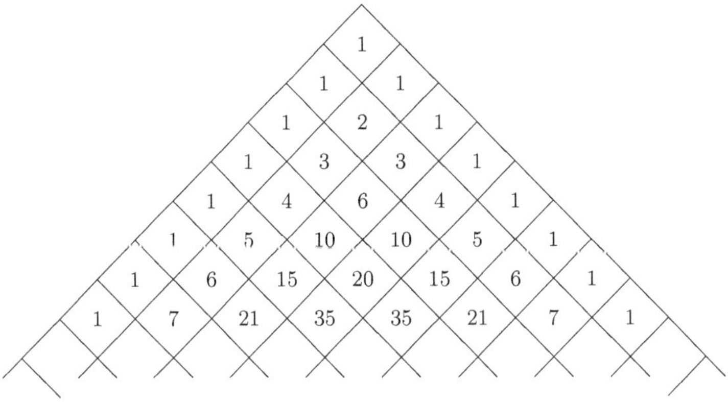

Fig. 6.2. The entries in Pascal's triangle: each entry is the sum of the upper-left and upper-right entries.

In Pascal's triangle, how are the cells indexed? Recall: the levels are indexed top to bottom: $n = 0, 1, 2, 3, \ldots$. In each particular level, the cells are indexed left to right: $k = 0, 1, 2, 3, \ldots, n$. This way, each cell is indexed by two indices: n and k $(n \geq k \geq 0)$. Let's use these indices.

In the nth level, let's focus on the kth entry (from the left). How does it look like algebraically? Let's prove that it looks like this:

$$\binom{n}{k} \equiv \frac{n!}{k!(n-k)!}.$$

This is the binomial coefficient.

How to prove this? By mathematical induction, of course! Indeed, at the zeroth level $(n = 0)$, we must have $k = 0$ too, so our formula indeed works:

$$\binom{0}{0} = \frac{0!}{0! \cdot (0-0)!} = \frac{1}{1 \cdot 1} = 1,$$

as required. Now, for $n = 1, 2, 3, \ldots$, assume that our formula works at the $(n-1)$st level (the induction hypothesis). Does it also work at the nth level below? Well, in the nth level, check cell by cell. In the leftmost cell $(k = 0)$, we indeed have

$$\binom{n}{0} = \frac{n!}{0!(n-0)!} = \frac{n!}{1 \cdot n!} = 1,$$

as required. Likewise, in the rightmost cell ($k = n$),

$$\binom{n}{n} = \frac{n!}{n!(n-n)!} = \frac{n!}{n! \cdot 1} = 1,$$

as required. And in between ($0 < k < n$), thanks to the induction hypothesis,

$$\begin{aligned}\binom{n-1}{k-1} + \binom{n-1}{k} &= \frac{(n-1)!}{(k-1)!(n-k)!} + \frac{(n-1)!}{k!(n-k-1)!} \\ &= \frac{k}{n} \cdot \frac{n!}{k!(n-k)!} + \frac{n-k}{n} \cdot \frac{n!}{k!(n-k)!} \\ &= \frac{k+n-k}{n} \cdot \binom{n}{k} \\ &= \binom{n}{k},\end{aligned}$$

as required. This completes the induction step.

6.1.3 Geometrical Paths in Pascal's Triangle

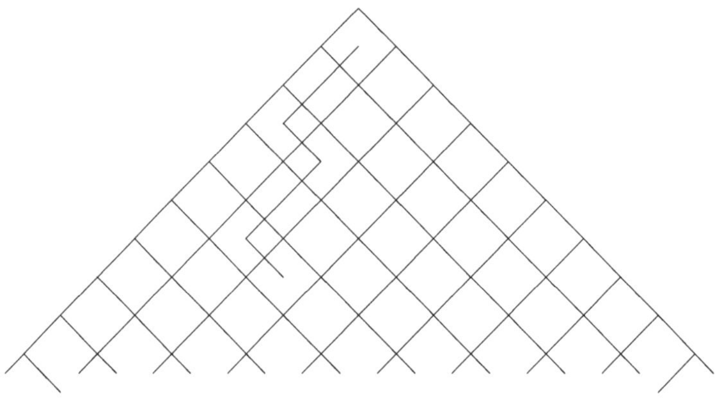

Fig. 6.3. How to get to level $n = 6$, and to cell $k = 2$ in it? Here is one option: the path $(0, 0, 1, 0, 0, 1)$: the third and sixth moves are down-right, and the rest are down-left.

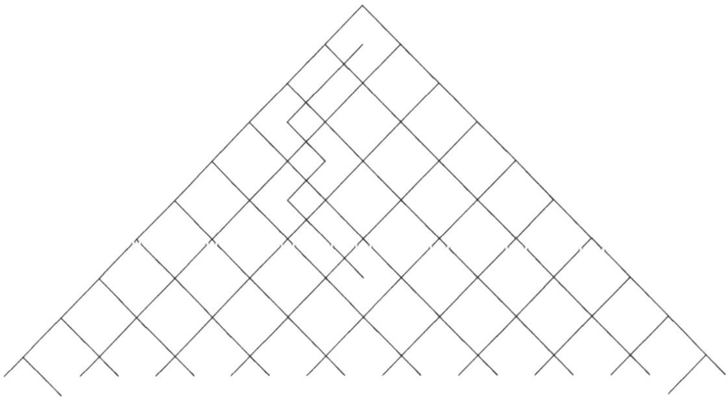

Fig. 6.4. How to get to level $n = 6$, and to cell $k = 3$ in it? Here is one option: the path $(0, 0, 1, 0, 1, 1)$: the third, fifth, and sixth moves are down-right, and the rest are down-left.

In Pascal's triangle, draw a path: start from the top, and "jump" from level to level, down the triangle. In each "jump," move to the next lower level: either down-left, or down-right (Figures 6.3–6.4).

Now, in the nth level, look at the kth cell ($0 \le k \le n$). How many paths lead to it? Well, the answer is already written in it! To prove this, use mathematical induction. Indeed, to reach our cell, the path must come from somewhere: either from the upper-left, or from the upper-right.

How many paths come from the upper-left, and how many from the upper-right? The answers are already written there (thanks to the induction hypothesis). Sum them up (if exist), and get your answer, which is already written here (the entry, in its original definition). Not convinced? Let's see this in yet another way.

6.2 Combinatorial Point of View

6.2.1 Binary Vectors

This was quite geometrical. Let's see another approach, with more combinatorial insight. Our path contains n moves: either down-right (1), or down-left (0). Let's list them:

$$v_1, \ v_2, \ v_3, \ \ldots, \ v_n,$$

where

$$v_i \equiv \begin{cases} 1 & \text{if the } i\text{th move is down-right} \\ 0 & \text{if the } i\text{th move is down-left} \end{cases}$$

$(1 \le i \le n)$. This is v: a binary (n-dimensional) vector, to model (and mirror) our path. In v, a zero component stands for a down-left move, and 1 for a down-right move (Figures 6.3–6.4). Don't forget: there is one degenerate case: $n = 0$: the empty vector $\emptyset$. It stands for the static "path" that goes nowhere: no move at all! Still, we are more interested in real paths, with $n > 0$.

6.2.2 Subsets

What is a binary vector? It is actually a discrete function: it maps an index i to a binary component v_i (either 0 or 1). Thus, v is actually a mapping: from a set of n numbers to a set of two:

$$v: \ \{1, 2, 3, \ldots, n\} \to \{0, 1\}.$$

(In set theory, this is called a characteristic function.) Which numbers are mapped to 1? In other words, what is the origin of 1? This makes a new subset:

$$S \subset \{1, 2, 3, \ldots, n\},$$

containing those i's for which $v_i \neq 0$:

$$i \in S \ \Leftrightarrow \ v_i = 1.$$

In Figure 6.3, for example,

$$v = (0, 0, 1, 0, 0, 1),$$

so

$$S = \{3, 6\} \subset \{1, 2, 3, 4, 5, 6\}.$$

In Figure 6.4, on the other hand,

$$v = (0, 0, 1, 0, 1, 1),$$

so S is a little bigger:

$$S = \{3, 5, 6\} \subset \{1, 2, 3, 4, 5, 6\}.$$

Our original question was geometrical: how many paths lead to the kth cell in the nth level? Now, it gets combinatorial: how many binary vectors have k 1's and $n-k$ 0's? Better yet, put it in terms of set theory: what is the cardinality (size) of the set

$$\{v \equiv (v_1, v_2, \ldots, v_n) \in \{0, 1\}^n \ | \ v_1 + v_2 + \cdots + v_n = k\}?$$

Better yet, put it in terms of subsets: what is the cardinality of the set

$$\{S \subset \{1, 2, 3, \ldots, n\} \ | \ |S| = k\}?$$

6.2.3 Subsets and The Binomial Coefficient

In Section 6.1.3, we already studied Pascal's triangle, and counted the geometrical paths in it. Thanks to subsets, we can now do this in yet another way, and obtain the same result back again: the binomial coefficient. For this, we rephrase our question combinatorially: how many different subsets of the form

$$S \equiv \{s_1, s_2, s_3, \ldots, s_k\}$$

could be extracted from $\{1, 2, 3, \ldots, n\}$? Well, to design such a subset, we need to pick k different numbers. How? Let's pick s_1. It could be either 1, or 2, or 3, or ..., or n. So, there are n options.

For each such an option, how to pick s_2? Well, it could be either smaller or bigger than s_1, but not the same. So, in total, there are $n(n-1)$ options to pick s_1 and s_2.

But wait a minute: in Figure 6.3,

$$\{3, 6\} = \{6, 3\}$$

are the same subset. No need to count it twice! So, our count was duplicate. To avoid this, divide by 2: count only well-ordered $\{s_1, s_2\}$'s, in which $s_1 < s_2$:

$$\{3, 6\} \quad \text{rather than} \quad \{6, 3\}.$$

With no duplication, there are only

$$\frac{n(n-1)}{2} = \binom{n}{2}$$

different options to pick

$$s_1 < s_2,$$

as required. For each such an option, there are $n-2$ options to pick s_3. After all, s_3 could be either smaller than s_1, or bigger than s_2, or in between. But there is still some duplication. For example, to design the path in Figure 6.4, you should pick

$$\{s_1, s_2, s_3\} = \{3, 6, 5\} = \{3, 5, 6\} = \{5, 6, 3\}$$

(in this order, satisfying $s_1 < s_2$). In other words, you could pick

$$s_1 = 3 \quad \text{and} \quad s_2 = 6$$

(as in Figure 6.3), and then

$$s_3 = 5$$

(in between). But this is not the only way: you could equally well pick

$$s_1 = 3, \; s_2 = 5, \quad \text{and then} \quad s_3 = 6,$$

or even

$$s_1 = 5, \; s_2 = 6, \quad \text{and then} \quad s_3 = 3.$$

(After all, in all these three options, $s_1 < s_2$, as required.) How to avoid this new duplication? Force

$$s_1 < s_2 < s_3.$$

This will divide your count by 3. Leaving only

$$\frac{n(n-1)}{2} \cdot \frac{n-2}{3} = \frac{n!}{(n-3)!} \cdot \frac{1}{3!} = \binom{n}{3}$$

different options to design $\{s_1, s_2, s_3\}$.

This goes on and on (by mathematical induction on $k = 1, 2, 3, \ldots, n$). In the end, there are

$$\binom{n}{k-1} \cdot \frac{n-k+1}{k} = \frac{n!}{(k-1)!(n-k+1)!} \cdot \frac{n-k+1}{k} = \frac{n!}{k!(n-k)!} = \binom{n}{k}$$

different options to pick

$$s_1 < s_2 < s_3 < \cdots < s_k,$$

to help design each subset of the form $\{s_1, s_2, \ldots, s_k\}$ exactly once: uniquely, as required.

6.2.4 Monotonically Increasing Sequences

By now, we counted how many sequences of the form

$$s_1 < s_2 < s_3 < \cdots < s_k$$

could be extracted from $\{1, 2, 3, \ldots, n\}$. Still, these inequalities are quite strong. To satisfy them, our s_i's must lie in different intervals:

$$\begin{aligned} 1 \le s_1 &\le n-k+1 \\ 2 \le s_2 &\le n-k+2 \\ 3 \le s_3 &\le n-k+3 \\ &\vdots \\ k \le s_k &\le n. \end{aligned}$$

Better define new numbers:

$$\begin{aligned} t_1 &\equiv s_1 - 1 \\ t_2 &\equiv s_2 - 2 \\ t_3 &\equiv s_3 - 3 \\ &\vdots \\ t_k &\equiv s_k - k, \end{aligned}$$

all in the same interval:

$$\begin{aligned} 0 \le t_1 &\le n-k \\ 0 \le t_2 &\le n-k \\ 0 \le t_3 &\le n-k \\ &\vdots \\ 0 \le t_k &\le n-k. \end{aligned}$$

Why are they better? Because they only increase weakly, not strongly:

$$0 \le t_1 \le t_2 \le t_3 \le \cdots \le t_k \le n - k.$$

This is invertible: given t_i's like these, we can always define s_i's from them:

$$\begin{aligned} s_1 &\equiv t_1 + 1 \\ s_2 &\equiv t_2 + 2 \\ s_3 &\equiv t_3 + 3 \\ &\vdots \\ s_k &\equiv t_k + k. \end{aligned}$$

To simplify, let's introduce a new number:

$$m = n - k.$$

(What is its geometrical meaning? See exercises below.) Since $n \ge k \ge 0$,

$$m, k \ge 0.$$

(This is more nice and symmetric: both m and k have equal status.) This puts our t_i's in the new interval $[0, m]$:

$$0 \le t_1 \le t_2 \le t_3 \le \cdots \le t_k \le m.$$

How many such sequences could be extracted from $\{0, 1, 2, 3, \ldots, m\}$? We already know the answer:

$$\binom{m+k}{k}.$$

Is this true even in degenerate cases? Let's check: for $k = 0$, there is just one possible sequence: the empty sequence $\emptyset$. This agrees with our formula. Next, for $m = 0$, there is just one possible sequence:

$$0 \le 0 \le 0 \le \cdots \le 0.$$

This agrees with our formula as well.

6.3 Geometrical Problem

6.3.1 Discrete Simplex

By now, we avoided strong (or strict) inequalities, and replace them by weak ones. This way, our new t_i's lie in a more uniform interval: $[0, m]$. Still, by now, our discussion got a bit too combinatorial. Let's go geometrical back again. For this purpose, define new differences:

$$\begin{aligned} d_1 &\equiv t_1 \\ d_2 &\equiv t_2 - t_1 \\ d_3 &\equiv t_3 - t_2 \\ &\vdots \\ d_k &\equiv t_k - t_{k-1}. \end{aligned}$$

How should the d_i's look like? First, they must be nonnegative:

$$d_i \geq 0, \quad 1 \leq i \leq k.$$

Next, they must sum to t_k, which can't exceed m:

$$\sum_{i=1}^{k} d_i = d_1 + d_2 + d_3 + \cdots + d_k = t_k \leq m.$$

So, you can now forget about the t_i's: they were transformed to the new d_i's. This is invertible: given d_i's like these, you can always calculate their partial sums, and obtain new t_i's, if you like:

$$\begin{aligned} t_1 &\equiv d_1 \\ t_2 &\equiv d_1 + d_2 \\ t_3 &\equiv d_1 + d_2 + d_3 \\ &\vdots \\ t_k &\equiv d_1 + d_2 + d_3 + \cdots + d_k. \end{aligned}$$

What's so good about our new d_i's? They are no longer extracted from any existing set, but just satisfy two conditions:

$$d_i \geq 0, \quad 1 \leq i \leq k,$$

and

$$\sum_{i=1}^{k} d_i \leq m.$$

Place them in a new k-dimensional vector:

$$(d_1, d_2, d_3, \ldots, d_k) \in \left(\mathbb{Z}^+\right)^k,$$

where

$$\mathbb{Z}^+ \equiv \mathbb{N} \cup \{0\}$$

contains the natural numbers and zero. In this new vector, the components are nonnegative, and their sum is at most m. In other words, our new vector is in a discrete simplex of size m:

$$\left\{ (d_1, d_2, \ldots, d_k) \in \left(\mathbb{Z}^+\right)^k \mid \sum_{i=1}^{k} d_i \leq m \right\}.$$

6.3.2 Discrete Triangle

Let's start from a simple example: a planar v_1-v_2 grid (dimension: $k = 2$). In it, our simplex is just a triangle, confined to $v_1 \geq 0$, $v_2 \geq 0$, and $v_1 + v_2 \leq m$. In Figure 6.5, we draw a discrete triangle of size $m = 3$: each edge contains four grid points.

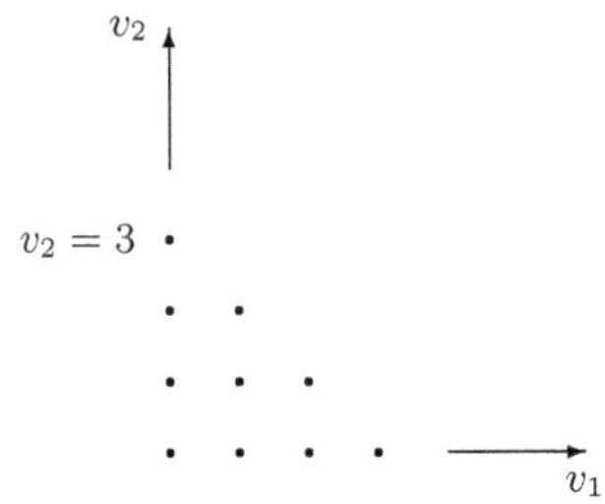

Fig. 6.5. A planar v_1-v_2 grid ($k = 2$). In it, our discrete triangle is confined to $v_1 \geq 0$, $v_2 \geq 0$, and $v_1 + v_2 \leq m = 3$.

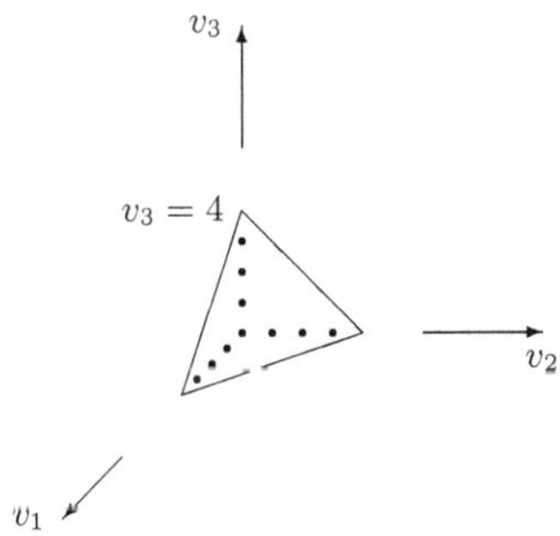

Fig. 6.6. A cubic v_1-v_2-v_3 grid (dimension $k = 3$). In it, our discrete tetrahedron is confined to $v_1 \geq 0$, $v_2 \geq 0$, $v_3 \geq 0$, and $v_1 + v_2 + v_3 \leq m = 4$.

6.3.3 Discrete Tetrahedron

For $k = 3$, this is still simple: a cubic v_1-v_2-V_3 grid. In this case, our simplex is a tetrahedron, of size m. In Figure 6.6, we draw a discrete tetrahedron of size $m = 4$: each edge contains five grid points.

6.3.4 How Many Points?

Next, look at the general case: a discrete k-dimensional simplex of size m. How many grid points are there in it? We already know the answer:

$$\left|\left\{(d_1, d_2, \ldots, d_k) \in \left(\mathbb{Z}^+\right)^k \mid \sum_{i=1}^{k} d_i \le m\right\}\right| = \binom{m+k}{k}.$$

Is this true even in degenerate cases? Let's check on this geometrically: for $k = 0$, there is just one possible vector: the empty vector $\emptyset$, with no component at all. This agrees with our formula. Next, for $m = 0$, there is just one possible vector – the origin:

$$(0, 0, 0, \ldots, 0).$$

This agrees with our formula as well.

6.3.5 Subsimplex of Lower Dimension

Look at a nontrivial simplex: $k \ge 1$. In it, the d_i's must sum to m or less. Now, look at its upper face. Here, the d_i's sum to m exactly:

$$d_1 + d_2 + d_3 + \cdots + d_k = m.$$

For this, d_k is no longer free, but must be defined as

$$d_k \equiv m - \sum_{i=1}^{k-1} d_i.$$

Fortunately, $d_1, d_2, \ldots, d_{k-1}$ are still free, and should satisfy two conditions:

$$d_i \ge 0, \quad 1 \le i \le k-1,$$

and

$$\sum_{i=1}^{k-1} d_i \le m.$$

Is this familiar? This is a discrete $(k-1)$-dimensional simplex of size m. How many points are there in it? We already know the answer:

$$\begin{aligned}
&\left|\left\{(d_1, d_2, \ldots, d_k) \in \left(\mathbb{Z}^+\right)^k \mid \sum_{i=1}^{k} d_i = m\right\}\right| \\
&= \left|\left\{(d_1, d_2, \ldots, d_{k-1}) \in \left(\mathbb{Z}^+\right)^{k-1} \mid \sum_{i=1}^{k-1} d_i \le m\right\}\right| \\
&= \binom{m+k-1}{k-1}.
\end{aligned}$$

6.3.6 Splitting The Simplex

This can go on, slice by slice. This splits the original simplex into $m+1$ disjoint subsimplices (parallel oblique slices). This gives us a new combinatorial formula:

$$\begin{aligned}
&\binom{m+k}{k} \\
&= \left|\left\{(d_1, d_2, \ldots, d_k) \in \left(\mathbb{Z}^+\right)^k \mid \sum_{i=1}^{k} d_i \le m\right\}\right| \\
&= \left|\cup_{j=0}^{m} \left\{(d_1, d_2, \ldots, d_k) \in \left(\mathbb{Z}^+\right)^k \mid \sum_{i=1}^{k} d_i = j\right\}\right| \\
&= \sum_{j=0}^{m} \left|\left\{(d_1, d_2, \ldots, d_k) \in \left(\mathbb{Z}^+\right)^k \mid \sum_{i=1}^{k} d_i = j\right\}\right| \\
&= \sum_{j=0}^{m} \binom{j+k-1}{k-1}.
\end{aligned}$$

Next, let's see a few interesting applications in physics.

6.4 Application in Statistical Physics

6.4.1 Brownian Motion

Consider a particle, moving step by step on the real axis. In each step, it "jumps" by ± 1, along the real axis: either rightwards, or leftwards.

In the beginning, the particle starts from the origin: 0. In each step, it moves exactly one unit: either rightwards (from some l to $l+1$), or leftwards (to $l-1$).

The process is random (or stochastic, or nondeterministic): in each individual step, we don't know where the particle will jump. Fortunately, we still know the odds. To move leftwards, the probability is $0 \le a \le 1$. To move rightwards, on the other hand, the probability is $0 \le b \le 1$. There is no other option:

$$a + b = 1.$$

(After all, the particle must go somewhere.) Now, what would happen after n steps? Where would the particle be? We can never tell this for sure. Fortunately, we can still tell the odds.

In n steps, the particle must make k leftward moves, and $n-k$ rightward moves (for some $0 \le k \le n$). Thus, after n steps, it could reach the point

$$-k + (n-k) = n - 2k$$

on the real axis. How likely is the particle to get there? We already know the answer: there are $\binom{n}{k}$ distinct ways to pick k leftward moves and $n-k$ rightward moves (Section 6.1.3). Furthermore, since the steps are independent of each other, all these ways have the same probability: $a^k b^{n-k}$. Thus, after n steps, the particle would reach $n-2k$ at probability

$$\binom{n}{k} a^k b^{n-k}.$$

These probabilities must sum to 1. (After all, after n steps, the particle must be somewhere.) Later on, we'll also prove this algebraically.

6.4.2 Symmetric Diffusion

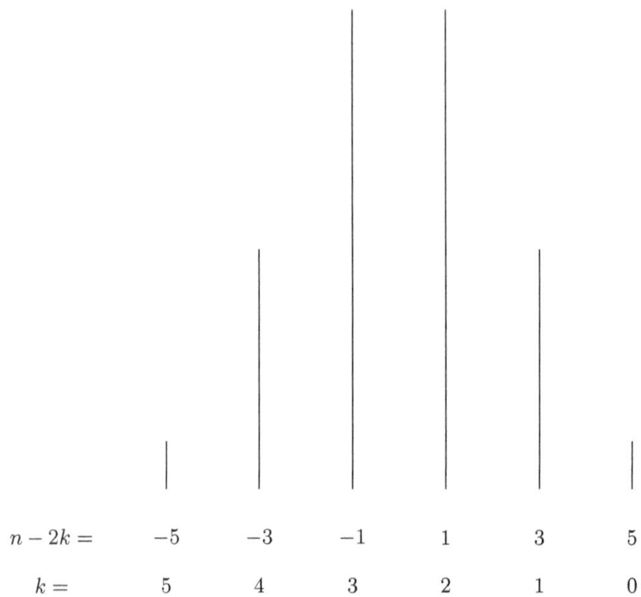

Fig. 6.7. Symmetric Brownian motion ($a = b = 1/2$, $n = 5$ steps). The columns tell us the probability to reach $n - 2k$ (k leftward and $n - k$ rightward moves).

Look at the symmetric case:

$$a = b = \frac{1}{2}.$$

This is a symmetric diffusion (in 1-D). Now, after n steps, where would the particle be? How likely is it to reach $n - 2k$? Thanks to our formula, the answer is

$$\binom{n}{k}\left(\frac{1}{2}\right)^k\left(\frac{1}{2}\right)^{n-k} = \frac{1}{2^n}\binom{n}{k}$$

(Figures 6.7–6.9).

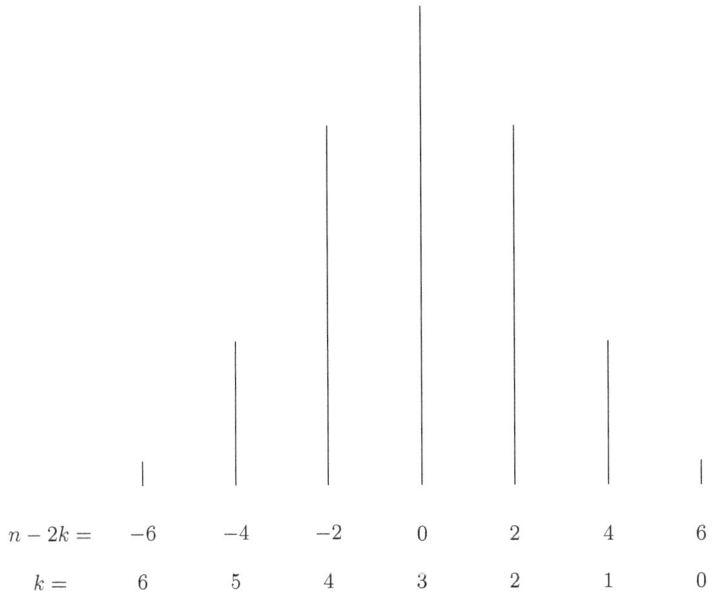

Fig. 6.8. Symmetric Brownian motion: $n = 6$. The columns tell us the probability to reach $n - 2k$ after $n = 6$ steps.

The nonsymmetric case $a < b$, on the other hand, models convection-diffusion, with a slight wind, blowing rightwards. Likewise, the case $a > b$ models a wind blowing leftwards.

6.4.3 Closed System: Entropy Must Increase

In a closed physical system, entropy means uniformity: to what extent everything is the same. In our model, assume that many particles start from the origin: 0. This is low entropy: the origin is quite special. But then again, thanks to diffusion, our particles move at random (independent of each other). Their distribution gets more and more spread out (Figures 6.7–6.9). Entropy increases.

6.4.4 Second Law of Thermodynamics

This is the second law of thermodynamics: on average, entropy must increase. At the very beginning of the universe, at the big bang, entropy was at its minimum: all

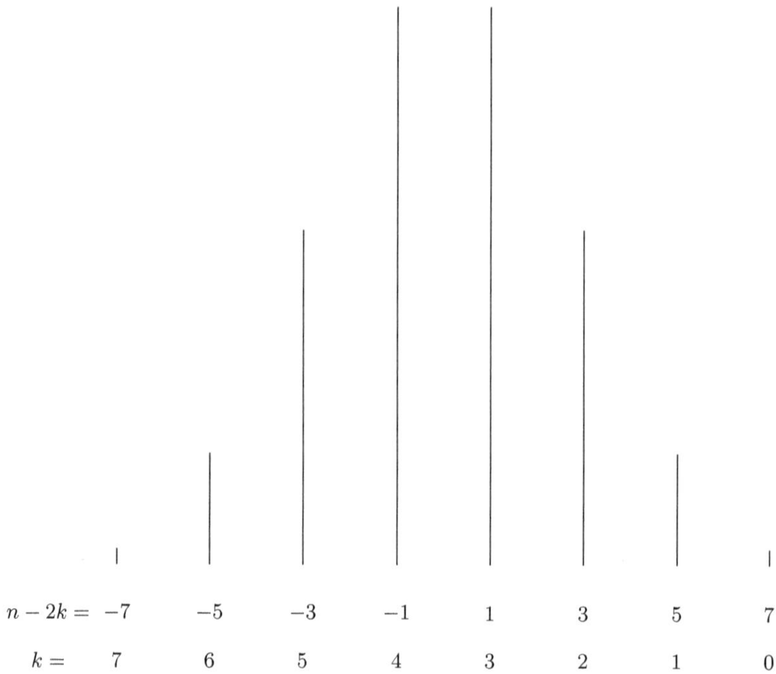

Fig. 6.9. Symmetric Brownian motion: $n = 7$. The columns tell us the probability to reach $n - 2k$ after $n = 7$ steps.

energy was concentrated at one singular point. Since then, entropy only increased.

In the inflation that followed, the universe expanded exponentially fast: even faster than light! Afterwards, matter and energy continued to spread out, increasing entropy even more.

Still, not everything expanded. Thanks to gravity, when the universe cooled down, stars and galaxies formed. Still, this didn't decrease entropy. On the contrary: this generated more and more heat, spreading energy around, and increasing entropy even more (on average).

6.5 The Binomial Formula: Mathematical Induction

6.5.1 Newton's Binomial Formula

Let n be a nonnegative integer number. Let a and b be some numbers. Look at their sum: $a + b$. Raise it to power n:

$$(a+b)^n \equiv (a+b)(a+b)(a+b) \cdots (a+b) \qquad (n \text{ times}).$$

In this product, there are n factors of the same form: $a + b$. In other words, the same factor $a + b$ multiplies n times.

How to open parentheses? From each factor of the form $a + b$, pick either a or b. This way, you'll pick k a's and $n - k$ b's (for some $0 \le k \le n$). Upon opening parentheses, this will produce a new product: $a^k b^{n-k}$. How many times? Mirror this by our good old binary vectors:

$$\left| \left\{ v \in \{0,1\}^n \mid \sum_{i=1}^{n} v_i = k \right\} \right| = \binom{n}{k}.$$

Thus, once the parentheses open up, each product of the form $a^k b^{n-k}$ will appear $\binom{n}{k}$ times:

$$(a+b)^n = \sum_{k=0}^{n} \binom{n}{k} a^k b^{n-k}.$$

This is Newton's binomial formula. Not convinced? In a moment, we'll prove this more directly.

To use the formula in practice, you need the binomial coefficients. How to calculate them efficiently? Better use no multiplication, but only addition: fill Pascal's triangle, level by level, until level n.

6.5.2 Mathematical Induction

Want to prove the binomial formula more directly? Use mathematical induction! For $n = 0$, this is trivial:

$$(a+b)^0 = 1 = \binom{0}{0} a^0 b^0 = \sum_{k=0}^{0} \binom{0}{k} a^k b^{0-k}.$$

More seriously, for $n = 1, 2, 3, \ldots$, assume the induction hypothesis:

$$(a+b)^{n-1} = \sum_{k=0}^{n-1} \binom{n-1}{k} a^k b^{n-1-k}.$$

Thanks to this (and the formula in Section 6.1.2),

$$\begin{aligned}(a+b)^n &= (a+b)(a+b)^{n-1} \\ &= (a+b) \sum_{k=0}^{n-1} \binom{n-1}{k} a^k b^{n-1-k}\end{aligned}$$

$$
\begin{aligned}
&= a\sum_{k=0}^{n-1}\binom{n-1}{k}a^k b^{n-1-k} + b\sum_{k=0}^{n-1}\binom{n-1}{k}a^k b^{n-1-k}\\
&= \sum_{k=0}^{n-1}\binom{n-1}{k}a^{k+1}b^{n-(k+1)} + \sum_{k=0}^{n-1}\binom{n-1}{k}a^k b^{n-k}\\
&= \sum_{k=1}^{n}\binom{n-1}{k-1}a^k b^{n-k} + \sum_{k=0}^{n-1}\binom{n-1}{k}a^k b^{n-k}\\
&= a^n + \sum_{k=1}^{n-1}\binom{n-1}{k-1}a^k b^{n-k} + \sum_{k=1}^{n-1}\binom{n-1}{k}a^k b^{n-k} + b^n\\
&= a^n + \sum_{k=1}^{n-1}\left(\binom{n-1}{k-1} + \binom{n-1}{k}\right)a^k b^{n-k} + b^n\\
&= a^n + \sum_{k=1}^{n-1}\binom{n}{k}a^k b^{n-k} + b^n\\
&= \sum_{k=0}^{n}\binom{n}{k}a^k b^{n-k},
\end{aligned}
$$

as required. This completes the induction step, and indeed the entire proof.

6.6 Extended Formulas

6.6.1 Extended Formula: Mathematical Induction

Next, let a take a new status. Like n, a is now a nonnegative integer number:

$$a, n \geq 0.$$

(This new a has nothing to do with the old a used in $a + b$.) Now, how to divide two factorials? Want to be naive? Calculate each factorial, and then divide. Say,

$$C_{a,n} \equiv \begin{cases} \frac{a!}{(a-n)!} & \text{if } a \geq n \\ 0 & \text{if } a < n. \end{cases}$$

But isn't this a bit silly? Why calculate the entire numerator explicitly? After all, it contains a lot of factors that appear in the denominator as well, and are going to drop soon anyway! Better use recursion:

$$C_{a,n} \equiv a(a-1)(a-2)\cdots(a-(n-1)) = \begin{cases} 1 & \text{if } n = 0 \\ aC_{a-1,n-1} & \text{if } n \geq 1. \end{cases}$$

This way, we also get a bonus: a can now be just any number: integer or not, positive or not. Furthermore, for every two numbers a and b, we can now mirror Newton's binomial formula:

$$C_{a+b,n} = \sum_{k=0}^{n}\binom{n}{k}C_{a,k}C_{b,n-k}.$$

To prove this, use mathematical induction. Indeed, for $n = 0$, this is trivial. Now, for $n \geq 1$, assume the induction hypothesis: for every a and b,

$$C_{a+b,n-1} = \sum_{k=0}^{n-1} \binom{n-1}{k} C_{a,k} C_{b,n-1-k}.$$

Thanks to this (and the formula in Section 6.1.2),

$$\begin{aligned}
C_{a+b,n} &= (a+b)C_{a+b-1,n-1} \\
&= aC_{(a-1)+b,n-1} + bC_{a+(b-1),n-1} \\
&= a\sum_{k=0}^{n-1} \binom{n-1}{k} C_{a-1,k} C_{b,n-1-k} + b\sum_{k=0}^{n-1} \binom{n-1}{k} C_{a,k} C_{b-1,n-1-k} \\
&= \sum_{k=0}^{n-1} \binom{n-1}{k} C_{a,k+1} C_{b,n-(k+1)} + \sum_{k=0}^{n-1} \binom{n-1}{k} C_{a,k} C_{b,n-k} \\
&= \sum_{k=1}^{n} \binom{n-1}{k-1} C_{a,k} C_{b,n-k} + \sum_{k=0}^{n-1} \binom{n-1}{k} C_{a,k} C_{b,n-k} \\
&= C_{a,n} + \sum_{k=1}^{n-1} \binom{n-1}{k-1} C_{a,k} C_{b,n-k} + \sum_{k=1}^{n-1} \binom{n-1}{k} C_{a,k} C_{b,n-k} + C_{b,n} \\
&= C_{a,n} + \sum_{k=1}^{n-1} \left(\binom{n-1}{k-1} + \binom{n-1}{k} \right) C_{a,k} C_{b,n-k} + C_{b,n} \\
&= C_{a,n} + \sum_{k=1}^{n-1} \binom{n}{k} C_{a,k} C_{b,n-k} + C_{b,n} \\
&= \sum_{k=0}^{n} \binom{n}{k} C_{a,k} C_{b,n-k},
\end{aligned}$$

as required. This completes the proof.

6.6.2 The Extended Binomial Coefficient

So far, the binomial coefficient was defined for natural numbers or zero (Section 6.1.2). Let's extend it to every number a: integer or not, positive or not:

$$\binom{a}{k} \equiv \frac{C_{a,k}}{k!}.$$

As in the original definition, k still has to be a nonnegative integer number. a, on the other hand, can now be just any number: even complex!

Now, look at this new definition. Is it consistent with the old one? Let's check: what happens when a is an integer? Well, there are two options: if $a \geq k$, then both definitions give us the same good old binomial coefficient. If, on the other hand, $0 \leq a < k$, then there is no old definition at all, and the new definition gives zero.

Now, look again at our new formula (Section 6.6.1). To make it look better, divide it by $n!$:

$$
\begin{aligned}
\binom{a+b}{n} &= \frac{1}{n!} C_{a+b,n} \\
&= \frac{1}{n!} \sum_{k=0}^{n} \binom{n}{k} C_{a,k} C_{b,n-k} \\
&= \sum_{k=0}^{n} \frac{1}{k!(n-k)!} C_{a,k} C_{b,n-k} \\
&= \sum_{k=0}^{n} \frac{C_{a,k}}{k!} \cdot \frac{C_{b,n-k}}{(n-k)!} \\
&= \sum_{k=0}^{n} \binom{a}{k} \binom{b}{n-k}.
\end{aligned}
$$

This mirrors the original binomial formula. In particular, set $a = b = n$:

$$
\binom{2n}{n} = \sum_{k=0}^{n} \binom{n}{k} \binom{n}{n-k} = \sum_{k=0}^{n} \binom{n}{k}^2.
$$

This is the middle binomial coefficient (the tallest column in Figure 6.8).

6.6.3 The Trinomial Formula

In the binomial formula, there are two numbers: a and b. In the trinomial formula, there is a third number: c. Apply the binomial formula twice:

$$
\begin{aligned}
(a+b+c)^n &= ((a+b)+c)^n \\
&= \sum_{k=0}^{n} \binom{n}{k} (a+b)^k c^{n-k} \\
&= \sum_{k=0}^{n} \binom{n}{k} \left(\sum_{l=0}^{k} \binom{k}{l} a^l b^{k-l} \right) c^{n-k} \\
&= \sum_{k=0}^{n} \sum_{l=0}^{k} \binom{n}{k} \binom{k}{l} a^l b^{k-l} c^{n-k} \\
&= \sum_{k=0}^{n} \sum_{l=0}^{k} \frac{n!}{k!(n-k)!} \cdot \frac{k!}{l!(k-l)!} a^l b^{k-l} c^{n-k} \\
&= \sum_{k=0}^{n} \sum_{l=0}^{k} \frac{n!}{l!(k-l)!(n-k)!} a^l b^{k-l} c^{n-k} \\
&= \sum_{0 \le l,j,m \le n,\ l+j+m=n} \frac{n!}{l!j!m!} a^l b^j c^m.
\end{aligned}
$$

In the latter sum, we introduced new indices: $j \equiv k - l$, and $m \equiv n - k$. These are more nice and symmetric: they have equal status (as in Section 6.3.1). This is the trinomial formula. Let's mirror it. Apply the formula in Section 6.6.1 twice:

$$\begin{aligned}
C_{a+b+c,n} &= C_{(a+b)+c,n} \\
&= \sum_{k=0}^{n} \binom{n}{k} C_{a+b,k} C_{c,n-k} \\
&= \sum_{k=0}^{n} \binom{n}{k} \left(\sum_{l=0}^{k} \binom{k}{l} C_{a,l} C_{b,k-l} \right) C_{c,n-k} \\
&= \sum_{k=0}^{n} \sum_{l=0}^{k} \binom{n}{k} \binom{k}{l} C_{a,l} C_{b,k-l} C_{c,n-k} \\
&= \sum_{k=0}^{n} \sum_{l=0}^{k} \frac{n!}{k!(n-k)!} \cdot \frac{k!}{l!(k-l)!} C_{a,l} C_{b,k-l} C_{c,n-k} \\
&= \sum_{k=0}^{n} \sum_{l=0}^{k} \frac{n!}{l!(k-l)!(n-k)!} C_{a,l} C_{b,k-l} C_{c,n-k} \\
&= \sum_{0 \le l,j,m \le n,\ l+j+m=n} \frac{n!}{l!j!m!} C_{a,l} C_{b,j} C_{c,m}.
\end{aligned}$$

To make it look better, divide it by $n!$:

$$\binom{a+b+c}{n} = \sum_{0 \le l,j,m \le n,\ l+j+m=n} \binom{a}{l} \binom{b}{j} \binom{c}{m}.$$

In the latter sum, how many terms are there? Thanks to Section 6.3.5, we already know the answer:

$$\binom{n+3-1}{3-1} = \binom{n+2}{2} = \frac{(n+1)(n+2)}{2}.$$

6.7 Exercises: Pascal's Triangle

6.7.1 Oblique Lines in Pascal's Triangle

1. Let $n \ge 0$ be an integer number. Show that

$$\binom{n}{0} = \binom{n}{n} = 1.$$

2. Let $n \ge k \ge 0$ be two integer numbers. Look at their binomial coefficient. Show that it is symmetric:

$$\binom{n}{k} = \binom{n}{n-k}.$$

3. What does this mean geometrically? Hint: Figures 6.7–6.9 are symmetric.
4. Let $n > k > 0$ be two natural numbers. Show that

$$\binom{n}{k} = \binom{n-1}{k} + \binom{n-1}{k-1}.$$

Hint: see Section 6.1.2.

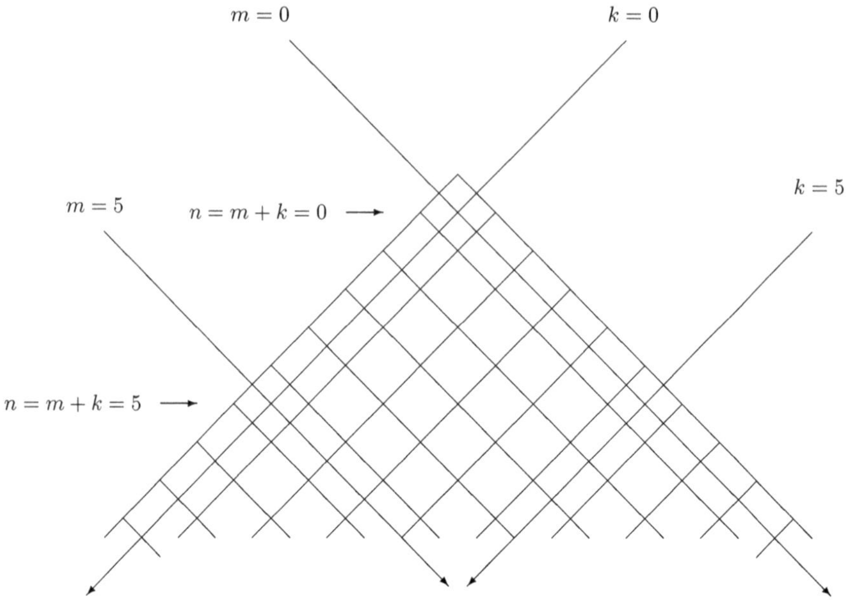

Fig. 6.10. How to index Pascal's triangle? $k = 0, 1, 2, \ldots$ indexes the oblique down-left lines. $m = 0, 1, 2, \ldots$ indexes the oblique down-right lines. This makes an oblique m-k grid of cells. In it, $n = m + k = 0, 1, 2, \ldots$ indexes the horizontal levels.

5. In Pascal's triangle, draw the lines n =const. How do they look like geometrically? Hint: these are horizontal rows (Figure 6.1).
6. Next, draw the lines k =const. How do they look like geometrically? Hint: oblique, pointing down-left (Figures 6.1 and 6.10).
7. Define the new index $m \equiv n - k$. What is its geometrical meaning? Hint: it indexes the oblique down-right lines.
8. Use this new index to draw the lines m =const. How do they look like geometrically? Hint: oblique, pointing down-right (Figure 6.10).
9. View Pascal's triangle as an oblique m-k grid of cells.
10. In it, draw the lines $m + k$ =const. How do they look like geometrically? Hint: horizontal (Figure 6.10).

6.7.2 Binary Vectors and Characteristic Functions

1. Define the set of n-dimensional binary vectors:

$$V \equiv V(n) \equiv \{(v_1, v_2, \ldots, v_n) \mid v_i = 0 \text{ or } 1,\ 1 \leq i \leq n\} .$$

2. View a binary vector as a characteristic function on the set $\{1, 2, 3, \ldots, n\}$. Hint: just define the new function:
$$v(i) \equiv v_i.$$
3. Write V as the set of all characteristic functions on the set $\{1, 2, 3, \ldots, n\}$. Hint: in set theory, V is written as
$$V \equiv V(n) \equiv \{0, 1\}^{\{1,2,\ldots,n\}}.$$
4. Note that V depends on n. Still, we often drop the '(n).'
5. What happens in the degenerate case $n \equiv 0$? How to define $V(0)$? Hint: $V(0)$ contains just one "vector:" the empty "vector" $\emptyset$, which contains no component at all:
$$V(0) \equiv \{\emptyset\}.$$
6. What is the cardinality of $V(n)$? In other words, how many vectors are there in $V(n)$?
7. Use mathematical induction on $n \geq 0$ to show that $V \equiv V(n)$ contains
$$|V| \equiv |V(n)| = 2^n$$
distinct vectors.
8. Let $n \geq 1$ be fixed. Let $V_k \subset V$ contain those vectors with exactly k nonzero components:
$$V_k \equiv \left\{ v \in V \mid \sum_{i=1}^{n} v_i = k \right\}.$$
9. Use mathematical induction on $k = 0, 1, 2, \ldots, n$ to show that V_k contains
$$|V_k| = \binom{n}{k}$$
distinct vectors. Hint: see Section 6.2.3.
10. Show that, for $k \neq l$, V_{l_0} and V_l are disjoint:
$$V_k \cap V_l = \emptyset.$$
11. Conclude that V could be written as the union of n disjoint subsets:
$$V = \cup_{k=0}^{n} V_k.$$
12. Conclude that
$$2^n = |V| = \sum_{k=0}^{n} |V_k| = \sum_{k=0}^{n} \binom{n}{k}.$$
13. Does this agree with the binomial formula?
14. In Pascal's triangle, how many distinct paths lead from the top cell (the zeroth level) to the kth cell in the nth level ($0 \leq k \leq n$)? Hint: see Section 6.1.3.
15. Why must this number be the same as $|V_k|$? Hint: mirror each such path by a unique binary vector in V_k.

16. Prove more directly (by mathematical induction on n rather than k) that V_k indeed contains
$$|V_k| = \binom{n}{k}$$
distinct binary vectors. Hint: split V_k into two disjoint subsets: those binary vectors ending with 0, and those ending with 1. Then, use the induction hypothesis, and the formula in Section 6.1.2.
17. Look at the algebraic expression: $(a+b)^n$, Open parentheses. This gives a sum. In it, look at one particular term: $a^k b^{n-k}$ (where k is a fixed number between 0 and n). How many duplicate copies are there? Hint: pick exactly k a's from exactly k factors of the form $a+b$. Mirror each one by a unique binary vector in V_k.
18. Conclude that there are
$$|V_k| = \binom{n}{k}$$
duplicate copies of this term.
19. Conclude that, in Newton's binomial formula, the term $a^k b^{n-k}$ has coefficient
$$\binom{n}{k}.$$

7

Applications in Quantum Statistical Mechanics

The binomial formula is particularly useful in statistical mechanics. In Sections 6.4.1–6.4.3, we already saw one example: Brownian motion. Here, we'll see more. For this purpose, we need to extend the binomial and trinomial formulas, to help model a system of molecules. This is related to the Maxwell-Boltzmann statistics.

Molecules are distinguishable from one another. Bosons, on the other hand, are not. Thanks to our discrete simplex, we can now model bosons too. This is related to the Bose-Einstein statistics.

Electrons are indistinguishable as well. Still, they are different from bosons: they obey Pauli's exclusion principle. Thanks to our formulas, we can still model them too. This is related to the Fermi-Dirac statistics [7, 19, 29, 40].

7.1 The Quadrinomial Formula

7.1.1 The Trinomial Coefficient

To obtain the present applications in quantum statistical mechanics, let's use our formulas. The trinomial formula tells us that

$$(a+b+c)^n = \sum_{0\le l,j,m\le n,\ l+j+m=n} \frac{n!}{l!j!m!} a^l b^j c^m$$

(Section 6.6.3). What is the combinatorial meaning of this? Well, upon opening parentheses on the left-hand side, we obtain a sum of many products. Each product may contain l a's, j b's, and m c's, where l, j, and m are some nonnegative integers that sum to n:

$$l+j+m=n.$$

This produces a product of the form

$$a^l b^j c^m,$$

as in the right-hand side.

How many duplicate copies of this product are there? In other words, how many different ways are there to pick l a's, j b's, and m c's? The answer is the trinomial coefficient:

$$\frac{n!}{l!j!m!}.$$

Let's go ahead and extend this.

7.1.2 Extending The Trinomial Formula

Let's extend this yet more:

$$(a+b+c+d)^n = \sum_{0\le l,j,m,k\le n,\ l+j+m+k=n} \frac{n!}{l!j!m!k!} a^l b^j c^m d^k.$$

This is the quadrinomial formula. What is its combinatorial meaning? Well, upon opening parentheses on the left-hand side, we obtain a sum of products of the form

$$a^l b^j c^m d^k,$$

where l, j, m, and k are some nonnegative integers that sum to n:

$$l+j+m+k=n.$$

To obtain such a product, how many different ways are there to pick l a's, j b's, m c's, and k d's? The answer is the quadrinomial coefficient:

$$\frac{n!}{l!j!m!k!}.$$

Let's go ahead and use this in physics.

7.2 Mass and Energy

7.2.1 Closed System: Constant Energy

To use this in physics, we need four new indices:

$$\begin{aligned} n_1 &\equiv l \\ n_2 &\equiv j \\ n_3 &\equiv m \\ n_4 &\equiv k. \end{aligned}$$

Consider now a system of n molecules. Assume that the system is closed: isolated and insulated. This way, it has constant energy.

7.2.2 Energy Levels

In quantum mechanics, energy is not continuous but discrete: it comes in discrete energy levels:

$$0 < E_1 < E_2 < E_3 < E_4 < E_5 < \cdots.$$

Only these E_i's are allowed: the system can never have any other energy in between. In this sense, energy is different from position and time, which remain continuous: at each individual time, the molecule can still lie in any position in space.

7.2.3 Conservation of Mass

Now, what are the a's? Let's interpret them as molecules of energy E_1. Likewise, let's interpret the b's, the c's, and the d's as molecules of energy E_2, E_3, and E_4, respectively. This way, in our original system of n molecules, n_i molecules have energy E_i $(1 \le i \le 4)$. This splits the original system into four disjoint subsets:

$$\sum_{i=1}^{4} n_i = n_1 + n_2 + n_3 + n_4 = n.$$

Actually, this is conservation of mass: together, the subsets produce the original system back again.

7.2.4 Configuration – Distribution

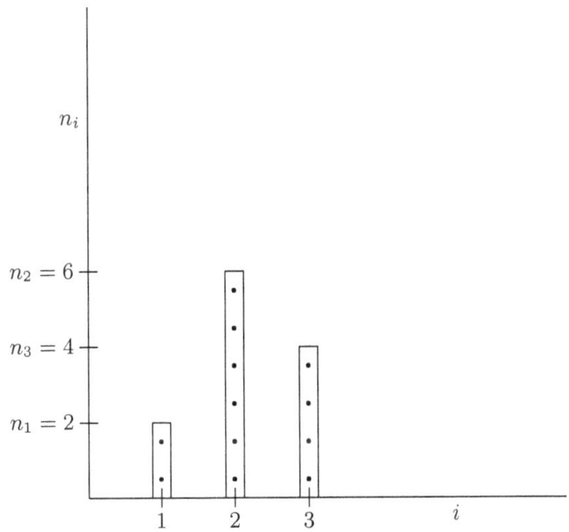

Fig. 7.1. Distribution of molecules: there are n_1 molecules of energy E_1, n_2 of energy E_2, and so on.

As a matter of fact, why stop at 4? After all, there could be many more subsets of yet higher energy levels: n_5 molecules of energy E_5, n_6 of energy E_6, and so on. This way, our sum takes a new form:

$$\sum_{i=1}^{\infty} n_i = n.$$

This is not really infinite: only a few n_i's are nonzero, and all the rest vanish. (Since we don't know which, we leave them all in.) They make a new configuration (or distribution), illustrated in a column diagram (Figure 7.1): the ith column contains n_i molecules of energy E_i. This is how our n molecules split into separate energy levels.

7.2.5 Conservation of Energy

So far, we used one constraint only: conservation of mass. This law is more mathematical than physical. After all, the total number of molecules must be n. Still, we also have a more fundamental (and really physical) law: conservation of energy:

$$\sum_{i=1}^{\infty} n_i E_i = E_{\text{total}}$$

(the total energy in the entire system). This is indeed the first law of thermodynamics (Sections 1.7.2–1.9.4).

Let's use both constraints to design an optimization problem. For this purpose, we need one more formula.

7.3 The Multinomial Coefficient

7.3.1 Random Variables

In our original system, n and E_{total} are given. Likewise, the individual E_i's are available as well. The n_i's, on the other hand, are not: they are random variables, not yet known. Furthermore, they depend on each other, as in our constraints. Still, let's focus on some particular configuration, with some specific n_i's.

7.3.2 Extending The Trinomial Coefficient

To design this particular configuration, how many different ways are there to pick the molecules and group them in subsets? Well, there are many ways to pick the first n_1 molecules of energy E_1. On top of these, there are also many ways to pick n_2 molecules of energy E_2, and so on. In total, there are

$$n! \frac{1}{n_1!} \cdot \frac{1}{n_2!} \cdot \frac{1}{n_3!} \cdots$$

different ways. This is just a finite product. After all, only a few n_i's are nonzero. Still, this product could be quite long. This is indeed the multinomial coefficient.

7.3.3 Degeneracy

In our configuration, for a particular $i \geq 1$, there are n_i molecules of energy E_i. Still, they are not in the same state, but split into d_i different states (or degeneracies), where $d_i \geq 1$ is a constant natural number. This is illustrated in Figure 7.2: the ith column splits into d_i cells.

How many different ways are there to split our n_i molecules into d_i disjoint cells? Clearly, the first molecule could be placed in either cell. On top of this, the second

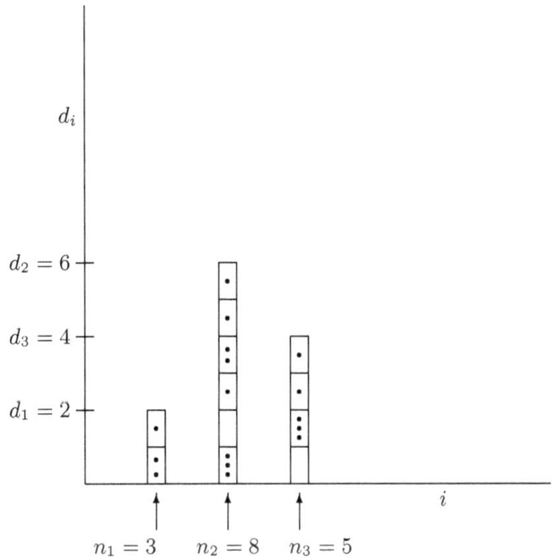

Fig. 7.2. At energy level E_i, our n_i molecules split now into d_i different states (degeneracies), each confining a few molecules to a cell.

molecule could be placed in either cell as well, and so on. Thus, the answer is $d_i^{n_i}$. Thus, the product in Section 7.3.2 should be modified to read

$$n!\frac{d_1^{n_1}}{n_1!} \cdot \frac{d_2^{n_2}}{n_2!} \cdot \frac{d_3^{n_3}}{n_3!} \cdots .$$

This is the total number of different ways to design our configuration.

7.3.4 Configuration and Its Entropy

This product is called the entropy of the configuration. It tells us how likely the configuration is to take place: the higher it is, the more options the configuration has to form, and the more likely it is to appear in reality. As time goes by, the system may change, but only in one direction: to a yet "better" configuration of higher entropy (and higher probability to occur). This is the second law of thermodynamics: entropy can only increase, but never decrease (Section 6.4.3). As a result, in the physical process in our closed system, a configuration can only transform into a new configuration of higher entropy and likelihood. In the end, the process will approach an equilibrium: a configuration of maximum entropy (among those that are allowed).

7.4 Distinguishable Molecules

7.4.1 Equilibrium: Maximum Entropy and Likelihood

Let's use the entropy to obtain a new useful quantity: the probability of the configuration. For this purpose, we only need to normalize: make sure that all probabilities of all configurations indeed sum to 1. How to do this? Easy: sum up all entropies of all possible configurations. Then, look at one particular configuration. To obtain its probability, just look at its entropy, and divide it by the above sum. This will tell us how likely the configuration is to take place in the real world.

Actually, in practice, there is no need to normalize at all. After all, this makes no difference to our practical question: what is the most probable configuration? Clearly, this is the configuration of maximum entropy (among those that are allowed). How to uncover it? For this purpose, better work not with the entropy but with its (natural) logarithm. After all, the logarithm is a monotonically-increasing function (Figure 26.17). Thus, maximizing it is the same as maximizing the entropy itself.

We must also work under two constraints: conservation of mass and energy. This will uncover the solution: the configuration that maximizes entropy (among those that preserve mass and energy). This is the true equilibrium, approached by the real physical process in our closed system. This is the Maxwell-Boltzmann distribution (or statistics).

7.4.2 Maxwell-Boltzmann Statistics

To obtain the probability of a specific configuration, we must normalize: looked at its entropy, and divide it by the sum of all entropies of all possible configurations. Fortunately, this is just theoretical: in practice, it is never done.

As time goes by, the system will eventually arrive at its equilibrium: the optimal configuration. How does it look like? Well, this is just a column diagram: at energy level i, there are n_i molecules of energy E_i (Figure 7.1). In other words, this is a discrete distribution. To make it continuous, we must first normalize. This, however, is a different kind of normalization. (After all, by now, we already have only one configuration, not many.) This new normalization has one purpose only: to make the columns sum to 1, to make a proper probability.

Thus, in the Maxwell-Boltzmann distribution, look at the ith column, and normalize it. In other words, look at the number of molecules of energy E_i, and divide by the total number of molecules. This defines a new probability function:

$$p(E_i) \equiv \frac{n_i(E_i)}{n}.$$

Now, if you pick a molecule from the system at random, how likely is it to have energy E? We now have the answer: the probability for this is $p(E)$. Fortunately, $p(E)$ doesn't depend on the discrete index i any more, but on the (continuous) energy E only.

Later on, in quantum mechanics, we'll see that energy is not really continuous, but only discrete: it comes in discrete lumps only. Still, these lumps are so tiny that they could make a continuous graph, to illustrate the behavior of p as a function of E. For this purpose, however, p must renormalize, to make sure that the area under the graph is 1, as required.

7.5 Indistinguishable Bosons

7.5.1 Indistinguishable Bosons

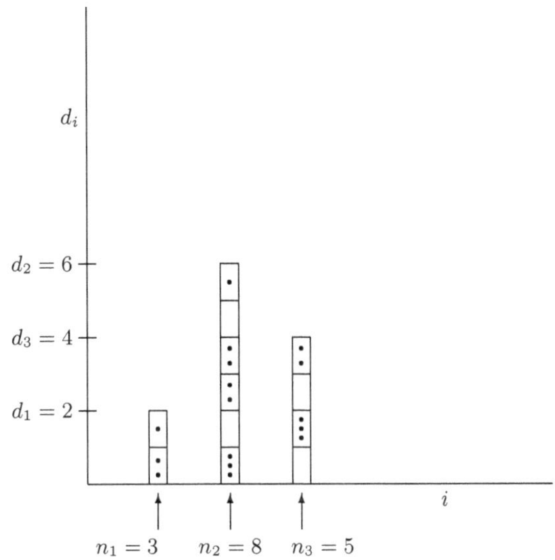

Fig. 7.3. Bosons, on the other hand, are indistinguishable: it doesn't matter which bosons to place where.

Instead of molecules, consider now n bosons, in one particular configuration. These are indistinguishable particles: there is no way to distinguish one boson from another! Therefore, it doesn't matter which bosons are placed in which energy level.

Still, there are many ways to split n_1 bosons among d_1 possible states in the first energy level (Figure 7.3). How many? Well, to split n_1 bosons into d_1 disjoint states, we need to decide how many are in state 1, how many in state 2, and so on. (After all, since the bosons are indistinguishable, it doesn't matter which ones, but only how many.) For this purpose, we need to pick d_1 nonnegative integers that sum to n_1. How many ways are there to do this? We already know the answer:

$$\binom{n_1 + d_1 - 1}{d_1 - 1}$$

(Sections 6.3.5–6.3.6). The same could be done in the other energy levels as well. Thus, to have our original configuration, there are

$$\binom{n_1 + d_1 - 1}{d_1 - 1} \binom{n_2 + d_2 - 1}{d_2 - 1} \binom{n_3 + d_3 - 1}{d_3 - 1} \cdots$$

possible ways. This is its entropy.

7.5.2 Bose-Einstein Statistics

What is the meaning of this product? Well, it is proportional to the probability of our configuration to appear in nature.

What is the most probable configuration? As before, take the logarithm of the above product, and maximize it, subject to a constraint: conservation of energy. (Conservation of mass is often irrelevant – the system may contain arbitrarily many bosons.) The configuration of maximum probability is called the Bose-Einstein distribution (or statistics).

The Bose-Einstein distribution is a discrete column diagram. To make it continuous, look at the number of bosons in the ith column:

$$n_i\left(E_i\right).$$

This is written as a continuous function of E_i. It is proportional to the probability that a boson picked from the system at random would indeed have energy E_i.

Later on, in quantum mechanics, we'll see that energy comes in discrete lumps only: not every value is allowed. Still, for our purpose, we assume that E_i may take just any value. This way, the Bose-Einstein statistics indeed takes its continuous face.

7.6 Indistinguishable Electrons

7.6.1 Pauli's Exclusion Principle

Instead of bosons, consider now n electrons, in our configuration. Like bosons, electrons are indistinguishable. But unlike bosons, they must obey Pauli's exclusion principle: in each energy level, each state may contain one electron at most. For this reason, the number of electrons in the level can never exceed the number of states:

$$n_i \le d_i, \quad i \ge 1.$$

Under these circumstances, how to split n_i electrons into d_i disjoint states? For this purpose, look at the ith column in Figure 7.4. It contains d_i cells. Out of these, pick just n_i cells to contain an electron, leaving $d_i - n_i$ cells empty. (Since electrons are indistinguishable, it doesn't matter which ones are picked, but only which states are picked.) How many ways are there to do this? We already know the answer:

$$\binom{d_i}{n_i}.$$

Thus, to produce our configuration fully, there are

$$\binom{d_1}{n_1}\binom{d_2}{n_2}\binom{d_3}{n_3}\cdots$$

different ways. This is its entropy.

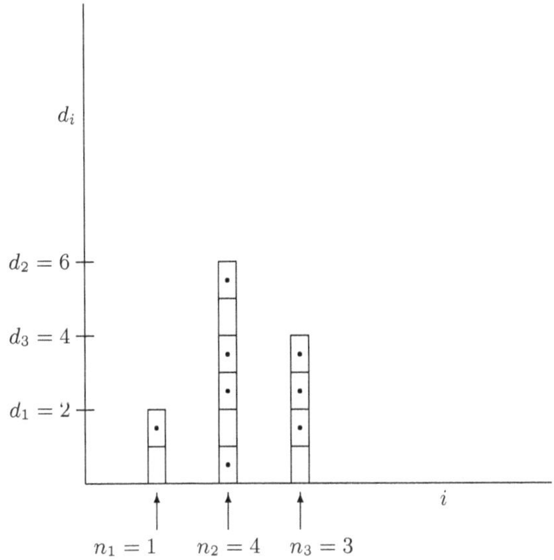

Fig. 7.4. Electrons, on the other hand, must obey Pauli's exclusion principle: each state (degeneracy) may contain one electron at most, so $n_i \leq d_i$.

7.6.2 Fermi-Dirac Statistics

What is the meaning of this product? Well, it is proportional to the probability of our configuration to take place in reality.

What is the most probable configuration? As before, take the logarithm of the above product, and maximize it, subject to two constraints: conservation of mass and energy. The configuration of maximum probability is called the Fermi-Dirac distribution (or statistics).

The Fermi-Dirac distribution is a discrete column diagram. To make it continuous, look at the number of electrons in the ith column, and normalize it. This defines a new probability function:

$$p(E_i) \equiv \frac{n_i(E_i)}{n}.$$

This could also be written as a continuous function of E_i. (For this purpose, however, there is a need to renormalize, as discussed below.) How to use it? Well, if you pick an electron from the system at random, how likely is it to have energy E? The probability for this is $p(E)$. Let's see how this function looks like.

7.6.3 Fermi Level and Fermi Energy

It turns out that the probability $p(E)$ looks like this [19]:

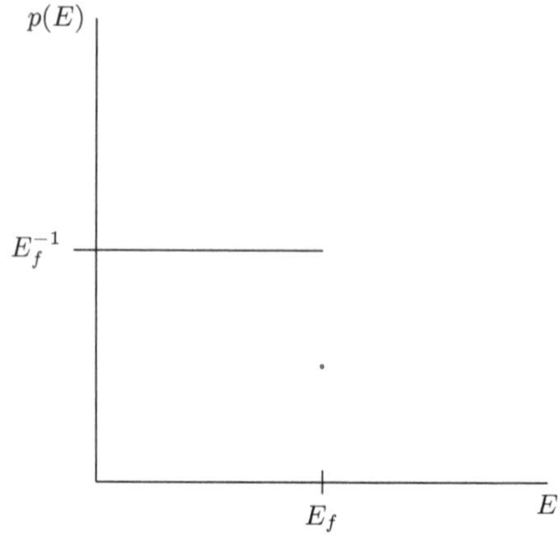

Fig. 7.5. E_f is the Fermi energy. Assume that the temperature T approaches the absolute zero. This way, $p(E)$ gets discontinuous, with two possible values: either E_f^{-1} (at $E < E_f$: full levels), or 0 (at $E > E_f$: empty levels). This way, E_f is maximal: no electron can pass it.

$$p(E) = \frac{E_f^{-1}}{\exp\left(\frac{E-E_f}{K_B T}\right) + 1},$$

where

- f is a fixed natural number: the Fermi level.
- At this level, E_f is the Fermi energy.
- T is the temperature (in Kelvin's scale).
- K_B is Boltzmann's constant.

This is the probability density, used weakly: not pointwise, but only in an integral. How likely is an electron to have $a <$energy$< b$? The probability for this is

$$\int_a^b p(E)dE.$$

This is no longer a sum, but an integral. Thus, we need a new normalization, to make sure that

$$\int_{-\infty}^{\infty} p(E)dE = 1.$$

How does $p(E)$ look like? Let's start from the limit case, in which E_{total} is minimal, and T approaches the absolute zero:

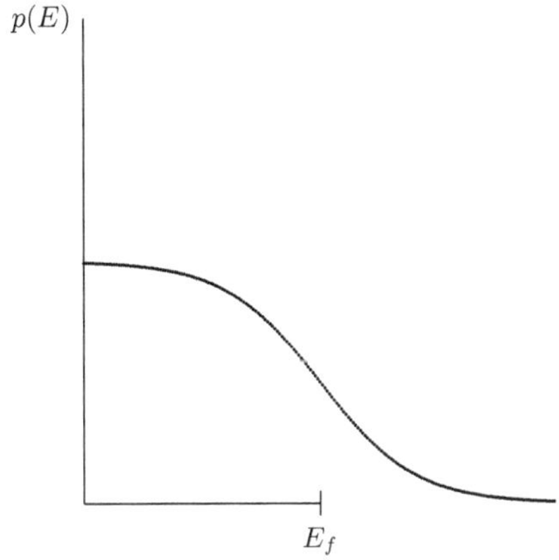

Fig. 7.6. As T increases, E_{total} increases too, so some electron may pass E_f. This way, $p(E)$ gets smooth (but still symmetric around E_f).

$$T \to 0^+ = -273°C.$$

At this limit,

$$p(E) \to \begin{cases} E_f^{-1} & \text{if} \quad E < E_f \\ 0 & \text{if} \quad E > E_f \\ E_f^{-1}/2 & \text{if} \quad E = E_f \end{cases}$$

(Figure 7.5). What is the Fermi level? This is the fth level. Up to it, all levels are full. Beyond it, all levels are empty. It could be half-full: the first nonfull level, and the last nonempty level. Its energy is called the Fermi energy: E_f. This is maximal: no electron can have more:

$$\int_0^{E_f} p(E)dE = E_f^{-1}E_f = 1.$$

But this is just a theoretical case. Next, let's heat the system a little, and pump some energy into it. This way, both E_{total} and T increase. What happens quantum-mechanically? An electron must "jump" from some nonempty level (say, level f) to the next higher (nonfull) level (say level $f+1$). This increases E_{total} by

$$E_{f+1} - E_f > 0.$$

As T increases, more electrons may "jump" too (say from level $f-1$ to level $f+2$), and so on. Still, $p(E)$ remains symmetric around E_f (Figure 7.6).

7.7 Exercises: Spin

7.7.1 Spin-Up-or-Down

1. Extend the trinomial formula, and obtain the quadrinomial formula:

$$(a+b+c+d)^n = \sum_{0\le l,j,m,k\le n,\ l+j+m+k=n} \frac{n!}{l!j!m!k!} a^l b^j c^m d^k.$$

   Hint: extend the proof in Section 6.6.3.
2. How many terms are there in this sum? Hint:

$$\binom{n+4-1}{4-1} = \binom{n+3}{3} = \frac{(n+1)(n+2)(n+3)}{6}$$

   (Sections 6.3.5–6.3.6).
3. In the above sum, look at the quadrinomial coefficient:

$$\frac{n!}{l!j!m!k!}.$$

   What is its combinatorial meaning? Hint: this is the total number of different ways to pick l a's, j b's, m c's, and k d's.
4. What is its physical meaning? Hint: this is the total number of different ways to pick l molecules of energy E_1, j of energy E_2, m of energy E_3, and k of energy E_4 from a system of $n = l + j + m + k$ molecules.
5. What would happen if, instead of molecules, we had bosons? Hint: unlike molecules, bosons are indistinguishable. Therefore, all the above ways would coincide, and become one and the same.
6. Instead of a boson, think of an electron (or any other fermion). What would happen now? Hint: electrons are indistinguishable as well. Thus, there would be no change: all the above ways still coincide.
7. Consider a system of n electrons. In it, look at a specific energy level (say, the first one). In it, there are just two possible states (degeneracies): either spin-up, or spin-down. In other words,

$$d_1 = 2.$$

   How many ways are there to have $n_1 = 2$ electrons in this level? Hint: there is just one way: one electron spins up, and the other spins down. (After all, thanks to Pauli's exclusion principle, they must differ in some aspect.) It doesn't matter which: they are indistinguishable anyway.
8. Is this as in our formula? Hint: yes:

$$\binom{d_1}{n_1} = \binom{2}{2} = 1.$$

9. One student said a different answer: "there are two ways:
   - either the first electron spins up and the second spins down,
   - or the first electron spins down and the second spins up."

   What's wrong? Hint: these ways coincide: there is no way to distinguish between the "first" and "second" electrons.

10. On the other hand, assume now that the first level contains just one electron: $n_1 = 1$. How many different ways are there? Hint: two: either spin-up, or spin-down.
11. Is this as in our formula? Hint: yes:

$$\binom{d_1}{n_1} = \binom{2}{1} = 2.$$

12. Conclude that $n_1 = 1$ has more entropy, and is thus more likely than $n_1 = 2$.
13. Does this mean that, in the Fermi-Dirac distribution, the first energy level must always contain just one electron? Hint: no! Like every configuration, it must preserve mass and energy:

$$\begin{aligned} n &= n_1 + n_2 + n_3 + \cdots \\ E_{\text{total}} &= n_1 E_1 + n_2 E_2 + n_3 E_3 \cdots \end{aligned}$$

(for some given n and E_{total}). Under these constraints, if the high-entropy configuration has $n_1 = 2$, then the first level is full:

$$n_1 (E_1) = 2.$$

In practice, this is often the case: the first energy level is usually full (Figures 7.5–7.6).
14. Look at a specific example, in which

$$E_{\text{total}} = E_1 + E_3 = 2E_2.$$

How to arrange $n = 2$ electrons? Hint: place one in the first level (spinning up or down), and the other in the third level (spinning up or down). This can be done in four different ways. This is better than both in the second level, which can be done in one way only (less entropy).
15. Consider now a system of bosons rather than electrons. Look at a specific energy level again (say, the first one). In it, there are now three possible states (degeneracies): either spin-up, or spin-down, or spin-zero:

$$d_1 = 3.$$

How many ways are there to have n_1 bosons in this level? Hint:

$$\binom{n_1 + 3 - 1}{3 - 1} = \binom{n_1 + 2}{2} = \frac{(n_1 + 1)(n_1 + 2)}{2}.$$

16. For example, assume that $n_1 = 2$ (just two bosons). How many ways are there to place them in the first level? Hint: six: either up-up or zero-zero or down-down or up-zero or zero-down or up-down.
17. Later on, in the context of Lie algebras, we'll learn more about bosons and electrons and their spin.

Part IV

Introduction to Special Relativity

8

Introduction to Special Relativity

A physical phenomenon should better be independent of the mathematical coordinates used to describe it. After all, the laws of nature must be invariant. You should observe the same laws, no matter what coordinates you use to carry out your experiment.

But what if you are on a spaceship, traveling away from the Earth (at a constant speed)? In your spaceship, you may have a lamp. Look at its light, and measure its speed: $300,000$ kilometer per second.

I, here on the Earth, also want to calculate the speed of light. But I'm a little naive: I take the speed that you just measured, and add to it yet another speed: the speed of the spaceship (with respect to the Earth). So, I get a higher speed!

Fortunately, we know better: light travels at the same speed at all inertial systems (moving at a constant velocity with respect to one another). Indeed, this was already proved in many experiments. How to solve this paradox? Well, who says that speeds are added linearly?

This is how special relativity improves on Galileo's original laws. In practical engineering problems, this effect is negligible: Newtonian mechanics is good enough. In modern technology, on the other hand, at a very high speed, this can no longer be ignored.

Relativity designs a new geometry: it takes the time axis, and sticks it to space, to form spacetime. In it, we get to see two new effects: time dilation, and length contraction. Together, they join forces, and may even solve the twin paradox.

Together, time-space make a new relativistic pair. Want to know them in another (inertial) system? Transform them, as a pair. Likewise, momentum-energy make a relativistic pair as well. This leads to Einstein's famous formula: $E = mc^2$ (where the mass m is invariant: the same at all inertial systems). But this holds only in the self-system, where the particle is at rest, and only has its potential (nuclear) energy. In any other system, it also has velocity, and kinetic energy.

In special relativity, spacetime is a linear hyperspace. In it, we have a constant metric: Minkowski's metric. General relativity, on the other hand, is more general. In it, spacetime gets curved: a 4-D differentiable manifold, with a nonconstant metric. How to design it? Around each event in spacetime, draw a linear chart: a linear t-x-y-z coordinate system, to help carry out an experiment in your local "lab."

Thanks to Zorn's lemma, there is a maximal atlas of charts. At each event in spacetime, use the axiom of choice to pick one chart, and span a tangent hyperspace in terms of Riemann's normal coordinates. These are only local: they can never be "tied" into a global coordinate system. Still, they "envelope" spacetime with tangent hyperspaces all over, and support Einstein's equivalence principle, geometrically and locally. Here, we only touch on this algebraically: a Lie algebra, tangent to a Lie group.

8.1 Michelson-Morley's Experiment

8.1.1 Light: An Electromagnetic Wave

In the 19th century, Maxwell unified two physical phenomena: electricity and magnetism. This formed a new phenomenon: an electromagnetic field. In vacuum, this is an electromagnetic wave, traveling at the speed of light: $c \equiv 300,000$ kilometer per second.

Nowadays, we know that light is just an example of an electromagnetic wave: while traveling forward, it also oscillates at a frequency that we can see. X-rays, on the other hand, make a different kind of radiation, too frequent for us to see. Radio waves, on the other hand, make yet another kind of wave, not frequent enough for us to see. Still, all these are in principle the same: an oscillating disturbance in the electromagnetic field. They only differ in frequency and wavelength, and occupy different parts of the spectrum.

8.1.2 Wave in Its Medium

In Maxwell's day, everybody assumed that a wave must oscillate in some medium. In what medium could the electromagnetic wave oscillate? After all, it could travel and oscillate even in vacuum!

To answer this, everyone believed that space is full of an invisible substance: aether. This is the medium in which light travels.

8.1.3 Aether: An Absolute Reference Frame

Our Earth may rotate, day and night, but the aether doesn't: it stands still, and makes an absolute reference frame. More precisely, the Earth rotates eastwards (at a constant speed $v > 0$ with respect to the aether). This is why the sun rises from the east. But we humans look at things the other way around: we think that we stand still, and that the aether blows at velocity v westwards. After all, we look at things from our own private perspective: our own reference frame. So, we should feel a slight wind, blowing at our faces: the aether wind, blowing westwards at velocity v, all the time. Fortunately, the aether is too fine to feel.

8.1.4 Inertial Reference Frames

So, the aether can serve as an absolute reference frame. With respect to it, the Earth is an inertial reference frame, moving at the constant velocity v eastwards. For simplicity, assume that this motion is linear: in a straight line. This way, these are inertial reference frames: they never accelerate with respect to one another, and never change direction. This is the subject of special relativity. Acceleration, on the other hand, is the subject of general relativity.

8.1.5 Michelson-Morley's Experiment

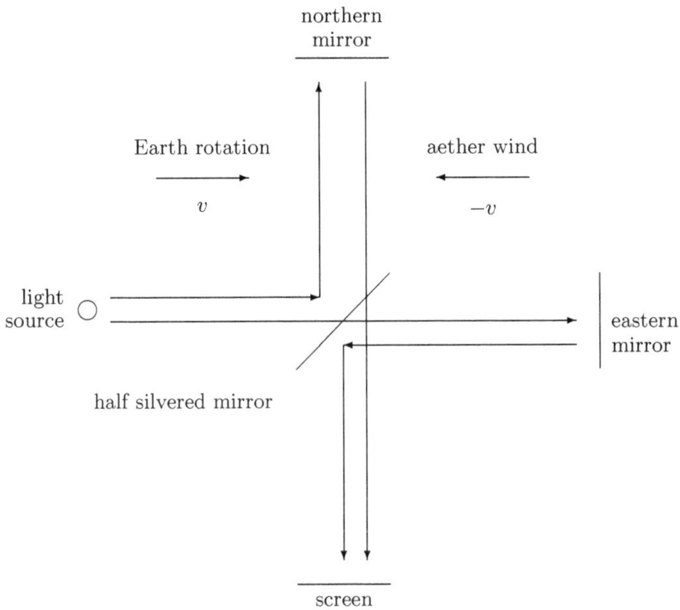

Fig. 8.1. Light comes from the left, and splits at the center: half goes through, reflects on the far east, and returns to the center, where it turns left. The other half turns left, reflects at the far north, and returns all the way south, to hit the screen at the far south. Both halves match! No aether wind!

To measure v, Michelson and Morley carried out their experiment (Figure 8.1). At the center, they placed an (oblique) half-mirror: half-transmitting, and half-reflecting. From the left, they fired a light beam, straight eastwards. Upon hitting the half-mirror at the center, it splits: half goes straight ahead, reflects at the far east, and returns to the center, where (half of it) turns south. The other half doesn't go through the center, but turns north, where it reflects and returns all the way south, to the screen.

Now, look at the light from the perspective of the Earth. What happens after the split? One half goes to the north mirror (at distance d) and back. On its way, it feels a little wind from the east. To overcome it, it must go a little obliquely north-east. Thanks to Pythagoras' theorem, its net velocity northwards is only

$$\sqrt{c^2 - v^2} < c.$$

Thus, the round trip from the center to the northern mirror and back should last

$$\triangle t = \frac{2d}{\sqrt{c^2 - v^2}}$$

seconds (distance divided by velocity).

And how about the other half? It transmits straight eastwards, feeling the wind at its face, or (later) at its back. Assuming that velocity is additive, this should last

$$\begin{aligned}
\frac{d}{c-v} + \frac{d}{c+v} &= d\frac{c+v+c-v}{(c-v)(c+v)} \\
&= 2d\frac{c}{c^2 - v^2} \\
&= \frac{2d}{\sqrt{c^2 - v^2}} \cdot \frac{c}{\sqrt{c^2 - v^2}} \\
&> \frac{2d}{\sqrt{c^2 - v^2}}
\end{aligned}$$

seconds, which is a little slower. So, the two halves won't match, but miss, and interfere: constructively, then destructively, then constructively again, and so on. This is how Michelson and Morley had hoped to measure v. They must be quick: in six hours, the wind will change, and blow from the sky down at our heads, with no effect on our experiment any more!

Michelson and Morley were very sad. They failed completely: at all versions (and angles), there was constructive interference only! Maybe there is no aether after all?

8.1.6 No Aether!

Nowadays, we know better: there is no aether at all! The electromagnetic wave is just a series of periodic fluctuations in the electromagnetic field.

Now, look at this electromagnetic field. From where? From the lab? Or from any other (moving) point of view? So, the electromagnetic field may actually play the role of the aether. Why is there no Michelson-Morlet effect? This is because the speed of light is constant: c (from any inertial reference frame). This is Einstein's law. Light obeys a strange equation:

$$c \text{ ``}+\text{''} \; v \; = \; c \text{ ``}-\text{''} \; v \; = \; c.$$

Why? Because velocity is nonlinear.

8.1.7 Einstein's Law: Constant Speed of Light

Each reference frame has its own (relativistic) time scale (measured in its own private clock) and length scale (measured in its own private ruler). These scales match. In what sense? They have the same ratio. This way, the speed of light remains c. This is Einstein's law: the speed of light is constant (invariant), independent of the reference frame (and the coordinates) used to measure it.

8.1.8 Synchronized Reference Frames

To see this, our (inertial) reference frames should better be synchronized: both clocks should start ticking at the same initial time: zero. At this time, both should also share the same origin: 0 (from which distance is measured). In spacetime, these will form an event: the origin $(0, 0)$, at which both time and space vanish.

8.2 Adding Velocities Accurately

8.2.1 How to Add Velocities?

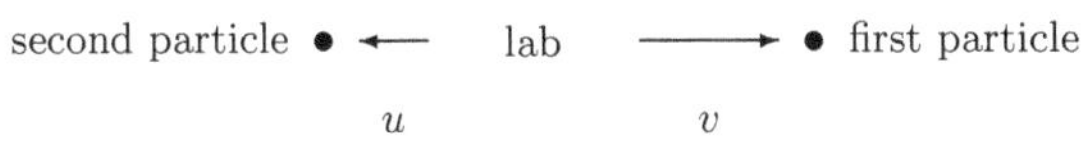

Fig. 8.2. In our lab, the first particle moves rightwards at velocity v, and the second moves leftwards at velocity u.

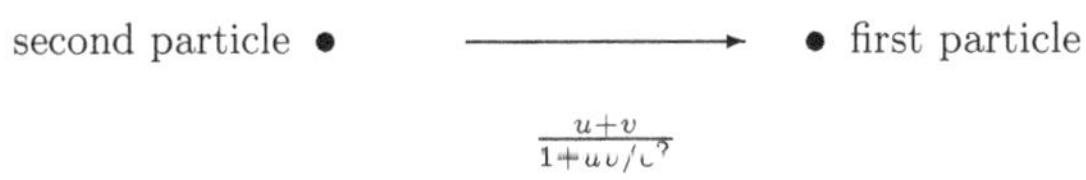

Fig. 8.3. And how fast does the first particle get away from the second one? Not quite $u + v$, but $(u + v)/(1 + uv/c^2)$.

In Newtonian mechanics, velocities are added linearly. Consider, for example, a particle that travels at constant velocity v (at some constant direction). At the same time, another particle travels at constant velocity u in the opposite direction (Figure 8.2). These velocities are not absolute: they are measured in our lab.

Still, our lab is not an absolute reference frame: it is not quite static, but dynamic. After all, it sits on the Earth, which orbits the sun, which orbits the center of our galaxy: the Milky Way. Fortunately, this underlying motion can be ignored.

After all, we don't feel it at all. In fact, we are only interested in the velocities with respect to our lab. For us, the lab is at rest.

Better yet, disregard the lab altogether, and focus on the relation between the particles: how do they get away from one another? Forget about the lab, and look at one particle from the perspective of the other. After all, this is the true physical quantity here: the lab is not really part of the game.

How fast does the first particle get away from the second one? In Galileo-Newtonian mechanics, velocities add up: $u + v$. For small velocities, this makes sense. But what about a particle (nearly) as fast as light (say, $v = c$)? After all, $c + u$ would exceed the speed of light, which is impossible (as is evident from many experiments)! Fortunately, velocities are nonlinear: the accurate answer is not $u + v$ but

$$\frac{u+v}{1+\frac{uv}{c^2}}$$

(Figure 8.3). So long as both u and v are moderate, this is nearly the same as $u + v$. This is why Galileo-Newtonian mechanics works well in practical engineering problems. Still, at very high speeds, our fix gets crucial, and can't be ignored any more.

8.2.2 Never Exceed The Speed of Light!

Thanks to our new formula, we can now keep our law: never exceed the speed of light! Indeed, assume that, with respect to the lab, both particles don't exceed the speed of light:

$$|u| \leq c \quad \text{and} \quad |v| \leq c.$$

Still, with respect to one another, could they ever exceed the speed of light? No! To prove this algebraically, rescale the speeds. For this, define

$$\beta_u \equiv \frac{u}{c} \quad \text{and} \quad \beta_v \equiv \frac{v}{c}.$$

Thanks to Einstein's law,

$$|\beta_u| \leq 1 \quad \text{and} \quad |\beta_v| \leq 1.$$

Assume first that our particles are heavier than photons, so

$$|u| < c \quad \text{and} \quad |v| < c,$$

or

$$|\beta_u| < 1 \quad \text{and} \quad |\beta_v| < 1.$$

In this case, we can estimate:

$$1 - \beta_u - \beta_v + \beta_u\beta_v = (1-\beta_u)(1-\beta_v) > 0,$$

so

$$\beta_u + \beta_v < 1 + \beta_u\beta_v,$$

so

$$\frac{\beta_u + \beta_v}{1 + \beta_u \beta_v} < 1.$$

For the same reason,

$$-\beta_u - \beta_v < 1 + \beta_u \beta_v$$

as well. Thus,

$$\left| \frac{\beta_u + \beta_v}{1 + \beta_u \beta_v} \right| = \frac{|\beta_u + \beta_v|}{1 + \beta_u \beta_v} < 1.$$

Multiply this by c:

$$\left| \frac{u + v}{1 + \frac{uv}{c^2}} \right| < c.$$

This is what we wanted: even with respect to one another, the particles never exceed the speed of light.

8.2.3 Photon

But what if (at least) one particle was a photon, as fast as light? In this case, things may get quite strange. Let's start with the case in which our first particle wins:

$$v = c \quad \text{and} \quad -c < u \le c.$$

In this case,

$$\frac{u + c}{1 + \frac{uc}{c^2}} = \frac{u + c}{1 + \frac{u}{c}} = c.$$

What is the physical meaning of this? Well, the first particle travels rightwards as fast as light, leaving the lab behind. Being so fast, it can no longer distinguish between the lab and the second particle: both look as one and the same thing, left behind at the speed of light.

8.2.4 Singularity: Indistinguishable Photons

Still, there is a yet stranger case:

$$v = c \quad \text{and} \quad u = -c.$$

In this (singular) case, the second particle is a photon too: it follows the first one, rightwards, as fast as light too. In this case, our formula is no good any more: it divides zero by zero. But we can still use our logic: at the initial time $t = 0$, both photons are at 0, with the same speed: c. Thus, they must be one and the same. Later on, in quantum mechanics, photons will indeed get indistinguishable. Moreover, like every (subatomic) particle, a photon has only uncertain position-momentum-energy (and speed-mass?). But here, we assume no uncertainty as yet.

8.3 Systems and Their Time

8.3.1 Motion in 1-D

In general, the motion is in 3-D. But here, things are even simpler: u and v span a 2-D plane, where the motion really takes place. Better yet, they are often on the same line: the x-axis. (This direction never changes: after all, there is no force.) The perpendicular coordinates, y and z, remain idle. They never take part in the motion, and can be ignored. This is how our spacetime reduces to the x-t plane, easy to work with. These coordinates will often be used in our lab.

8.3.2 How to Measure Time?

How to measure time? Usually, we use an angle: in our clock, in one second, the long hand makes an angle of 6° (1/60th of a complete circle). But here, we often "use" a linear (theoretical) clock: in one second, our light beam makes one light second, telling us that one more second had passed.

Why is this better? Well, this scales time by c: instead of the original t (measured in seconds), we can now use a new length variable: ct (measured in light seconds). After all, while t grows by one second, ct advances by one light second: $300,000$ kilometer.

This way, instead of the original t-axis, we now have the new ct-axis. Why is this attractive? Because ct is length too, producing a more symmetric spacetime.

8.3.3 The Self System

In our lab, we use our x-t coordinates. But what about another lab, moving at a constant speed v away from us? In it, our first particle is at rest. This is its self system. To describe it, use primed coordinates: x'-t'. (Don't get confused: this has nothing to do with derivative.) These are measured in new clock and ruler, moving with the self system. In them, you can read the primed coordinates. At $x' = 0$, you'll always find our first particle. After all, in its self system, it must be static. At $x' > 0$, you'll see what's to its right. At $x' < 0$, on the other hand, you'll see what's to its left. Likewise, want to read t'? Look at the moving clock. Better yet, work with ct', as before.

8.3.4 Synchronization

As before, the systems are synchronized: initially, at $t' = t = 0$, the particle is at the origin in both systems:

$$x' = x = 0.$$

In spacetime, this is the same event: the origin (in both systems):

$$\begin{pmatrix} x' \\ ct' \end{pmatrix} = \begin{pmatrix} x \\ ct \end{pmatrix} = \begin{pmatrix} 0 \\ 0 \end{pmatrix}.$$

This is the initial event in spacetime. At any later time, on the other hand, the systems may start to differ, not only in position but also in time: t' may differ from t. In this sense, time is relativistic too: it depends on the system where it is measured.

8.4 Spacetime: Two Inertial Systems

8.4.1 Our Coordinate System

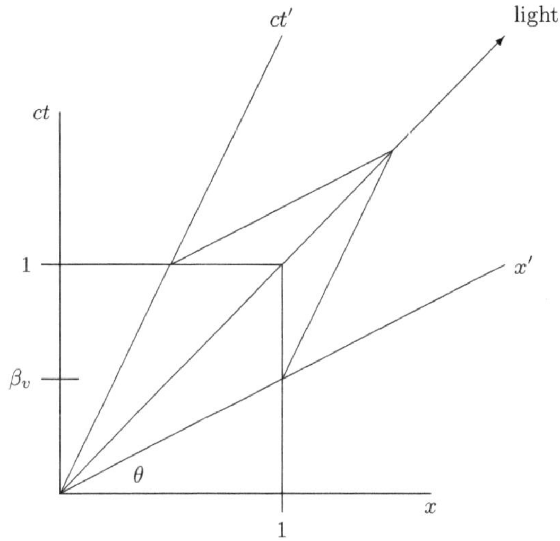

Fig. 8.4. Spacetime: the (perpendicular) x-ct axes span our lab system. The (nonperpendicular) x'-ct' axes, on the other hand, span another system, traveling rightwards at the constant speed v (away from us). In both systems, light travels at the same speed: c. This draws an oblique line, issuing from the origin (at angle 45°).

In special relativity, spacetime is easy to draw. After all, it is linear: no gravity, no force, and no curvature. (Constant flat metric: Minkowski's metric.) In our lab, use the x-ct coordinates. They are even and symmetric: both x and ct take the same status. How? Look at the oblique lines: $x = \pm ct$. Along them, light travels. This splits spacetime into four quarters. Only two are real. These are the upper and lower quarters: time-like, and physical. The left and right quarters, on the other hand, are space-like: nonphysical: only science fiction can go on.

8.4.2 Traveling Coordinate System

And what about the x'-ct' coordinate system, moving away from us at a constant speed v? It must keep the same oblique lines (Figure 8.4). After all, light travels at the same speed, at all (inertial) systems. Thus, although the new x'-ct' axes are no longer perpendicular, they are still even and symmetric: mirrored by the oblique lines $x' = \pm ct'$ (same as $x = \pm ct$).

8.4.3 Area in Spacetime

How to draw both systems in the same diagram? This is done in Figure 8.4. In it, look at the x- and x'-axes. What is the angle between them? This is θ, satisfying

$$\tan(\theta) = \beta_v.$$

Thanks to symmetry, this is also the angle between the ct- and ct'-axes. (After all, in special relativity, time and space take equal status.) Now, look at the x'-ct' axes. They issue from the origin, but are nonperpendicular: the angle between them is not $\pi/2$ but only

$$\frac{\pi}{2} - 2\theta.$$

This is the angle in a new diamond of edges of length $\sqrt{1+\beta_v^2}$ and area

$$\begin{aligned}
\text{area of diamond} &= \left(1+\beta_v^2\right)\sin\left(\frac{\pi}{2} - 2\theta\right) \\
&= \left(1+\beta_v^2\right)\cos(2\theta) \\
&= \left(1+\beta_v^2\right)\left(\cos^2(\theta) - \sin^2(\theta)\right) \\
&= \left(1+\beta_v^2\right)\left(1 - 2\sin^2(\theta)\right) \\
&= \left(1+\beta_v^2\right)\left(1 - \frac{2\beta_v^2}{1+\beta_v^2}\right) \\
&= \left(1+\beta_v^2\right)\frac{1-\beta_v^2}{1+\beta_v^2} \\
&= 1-\beta_v^2 \\
&= \gamma^{-2}
\end{aligned}$$

(Section 26.4.1), where γ is a new parameter, defined by

$$\gamma \equiv \gamma\left(\beta_v\right) \equiv \frac{1}{\sqrt{1-\beta_v^2}} \geq 1.$$

8.4.4 Time Dilation

To have area 1 (like the original square in Figure 8.4), our diamond must get bigger: its edges (and indeed the entire x'-ct' axes) must stretch by factor γ. This will help map the entire x-ct system onto the (stretched) x'-ct' system by a linear area-preserving mapping (of determinant 1).

Thanks to this stretching, the x'-ct' scales get sparser. This leads to time dilation (Figure 8.5): project the ct'-axis leftwards, onto the vertical ct-axis, and you'll see that one t'-second (measured in the moving clock) is slower: it takes as many as γ t-seconds (measured in the static clock).

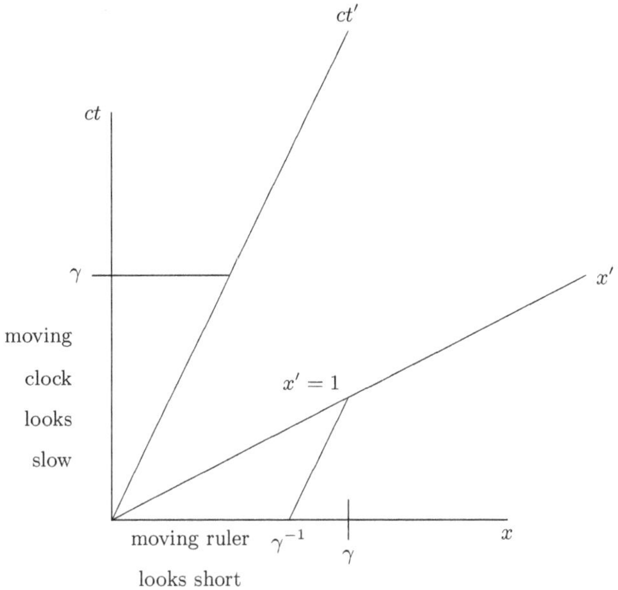

Fig. 8.5. The stretched x'-ct' axes. One second in the moving clock looks as long as γ seconds in the static clock. This is time dilation (illustrated by the short horizontal line). From the lab, the moving ruler looks as short as γ^{-1} static rulers. This is length contraction (illustrated by the short oblique line).

8.4.5 Length Contraction

And what about the moving ruler? From the lab, it looks γ times as short. Why? Well, assume that the moving ruler goes from $x' = 0$ to $x' = 1$. In the moving reference frame, this is static: it holds for all t'-time. This draws the (short) oblique line

$$x' \equiv 1$$

(Figure 8.5). What slope does it have? Well, it is parallel to the ct'-axis, so it makes angle θ with the ct-axis. Thanks to triangle similarity, it hits the x-axis at

$$\gamma - \frac{(\gamma\beta_v)^2}{\gamma} = \gamma\left(1 - \frac{(\gamma\beta_v)^2}{\gamma^2}\right) = \gamma\left(1 - \beta_v^2\right) = \gamma \cdot \gamma^{-2} = \gamma^{-1}.$$

Thus, from the lab, the ruler looks γ times as short. This is length contraction. Later on, we'll see this algebraically as well.

8.4.6 Singularity

And what if:

$$v \to c?$$

In this limit, our diamond gets degenerate: too thin, with no width at all. This is a singularity: space and time shrink to zero. Time is too lazy to move. Motion gets nonphysical: it takes infinite energy and momentum. This is a complete chaos: time stops!

In special relativity, this is only theoretical. In general relativity, on the other hand, this may happen at the center of a black hole. There, only quantum mechanics could save us.

8.5 Lorentz Matrix and Transformation

8.5.1 Coordinates and Their Equal Status

In terms of the self system, the particle is static. Indeed, its position is always

$$x' = 0.$$

In terms of the lab, on the other hand, the particle is dynamic: it moves at speed v. Therefore, at time t, it is at

$$x = \frac{dx}{dt}t = vt = \frac{v}{c}ct = \beta_v ct,$$

where

$$\beta_v \equiv \frac{v}{c} \equiv \frac{dx}{d(ct)}$$

is the scaled velocity (obtained from differentiation with respect to ct).

Consider now a more general point: at a fixed distance x' from the particle. In the self system, this marks a fixed position: x'. If $x' > 0$, then this is to the right of the particle. If, on the other hand, $x' < 0$, then this is to its left. In either case, this remains static (relative to the particle).

In the lab, on the other hand, this point is dynamic: it moves at speed v. At the initial time of $t = 0$, it starts from $x = x'$. Then, at time $t \geq 0$, it advances to

$$x = x' + vt = x' + \beta_v ct.$$

In other words,

$$x' = x - \beta_v ct.$$

Still, this is not the whole story. After all, x' and ct' must take an equal status, and form a new (relativistic) pair, preserving the speed of light: c (which is a true physical quantity, and mustn't depend on the mathematical coordinates). By now, we already did this geometrically. It is time to get algebraic. This will prepare us for the wider context of Lie groups, later on.

8.5.2 Lorentz Transformation

Let's transform the original x-ct lab coordinates to the new x'-ct' self coordinates. This must be insensitive to interchanging x and ct. After all, both measure distance: x measures distance from the lab's origin, and ct measures distance in our linear clock (Section 8.3.2). This is the Lorentz transformation:

$$\begin{pmatrix} x \\ ct \end{pmatrix} \to \begin{pmatrix} x' \\ ct' \end{pmatrix} \equiv \gamma(\beta_v) \begin{pmatrix} 1 & -\beta_v \\ -\beta_v & 1 \end{pmatrix} \begin{pmatrix} x \\ ct \end{pmatrix},$$

where

$$\gamma \equiv \gamma(\beta_v) \equiv \frac{1}{\sqrt{1-\beta_v^2}} \geq 1$$

is picked to make sure that the determinant is 1 (Figure 2.11).

8.5.3 Lorentz Matrix and The Infinity Point

Let's look at an interesting case. What if $|v| = c$? In this case, γ is no longer a number: it is the infinity point ∞ (Chapter 3). Still, it is assumed that

$$0 \cdot \infty = 0.$$

We'll come back to this later. Let's look at another interesting case: $v = 0$. In this case, the Lorentz matrix is just the identity matrix:

$$I \equiv \begin{pmatrix} 1 & 0 \\ 0 & 1 \end{pmatrix}.$$

This may look silly, but is not: it will help rewrite the Lorentz matrix better. For this purpose, define the new 2×2 matrix

$$J \equiv \begin{pmatrix} 0 & 1 \\ 1 & 0 \end{pmatrix}.$$

The original Lorentz matrix can now be written as

$$\gamma(\beta_v)(I - \beta_v J).$$

This will be useful later.

8.5.4 Interchanging Coordinates

Fortunately, the Lorentz matrix commutes not only with I but also with J:

$$J\gamma(\beta_v)(I - \beta_v J) = \gamma(\beta_v)(I - \beta_v J) J.$$

This is what we wanted: the Lorentz transformation is completely blind (insensitive) to interchanging x and ct:

$$\begin{aligned}
\begin{pmatrix} ct \\ x \end{pmatrix} &= J \begin{pmatrix} x \\ ct \end{pmatrix} \\
&\to \gamma(\beta_v)(I - \beta_v J) J \begin{pmatrix} x \\ ct \end{pmatrix} \\
&= J\gamma(\beta_v)(I - \beta_v J) \begin{pmatrix} x \\ ct \end{pmatrix} \\
&= J \begin{pmatrix} x' \\ ct' \end{pmatrix} \\
&= \begin{pmatrix} ct' \\ x' \end{pmatrix}.
\end{aligned}$$

8.5.5 Composite Transformation

How to add velocities? Let's prove our formula (Section 8.2.1). For this purpose, note that all Lorentz matrices commute. Consider now a composite Lorentz transformation. What does this mean physically? Look at our first particle, moving rightwards at speed v. The second particle, on the other hand, moves leftwards at speed u (or rightwards at speed $-u$). This is with respect to our (static) lab.

Now, look at things the other way around: from the perspective of the second particle, the entire lab moves rightwards at speed u. This way, the second particle is now assumed to be at rest. How does the first particle travel away from it? To calculate this, we need to compose two motions: the motion of the first particle with respect to the lab, on top of the motion of the lab away from the second particle. (Thanks to commutativity, order doesn't matter.) Since

$$J^2 = I,$$

the product is

$$\begin{aligned}
\gamma(\beta_u)(I - \beta_u J)\gamma(\beta_v)(I - \beta_v J) &= \gamma(\beta_u)\gamma(\beta_v)(I - \beta_u J)(I - \beta_v J) \\
&= \gamma(\beta_u)\gamma(\beta_v)\left(I - \beta_u J - \beta_v J + \beta_u \beta_v J^2\right) \\
&= \gamma(\beta_u)\gamma(\beta_v)\left((1 + \beta_u \beta_v) I - (\beta_u + \beta_v) J\right) \\
&= \gamma(\beta_u)\gamma(\beta_v)(1 + \beta_u \beta_v)\left(I - \frac{\beta_u + \beta_v}{1 + \beta_u \beta_v} J\right) \\
&= \gamma\left(\frac{\beta_u + \beta_v}{1 + \beta_u \beta_v}\right)\left(I - \frac{\beta_u + \beta_v}{1 + \beta_u \beta_v} J\right)
\end{aligned}$$

(since the determinant must remain 1). This composition describes the total motion of the first particle away from the second one. The total velocity is, thus, not $u + v$ but

$$c\frac{\beta_u + \beta_v}{1 + \beta_u \beta_v} = \frac{u + v}{1 + \frac{uv}{c^2}},$$

as asserted in Section 8.2.1.

8.5.6 The Inverse Transformation

In particular, look at the special case in which $u = -v$: the second particle coincides with the first one. What is the total "motion?" None! Indeed, in our composition, set

$$\begin{aligned}\gamma\left(\beta_{-v}\right)\left(I-\beta_{-v}J\right)\gamma\left(\beta_{v}\right)\left(I-\beta_{v}J\right) &= \gamma\left(\frac{\beta_{-v}+\beta_{v}}{1+\beta_{-v}\beta_{v}}\right)\left(I-\frac{\beta_{-v}+\beta_{v}}{1+\beta_{-v}\beta_{v}}J\right) \\ &= \gamma(0)I \\ &= I.\end{aligned}$$

This gives us the inverse transformation, represented by the inverse matrix:

$$\gamma\left(\beta_{-v}\right)\left(I-\beta_{-v}J\right),$$

or

$$\gamma\left(\beta_{v}\right)\left(I+\beta_{v}J\right).$$

This is a legitimate Lorentz matrix, with determinant 1. What does it mean physically? It transforms back to the lab. Indeed, with respect to the particles (which are now one and the same), the entire lab moves leftwards at velocity v (or rightwards at velocity $-v$). Thus, the inverse transformation works the other way around: it considers the particles to be at rest, and the entire lab to move (at velocity $-v$) away from them. This is why it picks a minus sign: it uses $-v$ rather than v.

Topologically, mirror all Lorentz matrices by the open interval $(-1,1)$:

$$\gamma\left(\beta_{v}\right)\left(I-\beta_{v}J\right) \leftrightarrow \beta_{v} \in (-1,1).$$

8.6 Proper Time in The Self System

8.6.1 Proper Time: Invariant

In its self system, the particle is static: it is always at $x'=0$. In its "pocket," it carries a tiny clock. What time does it show? This is the proper time:

$$s \equiv t'.$$

Once s is determined in the self system, it could also be calculated from any other inertial system (including the lab), and always give the same result: s. (Soon, we'll see how.) Thus, the proper time is not relativistic but invariant.

In the x'-ct' self system, the tiny clock is at

$$(x',ct') = (0,cs).$$

Let's use this to form the matrix

$$\begin{pmatrix} ct' & x' \\ x' & ct' \end{pmatrix} = \begin{pmatrix} cs & 0 \\ 0 & cs \end{pmatrix} = csI$$

(of determinant c^2s^2). To it, apply the inverse Lorentz matrix (column by column). The second column is easy: it goes back to the x-ct lab coordinates:

$$\begin{pmatrix} x' \\ ct' \end{pmatrix} \to \gamma\left(\beta_{v}\right)\left(I+\beta_{v}J\right)\begin{pmatrix} x' \\ ct' \end{pmatrix} = \begin{pmatrix} x \\ ct \end{pmatrix}.$$

Since the inverse matrix commutes with J (Section 8.5.4), the first column transforms in the same way:

$$\begin{pmatrix} ct' \\ x' \end{pmatrix} = J \begin{pmatrix} x' \\ ct' \end{pmatrix} \to J\gamma(\beta_v)(I + \beta_v J) \begin{pmatrix} x' \\ ct' \end{pmatrix} = J \begin{pmatrix} x \\ ct \end{pmatrix} = \begin{pmatrix} ct \\ x \end{pmatrix}.$$

In summary, the entire matrix transforms to

$$\begin{pmatrix} ct & x \\ x & ct \end{pmatrix} = \gamma(\beta_v)(I + \beta_v J) \begin{pmatrix} ct' & x' \\ x' & ct' \end{pmatrix} = cs\gamma(\beta_v)(I + \beta_v J).$$

Now, both sides have the same determinant:

$$c^2t^2 - x^2 = c^2s^2.$$

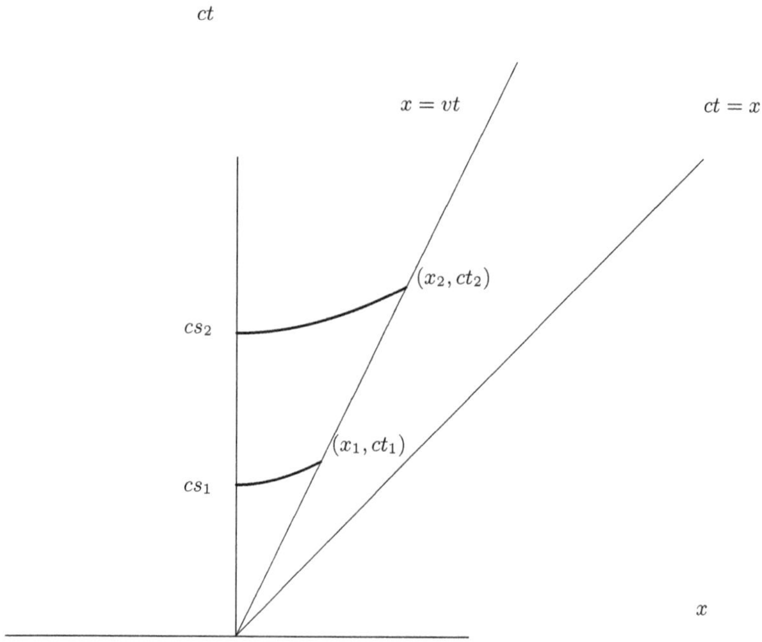

Fig. 8.6. In the lab, the proper time is read from a static clock: $t = t_1, t_2, t_3, \ldots$. This is maximal: a clock that moves at the constant speed of $x/t = v$ ticks slower: its proper time is $t_1' = s_1 < t_1$, then $t_2' = s_2 < t_2$, and so on.

So, we managed to calculate the proper time not only from the self system but also from the lab:

$$s \equiv \sqrt{t^2 - \frac{x^2}{c^2}}.$$

Here, what is x? This is the initial position of the particle in the lab. On the other hand, what is t? This is the proper time of the lab: the time in a static clock. Still, it has yet another (mathematical) meaning: the maximal proper time of any particle (Figure 8.6). After all, at time t, a moving particle would be at $x = vt$, with proper time

$$s \equiv \sqrt{t^2 - \frac{x^2}{c^2}} = \sqrt{t^2 - \beta_v^2 t^2} = \frac{t}{\gamma\left(\beta_v\right)} \le t.$$

Thus, if you want to think that a lot of time had passed, then look at your own (static) clock: it ticks fast, telling you that many seconds passed. If, on the other hand, you look at another clock (moving towards or away from you), then it would tick slower, telling you that fewer seconds passed. This is time dilation: your "penalty" for looking at someone else's clock...

8.6.2 The Twin Paradox

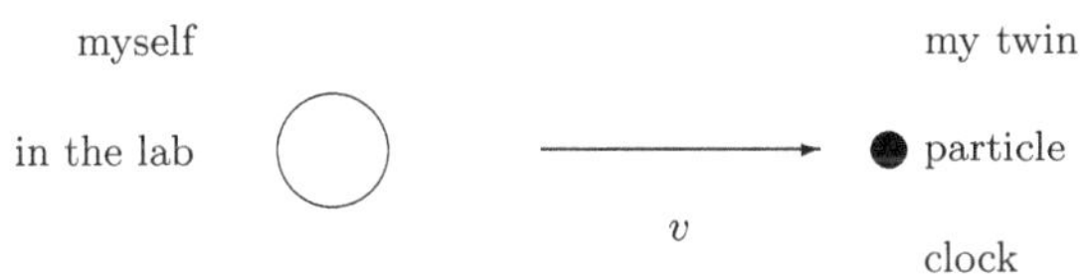

Fig. 8.7. I live in the lab. My twin, on the other hand, lives in the system of the particle, getting away at speed v. I say: "my time ticks faster, so I'm older!" My twin sees things the other way around: "my time ticks faster, so I'm older!"

Suppose that I live (happily) in the lab. With me, I have a static clock to show me my proper time: t. I also have a twin brother who lives in the system of the particle, moving at the constant velocity v away from me (Figure 8.7). With him, he carries his own clock that shows him his own proper time: $s = t'$.

As discussed above, I think that my own proper time ticks faster, so I'm older. My twin, on the other hand, sees things the other way around. He thinks that he is static, and that the entire lab travels at velocity $-v$ away from him. Thus, he believes that his own proper time ticks faster, so he is older. Who is right?

Well, thanks to length contraction, I see the moving system shrink. In my eyes, I'm bigger than my twin: my veins run longer, and my blood needs more time to get to my entire body. Thanks to time dilation, I do have more time. Both effects cancel out: both twins have the same metabolism, and will age at the same rate.

Why did we have a paradox in the first place? Because we only looked at one position: $x' = 0$. Still, in the x'-ct' plane, $x' = 0$ is not only a point but a complete line: the whole ct'-axis. Once transformed from system to system, it changes scale, leading to time dilation, and indeed to the twin paradox. To avoid this, transform not 1-D lines but 2-D areas. This should scale right, thanks to the (area-preserving) Lorentz transformation.

And what about the (idle) y-z dimensions, perpendicular to both x-ct and x'-ct'? There, we still have a paradox. But don't worry: to keep the systems inertial, the twins must never meet anyway...

8.6.3 Back To Length Contraction

In its self system, the particle is always at $x' = 0$. In the lab, on the other hand, it is at x. Where is this? Start from 0 in the lab, measure x rightwards, and you arrive at x.

Thus, the location has no meaning on its own, but only relative to a reference point: 0. Only the distance between two locations is meaningful.

Consider, for example, a horizontal stick that moves at velocity v away from us. In its self system, it is at rest: one endpoint at x'_1, and the other at x'_2, so its length is

$$\triangle x' \equiv x'_2 - x'_1.$$

And what is the view from the lab? Well, let's use Lorentz transformation:

$$\begin{pmatrix} x'_1 \\ ct'_1 \end{pmatrix} = \gamma(\beta_v) \begin{pmatrix} 1 & -\beta_v \\ -\beta_v & 1 \end{pmatrix} \begin{pmatrix} x_1 \\ ct_1 \end{pmatrix}$$

and

$$\begin{pmatrix} x'_2 \\ ct'_2 \end{pmatrix} = \gamma(\beta_v) \begin{pmatrix} 1 & -\beta_v \\ -\beta_v & 1 \end{pmatrix} \begin{pmatrix} x_2 \\ ct_2 \end{pmatrix}.$$

Now, subtract the former equation from the latter:

$$\begin{pmatrix} \triangle x' \\ c\triangle t' \end{pmatrix} = \gamma(\beta_v) \begin{pmatrix} 1 & -\beta_v \\ -\beta_v & 1 \end{pmatrix} \begin{pmatrix} \triangle x \\ c\triangle t \end{pmatrix}.$$

To measure the length of the moving stick, a viewer who sits in the lab has no access to the self system: he/she must use the x-ct lab coordinates. For this purpose, he/she must have both endpoints x_1 and x_2 at the same time $t_1 = t_2$:

$$\begin{aligned} \begin{pmatrix} \triangle x' \\ c\triangle t' \end{pmatrix} &= \gamma(\beta_v) \begin{pmatrix} 1 & -\beta_v \\ -\beta_v & 1 \end{pmatrix} \begin{pmatrix} \triangle x \\ 0 \end{pmatrix} \\ &= \gamma(\beta_v) \begin{pmatrix} \triangle x \\ -\beta_v \triangle x \end{pmatrix}. \end{aligned}$$

We only need the top equation:

$$\triangle x' = \gamma(\beta_v) \triangle x,$$

or

$$\triangle x = \frac{\triangle x'}{\gamma(\beta_v)}.$$

Since $\gamma \geq 1$, $|\triangle x| \leq |\triangle x'|$. This is length contraction. From its own self system, the stick looks longer than from any other system (including the lab).

Moreover, the above length $\triangle x$ (observed from the lab) decreases monotonically as $|v|$ increases. In the extreme case of $|v| = c$ and $\gamma = \infty$, $\triangle x = 0$. This means that, from the lab, the stick seems like one point, passing as fast as light.

This makes sense: two photons (traveling together at the speed of light) are indistinguishable, and could be considered as one and the same (Section 8.2.4). Here too, both endpoints shrink into one point. This is also why, in a system as fast as light, no change can be observed. Its proper time is stuck, and doesn't tick at all!

8.6.4 Simultaneous Events

In the lab, both endpoints of the stick were measured at the same time: $t_1 = t_2$. (After all, this is the right way to measure length.) These are simultaneous events. Still, in the x-t plane, they are not identical, but differ in their first coordinate: $x_1 \neq x_2$ (two different endpoints).

But in the self system these events are not simultaneous any more. Indeed, in the equation in Section 8.6.3, look now at the bottom:

$$c\triangle t' = -\gamma\left(\beta_v\right)\beta_v \triangle x = -\beta_v \triangle x' \neq 0.$$

Thus, the events are simultaneous in one system only: the lab. In any other system, on the other hand, they are not.

8.6.5 Back To Time Dilation

Thus, in spacetime, two events could differ in time or location or both. So far, we discussed simultaneous events that happen at the same time. Next, let's consider events that take place at the same location, but at different times.

Consider again our particle that moves at velocity v away from us (Figure 8.6). Consider also a clock, moving at the same speed. This way, in the self system, the clock is static.

In this clock, consider two successive times: $t'_2 > t'_1$. (In both, the clock is still at the same place in the self system: $x'_2 = x'_1$.) This way, in the self system, the time difference is

$$\triangle t' \equiv t'_2 - t'_1.$$

(This difference is in terms of proper time, as in Section 8.6.1.) From the lab, on the other hand, we measure the time difference

$$\triangle t \equiv t_2 - t_1.$$

Is this the same? Well, thanks to the inverse transformation,

$$\begin{pmatrix} x_1 \\ ct_1 \end{pmatrix} = \gamma\left(\beta_v\right)\begin{pmatrix} 1 & \beta_v \\ \beta_v & 1 \end{pmatrix}\begin{pmatrix} x'_1 \\ ct'_1 \end{pmatrix}$$

and

$$\begin{pmatrix} x_2 \\ ct_2 \end{pmatrix} = \gamma\left(\beta_v\right)\begin{pmatrix} 1 & \beta_v \\ \beta_v & 1 \end{pmatrix}\begin{pmatrix} x'_2 \\ ct'_2 \end{pmatrix}.$$

Let's subtract the former equation from the latter:

$$\begin{pmatrix} \Delta x \\ c\Delta t \end{pmatrix} = \gamma(\beta_v) \begin{pmatrix} 1 & \beta_v \\ \beta_v & 1 \end{pmatrix} \begin{pmatrix} \Delta x' \\ c\Delta t' \end{pmatrix}$$
$$= \gamma(\beta_v) \begin{pmatrix} 1 & \beta_v \\ \beta_v & 1 \end{pmatrix} \begin{pmatrix} 0 \\ c\Delta t' \end{pmatrix}$$
$$= \gamma(\beta_v) \begin{pmatrix} \beta_v c\Delta t' \\ c\Delta t' \end{pmatrix}.$$

We only need the bottom equation:

$$\Delta t = \gamma(\beta_v) \Delta t'.$$

Since $\gamma \geq 1$, $\Delta t \geq \Delta t'$. This is time dilation.

Now, let $|v|$ get greater and greater. This way, time dilation gets stronger and stronger. In the extreme case of $|v| = c$ and $\gamma = \infty$, $\Delta t = \infty$ as well. This is why, in a system as fast as light, even a tiny change seems to take forever.

8.7 Spacetime and Velocity

8.7.1 Doppler's Effect

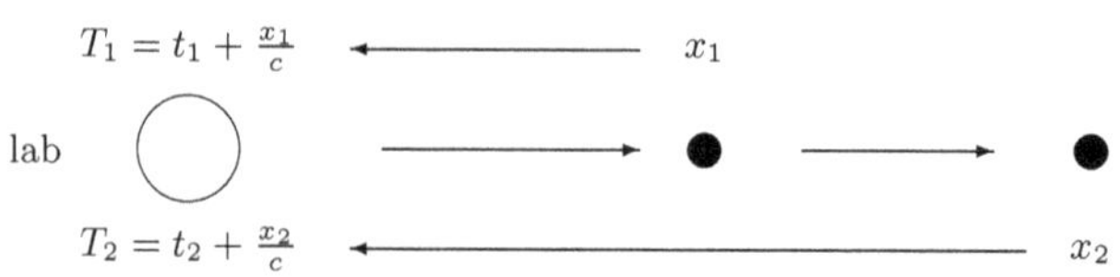

Fig. 8.8. Doppler's effect (a view from the lab). At time $t_1 > 0$, the particle is at $x_1 = vt_1$. At this time, a signal issues back to the lab, to arrive at $T_1 = t_1 + x_1/c$. At $t_2 > t_1$, on the other hand, the particle is already at $x_2 = vt_2$. At this time, another signal issues back to the lab, to arrive at $T_2 = t_2 + x_2/c$.

In our setting, the particle gets away from us at speed v. In its "pocket," it carries a tiny clock, to show its proper time. If we read it, then we see time dilation.

This is still quite theoretical. In practice, this information must travel back to the lab, at a finite speed, not exceeding the speed of light. How?

In its own self system, the tiny clock shows the proper time: $s = t'$. Once read from the lab, this must transform to t. For example, as things look from the lab, at time $t_1 > 0$, the particle gets as far as $x_1 = vt_1$. At this time, a signal (as fast as light) issues from the particle, to carry the news all the way back to the lab. To arrive, it needs some more time: $x_1/c = vt_1/c$. (Here, we assume for simplicity that $v > 0$, as in Figure 8.8.) Denote the arrival time by T_1. Later on, at time $t_2 > t_1$, the next signal will issue as well, to arrive at $T_2 > T_1$, and so on.

How to write the arrival-time difference $T_2 - T_1$ in terms of proper-time difference $t_2' - t_1'$? Thanks to time dilation (Section 8.6.5) and the above discussion,

$$\begin{aligned}
\triangle T &\equiv T_2 - T_1 \\
&= t_2 + \frac{x_2}{c} - \left(t_1 + \frac{x_1}{c}\right) \\
&= t_2 + \frac{vt_2}{c} - \left(t_1 + \frac{vt_1}{c}\right) \\
&= t_2 - t_1 + \frac{v}{c}(t_2 - t_1) \\
&= (\triangle t)(1 + \beta_v) \\
&= (\triangle t')\gamma(\beta_v)(1 + \beta_v) \\
&= (\triangle t')\frac{1 + \beta_v}{\sqrt{1 - \beta_v^2}} \\
&= (\triangle t')\frac{1 + \beta_v}{\sqrt{1 - \beta_v}\sqrt{1 + \beta_v}} \\
&= (\triangle t')\sqrt{\frac{1 + \beta_v}{1 - \beta_v}}.
\end{aligned}$$

What is the practical effect? Well, inside the particle, place a tiny camera, to take a movie. We, in the lab, will watch the movie in slow motion. Indeed, since $v > 0$, what takes $\triangle t'$ seconds in the movie will seem to us to take as many as $\sqrt{(1 + \beta_v)/(1 - \beta_v)}\triangle t'$ seconds. This is Doppler's effect: lower frequency, and red-shift.

8.7.2 Velocity in Spacetime

And how do things look like from the second particle, which travels at speed $-u$ away from us? To describe its self system, let's use now the x-t coordinates (with no prime). Why? Because this will be our main system. From it, we'll look at the world. In fact, the second particle will be just theoretical and nonphysical: a reference point. This is why it deserves our main x-t coordinates. This will be our new spacetime. Our lab, on the other hand, will be intermediate and unimportant, and will have primed coordinates: x'-t' (traveling at speed u with respect to our new spacetime). In them, the first particle moves at speed v:

$$\frac{dx'}{dt'} = v,$$

or

$$\frac{dx'}{cdt'} = \frac{v}{c} = \beta_v.$$

Better yet, write this in a column vector:

$$\begin{pmatrix} dx' \\ cdt' \end{pmatrix} = \begin{pmatrix} \beta_v \\ 1 \end{pmatrix}.$$

Indeed, the left-hand side is just a differential, which could be divided by cdt', as in projective geometry (Chapter 6 in [49]). And how do things look like from our new spacetime (the x-t self system of the second particle)? To see this, apply an inverse Lorentz matrix:

$$\begin{aligned} \begin{pmatrix} dx \\ cdt \end{pmatrix} &= \gamma(\beta_u) \begin{pmatrix} 1 & \beta_u \\ \beta_u & 1 \end{pmatrix} \begin{pmatrix} dx' \\ cdt' \end{pmatrix} \\ &= \gamma(\beta_u) \begin{pmatrix} 1 & \beta_u \\ \beta_u & 1 \end{pmatrix} \begin{pmatrix} \beta_v \\ 1 \end{pmatrix} \\ &= \gamma(\beta_u) \begin{pmatrix} \beta_v + \beta_u \\ \beta_u\beta_v + 1 \end{pmatrix}. \end{aligned}$$

Now, in this vector, divide the top component by the bottom component:

$$\frac{dx}{cdt} = \frac{\beta_v + \beta_u}{\beta_u\beta_v + 1},$$

or

$$\frac{dx}{dt} = c\frac{\beta_v + \beta_u}{\beta_u\beta_v + 1}, = \frac{u+v}{1+\beta_u\beta_v}.$$

Is this familiar? This is our law of adding velocities (Section 8.2.1). This is how fast the first particle gets away from the second one.

8.7.3 Moebius Transformation

The Moebius transformation is even more direct. It transforms ratio to ratio: the old velocity (away from the lab) to the new one (away from the second particle):

$$\beta_v = \frac{dx'}{cdt'} \to \frac{dx}{cdt} = \frac{\beta_u + \beta_v}{1 + \beta_u\beta_v}.$$

Let's use the same trick to calculate the perpendicular velocity as well.

8.7.4 Perpendicular Velocity

Next, consider a two-dimensional motion: not only in x' but also in y'. For this purpose, our lab still uses primes: x'-y'-t'. In it, the first particle moves now at an oblique velocity $(v_{x'}, v_{y'})$: velocity $v_{x'}$ in the x'-direction, and velocity $v_{y'}$ in the perpendicular y'-direction (Figure 8.9). (Don't get confused: these are not partial derivatives, but just velocity components.) The second particle, on the other hand, still moves at velocity $(-u, 0)$ in the x'-direction only. This is the view from the lab.

And how do things look like from the second particle? Well, this is now our new spacetime: the x-y-t self system of the second particle. To observe things from there, use the same trick, but with a few changes.

In its own self system, the second particle is at rest, while the entire lab moves at velocity $(u, 0)$. In these terms, how does the first particle move? Well, to transform

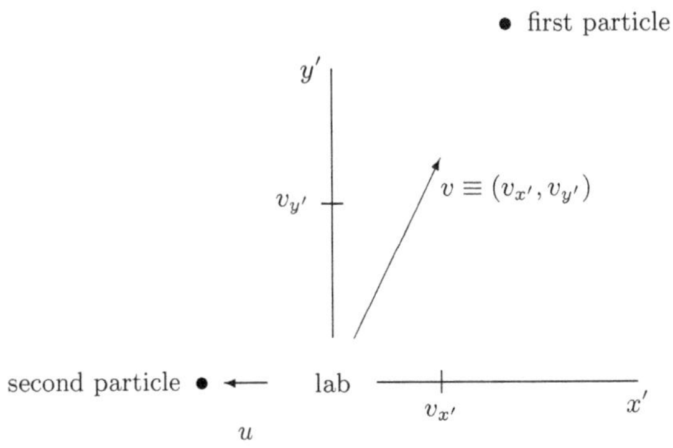

Fig. 8.9. View from the lab: the first particle travels at velocity $(v_{x'}, v_{y'})$, while the second particle travels at velocity $(-u, 0)$.

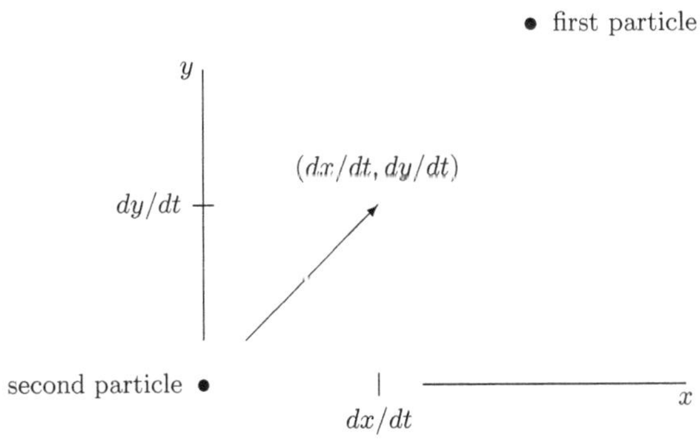

Fig. 8.10. View from the second particle: the first particle gets away at a new velocity: $(dx/dt, dy/dt)$ in the x-y-t system.

back to the self system of the second particle, apply an extended 3×3 Lorentz matrix, which leaves the second component idle. As usual, blank spaces stand for zeroes:

$$\begin{pmatrix} dx \\ dy \\ cdt \end{pmatrix} = \begin{pmatrix} \gamma(\beta_u) & & \\ & 1 & \\ & & \gamma(\beta_u) \end{pmatrix} \begin{pmatrix} 1 & & \beta_u \\ & 1 & \\ \beta_u & & 1 \end{pmatrix} \begin{pmatrix} dx' \\ dy' \\ cdt' \end{pmatrix}$$
$$= \begin{pmatrix} \gamma(\beta_u) & & \\ & 1 & \\ & & \gamma(\beta_u) \end{pmatrix} \begin{pmatrix} 1 & & \beta_u \\ & 1 & \\ \beta_u & & 1 \end{pmatrix} \begin{pmatrix} \beta_{v_{x'}} \\ \beta_{v_{y'}} \\ 1 \end{pmatrix}$$
$$= \begin{pmatrix} \gamma(\beta_u) & & \\ & 1 & \\ & & \gamma(\beta_u) \end{pmatrix} \begin{pmatrix} \beta_{v_{x'}} + \beta_u \\ \beta_{v_{y'}} \\ \beta_u \beta_{v_{x'}} + 1 \end{pmatrix}.$$

Next, divide by dt, and obtain the new velocities dx/dt and dy/dt (as observed from the second particle). (This eliminates the unimportant x'-y'-t' lab coordinates, and drops them altogether.) As observed from the second particle, the x-velocity of the first particle is still

$$\frac{dx}{dt} = c\frac{\beta_u + \beta_{v_{x'}}}{1 + \beta_u \beta_{v_{x'}}} = \frac{u + v_{x'}}{1 + \beta_u \beta_{v_{x'}}}$$

(Sections 8.7.2–8.7.3). The y-velocity, on the other hand, is

$$\frac{dy}{dt} = c\frac{\beta_{v_{y'}}}{\gamma(\beta_u)\left(1 + \beta_u \beta_{v_{x'}}\right)} = \frac{v_{y'}}{\gamma(\beta_u)\left(1 + \beta_u \beta_{v_{x'}}\right)}.$$

Thus, as observed from the second particle, the first particle indeed follows the arrow in Figure 8.10. To draw this arrow, we now have the new velocities dx/dt and dy/dt in terms of three known parameters: u, $v_{x'}$, and $v_{y'}$.

8.8 Relativistic Momentum and Its Conservation

8.8.1 Invariant Mass

A particle has mass: m. How to measure it (from any inertial system)? Well, if it is at rest, then just measure its mass: m. (This is why it is also called rest mass. But there is no need: better call it just mass.) And what if it is moving? No problem: transform to its self system, where it is at rest. There, calculate its mass: m. This is still invariant: same m in all (inertial) systems.

8.8.2 Momentum: Old Definition

Consider now a new particle (of mass m), traveling rightwards at a new velocity: not v, but u. Suddenly, with no outer force or energy, it explodes, and splits into two equal subparticles (Figure 8.11). How should they fly away (with respect to the original particle)? Well, thanks to symmetry, they should fly sideways: one rightwards (at the extra velocity v), and the other leftwards at the extra velocity v (or rightwards at the extra velocity $-v$).

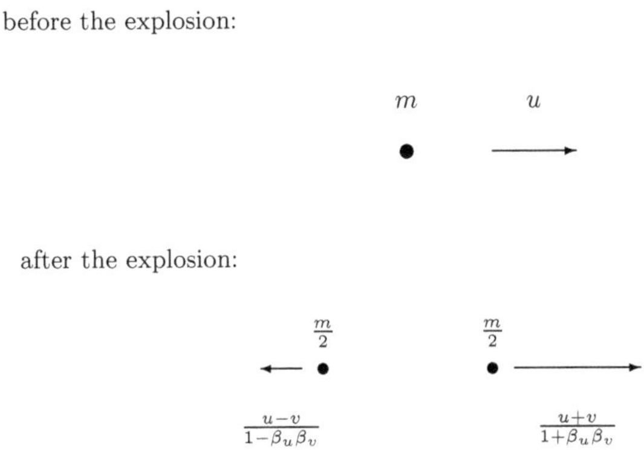

Fig. 8.11. In the beginning, the original particle travels rightwards at velocity u (top picture). Suddenly, it breaks into two new subparticles (bottom picture). But the total momentum remains the same: $mu\gamma(\beta_u)$.

In Newtonian mechanics, the momentum is mass times velocity (Section 2.3.1). (We are talking about linear momentum in 1-D.) In these terms, the momentum remains the same:

$$\frac{m}{2}(u+v) + \frac{m}{2}(u-v) = \frac{m}{2}(u+v+u-v) = mu.$$

8.8.3 Relativistic Momentum

Thanks to special relativity, we already know better: this is *not* the correct way to add velocities (Section 8.2.1)! To fix this, redefine momentum better. In the lab, the initial velocity is

$$u \equiv \frac{dx}{dt}.$$

But space and time are both relativistic, so their ratio is no good. To define our new momentum, better differentiate with respect to an invariant (nonrelativistic) quantity: s (the proper time of the particle):

$$p \equiv m\frac{dx}{ds} = m\frac{dx}{dt} \cdot \frac{dt}{ds} = mu\gamma(\beta_u)$$

(because, in the lab, t is faster than s, thanks to time dilation). This is our new momentum p: not absolute, but only relative to our lab. In particular, as $|u| \to c$, it gets infinite, resisting any change most strongly. (It is also called relativistic momentum, but there is no need: better call it just momentum.) Thanks to it, we can even uncover the location of the particle at proper time s:

$$mx(s) = m\int_0^s \frac{dx}{ds}ds = \int_0^s pds = ps$$

(since p is constant). Thus, since m and s are both invariant, $p/x(s)$ is invariant as well.

8.8.4 Relativistic Momentum and Energy

By now, we have one relativistic pair: x-ct, transformed together by the Lorentz transformation. Why? Because they are proportional to one another:

$$x = \frac{dx}{dt}t = \frac{dx}{cdt}ct = \frac{u}{c}ct = \beta_u ct.$$

Later on, we'll see yet another relativistic pair: momentum-energy/c. Indeed, they'll keep the same proportion:

$$p = mu\gamma(\beta_u) = \frac{u}{c}mc\gamma(\beta_u) = \beta_u\frac{mc^2\gamma(\beta_u)}{c} = \beta_u\frac{\text{energy}}{c}.$$

This will be proved later (Einstein's formula).

In a sense, momentum mirrors space. It could even replace space, and play the same role, in a more fundamental way. This is why momentum is often constant in space: invariant under any spatial translation. Likewise, energy mirrors time, and plays the same role, in a more fundamental way. This is why energy is conserved (time-invariant): unchanged under any time translation. This marks a new symmetry line in spacetime. We'll come back to this in quantum and Hamiltonian mechanics.

8.8.5 Relativistic Mass

To define momentum, we used not m but the product

$$m\gamma(\beta_u).$$

This is also called relativistic mass. It resists any outer force, or any change to momentum. (In this sense, it is like inertial mass in classical mechanics.) As $|u| \to c$, it gets infinite, resisting an outer force most strongly. This is why a particle of a positive mass $m > 0$ could never reach the speed of light: its relativistic mass would get too big:

$$m\gamma(\beta_u) \to_{u\to c} \infty.$$

It would resist an outer force ever so strongly, and win! The final bit of outer force required to accelerate to c will remain missing. Only a massless photon could reach the speed of light.

Still, our new momentum could do the same kind of resistance equally well. This is why we don't use the concept of relativistic mass any more, but only (rest) mass: m.

8.8.6 Moderate Velocity

Is our new momentum good? Well, look at a moderate velocity $|u| \ll c$ (or $\beta_u \ll 1$), or $\gamma(\beta_u) \sim 1$):

$$mu\gamma(\beta_u) \sim mu$$

(the good old Newtonian momentum). So, our new momentum can only improve on the classical one.

8.8.7 Closed System: Lose Mass – Gain Motion

In Figure 8.11, to make the explosion happen, some energy must come from somewhere. From where? Well, our system is isolated: no force or energy could come from the outside. Thus, the extra energy must come from mass: at the explosion, the subparticles must lose some of their original mass. Indeed, before the explosion, each had mass $m/2$. After it, on the other hand, each has mass

$$\frac{m}{2\gamma(\beta_v)} < \frac{m}{2\gamma(0)} = \frac{m}{2}.$$

(Later on, we'll see why.) What is the physical meaning of this? Well, after the explosion, the mass must be less than in the beginning (when the subparticle was still "inside" the original particle, and had velocity $v = 0$ with respect to it). Still, no mass was lost for nothing: it supplied the extra energy required to make the explosion happen.

8.8.8 The Momentum Matrix

Let's define the (so-called) momentum matrix. This will be a Lorentz matrix, multiplied by mass. In it, one element will be the momentum itself. The other elements will be useful too.

For this purpose, look at things from the lab again (Figure 8.11). For each subparticle, compose two Lorentz matrices (as in Section 8.5.5, but go the other way around: from bottom to top):

$$\begin{aligned}
&\text{mass} \cdot \text{Lorentz matrix} \quad + \quad \text{mass} \cdot \text{Lorentz matrix} \\
&= \frac{m}{2\gamma(\beta_v)}\gamma\left(\frac{\beta_u+\beta_v}{1+\beta_u\beta_v}\right)\left(I+\frac{\beta_u+\beta_v}{1+\beta_u\beta_v}J\right)+\frac{m}{2\gamma(\beta_v)}\gamma\left(\frac{\beta_u-\beta_v}{1-\beta_u\beta_v}\right)\left(I+\frac{\beta_u-\beta_v}{1-\beta_u\beta_v}J\right) \\
&= \frac{m}{2\gamma(\beta_v)}\gamma(\beta_v)(I+\beta_v J)\gamma(\beta_u)(I+\beta_u J)+\frac{m}{2\gamma(\beta_v)}\gamma(\beta_v)(I-\beta_v J)\gamma(\beta_u)(I+\beta_u J) \\
&= \frac{m}{2}(I+\beta_v J+I-\beta_v J)\gamma(\beta_u)(I+\beta_u J) \\
&= m\gamma(\beta_u)(I+\beta_u J).
\end{aligned}$$

Thus, after the explosion, the momentum matrix remains the same (with respect to the lab). This is indeed conservation of momentum (in matrix form).

8.8.9 Momentum and Its Conservation

This is a matrix equation (four scalar equations). Focus on one of them (say, the upper-right one):

$$\frac{m}{2\gamma\left(\beta_v\right)}\gamma\left(\frac{\beta_u+\beta_v}{1+\beta_u\beta_v}\right)\frac{\beta_u+\beta_v}{1+\beta_u\beta_v}+\frac{m}{2\gamma\left(\beta_v\right)}\gamma\left(\frac{\beta_u-\beta_v}{1-\beta_u\beta_v}\right)\frac{\beta_u-\beta_v}{1-\beta_u\beta_v}=m\gamma\left(\beta_u\right)\beta_u.$$

To simplify, multiply this by c:

$$\frac{m}{2\gamma\left(\beta_v\right)}\frac{u+v}{1+\beta_u\beta_v}\gamma\left(\frac{\beta_u+\beta_v}{1+\beta_u\beta_v}\right)+\frac{m}{2\gamma\left(\beta_v\right)}\frac{u-v}{1-\beta_u\beta_v}\gamma\left(\frac{\beta_u-\beta_v}{1-\beta_u\beta_v}\right)=mu\gamma\left(\beta_u\right).$$

This is indeed conservation of momentum: after the explosion, with respect to the lab, mass times velocity times γ still sums to the same: $mu\gamma(\beta_u)$.

8.9 Relativistic Energy and Its Conservation

8.9.1 Force: Derivative of Momentum

Mass is invariant, not relativistic. Momentum, on the other hand, is relativistic, not invariant (in its new definition). Once coupled with (relativistic) energy, it'll form a relativistic pair, transformed by the Lorentz transformation. And what is relativistic energy? To define it, we'll need to differentiate $\gamma(\beta_v)$:

$$\begin{aligned}\gamma'\left(\beta_v\right) &\equiv \left(\left(1-\beta_v^2\right)^{-1/2}\right)' \\ &= -\frac{1}{2}\left(1-\beta_v^2\right)^{-3/2}\left(-2\beta_v\right) \\ &= \beta_v\left(1-\beta_v^2\right)^{-3/2}.\end{aligned}$$

Let's use this to differentiate $\gamma(\beta_v)$ as a composite function of v:

$$\begin{aligned}\frac{d}{dv}\gamma\left(\beta_v\right) &= \gamma'\left(\beta_v\right)\frac{d\beta_v}{dv} \\ &= \frac{1}{c}\gamma'\left(\beta_v\right) \\ &= \frac{1}{c}\beta_v\left(1-\beta_v^2\right)^{-3/2}.\end{aligned}$$

Let's use this to differentiate the product $v\gamma(\beta_v)$:

$$\begin{aligned}\frac{d}{dv}\left(v\gamma\left(\beta_v\right)\right) &= \gamma\left(\beta_v\right)+v\frac{d}{dv}\gamma\left(\beta_v\right) \\ &= \gamma\left(\beta_v\right)+\frac{v}{c}\beta_v\left(1-\beta_v^2\right)^{-3/2} \\ &= \gamma\left(\beta_v\right)+\beta_v^2\left(1-\beta_v^2\right)^{-3/2} \\ &= \gamma^{-2}\left(\beta_v\right)\gamma^3\left(\beta_v\right)+\beta_v^2\left(1-\beta_v^2\right)^{-3/2} \\ &= \left(1-\beta_v^2\right)\left(1-\beta_v^2\right)^{-3/2}+\beta_v^2\left(1-\beta_v^2\right)^{-3/2} \\ &= \left(1-\beta_v^2\right)^{-3/2}.\end{aligned}$$

We are now ready to define relativistic energy. (This will be more accurate, even at high speeds.) Consider a particle of mass m, initially at rest (in the x-t lab coordinates). But then, an external force F is applied to it from time 0 until time $q > 0$, to increase its momentum and energy (without changing m). Thanks to this, the particle doesn't have to "pay," and can keep its mass m throughout the entire time interval $[0, q]$.

And how to calculate the force? For this, differentiate the relativistic momentum (Section 8.8.3) with respect to time. This will help define the relativistic energy:

$$\begin{aligned}
\int_{x(0)}^{x(q)} F(x)dx &= \int_0^q F(x(t))dx(t) \\
&= \int_0^q F(x(t))\frac{dx(t)}{dt}dt \\
&= \int_0^q F(x(t))v(t)dt \\
&= m\int_0^q \frac{d}{dt}\left(v\gamma\left(\beta_v\right)\right)v(t)dt \\
&= m\int_0^q \frac{d}{dv}\left(v\gamma\left(\beta_v\right)\right)\frac{dv}{dt}v(t)dt \\
&= m\int_{v(0)}^{v(q)} \frac{d}{dv}\left(v\gamma\left(\beta_v\right)\right)vdv \\
&= m\int_{v(0)}^{v(q)} \left(1-\beta_v^2\right)^{-3/2} vdv \\
&= mc^2\int_{v(0)}^{v(q)} \left(1-\beta_v^2\right)^{-3/2}\frac{1}{c}\beta_v dv \\
&= mc^2\left(\gamma\left(\beta_{v(q)}\right)-\gamma\left(\beta_{v(0)}\right)\right) \\
&= mc^2\left(\gamma\left(\beta_{v(q)}\right)-1\right).
\end{aligned}$$

8.9.2 Relativistic Energy: Kinetic plus Potential

This is indeed the new kinetic energy that the external force introduced into the particle from time 0 until time q (without changing m). The potential (nuclear) energy, on the other hand, is not relativistic but invariant. It was in the particle since the start, and never changed. This is the amount subtracted in the above formula:

$$E_{\text{potential}} \equiv E(0) \equiv mc^2.$$

Thanks to this part, we now have the total energy (as a smooth function of v):

$$E(v) \equiv E_{\text{potential}} + E_{\text{kinetic}}(v) = mc^2 + mc^2\left(\gamma\left(\beta_v\right)-1\right) = mc^2\gamma\left(\beta_v\right).$$

This is our new relativistic energy: kinetic plus potential. Is it familiar? Look at the momentum matrix:

$$m\gamma\left(\beta_v\right)\left(I+\beta_v J\right).$$

Look at the lower-right entry, and multiply by c^2. This is our relativistic energy back again. Better yet, call it just energy.

8.9.3 Moderate Velocity

In particular, what happens at a moderate velocity $|v| \ll c$ (or $\beta_v^2 \ll 1$)? Use a Taylor expansion around 0:

$$\gamma(\beta_v) = \left(1-\beta_v^2\right)^{-1/2} \sim 1 + \frac{1}{2}\beta_v^2.$$

Thus, our new definition agrees with the classical one:

$$E(v) = mc^2\gamma(\beta_v) \sim mc^2\left(1+\frac{1}{2}\beta_v^2\right) = mc^2 + \frac{1}{2}mv^2.$$

It can only improve on it.

8.9.4 Tradeoff: Kinetic Energy vs. Mass

With no external force, could the particle gain any new velocity $v \neq 0$? For this, it must pay cash: mass. This way, the new kinetic energy comes from its original potential energy (before it started to move). For this, it must lose some mass. Only a static particle ($v = 0$) can keep its original (maximal) mass m. This is why it is also called rest mass.

8.9.5 Energy Is Conserved, Mass Is Not

In Section 8.9.2, an outer force affects the particle, increasing its momentum and energy. This is how it can keep the same mass: m. This is an open system. In Figure 8.11, on the other hand, things are the other way around. The explosion takes place in a closed (isolated) system: no external force is allowed at all! This is why total momentum and energy remain the same. But mass is not.

During the explosion, where do the subparticles get their extra kinetic energy from? It must come from somewhere: from the original potential (nuclear) energy, stored in the original particle and its mass. There is a price to pay: to make the explosion happen, some mass must be lost.

As a matter of fact, this is true not only for an exploding particle but also for any particle that starts to move. Indeed, at the initial time, it accelerates: it changes its velocity from 0 to $v \neq 0$ (while preserving its total energy). Its new kinetic energy must come from somewhere: from its potential nuclear energy. As a price, its mass must decrease from $m(0) = m$ to

$$m(v) \equiv \frac{m(0)}{\gamma(\beta_v)} < m(0) = m.$$

This way, its total energy remains the same as at rest:

$$E(v) = m(v)c^2\gamma(\beta_v) = \frac{m(0)}{\gamma(\beta_v)}c^2\gamma(\beta_v) = m(0)c^2 = E(0).$$

This is how energy is conserved in a closed system.

8.9.6 Mass is Invariant, Energy Is Not

This is why $m(0)$ is also called rest mass: the original mass in the lab, before the particle started to move. In the lab, this is $m(0) = m$. But what about a moving reference frame? There, the particle may get kinetic energy from nothing! It can keep its mass m (as measured in the self system: the lab). It just has more energy. There is no problem: energy is relativistic, not invariant.

Back to our lab, here the new velocity comes from a real physical process: converting energy from potential to kinetic. To start moving at velocity v (in the lab), the particle must lose mass: from $m(0) = m$ to

$$m(v) = \frac{m(0)}{\gamma\left(\beta_v\right)} < \frac{m(0)}{\gamma(0)} = m(0) = m.$$

8.9.7 Closed System: Energy Can Only Convert

The decrease from $m(0)$ to $m(v)$ takes place in an isolated system: no external force is welcome. This is why the total energy must remain the same. Only mass can change: as $|v|$ increases, $m(v)$ decreases. This way, energy is never lost, but only converts from potential to kinetic. In this process, the total energy remains the same:

$$E = E_{\text{potential}}(v) + E_{\text{kinetic}}(v) = \frac{m(0)}{\gamma\left(\beta_v\right)}c^2 + \frac{m(0)}{\gamma\left(\beta_v\right)}c^2\left(\gamma\left(\beta_v\right) - 1\right) = m(0)c^2.$$

This is Einstein's famous formula. It tells us how powerful the original nuclear energy really is: it can produce a lot of kinetic energy.

8.9.8 Open System

This is quite different from the situation in Section 8.9.2, in which the system is open, and welcomes an outer force. In that case, the kinetic energy grows (without changing mass). This was necessary to help define the kinetic energy obtained from the work done by the outer force.

Mass, on the other hand, can change even in a closed system. In a particle that starts to move, as $|v|$ increases, $m(v)$ decreases. As $|v| \to c$, $\gamma \to \infty$, so the particle has no mass at all any more: all its potential energy has already been exploited, and converted into kinetic energy.

Still, even with no mass at all, the particle is not quite gone: it may still have momentum (as in a photon). This is why momentum and energy are more fundamental than mass: relativistic, not absolute.

8.10 Momentum–Energy and Their Transformation

8.10.1 New Mass

By now, both energy and momentum are relativistic: both depend on the velocity, which may change from system to system. In the lab, the particle travels rightwards at velocity v. As discussed above, its mass is $m(0)/\gamma(\beta_v)$. Still, who cares about its history? Who cares whether it was initially at rest or not? Better redefine m as it is now:

$$m \leftarrow \frac{m(0)}{\gamma(\beta_v)}.$$

This is its new mass, measured now (while it rests in its self system, traveling at velocity v away from the lab). Let's use it to calculate both energy and momentum in the lab.

8.10.2 Spacetime

To describe the lab, let's use the x'-t' coordinates again (Section 8.7.2). (After all, the lab won't be our main system, but only intermediate and unimportant.) Thus, in the lab, the momentum of the particle should be primed too:

$$p' \equiv mv\gamma(\beta_v).$$

Likewise, in the lab, the total energy of the particle should be primed too:

$$E' \equiv mc^2\gamma(\beta_v).$$

And what is the x-t system? This is our main system: the self system of the second particle, traveling at velocity u leftwards (or $-u$ rightwards) away from the lab. This will be our spacetime. In it, what are the energy E and momentum p of the first particle? the mom

8.10.3 A Naive Approach

To transform to our new spacetime, we could take a naive approach:

- Use the rule of adding velocities to calculate the velocity of the first particle away from the second one.
- Use it to calculate both E and p, as required.

Still, is there a more direct way?

8.10.4 The Momentum–Energy Vector

Fortunately, there is. For this purpose, note that p' and E'/c have a familiar ratio:

$$\frac{p'}{E'/c} = \frac{mv\gamma(\beta_v)}{mc\gamma(\beta_v)} = \frac{v}{c} = \beta_v.$$

So, let's put them in a new column vector, proportional to the column $(\beta_v, 1)^t$. In fact, this is just the second column in the momentum matrix (times c):

$$\begin{pmatrix} p' \\ E'/c \end{pmatrix} \equiv cm\gamma(\beta_v)(I + \beta_v J)\begin{pmatrix} 0 \\ 1 \end{pmatrix}.$$

8.10.5 Momentum Matrix in Spacetime

How do things look like from our new spacetime? Well, from the perspective of the second particle, the entire lab travels rightwards at velocity u. On top of this, the first particle also travels in the lab at velocity v. Thus, to transform the old momentum matrix to our new spacetime, we need to apply an inverse Lorentz matrix:

$$m\gamma(\beta_u)(I+\beta_u J)\gamma(\beta_v)(I+\beta_v J).$$

This is the new momentum matrix in spacetime.

8.10.6 Lorentz Transformation on Momentum–Energy

Now, look at the second column, and multiply by c:

$$\begin{aligned}\begin{pmatrix} p \\ E/c \end{pmatrix} &= cm\gamma(\beta_u)(I+\beta_u J)\gamma(\beta_v)(I+\beta_v J)\begin{pmatrix} 0 \\ 1 \end{pmatrix} \\ &= \gamma(\beta_u)(I+\beta_u J)\begin{pmatrix} p' \\ E'/c \end{pmatrix}.\end{aligned}$$

So, we managed to drop the primes! This will be quite useful below.

8.11 Energy and Mass

8.11.1 Invariant Nuclear Energy

So far, we put momentum-energy in the second column of the momentum matrix. But we could also work the other way around, and put them in the first column too:

$$\begin{pmatrix} E'/c & p' \\ p' & E'/c \end{pmatrix} = cm\gamma(\beta_v)(I+\beta_v J).$$

Fortunately, a Lorentz matrix must have determinant 1:

$$\begin{aligned}\frac{E'^2}{c^2} - p'^2 &= \det\left(\begin{pmatrix} E'/c & p' \\ p' & E'/c \end{pmatrix}\right) \\ &= \det\left(cm\gamma(\beta_v)(I+\beta_v J)\right) \\ &= c^2m^2\det\left(\gamma(\beta_v)(I+\beta_v J)\right) \\ &= m^2c^2.\end{aligned}$$

This is invariant (independent of system). Thus, in the left-hand side too, we can safely drop the primes. Indeed,

$$\begin{aligned}\frac{E^2}{c^2} - p^2 &= \det\left(\begin{pmatrix} E/c & p \\ p & E/c \end{pmatrix}\right) \\ &= \det\left(\gamma(\beta_u)(I+\beta_u J)\begin{pmatrix} E'/c & p' \\ p' & E'/c \end{pmatrix}\right) \\ &= \det\left(\begin{pmatrix} E'/c & p' \\ p' & E'/c \end{pmatrix}\right) \\ &= m^2c^2.\end{aligned}$$

To simplify, multiply by c^2:

$$E^2 - c^2p^2 = m^2c^4 = \left(mc^2\right)^2.$$

Is this familiar? This is the squared nuclear energy (stored in the particle). This is not relativistic, but invariant: same at all systems. On the left-hand side, you could put the primes back on, if you like.

This formula ties two relativistic quantities to an invariant one. Is this familiar? We already saw this structure before:

$$c^2t^2 - x^2 = c^2s^2,$$

where s is invariant: the proper time of the particle (Section 8.6.1).

In the exercises below, we'll extend this to 3-D: p will be not just a scalar, but a 3-D vector. For this, replace p^2 by the inner product $\|p\|^2 = (p, p)$. After all, in the direction of p, this is just the same.

8.11.2 How to Calculate Mass?

Thanks to our formula, we can now calculate mass (without using the self system):

- First, calculate the relativistic momentum and energy in your system.
- This will give you p and E (or p' and E').
- Fortunately, these primes don't matter at all: in either case, use our formula to calculate the same mass: m.

This will be useful below.

8.11.3 Einstein's Formula

In particular, why not work in the self system of the particle? In it, there is no velocity or momentum or kinetic energy at all, so our formula simplifies to read

$$E^2_{\text{potential}} = m^2c^4,$$

or

$$E_{\text{potential}} = mc^2.$$

This is Einstein's formula back again.

8.12 Center of Mass

8.12.1 Collection of Subparticles

In the lab, the first particle travels at velocity v. Assume that v is unknown. How to uncover it? Assume that we know the momentum p' and the energy E', and use them:

$$v = \frac{mv\gamma\left(\beta_v\right)}{mc^2\gamma\left(\beta_v\right)}c^2 = \frac{p'}{E'}c^2.$$

This is quite practical: p' and E' are more fundamental than v, which is often missing.

Throughout this chapter, our second particle was just theoretical: a reference point. The first particle, on the other hand, is more real and physical. In fact, it could even break into a collection of $k \ge 1$ subparticles, each with velocity v_i, momentum p'_i, and energy E'_i with respect to the lab ($1 \le i \le k$). What are the total momentum and energy? Well, these are fundamental (conserved) quantities, which sum up:

$$p' \equiv \sum_{i=1}^{k} p'_i \quad \text{and} \quad E' \equiv \sum_{i=1}^{k} E'_i.$$

Let's use them to define the total velocity as well.

8.12.2 Center of Mass

We are now ready to define the velocity of the entire collection:

$$v \equiv \frac{p'}{E'}c^2 = c^2 \frac{\sum_{i=1}^{k} p'_i}{\sum_{i=1}^{k} E'_i}.$$

(This is *not* the sum of velocities. After all, velocity is not conserved.) This new velocity refers to no physical object, but only a theoretical object: the center of mass of the entire collection. Where is this? To tell this, let's use the second particle. Besides, let's do our best to eliminate its velocity, and use fundamental quantities instead.

The above velocity is in terms of the lab. Next, let's look at things from our new spacetime: the self system of the second particle, which travels at velocity $-u$ in the lab. In this spacetime, what are the new momentum-energy of the collection? We already know what they are:

$$\begin{pmatrix} p \\ E/c \end{pmatrix} = \gamma(\beta_u)(I + \beta_u J) \begin{pmatrix} p' \\ E'/c \end{pmatrix}$$

(Section 8.10.6). In this formula, look at the top equation. Assume also that the second particle follows at the same speed:

$$u \equiv -v, \quad \text{so} \quad \beta_u = -\beta_v = -\frac{p'}{E'/c}.$$

This way, the top equation simplifies to read

$$p = 0.$$

Thus, with respect to the second particle, the entire collection has no momentum at all! It is at a complete rest. This is why the second particle marks the center of mass itself.

8.12.3 Mass of Collection

What is the mass m of the entire collection? Is it the same as the sum of individual masses? No! After all, mass is not conserved. To find m, let's use more fundamental quantities: momentum-energy.

In our spacetime (the self system of the second particle, traveling at velocity $-u = v$ in the lab), the collection has no momentum at all: $p = 0$. Therefore, the formula in Section 8.11.1 tells us that

$$E^2 = m^2c^4,$$

or

$$m \equiv \frac{E}{c^2}.$$

This is a good definition of the total mass of the entire collection.

What is the physical meaning of this? Well, we are now in our spacetime: both p and E have no prime any more. This is also the self system of the collection (or its center of mass, marked by the second particle). In it, the collection is at a complete rest, with no momentum or kinetic energy at all. Thus, the above formula actually defines its mass m (or its rest mass) in terms of its total energy E (which is conserved, and can be written as a sum):

$$\begin{aligned} m &\equiv \frac{E}{c^2} \\ &= \frac{1}{c^2}\sum_{i=1}^{k} E_i, \end{aligned}$$

where E_i is the energy of the ith subparticle in our spacetime: kinetic plus potential. After all, with respect to the center of mass, each subparticle could move in its own way.

8.13 Oblique Force and Momentum

8.13.1 Oblique Momentum in x'-y'

Let's add yet another dimension (Section 8.7.4): our lab is now in the x'-y'-t' coordinates. This way, in the lab, the entire collection moves obliquely, at the new velocity $v \equiv (v_{x'}, v_{y'})^t$. This means that its velocity has two components: $v_{x'}$ in the x'-direction, and $v_{y'}$ in the perpendicular y'-direction. In this setting, the momentum is oblique as well: $p' \equiv (p'_{x'}, p'_{y'})^t$, proportional to v. This is the view from the lab (Figure 8.12). (From the second particle, on the other hand, things look different: Figure 8.13.) Thus, in the formula in Section 8.11.1, p'^2 should be replaced by the inner product

$$\|p'\|^2 = (p', p') = p'^2_{x'} + p'^2_{y'}.$$

After all, in theory, we could always redefine x' to align with v and p' (see exercises below). Fortunately, there is no need to do this explicitly.

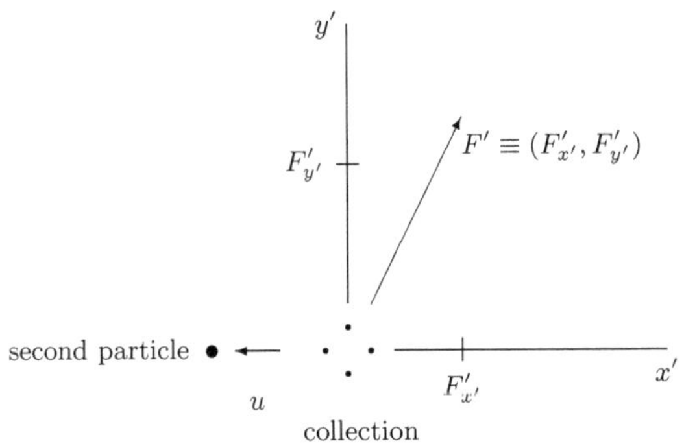

Fig. 8.12. View from the lab: initially, at time $t' = 0$, the collection is still at rest at $(x', y') = (0, 0)$. At $t' = 0$, an oblique external force $F' = (F'_{x'}, F'_{y'})$ starts to act upon it, to increase its momentum and kinetic energy (without changing its mass).

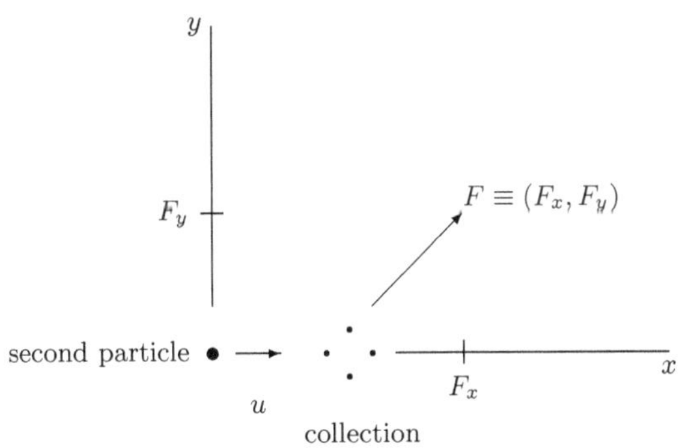

Fig. 8.13. View from the second particle: at the initial time $t = t' = 0$, a force F starts to act on the collection. In the horizontal x-direction, it looks the same as before. But in the perpendicular y-direction, it looks weaker than before.

8.13.2 View From Our Spacetime

The second particle, on the other hand, still moves in the x'-direction only: at velocity $(-u, 0) \neq (0, 0)$ away from the lab. Its self system will be our new spacetime. To transform to there, we must use an extended 3×3 Lorentz matrix:

$$\begin{pmatrix} p_x \\ p_y \\ E/c \end{pmatrix} = \begin{pmatrix} \gamma(\beta_u) & & \\ & 1 & \\ & & \gamma(\beta_u) \end{pmatrix} \begin{pmatrix} 1 & & \beta_u \\ & 1 & \\ \beta_u & & 1 \end{pmatrix} \begin{pmatrix} p'_{x'} \\ p'_{y'} \\ E'/c \end{pmatrix}.$$

Here, we don't assume $u = -v$ any more: the second particle doesn't mark the center of mass any more. This job is left to the lab itself.

8.13.3 The Lab: The New Self System

Indeed, assume now that the lab is initially the self system of the entire collection: at time $t' = 0$, the collection is at rest in the lab:

$$p' \equiv \begin{pmatrix} p'_{x'} \\ p'_{y'} \end{pmatrix} = \begin{pmatrix} 0 \\ 0 \end{pmatrix}, \quad E' = mc^2 > 0, \quad \text{and} \quad v \equiv \begin{pmatrix} v_{x'} \\ v_{y'} \end{pmatrix} \equiv \frac{p'}{E'}c^2 = \begin{pmatrix} 0 \\ 0 \end{pmatrix}.$$

This is true at time $t' = 0$ only. At $t' > 0$, on the other hand, things are going to change.

8.14 Force in an Open System

8.14.1 Force in an Open Passive System

Unlike before, assume now that the lab is an open system: from time $t = t' = 0$ onward, an external force is applied to the entire collection, to increase its momentum and kinetic energy in the lab (without changing its mass). In this sense, the lab is a passive system: the initial force acts directly on the static collection in it. What is this force? It is the time-derivative of the momentum (in the original x'-y'-t' coordinates):

$$F' \equiv \begin{pmatrix} F'_{x'} \\ F'_{y'} \end{pmatrix} \equiv \begin{pmatrix} \frac{dp'_{x'}}{dt'} \\ \frac{dp'_{y'}}{dt'} \end{pmatrix}.$$

Let's focus on the initial time $t' = 0$:

$$F' \equiv \begin{pmatrix} F'_{x'} \\ F'_{y'} \end{pmatrix} \equiv \begin{pmatrix} \frac{dp'_{x'}}{dt'}(0) \\ \frac{dp'_{y'}}{dt'}(0) \end{pmatrix}.$$

(After all, each fixed time $t' > 0$ could in theory be shifted back to zero.) How does this force look like from the second particle? In other words, how to transform it to our new spacetime (the x-y-t self system of the second particle)?

8.14.2 Force in Our Spacetime

Of course, we could take a naive approach: transform momentum-energy to our new spacetime, and define the force F there by differentiating momentum with respect to t. Still, is there a more direct way?

8.14.3 Proper Time in The Lab

Let's use the trick in Section 8.7.4. To start, let's differentiate t with respect to t'. This seems easy: after all, in the lab, t' is proper, isn't it? So, from our new spacetime, it should look slower (Section 8.6.1):

$$t' = \frac{t}{\gamma(\beta_u)},$$

shouldn't it? Unfortunately not: t' is proper only in an isolated lab, which welcomes no outer force. In our open lab, on the other hand, t' is only nearly proper: only at $t' = 0$, before the force had time to act, does t' behave like a proper time.

8.14.4 Nearly Proper Time in The Lab

To see this, transform the lab back to our spacetime, using the inverse Lorentz transformation:

$$ct = \gamma(\beta_u)(\beta_u x' + ct'),$$

or

$$t = \gamma(\beta_u)\left(\frac{\beta_u}{c}x' + t'\right).$$

Differentiate this with respect to t':

$$\frac{dt}{dt'} = \gamma(\beta_u)\left(\frac{\beta_u}{c}\cdot\frac{dx'}{dt'} + 1\right) = \gamma(\beta_u)\left(\frac{\beta_u}{c}v_{x'} + 1\right).$$

At $t' = 0$, in particular, $v_{x'} = 0$, so this simplifies to read

$$\frac{dt}{dt'} = \gamma(\beta_u).$$

Thus, at $t' = 0$, t' is indeed nearly proper: it behaves just like a proper time. Let's use this to look at the force from the second particle (Figure 8.13).

8.15 Perpendicular Force

8.15.1 Force: Time-Derivative of Momentum

Let's start from the perpendicular force in our spacetime: F_y. Thanks to the above 3×3 matrix, in the y-direction, the momentum is still the same:

$$p_y = p'_{y'}.$$

Thus, the differentiation is simple:

$$\begin{aligned} F_y &\equiv \frac{dp_y}{dt} \\ &= \frac{dp'_{y'}}{dt} \\ &= \frac{dp'_{y'}}{\gamma(\beta_u)\, dt'} \\ &= \frac{1}{\gamma(\beta_u)} \cdot \frac{dp'_{y'}}{dt'} \\ &= \frac{1}{\gamma(\beta_u)} F'_{y'}. \end{aligned}$$

8.15.2 Passive System – Strong Perpendicular Force

This is weaker! Thus, the passive system (our lab) feels the maximal perpendicular force. Any other system would feel a weaker perpendicular force. In particular, our spacetime (the self system of the second particle) feels a perpendicular force $\gamma(\beta_u)$ times as weak.

8.16 Nonperpendicular Force

8.16.1 Force: Time-Derivative of Momentum

What about the force in the x-direction? Does it also feel weaker? Well, let's use the same trick:

$$\begin{aligned} F_x &= \frac{dp_x}{dt} \\ &= \frac{d\left(\gamma(\beta_u)\left(p'_{x'} + \beta_u \frac{E'}{c}\right)\right)}{\gamma(\beta_u)\, dt'} \\ &= \frac{d\left(p'_{x'} + \beta_u \frac{E'}{c}\right)}{dt'} \\ &= \frac{dp'_{x'}}{dt'} + \beta_u \frac{dE'}{cdt'} \\ &= F'_{x'} + \frac{\beta_u}{c} \cdot \frac{dE'}{dt'}. \end{aligned}$$

8.16.2 Energy in an Open System

To simplify this, let's look at the latter term, and show that it contributes nothing. For this purpose, let's look at the original equation:

$$E'^2 = c^2 (p', p') + m^2 c^4.$$

This equation comes from the original definition of E' and p' in the lab. Therefore, it holds for every time $t' \geq 0$ (with the same m, but possibly with a different E', p', and v). In a moment, we'll come back to the initial time $t' = 0$.

8.16.3 Open System – Constant Mass

In this equation, the latter term remains constant. After all, thanks to the external force, mass (and potential energy) can remain unchanged. So, upon differentiating both sides with respect to t', the latter term drops:

$$2E'\frac{dE'}{dt'} = 2c^2\left(p', \frac{dp'}{dt'}\right) = 2c^2\left(p', F'\right).$$

8.16.4 Nearly Constant Energy in The Lab

This is good at every $t' \geq 0$. At $t' = 0$, in particular, we know even more: the momentum was zero:

$$2E'\frac{dE'}{dt'} - 2c^2(0,0)F' = 0.$$

Since $E' > 0$, we must have

$$\frac{dE'}{dt'} = 0,$$

as required.

8.16.5 Nonperpendicular Force: Same at All Systems

In summary, at $t = t' = 0$,

$$F_x = F'_{x'} + \frac{\beta_u}{c} \cdot \frac{dE'}{dt'} = F'_{x'}.$$

Thus, unlike the perpendicular force, the nonperpendicular force is the same at all systems.

8.16.6 The Photon Paradox

What is light? It has two faces. On one hand, it is a wave. On the other hand, it is also a particle: a massless photon, traveling at speed c with respect to any (inertial) reference frame.

The photon is in a kind of singularity. Indeed, from our perspective, it has no size at all (due to length contraction: Section 8.6.3). Still, this is only from our own (subjective) point of view. The photon may disagree: in its own self system, it is at rest, with positive mass and size, and the entire universe shrinks to a singular point, traveling at speed c in the opposite direction. Who is right?

Thanks to Einstein's law, c is invariant: same at all (inertial) systems. But, in its own self system, the photon has speed 0, not c. This system can't be inertial. This is why it can never be transformed: its Lorentz matrix gets singular.

8.17 Exercises: 3-D

8.17.1 Lorentz Matrix and Its Determinant

1. Consider a 2×2 matrix. It contains two column vectors. Geometrically, they make a parallelogram in 2-D. What is its area? Hint: same as the determinant (Figure 2.11).
2. Is this true for the row vectors too?
3. Consider two 2×2 matrices. Look at their product. What is its determinant? Hint: same as the product of determinants (Figure 2.11, or Section 13.4.3).
4. Look at a Lorentz matrix. What is its determinant? Hint: 1.
5. Consider a 2×2 matrix. Multiply it by a Lorentz matrix. What is the new determinant? Hint: same as before.
6. Conclude that the Lorentz transformation preserves area in 2-D.
7. Look at a Lorentz matrix. What is its inverse? Hint: just change sign: replace v by $-v$ (Section 8.5.6).
8. What is the physical meaning of this? Hint: from the moving system, our lab seems to move in the opposite direction.
9. Conclude that the inverse Lorentz matrix has determinant 1 as well.
10. Conclude that the inverse transformation preserves area as well.

8.17.2 Motion in 3-D

1. Let
$$v \equiv \begin{pmatrix} v_1 \\ v_2 \\ v_3 \end{pmatrix} \in \mathbb{R}^3$$
be some (nonzero) vector in 3-D.
2. Define the orthogonal 3×3 matrix O_v, whose columns are: v (normalized), $v^\perp$ (orthogonal to v and normalized), and their vector product:
$$O_v \equiv \left(\frac{v}{\|v\|} \;\middle|\; \frac{v^\perp}{\|v^\perp\|} \;\middle|\; \frac{v \times v^\perp}{\|v\| \cdot \|v^\perp\|} \right).$$
3. Show that O_v is orthogonal. Hint: see Section 28.3.4.
4. Show that O_v has determinant 1. Hint: use the Euler angles (Section 2.6.3). Alternatively, see Section 13.4.6.
5. Consider a particle that moves at velocity $v \in \mathbb{R}^3$ with respect to the lab. In other words, the particle moves in direction $v/\|v\|$ at speed $\|v\|$. Let (x', y', z', ct') denote the lab coordinates, and (x'', y'', z'', ct'') the self coordinates of the particle. We are now ready to define the more general Lorentz transformation
$$\begin{pmatrix} x' \\ y' \\ z' \\ ct' \end{pmatrix} \rightarrow \begin{pmatrix} x'' \\ y'' \\ z'' \\ ct'' \end{pmatrix} = L_v \begin{pmatrix} x' \\ y' \\ z' \\ ct' \end{pmatrix},$$
where L_v is the following 4×4 Lorentz matrix:

$$L_v \equiv \begin{pmatrix} O_v & \\ & 1 \end{pmatrix} \begin{pmatrix} \gamma\left(\beta_{\|v\|}\right) & & & \\ & 1 & & \\ & & 1 & \\ & & & \gamma\left(\beta_{\|v\|}\right) \end{pmatrix} \begin{pmatrix} 1 & & & -\beta_{\|v\|} \\ & 1 & & \\ & & 1 & \\ -\beta_{\|v\|} & & & 1 \end{pmatrix} \begin{pmatrix} O_v^t & \\ & 1 \end{pmatrix}.$$

(As usual, blank spaces stand for zero matrix elements.)

6. Show that this indeed transforms the lab system to the self system of the particle.
7. Show that L_v has determinant 1.
8. Consider also a second particle, moving at velocity $-u \in \mathbb{R}^3$ with respect to the lab. Denote its self coordinates by (x, y, z, ct). This will be our new spacetime.
9. With respect to this spacetime, how does the entire lab move? Hint: it moves at velocity $u \in \mathbb{R}^3$.
10. How could this spacetime transform to the lab? Hint: the transformation is

$$\begin{pmatrix} x \\ y \\ z \\ ct \end{pmatrix} \rightarrow \begin{pmatrix} x' \\ y' \\ z' \\ ct' \end{pmatrix} = L_u \begin{pmatrix} x \\ y \\ z \\ ct \end{pmatrix}.$$

11. Look at the composite transformation

$$\begin{pmatrix} x \\ y \\ z \\ ct \end{pmatrix} \rightarrow \begin{pmatrix} x'' \\ y'' \\ z'' \\ ct'' \end{pmatrix}$$

from our spacetime (the self system of the second particle) to the self system of the first particle. Show that it is represented by the matrix product $L_v L_u$:

$$\begin{pmatrix} x'' \\ y'' \\ z'' \\ ct'' \end{pmatrix} = L_v \begin{pmatrix} x' \\ y' \\ z' \\ ct' \end{pmatrix} = L_v L_u \begin{pmatrix} x \\ y \\ z \\ ct \end{pmatrix}.$$

12. Show that $L_v L_u$ has determinant 1 as well. Hint: see Section 13.4.3.
13. Does L_u commute with L_v? Hint: only if u is a scalar multiple of v.
14. Consider the inverse transformation

$$\begin{pmatrix} x'' \\ y'' \\ z'' \\ ct'' \end{pmatrix} \rightarrow \begin{pmatrix} x \\ y \\ z \\ ct \end{pmatrix}$$

from the self system of the first particle back to our spacetime. Show that it is represented by the inverse matrix

$$(L_v L_u)^{-1} = L_u^{-1} L_v^{-1} = L_{-u} L_{-v}.$$

15. Conclude that the last column in $L_{-u} L_{-v}$ describes the motion of the first particle away from the second one. Hint: see below.

16. Show that, in its own self system, the first particle is at rest:

$$x'' = y'' = z'' = 0.$$

17. Conclude that t'' is its proper time. Hint: see Section 8.6.1.
18. Write this in vector form. Hint:

$$\begin{pmatrix} x'' \\ y'' \\ z'' \\ ct'' \end{pmatrix} = \begin{pmatrix} 0 \\ 0 \\ 0 \\ ct'' \end{pmatrix}.$$

19. Write this in differential form. Hint:

$$\begin{pmatrix} dx'' \\ dy'' \\ dz'' \\ cdt'' \end{pmatrix} = \begin{pmatrix} 0 \\ 0 \\ 0 \\ 1 \end{pmatrix}.$$

20. What is the meaning of this? Hint: this is just a proportion. After all, upon dividing the left-hand side by dt'', you get zero velocity (a static particle in its own self system).
21. Transform this back to our spacetime. Hint:

$$\begin{pmatrix} dx \\ dy \\ dz \\ cdt \end{pmatrix} = L_{-u}L_{-v} \begin{pmatrix} dx'' \\ dy'' \\ dz'' \\ cdt'' \end{pmatrix} = L_{-u}L_{-v} \begin{pmatrix} 0 \\ 0 \\ 0 \\ 1 \end{pmatrix}.$$

22. How to do this efficiently? Hint: up to a scalar multiple,

$$L_{-v} \begin{pmatrix} 0 \\ 0 \\ 0 \\ 1 \end{pmatrix} = \begin{pmatrix} v \\ c \end{pmatrix}.$$

Therefore, we only need to calculate

$$\begin{pmatrix} dx \\ dy \\ dz \\ cdt \end{pmatrix} = L_{-u} \begin{pmatrix} v \\ c \end{pmatrix}.$$

23. Divide the left-hand side by dt, and obtain the velocity in space:

$$\begin{pmatrix} dx/dt \\ dy/dt \\ dz/dt \end{pmatrix} \in \mathbb{R}^3.$$

24. What is its physical meaning? Hint: it tells us how the first particle gets away from the second one.

25. Consider a special case, in which u aligns with the x-axis:

$$u \equiv \begin{pmatrix} \|u\| \\ 0 \\ 0 \end{pmatrix}.$$

26. Show that, in this case, one could design

$$O_u \equiv I$$

(the 3×3 identity matrix).
27. Show that, in this case,

$$L_u \equiv \begin{pmatrix} \gamma\left(\beta_{\|u\|}\right) & & & \\ & 1 & & \\ & & 1 & \\ & & & \gamma\left(\beta_{\|u\|}\right) \end{pmatrix} \begin{pmatrix} 1 & & & -\beta_{\|u\|} \\ & 1 & & \\ & & 1 & \\ -\beta_{\|u\|} & & & 1 \end{pmatrix}.$$

28. Show that, in this case,

$$L_{-u} \equiv \begin{pmatrix} \gamma\left(\beta_{\|u\|}\right) & & & \\ & 1 & & \\ & & 1 & \\ & & & \gamma\left(\beta_{\|u\|}\right) \end{pmatrix} \begin{pmatrix} 1 & & & \beta_{\|u\|} \\ & 1 & & \\ & & 1 & \\ \beta_{\|u\|} & & & 1 \end{pmatrix}.$$

29. Show that this is not just a special case, but a most general case. Hint: for a general u, pick the x-axis to align with u in the first place.
30. Interpret L_u as a projective mapping in the real projective space. Hint: see Chapter 6 in [49].
31. Interpret the above method as the three-dimensional extension of the methods in Sections 8.7.2–8.7.4.
32. Likewise, in Section 8.7.4, interpret the inverse Lorentz transformation (back to our spacetime) as a projective mapping in the real projective plane.
33. What does it do? Hint: it maps the original velocity $(dx'/dt', dy'/dt')$ of the first particle in the lab (Figure 8.9) to the new velocity $(dx/dt, dy/dt)$ of the first particle away from the second one (Figure 8.10).
34. In Figures 8.9–8.10, where is the time variable? Why is it missing? Hint: these are phase planes, as in Figure 1.8. They only show us the position that the particle may reach, no matter when. Time remains an implicit parameter that "pushes" the particle along the arrow. You are welcome to make a movie that shows this!
35. In these figures, how did we get rid of time? Hint: we just divided by dt. This eliminated time, leaving just the velocity arrow in space.

Part V

Introduction to Quantum Mechanics

Part V: Introduction to Quantum Mechanics

So far, we saw how Einstein improved on Galileo-Newtonian physics, and introduced a more accurate theory, good even for very high speeds. This is relativity: a complete switch of mind. This is macroscale: but what about microscale? There too, there is room for improvement. For this too, we need a leap of imagination.

In classical mechanics, the particle is pointwise: it occupies a single point: its position in space. It also has its own time. This makes spacetime. Moreover, differentiate position (with respect to time), and get velocity. Better yet, look at more fundamental quantities: momentum-energy. But there is still a problem: the particle is too big! After all, it has mass and size. How can it be pointwise, and occupy a dimensionless point?

In quantum mechanics, the particle is no longer pointwise. Its position gets blurred and uncertain. It only has a probability to be at a certain position. More precisely, it has a distribution (probability density), telling us how likely it is to be in some interval. This is the square of its wave function, which is its true physical state. Time, Momentum, and energy are also modeled in this way, leading to their own uncertainty principles. All are uncertain. Why? Because the particle can never occupy a dimensionless point: neither in position-momentum nor in time-energy plane.

So, the key is the wave function: it contains all physical information about the particle. This is nondeterministic: it only tells you the probability to have a certain physical property. For example, how to have the energy? Take the wave function, and apply the Hamiltonian to it (the energy operator). It has discrete eigenvalues: energy levels. Only these are allowed: any other energy is nonphysical. This is quantum mechanics for you: energy comes in discrete packets, or quanta.

In angular momentum, things get even more bizarre: only a few values are allowed. Spin is most bizarre: only two values are allowed: up or down. (We already touched on this in quantum statistical mechanics: Chapter 7.) This is quite discrete, and must be modeled algebraically, by three Pauli matrices (good to model polarization too). These will later span a new linear space: a Lie algebra.

9

Introduction to Quantum Mechanics

In microscale, classical physics doesn't work any more. Instead, quantum mechanics teaches us new laws of nature. For one, order matters: measuring position before momentum is not the same as measuring momentum before position. To model this, use matrices: multiplying from the left is not the same as multiplying from the right. How different could this be? To measure this, we have the commutator. Later on, this will make a new Lie algebra.

To our matrices, we can now apply the Fourier transform. This is duality: a particle is also a wave. This will give us new laws of nature for free (Planck, Schrodinger, Debroglie, and more).

One important matrix is the Hamiltonian: the energy operator. It has discrete eigenvalues: energy levels. Only these are allowed, not any other energy in between. This is quantum mechanics for you: energy comes in packets, or quanta.

9.1 The Double-Slit Experiment

9.1.1 Nature and Chaos

In Chapters 4-5, we already got a taste of the important subject of stability. This marks the start of modern physics: initial data are not quite enough to predict the future! One must look at nature with a little bit of salt: things are not always as they seem. Nature is more relative than absolute. In some cases, nature makes no sense at all. Better have a little doubt, and not always trust your own eyes. Nature could be quite mysterious: not quite physical, but more mathematical.

9.1.2 Light and Its Polarization

For decades, people knew about two curious phenomena: electricity and magnetism. Maxwell unified them into one and the same phenomenon: electromagnetism. In vacuum, this looks like a plane wave: propagating in the z-direction, and oscillating (or even rotating) in the perpendicular x-y plane. This is polarization (to be modeled later).

Light is an electromagnetic wave (in a certain range of frequency and wavelength). Like all electromagnetic waves, it travels at the maximal speed in nature: c

(300, 000 kilometer per second). This is in vacuum, in all inertial systems (Chapter 8).

9.1.3 Interference

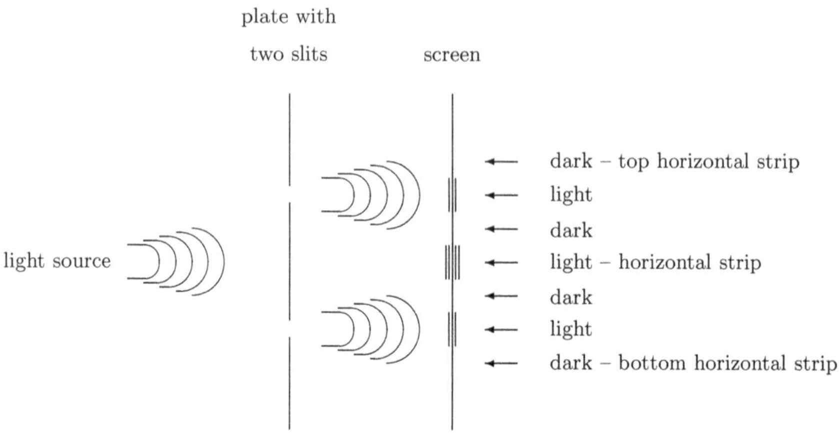

Fig. 9.1. At each (horizontal) slit, the light diffracts and restarts. On the screen on the far right, these waves interfere: enhance constructively, or cancel destructively, making alternating strips of light and dark.

As a wave, light exhibits yet another interesting phenomenon: interference. To see this, consider the double-slit experiment: a Laser beam (monochromatic light of a specific frequency and wavelength) comes from the left (Figure 9.1). On its way, it meets a plate, with two (narrow) horizontal slits in it. Through each slit, it passes and diffracts: restarts as a new source of light. These two light waves continue to travel rightwards, until they meet the screen on the far right, and shed their light on it.

What will we see on the screen? At the middle, we'll see a horizontal strip of bright light. Why? Thanks to symmetry, both light waves (from both slits) will arrive there at the same time: synchronically, in phase. This way, they'll enhance one another: crest meets a crest to make a yet higher peek, and trough meets a trough to make a yet deeper well. This is a constructive interference.

On top of the middle strip, however, symmetry breaks. The light wave from the top slit arrives a little too early: out of phase. As a result, things are not so benign any more: crest meets a trough, and trough meets a crest, canceling one another. This makes a flat (zero) wave: no light at all: a dark strip. This is a destructive interference.

On top of that, things get benign again: light from the top slit arrives yet earlier: crest meets the next crest, and trough meets the next trough. This makes a constructive interference again, and so on.

The same is mirrored from the bottom, below the middle strip. In summary, we get alternating (horizontal) strips: light, dark, light, dark, and so on.

9.1.4 Light: Wave or Particle?

Still, light is not only a wave but also a particle: photon. Thus, one could make the Laser beam dimmer and dimmer, eventually becoming a stream of individual photons, fired one at a time. But this doesn't matter: the same interference still forms (after many photons hit the screen). How come? Can a photon interfere with itself? Does it have a wave function that can interfere with itself?

Yes, it has! This is duality. The photon is not only a particle but also a wave. Its physical state is not just a point but a complete wave function, whose square is the probability density to be in some interval. This wave function can indeed interfere with itself.

9.1.5 Detector: No Interference

Not convinced? This experiment was carried out not only with photons but also with massive particles: electrons, or even atoms. The results were the same: interference.

How to avoid interference altogether? On the plate, place a detector, to find out which slit was picked. The particles would "feel" this, and behave like bullets, with only two strips of light: each slit would have its own strip in front of it (made of many particles that had hit there).

How to have interference back again? No good to delay detection. Even if you place the detector behind the plate, near the screen, there is still no interference. To have interference back again, you must be quick: before the particle hits the screen, drop the result from your detector, and store it nowhere (not even in your own brain). This is like no detection at all: upon hitting the screen, the particle will behave like a wave back again.

9.1.6 Wave Function and Superposition

To explain these strange results, Schrodinger came up with a new idea: nature is not quite unique! It is a superposition (sum) of two different outcomes: the top slit was picked (at some probability), and also the bottom slit was picked (at some probability). How big is the probability? This depends on the (squared) wave function. Where it is high, the particle is likely to be. This is constructive interference: bright light.

Thus, the particle is placed at all positions at the same time (each at its own probability). The wave in Figure 9.1 is just a mathematical wave function that tells us how likely the particle is to be there.

Only at the screen on the far right (where you measure energy for the first time) does the wave function "collapse" into a specific outcome: either constructive or destructive interference. If, however, you are too curious, and spy on the particle beforehand, then nature must decide earlier: the "collapse" must take place before any interference had a chance to form. This forces the particle to pass like a bullet, either through the top or the bottom slit, with no interference at all any more.

9.2 Nondeterminism

9.2.1 Relativistic Observation

In Newtonian mechanics, we often have a particle, or just any physical object. It must lie somewhere in space. This is its location or position.

At each individual time, the object may have a specific position. This is deterministic: we can measure it, and tell it for sure. We can even differentiate it, and tell how fast it changes: its (momentary) speed or velocity. Multiply by mass, and have the (momentary) momentum, and even calculate the kinetic energy. This gives us the complete picture of the dynamics or the motion.

In Chapter 8, we already saw that things are not so absolute, but more relativistic. The position I measure in my lab may differ from the position that you measure in your (moving) lab. Worse, position is meaningless on its own, unless it comes with some (fixed) reference point: the origin. What is meaningful is only the difference between two points.

The same is true for another important quantity: time. Time is only relativistic too: the time that I see in my clock may differ from the time that you see in your own clock, especially if you travel away from me fast.

Don't worry: no clock is wrong: both work well. Time is not absolute, but only relativistic: depends on the perspective from which it is measured. This is why time is never defined absolutely.

Thus, time is just an observation. To know it, you must look and observe: make an experiment, and measure. This measurement is relativistic: depends on the perspective it is made from. For this reason, two observers may see a different time in their different systems. For example, the time in my clock here on the Earth may differ from the time in some other clock, placed on a satellite.

9.2.2 Determinism

So, special relativity teaches us to be more humble, and not trust our own eyes. What you see is not quite the absolute truth: it is often different from what I see in my own (moving) system. This applies not only to time-position but also to energy-momentum: they too form a relativistic pair, transformed by the Lorentz transformation.

Fortunately, this is still deterministic. In your own system, you know what you see, with no doubt. This way, the physical quantities are still well-defined, uniquely and unambiguously. But this is true only in macroscale, not in microscale or nanoscale, used often to study molecules, atoms, and subatomic particles.

9.2.3 Nondeterminism: Observables

In quantum mechanics, things get even worse: nothing is certain any more! Even in your own system, you can't trust your observation any more. After all, a particle could be so small that its position has no physical meaning any more. Or even if it has, it comes at a price: if you measure it, then you can never know what the momentum was. In this context, position is no longer an observation, but only an observable (or measurable): in theory, you could observe and measure it, but better not.

This way, position is no longer a physical quantity, but only a random variable. At some probability, the particle *could* be there. Perhaps it is, and perhaps not. We'll never know, unless we're ready to measure and destroy the original state completely.

In fact, detecting the position of a very small particle requires a complicated experiment, with a lot of energy. This may change the original state forever, losing valuable information about how other quantities looked like (including momentum).

This also works the other way around: measuring the momentum of a very small particle is so difficult (and requires so much energy) that the original state gets lost, with a lot of vital information about how other observables looked like (including position). Better not measure at all, but only use matrices (or operators) to model things algebraically. Don't look or observe as yet: this could wait until later. In the meantime, we can still "play" with our matrices, and design more and more interesting observables.

9.3 State – Wave Function

9.3.1 State

What is a classical state? A particle travels along the x-axis, with position $x(t)$, velocity $x'(t)$, and momentum $mx'(t)$ (linear momentum, in the x-direction). This is true at any time t. But, in quantum mechanics, there is no determinism any more. In "true" physics, the particle is nowhere (or everywhere). The state only tells us where it *could* be. Perhaps it is there, and perhaps not.

Position is a random variable. Its state (its physical information) is its wave function:

$$v : \mathbb{R} \to \mathbb{C},$$

assumed to have norm 1:

$$\|v\|^2 \equiv (v, v) \equiv \int_{-\infty}^{\infty} v\bar{v}dx = \int_{-\infty}^{\infty} |v|^2 dx = 1.$$

In fact, v contains all physical information that nature tells us about the particle, including where it might be at time t. Thus, v isn't static, but dynamic: it depends on time too. Still, for now, we disregard this, and think of time as frozen, and of v as static.

9.3.2 Probability Density

Position is a random variable. What is its probability density? This is $|v|^2$. For instance, how likely is the particle to be between a and b? The probability for this is

$$\int_a^b |v|^2 dx.$$

Let's extend this to other observables as well.

9.4 Observable: Hermitian Operator

9.4.1 Position Operator

So, where might the particle be? For this purpose, we have the position operator: X. You can think of it as an infinite (diagonal) matrix. On its main diagonal, you'll find the possible positions: x. Thus,

$$X \equiv \text{diag}(x).$$

Now, think of v as a tall column vector (indexed by x). Apply X to v (from the left), and obtain a new column vector:

$$(Xv)(x) = xv(x)$$

(indexed by x). Algebraically, X is symmetric: real and Hermitian. This is why it has real eigenvalues: the hypothetical positions $x \in \mathbb{R}$.

9.4.2 Momentum: Fourier Expansion

Likewise, let P be the momentum operator. It is not diagonal. Still, we can diagonalize it. How? Superpose our particle in terms of waves (modes). How to do this algebraically? Transform to the eigenbasis (only implicitly). For this, you don't even need to know what P is. (Don't worry: later on, we'll define P explicitly.) Just remember that it is Hermitian, with real eigenvalues p (hypothetical momenta), and orthogonal eigenfunctions $\hat{p}$ (Fourier modes):

$$P\hat{p} = p\hat{p}.$$

How to write v in terms of p rather than x? Project it (orthogonally) on the $\hat{p}$'s:

$$v = \int_{-\infty}^{\infty} (v, \hat{p})\hat{p}dp,$$

where $(v, \hat{p})$ is the Fourier coefficient: the orthogonal projection on $\hat{p}$. (Here, no need to normalize $\hat{p}$: this is already incorporated in dp. This is how $\hat{p}$ approaches a Delta function in p. See Chapter 7 in [50].) Now, define

$$v(p) \equiv (v, \hat{p}).$$

This is the momentum wave function. Let's use it to define probability density.

9.4.3 Probability Density

What is the probability density? This is $|v(p)|^2$. Thanks to it, how likely is the particle to have momentum $a < p < b$? The probability for this is

$$\int_a^b |v(p)|^2 dp.$$

Later on, we'll extend this to another important observable: energy.

9.4.4 The Commutator

By now, we already have two observables (or measurables, or Hermitian operators): X and P. They don't commute, but have a nonzero commutator:

$$[X, P] \equiv XP - PX = i\bar{h}I,$$

where $i = \sqrt{-1}$, I is the identity operator (the infinite identity matrix), and

$$\bar{h} \doteq 10^{-34}$$

is Planck's constant. Later on, we'll define P explicitly, and obtain this commutator. Anyway, it is nonzero. Thus, measuring position before momentum is not the same as measuring momentum before position. This is what makes quantum mechanics more accurate than classical mechanics: at microscale and nanoscale, the commutator gets important, and can't be ignored any more. This will lead us to the uncertainty principle.

9.5 Heisenberg's Uncertainty Principle

9.5.1 Interference: Statistical Effect

In the double-slit experiment, we looked at one particle, passing through the top or bottom slit (we don't know which), and hitting the screen on the far right (Figure 9.1). There, we'll see a tiny spot of light. But where? For this, we need quantum mechanics. The particle is also a wave: it has a wave function, which splits into two, restarting from the slits, and interfering on the screen on the far right. After many particles did this, we can see a beautiful pattern: alternating strips of light and dark. This is a statistical effect: many particles are needed. Still, we can gather and keep the statistics, and use it to predict where a new particle will hit: probably where a bright strip was, not where a dark strip was.

9.5.2 Expectation

Let's return to our position operator: X. Assume that our particle has state v. What is its expected position? This is

$$e \equiv e(v) \equiv (v, Xv) = \int_{-\infty}^{\infty} x|v|^2 dx.$$

And what about its expected momentum? This is

$$f \equiv f(v) \equiv (v, Pv) = \int_{-\infty}^{\infty} p|v(p)|^2 dp.$$

9.5.3 Variance

Is the particle likely to deviate from e? This is its variance:

$$\|(X-eI)v\|^2=\int_{-\infty}^{\infty}(x-e)^2|v|^2dx.$$

And what about momentum? Is it likely to deviate from f? This is momentum variance:

$$\|(P-fI)v\|^2=\int_{-\infty}^{\infty}(p-f)^2|v(p)|^2dp.$$

9.5.4 Covariance

How likely are position-momentum to deviate together? This is their covariance:

$$((X-eI)v,(P-fI)v).$$

Let's estimate this. For simplicity, assume that

$$e=f=0.$$

After all, we can always shift the x-axis (and also the p-axis, by going to the self system). This way, X has variance $\|Xv\|^2$, and P has variance $\|Pv\|^2$. How small can they be? For this, we need an algebraic tool.

9.5.5 The Cauchy-Schwarz Inequality

Look at two wave functions:

$$u,v:\mathbb{R}\to\mathbb{C}.$$

Look at their inner product: $(u,v)\in\mathbb{C}$. How to estimate it (in magnitude)? For this, use their norms:

$$|(u,v)|\le\|u\|\cdot\|v\|.$$

This is the Cauchy-Schwarz inequality. (See exercises at the end of Chapter 28.) Let's go ahead and use it.

9.5.6 Hermitian and Anti-Hermitian Parts

Next, take our position operator X, and our momentum operator P. They are both Hermitian. Now, multiply them by one another. This will give you a new operator: the product XP. But this is not Hermitian any more. Split it into two parts: Hermitian and anti-Hermitian (Section 28.1.1):

$$\begin{aligned}XP&=\frac{XP+(XP)^*}{2}+\frac{XP-(XP)^*}{2}\\&=\frac{XP+PX}{2}+\frac{XP-PX}{2}\\&=\frac{XP+PX}{2}+\frac{[X,P]}{2}.\end{aligned}$$

Use this to calculate the expectation of XP at v:

$$(v, XPv) = \left(v, \frac{XP + PX}{2}v\right) + \left(v, \frac{[X, P]}{2}v\right).$$

What do we have here? On the left-hand side, we have a complex number. On the right-hand side, it splits into two parts: real and imaginary. Clearly, each part is smaller than (or equal to) the original complex number (in magnitude). In particular,

$$|(v, XPv)| \geq \frac{1}{2}|(v, [X, P]v)| = \frac{1}{2}|(v, i\bar{h}v)| = \frac{\bar{h}}{2}(v, v) = \frac{\bar{h}}{2}$$

(assuming that v has norm 1). Let's go ahead and use this.

9.5.7 Standard Deviation

What is the standard deviation? This is the square root of the variance. Now, take the standard deviation of X at v, times the standard deviation of P at v. How small can this be? How to estimate this from below? Thanks to the Cauchy-Schwarz inequality (and what we did so far),

$$\|Xv\| \cdot \|Pv\| \geq |(Xv, Pv)| = |(v, XPv)| \geq \frac{\bar{h}}{2}.$$

This is Heisenberg's uncertainty principle: at a particular v, X and P can't be known simultaneously. Indeed, if X was known, then its standard deviation would vanish. But then again, P would be forced to have an infinite standard deviation, meaning that P can't be known. The same works the other way around.

9.6 Duality: Wave or Particle?

9.6.1 Momentum: Differentiation

P is an operator, acting on any (differentiable) function of x. How? It differentiates it:

$$P \equiv -i\bar{h}\frac{d}{dx}.$$

For example, look at a Fourier mode:

$$\hat{p}(x) \equiv \exp(ikx),$$

where $k \in \mathbb{R}$ is the wave number. How does P act on it? Like this:

$$P\hat{p} = -i\bar{h}\hat{p}' = -i\bar{h} \cdot ik\hat{p} = k\bar{h}\hat{p}.$$

Thus, $\hat{p}$ is indeed an eigenvector, with eigenvalue

$$p \equiv k\bar{h}.$$

9.6.2 Debroglie: Momentum is Wave Number

This is momentum. why? Because, thanks to wave theory, the wave looks the same in t too: it oscillates at frequency $\omega = ck$. Better yet, in terms of ct (which is more comparable to x), frequency scales better, and becomes k exactly.

This is Debroglie's law: the particle is also a wave: it has a wave function, with a wave number proportional to momentum.

This is nondeterministic: k is a random variable too, proportional to p. Both have the same probability density: $|v(p)|^2$.

Heisenberg's uncertainty principle forces us to choose. Want to have a particle? Measure position. Want to have a wave? Measure momentum (and indeed wave number). You can't enjoy both worlds. This is duality for you.

We can't know both position and momentum at the same time. The better we know x, the worse we know k and p, and vice versa. Each picture has its own advantage. The particle picture helps explain the photo-electric effect: only a high-frequency photon has enough energy to knock out an electron. The wave picture, on the other hand, helps explain diffraction and interference in the double-slit experiment.

9.6.3 The Commutator

X and P don't commute. What is their commutator? Pick a (differentiable) function $f(x)$, and apply the commutator to it:

$$\begin{aligned}[X, P]f &= (XP - PX)f \\ &= -(PX - XP)f \\ &= i\bar{h}((xf)' - xf') \\ &= i\bar{h}(f + xf' - xf') \\ &= i\bar{h}f.\end{aligned}$$

Thus, the commutator is indeed

$$[X, P] = i\bar{h}I,$$

as required (Section 9.4.4).

9.6.4 An Electron in The Atom

So far, x could be any real number. But what if x was confined to a circle? In other words, what if $f(x)$ must be periodic? In this case, our Fourier mode must be periodic too. For this purpose, the wave number k must be an integer. This is quantum mechanics for you: nature is discrete, not continuous.

In the atom, the electron orbits the nucleus not only as a particle but also as a wave. For this, our Fourier mode should be interpreted as a (periodic) wave function. This is the state of the electron in its orbit. But k must now be an integer, leading to a discrete linear momentum: $p = \bar{h}k$. Likewise, the angular momentum takes discrete values only. In fact, the electron must orbit at a suitable radius, where the electrostatic attraction to the nucleus supplies the required centripetal force. This is Bohr's model: the nucleus is surrounded by more and more shells (energy levels), where the electron could orbit at more and more energy. Thanks to Debroglie, this is indeed stable: the periodic wave function makes a standing wave, from which the electron radiates no electromagnetic wave, and loses no mass or energy.

9.7 Time-Energy Uncertainty

9.7.1 Time-Energy and Space-Momentum

In special relativity, time-space make a relativistic pair. In it, both x and ct have the same status. Likewise, energy-momentum make a relativistic pair too. In this sense, time mirrors space, and energy mirrors momentum (Section 8.8.4).

In quantum mechanics too, time-energy mirror space-momentum. Thus, not only position-momentum but also time-energy have an uncertainty principle. You can't enjoy both worlds. You must choose. Want to have a particle? Measure time. Want to have a wave? Measure energy (or frequency, or color). The better you know energy, the worse you know time, and vice versa.

Like momentum, energy looks at the particle as a wave. Indeed, look at a Fourier mode in time: $\exp(-i\omega t)$. The energy operator, the Hamiltonian H, will pick the frequency ω as an eigenvalue. planck:

9.7.2 Schrodinger's Equation

For this purpose, H mirrors P: it differentiates in time:

$$H \equiv i\hbar h \frac{d}{dt}.$$

This is indeed Schrodinger's equation. Later on, we'll also rewrite it with a partial derivative.

9.7.3 Planck: Energy is Frequency

Thanks to this definition, a Fourier mode in time is indeed an eigenfunction:

$$H \exp(-i\omega t) = i\bar{h}(-i\omega)\exp(-i\omega t) = \omega\bar{h}\exp(-i\omega t).$$

This is Planck's law: energy is proportional to frequency:

$$E = \omega\bar{h}$$

(an eigenvalue of H). This was used by Einstein to explain the photo-electric effect: only a high-frequency photon has enough energy to knock out an electron.

Better yet, let's redefine both P and H in terms of a partial derivative:

$$P \equiv -i\bar{h}\frac{\partial}{\partial x}$$
$$H \equiv i\hbar h\frac{\partial}{\partial t}.$$

This way, P and H can now act even on a function of two variables, like the wave

$$\exp(-i\omega t)\exp(ikx) = \exp(i(kx - \omega t)).$$

Thanks to the left-hand side, this is indeed a standing wave: a function of x, precessing at frequency ω in time. In the atom, the true wave function is more complicated: a superposition (sum) of a few such waves, each with its own frequency and coefficient (which depend on the probability to be at this energy level). Later on, in quantum chemistry, we'll get deeper into this.

9.8 Eigenvalues

9.8.1 Shifting an Eigenvalue

Let's use linear algebra to study a general commutator. Let C and T be two operators (Hermitian or not). Assume that they don't commute, but have a nonzero commutator, proportional to T:

$$[C, T] \equiv CT - TC = \alpha T,$$

for some complex number $\alpha \neq 0$. What does this mean algebraically? In a sense, T is an "eigenvector" of the linear mapping $[C, \cdot]$. (We'll come back to this later, in Lie algebras.) Does this mean that T can also help design new eigenvectors for C itself? Let u be an eigenvector of C, with eigenvalue λ:

$$Cu = \lambda u.$$

How to find more eigenvectors? Just apply T to u. If Tu is a nonzero vector, then it is an eigenvector as well:

$$CTu = (TC + [C, T])\,u = (TC + \alpha T)\,u = (\lambda + \alpha)\,Tu.$$

What is its eigenvalue? $\lambda + \alpha$. Thus, we shifted λ by α, obtaining a new eigenvalue of C: $\lambda + \alpha$ (provided that Tu is not the zero vector). We can now repeat this, and apply T time and again (so long as we don't hit the zero vector). This will design more and more eigenvectors of C, with new eigenvalues, shifted more and more:

$$\lambda,\ \lambda + \alpha,\ \lambda + 2\alpha,\ \lambda + 3\alpha,\ \ldots.$$

This could continue forever, unless we hit the zero vector.

9.8.2 Shifting an Eigenvalue of a Product

Let's use this in a special case. Let A and B be two operators (Hermitian or not). Assume that they don't commute, but have a nonzero commutator, proportional to the identity operator:

$$[A, B] \equiv AB - BA = \beta I,$$

for some complex number $\beta \neq 0$. Now, let's look at the product BA. What is its commutator with B? We already know what it is:

$$[BA, B] \equiv BAB - BBA = B(AB - BA) = B[A, B] = \beta B.$$

Thanks to Section 9.8.1, we can now shift an eigenvalue of BA. Let u be an eigenvector of BA, with eigenvalue λ:

$$BAu = \lambda u.$$

If Bu is a nonzero vector, then it is an eigenvector as well:

$$BA(Bu) = (\lambda + \beta)Bu.$$

Likewise, one could also shift in the opposite direction. For this purpose, look again at the product BA. What is its commutator with A? It is just

$$[BA, A] = BAA - ABA = (BA - AB)A = -[A, B]A = -\beta A.$$

Thanks to Section 9.8.1, we can now design a new eigenvector of BA: take the original eigenvector u, and apply A to it:

$$BA(Au) = (\lambda - \beta)Au.$$

This way, if Au is a nonzero vector, then it is an eigenvector as well, with the new eigenvalue $\lambda - \beta$. Moreover, repeat this time and again (so long as you don't hit the zero vector), and design more and more eigenvectors of BA, with new eigenvalues:

$$\lambda,\ \lambda \pm \beta,\ \lambda \pm 2\beta,\ \lambda \pm 3\beta,\ \ldots.$$

This could continue forever, unless you hit the zero vector.

9.8.3 A Number Operator

Consider a special case, in which

$$B = A^* \quad \text{and} \quad \beta = 1.$$

This way, the assumption in the beginning of Section 9.8.2 takes the form

$$[A, A^*] = I,$$

and the product studied above is now Hermitian:

$$BA = A^*A.$$

We can now shift an eigenvalue of A^*A from λ to

$$\lambda \pm 1,\ \lambda \pm 2,\ \lambda \pm 3,\ \ldots$$

(until hitting the zero vector, which is going to be very soon, at least on the negative side). A^*A is called a number operator. Why? Because it has eigenvalues $0, 1, 2, 3, 4, \ldots$. In fact, our process is not so general, but quite limited: it must start from an integer $\lambda \geq 0$, and stop at 0. Let's see why.

9.8.4 Eigenvalue – Expectation

Indeed, let u be an eigenvector of A^*A, with eigenvalue λ:

$$A^*Au = \lambda u.$$

Without calculating them explicitly, what can we say about λ and u? Well, since A^*A is Hermitian, λ must be real (Section 28.4.4). Could λ be negative? No! Indeed, look at the expectation of A^*A at u:

$$\lambda(u, u) = (u, A^*Au) = (Au, Au) \geq 0.$$

Thus, $\lambda \geq 0$. Could λ be noninteger? No! To see this, let's design more eigenvectors.

9.8.5 Ladder Operator: Lowering an Eigenvalue

Recall that we consider now a special case: in Section 9.8.2, set $B \equiv A^*$, and $\beta \equiv 1$. Let's design a new eigenvector: Au. In fact, if Au is a nonzero vector, then it is an eigenvector of A^*A in its own right, with a lower eigenvalue: $\lambda - 1$. Here, A serves as a "ladder" operator. By applying it to u, we go down the ladder, to a lower eigenvalue.

What is the norm of Au? To find out, look again at the expectation of A^*A at u:

$$\|Au\|^2 = (Au, Au) = (u, A^*Au) = \lambda(u, u) = \lambda\|u\|^2.$$

In other words,

$$\|Au\| = \sqrt{\lambda}\|u\|.$$

Later on, we'll use this to normalize the eigenvectors.

9.8.6 Null Space

Lowering can't continue forever, or it would hit a negative eigenvalue, which is forbidden (Section 9.8.4). It must stop at some eigenvector w of eigenvalue 0. This w is in the null space of A^*A. At w, lowering stops. So, we see in retrospect that λ must have been integer. This is why lowering stopped, as required.

By now, what do we know about w? Well, at w, A^*A has zero expectation:

$$(Aw, Aw) = (w, A^*Aw) = (w, \mathbf{0}) = 0.$$

Thus,

$$Aw = \mathbf{0},$$

so w is actually in the null spaces of A too (Section 28.4.2). Let's go the other way around, and raise eigenvalues back again.

9.8.7 Raising an Eigenvalue

For this purpose, let's use Section 9.8.2 once again (with $B \equiv A^*$ and $\beta \equiv 1$). In fact, A^*u is an eigenvector of A^*A in its own right, with a bigger eigenvalue: $\lambda + 1$. Here, A^* serves as a new ladder operator, to help "climb" up the ladder.

What is the norm of A^*u? Well, since $[A, A^*] = I$,

$$\begin{aligned}
\|A^*u\|^2 &= (A^*u, A^*u) \\
&= (u, AA^*u) \\
&= (u, (A^*A + [A, A^*])\, u) \\
&= (u, (A^*A + I)\, u) \\
&= (u, (\lambda + 1)\, u) \\
&= (\lambda + 1)\, (u, u) \\
&= (\lambda + 1)\, \|u\|^2.
\end{aligned}$$

In other words,

$$\|A^*u\| = \sqrt{\lambda + 1}\|u\|.$$

Later on, we'll use this to normalize these eigenvectors.

9.9 Hamiltonian and Its Eigenvalues

9.9.1 Harmonic Oscillator and Its Hamiltonian

How to use this in physics? Look at a harmonic oscillator, or a spring. So far, we modeled it classically (Sections 1.6.1–1.9.4). How to model it quantum-mechanically? Thanks to our position-momentum operators, X and P, we can now define the energy operator: the Hamiltonian:

$$H \equiv \frac{m\omega^2}{2} \left(X^2 + \frac{1}{m^2\omega^2} P^2 \right),$$

where m is mass, and ω is frequency. This is the Hamiltonian observable: total energy: kinetic plus potential. This is nondeterministic: a random variable, telling us what the total energy *could* be: an eigenvalue of H. At what probability? This depends on the (normalized) eigenvector: take its inner product with the (normalized) state v, take the absolute value, and square it up.

9.9.2 Concrete Number Operator

In Section 9.8.3, A was a general operator. It is time to define it more concretely:

$$A \equiv= \sqrt{\frac{m\omega}{2\hbar}} \left(X + \frac{i}{m\omega} P \right).$$

This way, its Hermitian adjoint just picks a minus sign:

$$A^* = \sqrt{\frac{m\omega}{2\hbar}} \left(X - \frac{i}{m\omega} P \right).$$

Thus, their commutator is

$$[A, A^*] = \frac{m\omega}{2\bar{h}}\left[X + \frac{i}{m\omega}P, X - \frac{i}{m\omega}P\right] = -\frac{m\omega}{2\bar{h}} \cdot \frac{i}{m\omega}\left([X,P] - [P,X]\right) = I,$$

as required (Section 9.8.3). This allows us to design the eigenvectors of A^*A, and even normalize them: start from some w in the null space of A, normalize, apply A^* time and again, and normalize:

$$w,\ A^*w,\ \frac{1}{\sqrt{2}}\left(A^*\right)^2 w,\ \frac{1}{\sqrt{3!}}\left(A^*\right)^3 w,\ \ldots\ \frac{1}{\sqrt{k!}}\left(A^*\right)^k w,\ \ldots.$$

These are the orthonormal eigenvectors of A^*A, useful to expand any vector (Section 28.4.5).

9.9.3 Energy Levels

How to get some physical insight? Let's write our number operator explicitly:

$$\begin{aligned}
A^*A &= \frac{m\omega}{2\bar{h}}\left(X - \frac{i}{m\omega}P\right)\left(X + \frac{i}{m\omega}P\right) \\
&= \frac{m\omega}{2\bar{h}}\left(X^2 + \frac{1}{m^2\omega^2}P^2 + \frac{i}{m\omega}[X,P]\right) \\
&= \frac{m\omega}{2\bar{h}}\left(X^2 + \frac{1}{m^2\omega^2}P^2\right) + \frac{i}{2\bar{h}}i\bar{h}I \\
&= \frac{1}{\omega\bar{h}}H - \frac{1}{2}I.
\end{aligned}$$

Thus, thanks to our concrete number operator, the Hamiltonian takes the form

$$H = \omega\bar{h}\left(A^*A + \frac{1}{2}I\right).$$

What is the total energy? This is a random variable, which *could* take discrete values: eigenvalues of H:

$$\frac{\omega\bar{h}}{2},\ \frac{3\omega\bar{h}}{2},\ \frac{5\omega\bar{h}}{2},\ \frac{7\omega\bar{h}}{2},\ \ldots.$$

This is quantum mechanics for you: energy comes in packets, or quanta, or levels. (Position and time, on the other hand, remain continuous: at each individual time, the particle could occupy any position.) Each eigenvector of H (and of A^*A) is quite static: thanks to the Schrodinger equation, it can only precess by factor $\exp(-i\omega t)$, with no physical effect at all. It makes a static wave function: a standing wave that never moves. This is how it keeps the same energy: its eigenvalue with respect to H. This is an energy level.

9.9.4 Ground State: Zero-Point Energy

Our w is the eigenvector of minimal energy (Figure 9.2). Indeed, with respect to H, w has minimal eigenvalue: $\omega\bar{h}/2$. This is the ground state: zero-point (or vacuum)

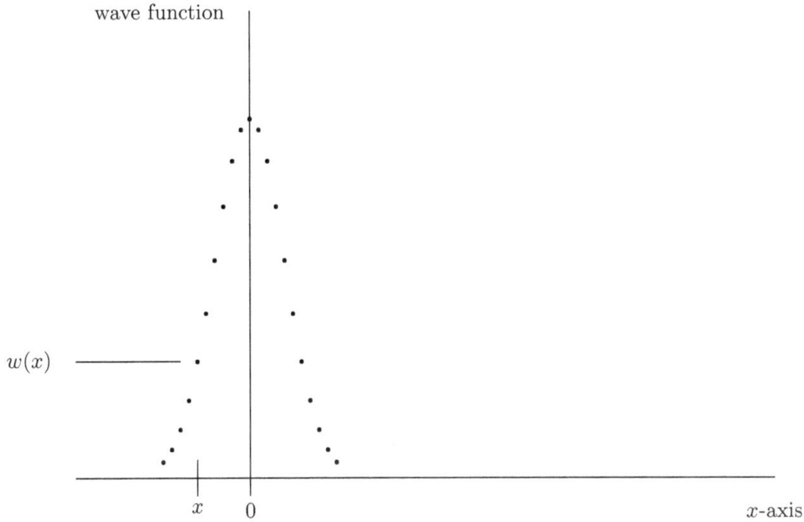

Fig. 9.2. Ground state: $w(x)$ is a (normalized) Gaussian. The position x has probability density $|w(x)|^2$. The most likely is $x = 0$ (the expectation).

energy, coming from virtual processes. We'll come back to this later, in Feynman diagrams.

Thanks to the Schrodinger equation, w never changes physically, but only precesses, with no effect on the probability density:

$$|\exp(-i\omega t/2)w(x)|^2 = |w(x)|^2,$$

or on the momentum probability density:

$$|(\exp(-i\omega t/2)w, \hat{p})|^2 = |(w, \hat{p})|^2.$$

Only a superposition of a few energy levels can have real dynamics, and really propagate in time.

9.10 Coherent States

9.10.1 Superposition: Dynamic Interference

Let's superpose our (orthonormal) eigenvectors, and expand v:

$$v = \sum_{k \geq 0} \left(v, \frac{1}{\sqrt{k!}} (A^*)^k w \right) \frac{1}{\sqrt{k!}} (A^*)^k w.$$

Each eigenvector has its own (deterministic) energy. But here, at state v, energy is nondeterministic again (just like position and momentum). In this sum, the probability to be at a certain energy level is the coefficient, squared and normalized. This doesn't change over time: each eigenvector can only precess (in its own frequency), with no effect on the probability. So, as a (discrete) random variable, energy is still conserved, as required. But position and momentum are not. After all, each eigenvector precesses at its own (private) frequency, leading to interference: constructive, then destructive, then constructive again, and so on. This is true dynamics: v is no longer standing, but really traveling.

9.10.2 Energy Conservation

How to see conservation of Energy, in its nondeterministic sense? So far, we only expanded $v(0)$. What happens later, at time $t > 0$? In our expansion, look at the kth coefficient. How does it change in time? It only precesses: it is multiplied by $\exp(-i\omega(k + 1/2)t)$:

$$v(t) = \sum_{k\geq 0} \exp\left(-i\omega\left(k + \frac{1}{2}\right)t\right)\left(v(0), \frac{1}{\sqrt{k!}}\left(A^*\right)^k w\right)\frac{1}{\sqrt{k!}}\left(A^*\right)^k w.$$

This new exponent has no effect on the probability to be in the kth energy level, which remains

$$\left|\left(v, \frac{1}{\sqrt{k!}}\left(A^*\right)^k w\right)\right|^2 = \frac{1}{k!}\left|\left(v, \left(A^*\right)^k w\right)\right|^2$$

(to be normalized soon). Still, this precession is not uniform: as a function of x, each eigenvector precesses in its own frequency (and energy). As functions of x, they interfere dynamically: constructive, then destructive, then constructive again, and so on. In x, this makes a dynamic wave function that really propagates in time. Its square (the probability density) propagates in the same way. The same holds for the momentum wave function and probability density. This is why, in the harmonic oscillator, momentum is not conserved (not even stochastically).

9.10.3 Coherent State

In physics, energy must be real. Fortunately, our energy levels are indeed real. Why? Because H is Hermitian. Why? Because A^*A is always Hermitian. But A is not. In fact, A may have a complex eigenvalue $\lambda \in \mathbb{C}$, with a nonzero imaginary part. Assume that v is its eigenvector:

$$Av = \lambda v.$$

This is a coherent state. How does it look like? It makes a Gaussian, shifted by λ (Figure 9.3).

9.10.4 Energy: Discrete Distribution

Our coherent state is *not* an eigenfunction of H, but superposed from many. Each precesses in its own frequency, producing a dynamic interference: constructive, then

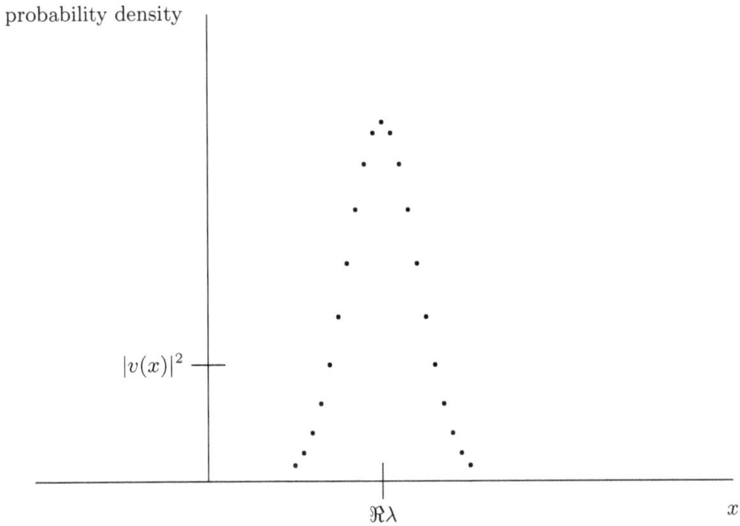

Fig. 9.3. A coherent state: a Gaussian, shifted by a complex number λ. The probability density is $|v(x)|^2$.

destructive, then constructive again, and so on. The (momentum) wave function and (momentum) probability density are really dynamic, and really travel in time. But what about energy? It has a discrete distribution that never changes in time. This is conservation of energy, in its stochastic face. In a coherent state, this is even easier to see: the formula in Section 9.10.2 simplifies to read:

$$\begin{aligned}\frac{1}{k!}\left|\left(v, (A^*)^k w\right)\right|^2 &= \frac{1}{k!}\left|\left(A^k v, w\right)\right|^2 \\ &= \frac{1}{k!}\left|\left(\lambda^k v, w\right)\right|^2 \\ &= \frac{1}{k!}|\lambda|^{2k}\left|(v, w)\right|^2 \\ &= \left|(v, w)\right|^2 \frac{|\lambda|^{2k}}{k!}.\end{aligned}$$

Here, we still have a mysterious factor: $|(v, w)|^2$. How to estimate it? Well, the probabilities must sum to 1:

$$1 = |(v, w)|^2 \sum_{k \geq 0} \frac{|\lambda|^{2k}}{k!} = |(v, w)|^2 \exp\left(|\lambda|^2\right).$$

As a result,

$$|(v, w)|^2 = \exp\left(-|\lambda|^2\right).$$

9.10.5 Poisson Distribution

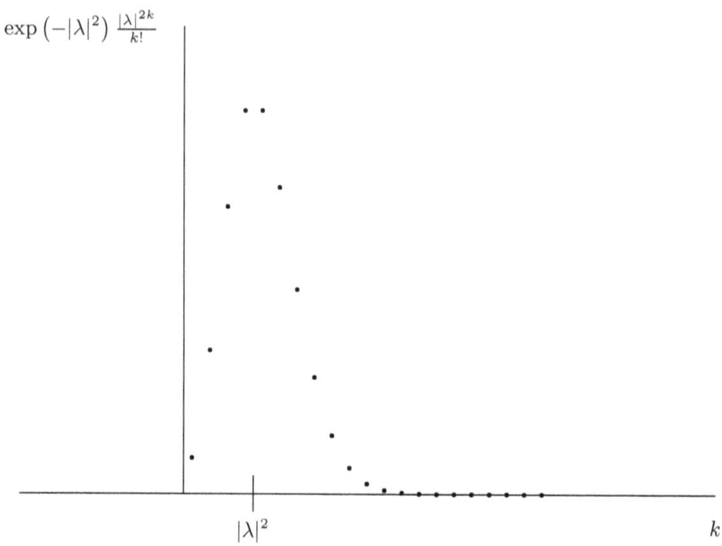

Fig. 9.4. The Poisson distribution. To have energy $\omega\bar{h}(k + 1/2)$, the probability is $\exp(-|\lambda|^2)|\lambda|^{2k}/k!$, where λ is the eigenvalue of the coherent state with respect to A. The maximal probability is at $k \doteq |\lambda|^2$.

This is the Poisson distribution (Figure 9.4). Unlike the Gaussian distribution, it doesn't tell us where the particle could be, but only what energy it could have. This is a discrete function of k: to have energy $\omega\bar{h}(k + 1/2)$, the probability is

$$\exp\left(-|\lambda|^2\right)\frac{|\lambda|^{2k}}{k!},$$

where λ is the eigenvalue of the coherent state with respect to A. Energy is nondeterministic, but can still be predicted quite well. Later on, in Lie groups, we'll see this in more general terms.

9.11 Exercises

9.11.1 Eigenvalues and Eigenvectors

1. What is an observable? Hint: same as measurable: Hermitian matrix (or operator).
2. What can you say about its eigenvalues? Hint: they are real (no imaginary part).

3. Consider two eigenvectors, corresponding to two distinct eigenvalues. What can you say about them? Hint: they are orthogonal to one another.
4. Can you make them orthonormal? Hint: normalize them (to have norm 1).
5. And what if they share the same eigenvalue? Assume they are linearly independent. How to orthogonalize them? Hint: pick one of them, and modify it: subtract its orthogonal projection on the other one (obtained from their inner product). This is the Gram-Schmidt process.
6. This may change their norm. Normalize them again.
7. What is a degenerate eigenvalue? Hint: there are at least two (linearly independent) eigenvectors that share it.
8. What is a degenerate observable? Hint: it has a degenerate eigenvalue.
9. Could the position operator X be degenerate? Hint: X must have distinct main-diagonal elements: possible positions.
10. Could the momentum operator P be degenerate? Hint: once diagonalized, it must have distinct main-diagonal elements: possible momenta.

9.11.2 Energy Levels

1. Consider the harmonic oscillator, and its Hamiltonian: H (Section 9.9.1). Is it Hermitian? Hint: X and P are, and so are X^2, P^2, and indeed H.
2. Is it a legitimate observable?
3. Can it have an eigenvalue with an imaginary part? Hint: a Hermitian matrix (or operator) can have real eigenvalues only.
4. Can it have a negative eigenvalue? Hint: no. The number operator can't have a negative expectation (Sections 9.8.3–9.8.4).
5. What is the minimal eigenvalue of H?
6. Can it be zero? Hint: it must be bigger than the minimal eigenvalue of the number operator, which is zero.
7. What is the physical meaning of this eigenvalue? Hint: this is vacuum energy: the minimal energy of the harmonic oscillator.
8. Can the harmonic oscillator have no energy at all? Hint: no! Its minimal energy is positive.

9.11.3 Ground State

1. At this minimal eigenvalue, what is the corresponding eigenvector? Hint: the ground state: the state of minimal energy.
2. How does it look like? Hint: a Gaussian (Figure 9.2).
3. At the ground state, what is the probability to have a particular amount of energy? Hint: at probability 1, it has the minimal energy. It can't have any more. This is deterministic.
4. Can the ground state change dynamically in time? Hint: it can only precess, with no physical effect at all.
5. Why? Hint: Energy must remain at its minimum.
6. How about any other eigenvector? Can it change dynamically in time? Hint: same again: it can only precess, with no physical effect at all.
7. Why? Hint: energy must remain the same, all the time.

8. How much energy can the harmonic oscillator have? Hint: the allowed energy levels are the eigenvalues of H: $\omega\bar{h}(k + 1/2)$ $(k \geq 0)$.
9. Is there a maximal energy? Hint: you can raise eigenvalues, time and again, forever. After all, you can never hit the zero vector (Section 9.8.7).

9.11.4 Coherent State

1. In a coherent state, what is the probability to have a particular amount of energy? Hint: this is the discrete Poisson distribution (Figure 9.4).
2. Is this deterministic? Hint: no! Each energy level has its own (nonzero) probability.
3. Is this probability constant in time? Hint: yes! Each eigenfunction only precesses, with no effect on its absolute value (whose square is the probability).
4. Is the coherent state constant in time? Hint: no! the (momentum) wave function is a superposition of many eigenfunctions (energy levels), each precessing at its own frequency. This leads to a dynamic interference: constructive, then destructive, then constructive again, and so on.
5. Is this physical? Hint: yes! This affects position-momentum. Over time, there is a real change in the (momentum) wave function, not only in phase but also in magnitude, whose square is the (momentum) probability density.
6. Does this affect the probability to have a particular amount of energy? Hint: no! The Poisson distribution remains constant. Indeed, each eigenfunction only precesses: changes phase, but not magnitude at all.
7. And what about any observable that commutes with the Hamiltonian? Does its (discrete) distribution change over time? Hint: no! It shares the same eigenvectors, which only precess, with no effect on probability.
8. What is the physical meaning of this? Hint: this observable behaves like total energy: it obeys the same conservation law (stochastically).
9. And what about an observable that doesn't commute with the Hamiltonian? Does it have any real dynamics? Hint: yes! Its wave function is superposed from many energy levels, each precessing at a different frequency. This leads to a dynamic interference again: constructive, then destructive, then constructive again, and so on.
10. Give an example. Hint: position-momentum. Indeed, in the harmonic oscillator, X and P don't commute with H.

10

Angular Momentum

How to study energy and angular momentum at the same time? For this, we already developed an algebraic tool: the commutator (which will later play a major role in Lie algebras). Thanks to it, we already designed ladder operators, to climb up or down the ladder, and obtain many energy levels, on which energy is conserved. They are discrete, not continuous. This is quantum mechanics for you: energy comes in packets, or quanta, or levels.

Likewise, angular momentum can also benefit from the same tools. In a sense, it is even more quantized, and even less classical: it has only a few levels. These are distributed symmetrically: positive ones, mirrored by negative ones.

How is this different from energy? Energy is unbounded. In theory, it could take very high values: as high as you like. In practice, however, this is quite unlikely, uncertain, and even unstable: easy to leak out, and fall down, back to a lower level, more steady and attractive.

Angular momentum, on the other hand, is strictly bounded: it could take only a few values (positive or negative). Spin, for example, has just two possible values: up or down. Later on, this will make a simple Lie algebra.

10.1 Particle in 3-D

10.1.1 Energy vs. Angular Momentum

How is energy different from angular momentum? Energy has infinitely many (positive) levels. Angular momentum, on the other hand, has only a few: positive ones, mirrored by their negative counterparts. Why? Because energy is unbounded: in theory, it can reach very high levels: as high as you like. (In practice, however, they are not very useful: highly unlikely, unstable, and uncertain.) In angular momentum, there are no high levels at all, but only a few low levels (positive or negative). Let's use our algebraic tools to prove this.

10.1.2 Wave Function

So far, our particle was confined to the x-axis. This was quite theoretical, and easy to model. It is time to move on to the real 3-D world. For this, we need a new wave function, defined in 3-D:

$$v : \mathbb{R}^3 \to \mathbb{C}.$$

Assume that v was already normalized in 3-D:

$$\|v\|^2 \equiv (v, v) \equiv \int_{-\infty}^{\infty} \int_{-\infty}^{\infty} \int_{-\infty}^{\infty} v\bar{v} dxdydz = \int_{-\infty}^{\infty} \int_{-\infty}^{\infty} \int_{-\infty}^{\infty} |v|^2 dxdydz = 1.$$

This way, the probability density is $|v|^2$. How likely is the particle to be in a given volume $V \subset \mathbb{R}^3$? The probability for this is

$$\int\int\int_V |v|^2 dxdydz.$$

Let's define new operators, each acting in one Cartesian direction, leaving the two others idle.

10.1.3 Position Operators

How to define position operators in 3-D (to act on functions like v)? X will now be x-position (regardless of y or z). Likewise, Y will multiply by y, and Z by z:

$$X = x\cdot, \ Y = y\cdot, \quad \text{and} \quad Z = z\cdot.$$

Clearly, they commute. Let's define momentum operators too.

10.1.4 Momentum Operators

How to define momentum operators? Well, P_x will be x-momentum (regardless of y or z, just like our good old 1-D P). Likewise, P_y will be y-momentum, and P_z z-momentum:

$$P_x \equiv -i\bar{h}\frac{\partial}{\partial x}, \ P_y \equiv -i\bar{h}\frac{\partial}{\partial y}, \quad \text{and} \quad P_z \equiv -i\bar{h}\frac{\partial}{\partial z}.$$

10.1.5 Commutator

Clearly, they commute. Only three pairs don't:

$$[X, P_x] = [Y, P_y] = [Z, P_z] = i\bar{h}I$$

(where I is the identity operator in 3-D). Let's use this.

10.2 Angular Momentum

10.2.1 Angular Momentum Component

In quantum mechanics, angular momentum is nondeterministic: three random variables (or observables, or measurables). (This will mirror the good old deterministic angular momentum in classical mechanics.) It will contain three (3-D) operators:

$$\begin{aligned} L_x &\equiv YP_z - ZP_y \\ L_y &\equiv ZP_x - XP_z \\ L_z &\equiv XP_y - YP_x. \end{aligned}$$

This is cyclic: the right-hand x-y-z coordinates shift cyclically. Now, look at these operators. Are they Hermitian? Thanks to commutativity, they sure are. For instance,

$$L_x^* = (YP_z - ZP_y)^* = P_z^* Y^* - P_y^* Z^* = P_z Y - P_y Z = YP_z - ZP_y = L_x$$

(Section 10.1.5).

10.2.2 The Commutator

By now, we have three new operators: our new angular-momentum components. Do they commute? No! On the contrary: they have a nonzero commutator:

$$\begin{aligned} [L_x, L_y] &= [YP_z - ZP_y, ZP_x - XP_z] \\ &= [YP_z, ZP_x] - [YP_z, XP_z] - [ZP_y, ZP_x] + [ZP_y, XP_z] \\ &= YP_x [P_z, Z] - YX [P_z, P_z] - P_y P_x [Z, Z] + P_y X [Z, P_z] \\ &= -i\bar{h} YP_x - 0 - 0 + i\bar{h} XP_y \\ &= i\bar{h} L_z. \end{aligned}$$

10.2.3 An Uncertainty Principle

What does this mean physically? They have their own uncertainty principle: assume that the particle is at a particular state. If you know L_x, then you can't know L_y any more (and vice versa). This is cyclic:

$$\begin{aligned} [L_x, L_y] &= i\bar{h} L_z \\ [L_y, L_z] &= i\bar{h} L_x \\ [L_z, L_x] &= i h L_y. \end{aligned}$$

Thus, our new uncertainty principle is cyclic too: assume that the particle is at a particular state. If you know one angular-momentum component, then you can't know the other two.

10.2.4 Up the Ladder

Let's use our formula to raise an eigenvalue. For example, let u be an eigenvector of L_z:

$$L_z u = \lambda u.$$

(Since L_z is Hermitian, λ must be real.) How to raise λ? Define a new operator:

$$T \equiv L_x + iL_y.$$

This way, T is not Hermitian, and doesn't commute with L_z. On the contrary: they have a nonzero commutator:

$$\begin{aligned}[L_z, T] &= [L_z, L_x + iL_y] \\ &= [L_z, L_x] + i\,[L_z, L_y] \\ &= i\bar{h}L_y + \bar{h}L_x \\ &= \bar{h}T.\end{aligned}$$

Thanks to Section 9.8.1, we can now raise λ: if $Tu \neq \mathbf{0}$, then it is an eigenvector of L_z in its own right, with eigenvalue $\lambda + \bar{h}$. This way, T serves as a new ladder operator, to help "climb" up. This is why T is often denoted by

$$A^+ \equiv T.$$

This can be done time and again, producing bigger and bigger eigenvalues, until hitting the zero vector, and reaching the maximal eigenvalue of L_z. This is quantum mechanics for you: angular momentum is not continuous, but can take only a few discrete values.

10.2.5 Down the Ladder

This also works the other way around: not only to raise an eigenvalue, but also to lower it. For this purpose, redefine T (this time with a minus sign):

$$T \equiv L_x - iL_y.$$

(This new T is the Hermitian adjoint of the old one.) As before, our new T doesn't commute with L_z, but has a nonzero commutator with it:

$$\begin{aligned}[L_z, T] &= [L_z, L_x - iL_y] \\ &= [L_z, L_x] - i\,[L_z, L_y] \\ &= i\bar{h}L_y - \bar{h}L_x \\ &= -\bar{h}T.\end{aligned}$$

Thanks to Section 9.8.1, we can now lower λ: if $Tu \neq \mathbf{0}$, then it is an eigenvector of L_z in its own right, with eigenvalue $\lambda - \bar{h}$. This way, our new T serves as a new ladder operator, to help go down the ladder. This is why our new T is often denoted by

$$A^- \equiv T.$$

Furthermore, this can be repeated time and again, producing lower and lower eigenvalues, until hitting the zero vector, and reaching the minimal eigenvalue of L_z.

So far, we worked with L_z. But there is nothing special about it: the same can be done with L_x and L_y as well. So, we have three observables, ready to serve as components in our new (quantum-mechanical) angular momentum.

10.2.6 Angular Momentum

Let's plug these operators as blocks in a new (rectangular) operator:

$$L \equiv \begin{pmatrix} L_x \\ L_y \\ L_z \end{pmatrix}.$$

This is our new (nondeterministic) angular momentum. It mirrors the good old (deterministic) angular momentum in classical mechanics. Physically, however, it is different: it obeys an uncertainty principle. Assume that our particle is at a particular state. At this state, you can only know one angular-momentum component, but not the other two. In the next chapter, we'll see an interesting example: spin.

10.3 Exercises

1. Look at an angular-momentum component (like L_z in Section 10.2.1). What is it? Hint: an operator.
2. On what function can it act? Hint: on a (3-D) function, like our wave function $v : \mathbb{R}^3 \to \mathbb{C}$ (Section 10.1.2).
3. Is it Hermitian?
4. Is it a legitimate observable?
5. What can you say about its eigenvalues? Hint: they are real.
6. What can you say about its eigenvectors? Hint: if they have different eigenvalues, then they are orthogonal.
7. And what if they share the same eigenvalue (but are still linearly-independent)? How to orthogonalize them? Hint: use a Gram-Schmidt process (Section 9.11.1).
8. Normalize them, so their square has integral 1 (in 3-D).
9. What are they now? Hint: orthonormal.
10. Consider some eigenvalue of L_z. Look at its eigenvector. By now, how does it look like? Hint: this is an eigenfunction, from $\mathbb{R}^3 \to \mathbb{C}$. Besides, it is already normal: its square has integral 1 (in 3-D).
11. Consider a state: a wave function $v : \mathbb{R}^3 \to \mathbb{C}$. Assume that it is normal too (its square has integral 1 in 3-D). What is the probability density (in 3-D)? Hint: $|v|^2$.
12. Consider a volume $V \subset \mathbb{R}^3$. How likely is the particle to be there? Hint: the probability for this is
$$\int\int\int_V |v|^2 dxdydz.$$
13. And how likely is the z-angular momentum to be the above eigenvalue? Hint: look at these two functions: the eigenfunction (corresponding to this eigenvalue), and v. By now, they are both normal (in 3-D). Take their inner product (in 3-D), take its absolute value, square it up, and normalize (over all eigenvectors). This is the required probability.
14. Does L_z have eigenvalue zero? Hint: look at the constant function (or any other function invariant under interchanging the x- and y-coordinates: $x \leftrightarrow y$).
15. Is L_z singular? Hint: yes: it has a nontrivial null space.
16. Next, look at a positive eigenvalue of L_z, and its eigenfunction.
17. How to modify it, and design a new eigenfunction, with the negative eigenvalue? Hint: interchange the x- and y-coordinates: $x \leftrightarrow y$. This way, the eigenvalue picks a minus sign.
18. Show that L_z may have discrete eigenvalues of the form
$$0,\ \pm\bar{h},\ \pm 2\bar{h},\ \pm 3\bar{h},\ \ldots$$
(a finite list). Hint: see Sections 10.2.4–10.2.5.

19. Show that L_z may also have eigenvalues of the form

$$\pm\frac{\hbar}{2},\ \pm\frac{3\hbar}{2},\ \pm\frac{5\hbar}{2},\ \pm\frac{7\hbar}{2},\ \ldots$$

(a finite list). Hint: use symmetry considerations to make sure that, in this (finite) list, the minimal and maximal eigenvalues have the same absolute value.

11

Spin and Pauli Matrices

In quantum mechanics, both energy and angular momentum are discrete: they come in packets, or quanta, or levels. In fact, angular momentum takes only a few (positive or negative) values, mirroring one another. The most elementary example is spin. It takes just two possible values: up or down. Geometrically, how do they look like? Spin-up means spinning counterclockwise, and spin-down clockwise.

But this is not the only spin. In our 3-D world, there are three: spin-up-or-down, right-or-left, and in-or-out. They are isomorphic: share the same algebraic structure. But they don't commute, and have a new uncertainty principle: for a particle at a given state, you can only know one, but not the other two.

Note that this is about spin, not about position at all. The particle can keep moving forward, while spinning in any direction. If you like, this is just like the Earth: orbiting the sun, while spinning around its own axis.

So far, we talked about the three basic spins: spin-up, spin-right, and spin-in. From these, you can now design a new spin, in any spatial direction in 3-D. This is nondeterministic as well: an uncertain random variable too. For example, spin-up-or-down has a (discrete) wave function $v \equiv (v_0, v_1) \in \mathbb{C}^2$. How likely is it to spin-up? $|v_0|^2/(|v_0|^2 + |v_1|^2)$. And how likely is it to spin-down? $|v_1|^2/(|v_0|^2 + |v_1|^2)$.

How to model our spins mathematically? For this, we have an algebraic tool: Pauli matrices. Like our spins, they don't commute. On the contrary: they even anti-commute. Later on, we'll see how they span a simple Lie algebra.

This chapter is written in a new style: questions, followed by their answers. This way, you can get more active, and discover physics on your own, before looking at the answers. Later on, we'll have a few more chapters like this.

11.1 Spin Matrices

11.1.1 Spin-One

1. Define three new 3×3 matrices:

$$S_x \equiv i\bar{h} \begin{pmatrix} 0 & 0 & 0 \\ 0 & 0 & -1 \\ 0 & 1 & 0 \end{pmatrix}$$

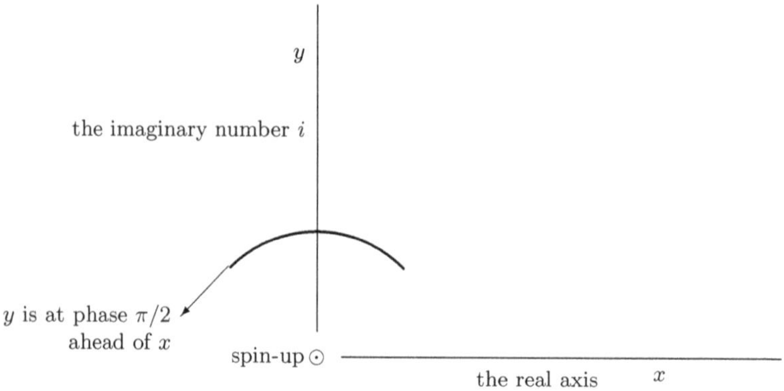

Fig. 11.1. The x-y plane: a view from above. The spin vector points in the positive z-direction: from the page towards your eyes (as indicated by '⊙' at the origin). Around it, the spin is counterclockwise, as in the right-hand rule. This is spin-up, represented by the eigenvector $(1, i, 0)^t/\sqrt{2}$ (y is at phase $\pi/2$ ahead of x).

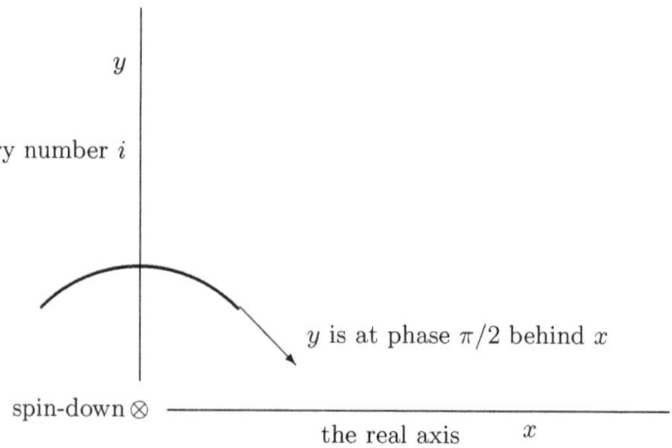

Fig. 11.2. In spin-down, on the other hand, the spin vector points down: into the page, away from your eyes (as indicated by '⊗' at the origin). Around it, the spin is counterclockwise again, as in the right-hand rule. This is represented by the other eigenvector: $(1, -i, 0)^t/\sqrt{2}$ (y is at phase $\pi/2$ behind x).

$$S_y \equiv i\bar{h} \begin{pmatrix} 0 & 0 & 1 \\ 0 & 0 & 0 \\ -1 & 0 & 0 \end{pmatrix}$$

$$S_z \equiv i\bar{h} \begin{pmatrix} 0 & -1 & 0 \\ 1 & 0 & 0 \\ 0 & 0 & 0 \end{pmatrix}$$

(where $i \equiv \sqrt{-1}$ is the imaginary number).

2. These are called spin matrices.
3. Show that these definitions are cyclic under the (right-hand) shift:

$$x \to y \to z \to x.$$

4. Are these matrices Hermitian? Hint: yes, thanks to their imaginary coefficient.
5. Are they legitimate observables?
6. What do they observe? Hint: they will observe the boson spin. S_x will observe spin-right, S_y spin-in, and S_z spin-up (Figures 11.1-11.2).
7. How? Hint: this comes from their geometrical meaning (see below).
8. What is their geometrical meaning? Hint: rotations (making a Lie group): S_x means $y \to z$, S_y $z \to x$, and S_z $x \to y$.
9. Compose the first with the second. What do you get? Hint: $y \to z \to x$, or simply $y \to x$, which is the third (but the other way around).
10. What is this algebraically? Hint: the commutator (see below).
11. Calculate their commutator. Hint:

$$\begin{aligned}
[S_x, S_y] &= i\hbar S_z \\
[S_y, S_z] &= i\hbar S_x \\
[S_z, S_x] &= i\hbar S_y.
\end{aligned}$$

12. Are these cyclic too?
13. Are these relations familiar? Hint: they mirror the formulas in Section 10.2.2.
14. Conclude that the spin matrices represent angular-momentum components.
15. Focus on S_z. What are its eigenvectors and eigenvalues?
16. Look at $(1, i, 0)^t$. Is it an eigenvector? Hint: yes (eigenvalue: $\hbar$).
17. What is its geometrical meaning?
18. Interpret it to point in the positive z-direction. Hint: its y-component is at phase 90° ahead of its x-component (Figure 11.1). Now, follow the right-hand rule: place your right hand with your thumb pointing in the positive x-direction, and your index finger in the positive y-direction. This way, your middle finger will point in the positive z-direction.
19. Next, look at $(1, -i, 0)^t$. Is it an eigenvector too? Hint: yes (eigenvalue: $-\hbar$).
20. What is its geometrical meaning?
21. Interpret it to point in the negative z-direction (Figure 11.2).
22. Normalize these eigenvectors to have norm 1 in $\mathbb{C}^3$. Hint: divide them by $\sqrt{2}$.
23. Next, look at $(0, 0, 1)^t$.
24. Is it an eigenvector too? Hint: yes (eigenvalue: 0).
25. Conclude that it is in the null space.
26. Conclude that S_z is singular.
27. Are these eigenvectors orthogonal? Hint: they are eigenvectors of a Hermitian matrix (of distinct eigenvalues).
28. Confirm this explicitly. Hint: calculate their inner product in $\mathbb{C}^3$. (Don't forget the complex conjugate!)
29. Are they orthonormal? Hint: they were already normalized.
30. Is this as in linear algebra? Hint: a Hermitian matrix indeed has real eigenvalues and orthogonal eigenvectors.
31. Is this also as expected from Sections 10.2.4–10.2.5? Hint: yes – an eigenvalue can be raised or lowered by $\hbar$.

11.1.2 The Boson Spin

1. Consider a new physical system: a boson. This is an elementary particle, with a new property: spin. This is a strange kind of angular momentum: it has a direction, but no value. As such, it is more mathematical than physical.
2. Interpret S_z as a new observable: spin around the z-axis. This is a strange kind of angular momentum: a new random variable, with only three possible results: either counterclockwise (spin-up), or clockwise (spin-down), or none (spin-zero).
3. How likely is the boson to spin up? Hint: take the state $v \in \mathbb{C}^3$, normalize it to have norm 1, calculate its inner product with $(1, i, 0)^t/\sqrt{2}$, take the absolute value, and square it up. This is the probability to spin up:

$$\frac{1}{2}\left|((1, i, 0)^t, v)\right|^2.$$

4. And how likely is the boson to spin down? Hint: do the same with $(1, -i, 0)^t/\sqrt{2}$.
5. And how likely is the boson to have spin-zero? Hint: do the same with $(0, 0, 1)^t$. The result is $|v_3|^2$.
6. Do the same with S_x as well. This makes a new observable: spin-right (or left), around the x-axis.
7. Do the same with S_y as well. This makes a new observable: spin-in (or out), around the y-axis.

11.2 Pauli Matrices

11.2.1 Spin-One-Half

1. This is called spin-one, because the maximal eigenvalue is 1 (times $\hbar$).
2. Consider now a new physical system: a fermion. (For example, an electron or a proton or a neutron.)
3. This is a simpler particle, with a simpler spin, called spin-one-half, because its maximal eigenvalue is 1/2 (times $\hbar$).
4. To model it, use a new state, of a lower dimension: $v \in \mathbb{C}^2$.
5. Normalize it to have norm 1 in $\mathbb{C}^2$.
6. To model this new spin, define the Pauli matrices:

$$\sigma_z \equiv \begin{pmatrix} 1 & 0 \\ 0 & -1 \end{pmatrix}$$
$$\sigma_x \equiv \begin{pmatrix} 0 & 1 \\ 1 & 0 \end{pmatrix}$$
$$\sigma_y \equiv \begin{pmatrix} 0 & -i \\ i & 0 \end{pmatrix}$$

   (where $i = \sqrt{-1}$).
7. Are these Hermitian?
8. Are they legitimate observables?
9. What is their trace (sum of main-diagonal elements)? Hint: zero.
10. In other words, they are trace-free (or traceless).

11. What is their determinant? Hint: -1.
12. What is their square? Hint: the 2×2 identity matrix.
13. Show that they are cyclic (in their product):

$$\sigma_x \sigma_y = i\sigma_z$$
$$\sigma_y \sigma_z = i\sigma_x$$
$$\sigma_z \sigma_x = i\sigma_y.$$

14. Conclude that they have an anti-Hermitian product.
15. Conclude that they anti-commute:

$$\sigma_x \sigma_y + \sigma_y \sigma_x = \sigma_y \sigma_z + \sigma_z \sigma_y = \sigma_z \sigma_x + \sigma_x \sigma_z = (0)$$

(the 2×2 zero matrix).

11.2.2 New Spin Matrices

1. Multiply the Pauli matrices by $\bar{h}/2$, and define new spin matrices:

$$\tilde{S}_z \equiv \frac{\bar{h}}{2}\sigma_z$$
$$\tilde{S}_x \equiv \frac{\bar{h}}{2}\sigma_x$$
$$\tilde{S}_y \equiv \frac{\bar{h}}{2}\sigma_y.$$

2. These are the new spin matrices (of a fermion, not a boson).
3. Calculate their commutator, and show that it is cyclic as well. Hint: recall that the original Pauli matrices anti-commute, and are cyclic (in their product). Thus,

$$\left[\tilde{S}_z, \tilde{S}_x\right] = i\bar{h}\tilde{S}_y$$
$$\left[\tilde{S}_x, \tilde{S}_y\right] = i\bar{h}\tilde{S}_z$$
$$\left[\tilde{S}_y, \tilde{S}_z\right] = i\bar{h}\tilde{S}_x.$$

4. Conclude that the new spin matrices mirror the angular-momentum components (Section 10.2.2).
5. What is their trace? Hint: zero. This will be useful below.
6. What is their determinant? Hint: $-\bar{h}/4$. This will be useful below.
7. Look at $\tilde{S}_z$. What are its eigenvectors (and eigenvalues)? Hint: the standard unit vectors: $(1,0)^t$ has eigenvalue $\bar{h}/2$, and $(0,1)^t$ eigenvalue $-\bar{h}/2$.
8. Show that σ_x flips the eigenvectors.
9. Are the eigenvectors orthonormal?
10. Why? Hint: this is a Hermitian matrix.
11. Are the eigenvalues real?
12. What is their sum? Is it indeed the same as the trace?
13. What is their product? Is it indeed the same as the determinant?

14. Next, look at $\tilde{S}_x$. What are its eigenvectors and eigenvalues? Hint: $(1,1)^t$ has eigenvalue $\bar{h}/2$, and $(1,-1)^t$ eigenvalue $-\bar{h}/2$.
15. Note that the eigenvectors differ by sign only.
16. Conclude that σ_z flips them.
17. Are the eigenvectors orthogonal?
18. Why? Hint: this is a Hermitian matrix.
19. Normalize them to have norm 1. Hint: divide them by $\sqrt{2}$.
20. Are the eigenvalues indeed real?
21. What is their sum? Is it the same as the trace?
22. What is their product? Is it the same as the determinant?
23. Are the eigenvalues as in Sections 10.2.4–10.2.5? Hint: yes – an eigenvalue can be raised or lowered by $\bar{h}$.
24. Next, look at $\tilde{S}_y$. What are its eigenvectors? Hint: $(1,i)^t$ and $(1,-i)^t$.
25. Note that they differ by sign only.
26. Conclude that σ_z flips them.
27. Are they indeed orthogonal? Hint: calculate their inner product in $\mathbb{C}^2$. (Don't forget the complex conjugate!)
28. Normalize them to have norm 1. Hint: divide them by $\sqrt{2}$.
29. What are their eigenvalues? Hint: $\pm\bar{h}/2$.
30. What is their sum? Is it indeed the same as the trace?
31. What is their product? Is it indeed the same as the determinant?
32. As in spin-one, interpret these eigenvectors to model spin. Hint: see Figures 11.1–11.2.
33. Look at the new spin matrices again. Square them up. What do you get? Hint:
$$\tilde{S}_z^2 = \tilde{S}_x^2 = \tilde{S}_y^2 = \frac{\bar{h}^2}{4} I$$
(where I is the 2×2 identity matrix).
34. Does this agree with the eigenvalues calculated above? Hint: take an eigenvector, and apply the matrix to it twice.

11.2.3 Pauli's Exclusion Principle

1. Consider now both spin and position at the same time. How should the state look like? Hint: it should have a vector wave function, containing two wave functions as components:
$$v \equiv (v_0(x,y,z), v_1(x,y,z))^t,$$
where $v_0 : \mathbb{R}^3 \to \mathbb{C}$ is the wave function if spin-up, and $v_1 : \mathbb{R}^3 \to \mathbb{C}$ is the wave function if spin-down.
2. What is the squared norm of the vector wave function? Hint:
$$\begin{aligned} \|v\|^2 &\equiv (v,v) \\ &\equiv (v_0,v_0) + (v_1,v_1) \\ &= \|v_0\|^2 + \|v_1\|^2 \\ &= \int_{-\infty}^{\infty}\int_{-\infty}^{\infty}\int_{-\infty}^{\infty} \left(|v_0|^2 + |v_1|^2\right) dxdydz. \end{aligned}$$

3. Normalize v to have norm 1.
4. What is the probability to spin up? Hint: $\|v_0\|^2$.
5. And what is the probability to spin down? Hint: $\|v_1\|^2$.
6. Do these two sum to 1? Hint: yes: thanks to normalization, $\|v\|^2 = 1$.
7. Consider two volumes: $V_0 \subset \mathbb{R}^3$ and $V_1 \subset \mathbb{R}^3$. What is the probability to be in V_0 with spin-up or in V_1 with spin-down? Hint:

$$\int\int\int_{V_0} |v_0|^2 dxdydz + \int\int\int_{V_1} |v_1|^2 dxdydz.$$

8. Consider a bigger system: two indistinguishable fermions. Could they share the same vector wave function (same wave function if spin-up, and also same wave function if spin-down)? Hint: no! This is Pauli's exclusion principle.

11.2.4 Towards Polarization

1. The Pauli matrices can help model not only spin-one-half but also a new physical property, in a new physical system: a photon. (See Section 12.2 below.)
2. The photon is a boson. As such, it has a state in $\mathbb{C}^3$ to model its spin-one.
3. This state gives rise to a yet simpler state in $\mathbb{C}^2$, to model another interesting property: polarization.
4. Indeed, the photon is not only a particle but also light: an electromagnetic wave. As such, it travels in some direction: say, upwards (in the positive z-direction). At the same time, it also oscillates (and even draws an ellipse) in the horizontal x-y plane.
5. To help model this, the Pauli matrices can also serve as new observables. Better yet, use their spin matrices. For example, $\tilde{S}_z$ tells us how likely the photon is to oscillate in the x- or y-direction. In a (normalized) state $v \equiv (v_1, v_2)^t$, the probability to oscillate in the x-direction is $|v_1|^2$, and in the y-direction is $|v_2|^2$.
6. How is this related to the eigenvectors of $\tilde{S}_z$? Hint: the inner product of v with $(1,0)^t$ is v_1, and with $(0,1)^t$ is v_2.
7. At the same time, $\tilde{S}_x$ tells us how likely the photon is to oscillate obliquely, at angle 45° in between the x- and y-axes. To calculate the probability for this, take the eigenvector $(1,1)^t/\sqrt{2}$, calculate its inner product with v, take the absolute value, and square it up.
8. Write the result as a formula in terms of v_1 and v_2.
9. And what is the probability to oscillate in the perpendicular direction? Hint: do the same with the perpendicular eigenvector: $(1,-1)^t/\sqrt{2}$.
10. Finally, $\tilde{S}_y$ tells us how likely the photon is to draw circles in the x-y plane (Figures 11.1–11.2). To calculate the probability to make circles counterclockwise, take the eigenvector $(1,i)^t/\sqrt{2}$, calculate its inner product with v in $\mathbb{C}^2$ (don't forget the complex conjugate), take its absolute value, and square it up.
11. Write the result as a formula in terms of v_1 and v_2.
12. To calculate the probability to make circles clockwise, on the other hand, do the same with the orthonormal eigenvector: $(1,-i)^t/\sqrt{2}$.

12

Spin and Polarization

How was quantum mechanics discovered in the first place? We already saw a major experiment: the double-slit experiment (Figure 9.1). But there was another one: the Stern-Garlech experiment. It looked at an electron, and how it behaves in a magnetic field. It used the fact that, while flying forward, the electron also "spins" around its own axis. This way, it actually has two kinds of motion at the same time. If you like, this is just like the Earth: orbiting the sun, while spinning around its own axis at the same time. But it turns out that the electron is a little different.

As a matter of fact, the electron is a little magnet: a magnetic dipole. In classical physics, it tends to accept (and adopt) the global magnetic field around it. If it spins parallel to the outer field, then it tends to stay this way. If, on the other hand, it spins anti-parallel, then it tends to flip. For example, if your magnetic field points up (in the positive z-direction), would a random electron flip up too? Only if hit by a boson (Figure 25.2). But here, we're only interested in its initial state: before entering the field, and before having time to get affected. How did it spin originally? Up or down? To find out, throw it into a nonuniform field.

The results were quite surprising. To explain them, we need quantum mechanics, in its algebraic face: Pauli matrices. In a photon, they will help explain not only spin but also polarization.

How to see spin? Well, we can't see it directly. Fortunately, spin leads to yet another interesting phenomenon: polarization of light. To see it, use a polarizer. Later on, we'll model this algebraically, in a Lie algebra. But here, we design a geometrical model: not only classical but also quantum-mechanical.

As in spin, in polarization too, we have three measurables: two linear (horizontal and vertical), and one circular. As in spin, they don't commute. For this reason, at a given state, we can only know one, but not the other two. This is the uncertainty principle.

Later on, we'll have a more powerful tool: Lie algebras. This will help model even a general polarizer: linear or circular or even both! Likewise, in an electron too, we can even model a general spin: not only up-or-down but also in any other direction in 3-D.

This is just an example of a general quantum-mechanical system. We already know the rules:

- The dynamics follows the Schrodinger equation.

- The eigenvectors of the Hamiltonian make energy levels (on which energy is conserved).
- An operator that commutes with the Hamiltonian models a conserved property too.

In our case, spin has magnetic energy (whose Hamiltonian is just a Pauli matrix). This is why only one spin is conserved (the one that is parallel to the magnetic field), but not the other two.

12.1 The Stern-Garlech Experiment

12.1.1 Potential in an Electrostatic Field

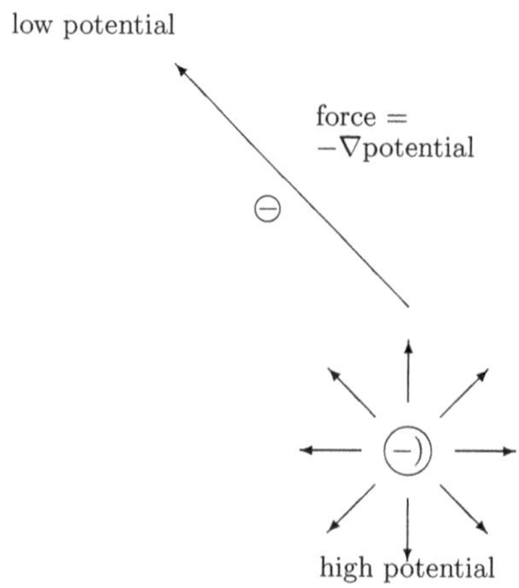

Fig. 12.1. The charged ball (at the middle) produces a radial electrostatic field: high potential near the middle, and low potential far away. In this field, an electron travels to the low-potential region, to make its (positive) energy minimal.

Let's start from classical electromagnetics, and highlight its shortcoming. For this, consider a static charged ball (at the middle of Figure 12.1). It produces an electrostatic field, whose flux is illustrated as outgoing arrows.

This is the force that the ball would exert on an electron. This comes from the ball only: there is no other source. This is why, farther and farther away from the ball, the arrows diverge, and spread out, more and more.

How to quantify this? Consider a big sphere, of radius $r > 0$. What is the total flux through it? Thanks to Green's divergence theorem, this is

$$\int\int_{\text{sphere}} \text{field} \cdot \mathbf{n}\, dxdy = \int\int\int \nabla \cdot \text{field}\, dxdydz = \text{charge of ball}$$

(where $\mathbf{n}$ is the outer normal vector). As a matter of fact, this is true for any closed surface around the ball. Why did we pick a sphere? Because we know its area: $4\pi r^2$. Thanks to symmetry, the flux per unit area (and indeed the field itself) behave like r^{-2}. Fortunately, the field is conservative: has no curl at all. Thus, it has a potential: its line integral (times -1). This behaves like r^{-1} (plus some constant): maximal near the ball, and monotonically decreasing farther and farther away. How would this affect an electron?

12.1.2 Force: Towards Minimal Energy

Into this field, throw an electron. What force would it feel? Well,

$$\text{force} = -\nabla \text{potential}.$$

This is radial: outwards. Better yet, look at things in terms of energy: an electron has a positive energy that agrees with the field. After all, it could "sit" on an arrow, and take a "free ride" to $r = \infty$. It has a positive potential energy: capacity to do work on its way.

Thus, the force could actually be interpreted in terms of energy: the electron tends to travel to a low-potential region:

$$r \to \infty, \quad \text{and} \quad r^{-1} \to 0.$$

A proton, on the other hand, behaves the other way around: it has a negative energy that disagrees with the field. This is why it wants to approach the ball, and even "kiss" it hello. After all, this is where its negative energy $-r^{-1}$ would get minimal. Let's use the same principle in a magnetic field as well.

12.1.3 An Electron in a Nonuniform Magnetic Field

So far, we looked at an electrostatic field. Consider now a magnetic field (Figure 12.2). This time, the arrows point upwards: from the north magnetic pole (at the bottom) to the south magnetic pole (at the top). This is nonuniform: on their way up, the arrows spread out, more and more. As a result, the potential gets lower and lower, all the way up.

Now, into this field, throw two kinds of electrons:

- A spin-up electron aligns with the field. Why? Because it makes a little magnet: a little arrow, whose north head is attracted to the south pole of the global magnetic field. This is why it has a positive magnetic energy. (Indeed, it has a positive eigenvalue with respect to the Hamiltonian, which is proportional to σ_z.) To minimize it, it tends to go up (to the low-potential region). In other words, it "feels" a magnetic force up.
- A spin-down electron, on the other hand, looks the other way around: it has a south head, and a north tail. It disagrees with the field, and has a negative magnetic energy. (Indeed, it has eigenvalue -1 with respect to σ_z.) To minimize it, it tends to go down, to the high-potential region (Figure 12.2). There, its energy will get minimal: negative, but with a big absolute value.

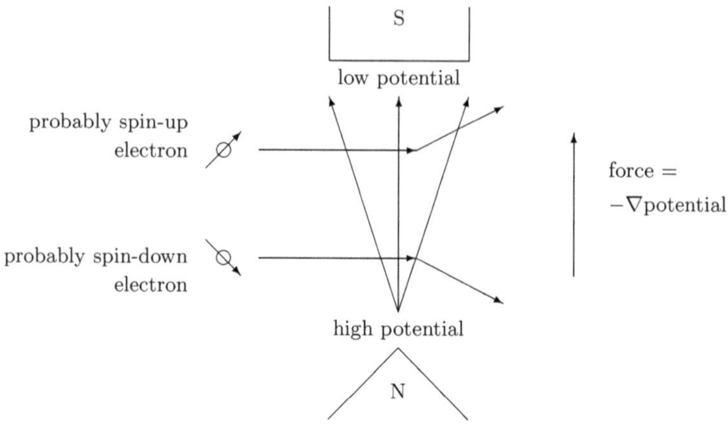

Fig. 12.2. Into a nonuniform magnetic field, electrons are coming from the left. A spin-up electron aligns with the field, and has positive energy. To minimize it, it tends to go up (to the low-potential region). A spin-down electron, on the other hand, has a negative energy. To minimize it, it tends to go down (to the high-potential region).

12.1.4 The Stern-Gerlech Experiment

This is the Stern-Gerlech experiment: into a nonuniform magnetic field, fire a stream of electrons (from the left). How should they go? They should go oblique: either up-right, or down-right, or somewhere in between.

Each electron spins (counterclockwise) around its own spin vector, which points at some angle (away from the z-axis): either $0°$ (spin-up) or $180°$ (spin-down) or in between. So, we expect all kinds: spin-up or spin-down or spin-in-between. On them, the magnetic force should act strongly (on spin-up or spin-down electrons) or weakly (on the rest). Some should go strongly up, some down, and the rest in between. Is this true?

Surprise surprise: no electron goes in between at all! No electron feels a weak magnetic force at all! All electrons feel a strong force, and go strongly up-right or down-right. Why?

This is quantum mechanics for you: discrete, not continuous. Spin can only be up or down, not in between. Algebraically, σ_z has just two eigenvalues: ± 1.

12.1.5 Nonphysical Stochastic Property

Next, look at those electrons that spin up. On them, carry out yet another experiment (rotated by $90°$), to check their spin-right-or-left (their eigenvalue with respect to σ_x). We expect them to be neutral and insensitive. After all, they spin up, don't they?

Surprise surprise: they are sensitive. Half spin right, and half left. Why? Because spin is nonphysical. It is only stochastic, born at measured. Until then, it doesn't "exist" at all! Only then does nature make its mind abut it. Indeed, look at those electrons that spin right, and carry out the original experiment on them. You'll find that only half spin up, but half spin down again!

This is an electron. What about a photon? Let's model it classically, then quantum-mechanically. For this, let's return to our good old active style: questions, followed by their answers.

12.2 Photon: Spin and Polarization

12.2.1 Spin Vector and Its Tail

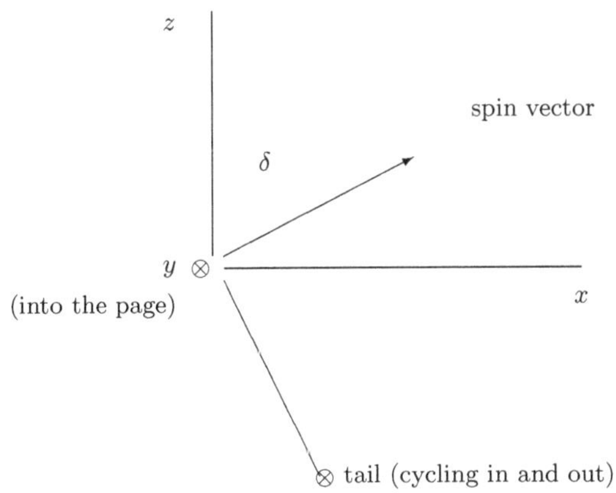

Fig. 12.3. The right-hand x-y-z system: z points upwards, x rightwards, and y into the page (away from your eyes, as indicated by '⊗' at the origin). The spin vector points up-right (at angle δ away from the z-axis). It has a (perpendicular) tail, which cycles around it: into the page, then out again, and so on.

1. Consider a laser beam, pointing upwards (in the positive z-direction). It contains many photons, flying upwards (at the speed of light).
2. On its way, the photon not only flies upwards but also spins around itself (like the Earth).
3. More precisely, it spins around a spin vector, issuing from it (Figure 12.3).

4. If you look from above, what would you see? Hint: a point of light, approaching your eyes from below. (Don't try this at home!) Spin is not about motion: the laser beam keeps coming from the same source, in the same direction: upwards (oscillating on its way, which you can never see).
5. As it propagates upwards, how does the electromagnetic wave oscillate? Hint: first of all, it cycles in the complex plane. This makes it look like a (co)sine in z. But the electromagnetic field is not just a scalar but a 3-D vector, cycling in the perpendicular x-y plane. You can never see this: this is no motion, but only behavior of the electromagnetic field (as a function of z).
6. In general, each photon may have a different spin vector, pointing in a different direction in 3-D.
7. In a polarized light, on the other hand, all photons share the same spin vector, pointing in parallel, at the same direction in 3-D.
8. The spin vector is constant, and always points in the same direction in 3-D, regardless of time or space.
9. Let's start from classical physics (which is inaccurate, but still interesting).
10. Draw the spin vector: from the photon, towards the up-right. (In our laser beam, assume that this is uniform: for all photons alike.)
11. In classical physics, this is real: the photon really spins around the spin vector.
12. Below the spin vector, draw the (horizontal) x-axis. This way, the spin vector is in the vertical z-x plane. (Later on, we'll extend this yet more.)
13. This way, in classical physics, the spin vector has no y-component at all. (In quantum mechanics, on the other hand, this will mean something else: spin-in is as likely as spin-out: see below.)
14. Moreover, the spin vector is at angle δ away from the z-axis.
15. For simplicity, normalize it to have norm 1. Hint: divide it by its own length.
16. In the vertical z-x plane, the spin vector also has a perpendicular "tail," pointing down-right.
17. For simplicity, normalize it to have norm 1 too. Hint: divide it by its own length.
18. But the tail is not static, but dynamic: it cycles around the spin vector. Thus, the tail depends on the time $t \geq 0$ (which is proportional to z), but not on x or y.
19. In Figure 12.3, we only see where the tail pointed initially. But this is going to change.
20. What is the electrical field E? Hint: project the tail onto the horizontal x-y plane, and obtain
$$E \equiv E(t) \equiv (E_1(t), E_2(t), 0)^t .$$
21. This way, E has no third component at all: it points horizontal, perpendicular to the laser beam, as required.

12.2.2 Classical Model

1. What happens at the beginning? Hint: at the beginning, the tail points down-right (and has norm 1, as always). Project it onto the horizontal x-axis. This is E: it points rightwards (in the positive x-direction), and has length
$$\|E\| = |\cos(\delta)|.$$

2. What will happen to E later? Hint: the tail will cycle around the spin vector, and E will be its (new) horizontal projection.
3. What will happen after a quarter of a period? Hint: E will be the same as the tail: both will point in the positive y-direction (into the page, away from your eyes).
4. What will their length be? Hint:

$$\|E\| = 1.$$

5. What will happen after half a period? Hint: the tail will point up-left, at angle δ from the negative part of the x-axis.
6. What will E be? Hint: project the tail onto the negative part of the x-axis.
7. What will its length be? Hint:

$$\|E\| = |\cos(\delta)|$$

again.

8. What will happen after three quarters of a period? Hint: E will be the same as the tail again: both will point in the negative y-direction (from the page towards your eyes).
9. And what will their length be? Hint:

$$\|E\| = 1$$

again, and so on.

10. In the horizontal x-y plane, place a (linear) y-polarizer. (It will let the photon go through only if E aligns with y. Later on, we'll model it algebraically.) In yet another experiment, place an x-polarizer instead. Which one is more likely to let the photon go through? Hint: y-polarization (the former) is more likely. More precisely, it is $1/|\cos(\delta)|$ times as likely. Indeed, look at $\|E\|$.
11. To verify this, look at a few simple examples.
12. What happens at $\delta = 0$ (or π)? Hint: this is spin-up (or down): the spin vector points up (or down, respectively). This way, E coincides with the tail (all the time): both cycle horizontally (and have length 1, all the time), so x-polarization is as likely as y-polarization.
13. Does this agree with our formula? Hint: yes:

$$|\cos(0)| = |\cos(\pi)| = 1.$$

14. On the other hand, what happens at $\delta = \pi/2$? Hint: this is spin-right: the spin vector points rightwards.
15. In this case, how does the tail behave? Hint: it cycles vertically: from the y-axis towards the z-axis: $y \to z$ (as in the right-hand rule).
16. Project it horizontally. How does E look like? Hint: E aligns with y, and oscillates along the y-axis: in and out.
17. This way, E is one-dimensional:

$$E = (0, E_2, 0)^t .$$

18. What is this? Hint: this is linear polarization: only in y, and never in x.
19. Does this agree with our formula? Hint: yes:

$$|\cos(\pi/2)| = 0.$$

20. As a matter of fact, this is correct even without normalization.

12.2.3 Circular Polarization

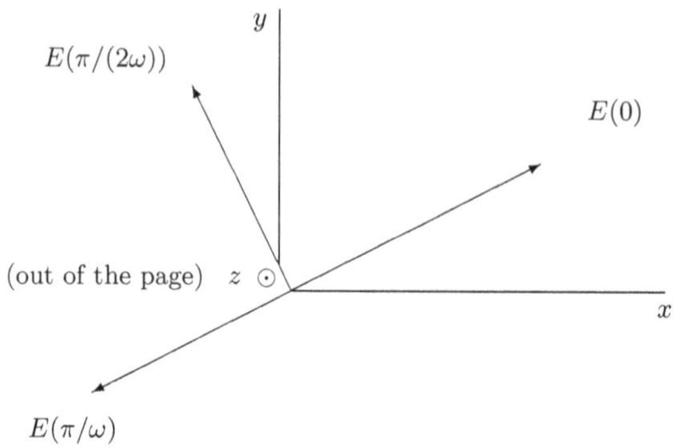

Fig. 12.4. Circular polarization. If you look from above, then you'd "see" E horizontal: (E_1, E_2) rotates counterclockwise: its real part like $E_1 \in \mathbb{C}$, and its imaginary part like $E_2 \in \mathbb{C}$.

1. This was spin-right.
2. Later on, we'll also see spin-in: the spin vector points in the positive y-direction (into the page, away from your eyes). In this case, how does the tail behave? Hint: it cycles vertically: $z \to x$ (as in the right-hand rule).
3. Better yet, multiply by $i = \sqrt{-1}$, to cycle $x \to -z$. (Spin-right plus spin-in will soon give us a new circular polarization: $x \to y$.)
4. Project it horizontally. How will E look like? Hint: E will oscillate along the x-axis: right and left.
5. This way, E will be one-dimensional again:
$$E = (E_1, 0, 0)^t .$$
6. What is this? Hint: this is linear polarization: only in x, and never in y.
7. Add both spin-right and spin-in. What is this physically? Hint: superposition or interference.
8. Mathematically, what do you get? Hint: E is now two-dimensional:
$$E = (E_1, E_2, 0)^t .$$
9. In the horizontal x-y plane, where does E point? Hint: (E_1, E_2).

10. Assume that E_1 is at phase $\pi/2$ ahead of E_2:
$$E_1 = iE_2.$$
11. Thanks to this, how does E look like? In particular, how does its real part look like? Hint:
$$(\Re E_1, \Re E_2) = (\Re E_1, \Im E_1).$$
This is how E_1 looks like in the complex plane (Figure 12.4).
12. And how does its imaginary part look like? Hint:
$$(\Im E_1, \Im E_2) = (\Re E_2, \Im E_2).$$
This is how E_2 looks like in the complex plane.
13. Give an example. Hint: horizontal, round and round (counterclockwise):
$$E(t) = (E_1(0)\exp(i\omega t), E_2(0)\exp(i\omega t), 0)^t.$$
14. Why not $\exp(-i\omega t)$? Hint: because we are looking at the x-y plane from above, not from below.
15. In this example, how to guarantee that $E_1 = iE_2$ (all the time)? Hint: it is sufficient to require that
$$E_1(0) = iE_2(0).$$
16. In this case, how does E look like? Hint:
$$E = E_2(0)\exp(i\omega t)(i, 1, 0)^t.$$
17. What will happen after a quarter of a period, at time $t = \pi/(2\omega)$? Hint: E_2 will be as E_1 was initially:
$$E_2\left(\frac{\pi}{2\omega}\right) = iE_2(0) = E_1(0).$$
E_1, on the other hand, will be as E_2 will be later (after half a period):
$$E_1\left(\frac{\pi}{2\omega}\right) = iE_1(0) = i^2E_2(0) = -E_2(0) = E_2\left(\frac{\pi}{\omega}\right)$$
(Figure 12.4).
18. What is this? Hint: circular polarization, in its general form (counterclockwise).
19. Modify this: after a quarter of a period, stretch $\|E\|$. Then, after half a period, shrink $\|E\|$ back again, and so on.
20. What is this? Hint: this is an elliptic polarization.

12.2.4 Superposition – Interference

1. This was spin-up: $E_1 = iE_2$. In spin-down, on the other hand, E_1 picks a minus sign: $E_1 = -iE_2$. How does the picture look like? Hint: the other way around: clockwise, rather than counterclockwise. (To see this, interchange: $E_1 \leftrightarrow E_2$, and $x \leftrightarrow y$.)

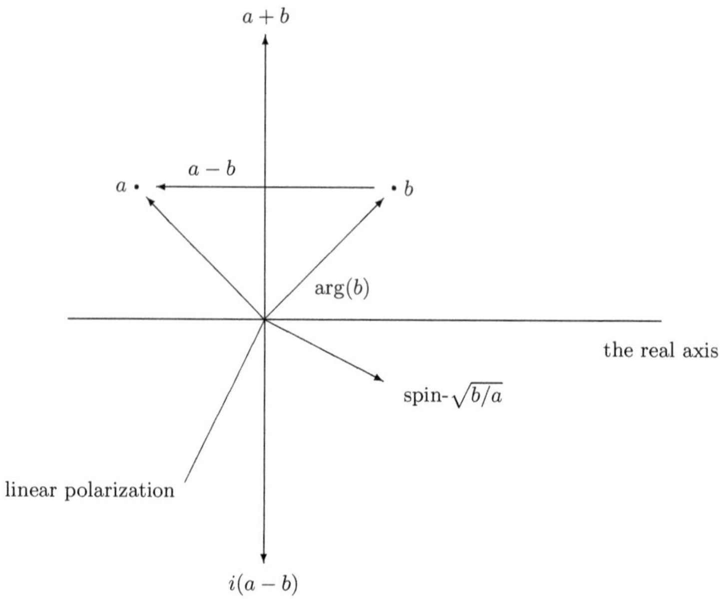

Fig. 12.5. If $a, b \in \mathbb{C}$ and $|a| = |b|$, then $a+b$ is perpendicular to $a-b$. Thus, $a+b$ points away from $i(a-b)$: they have a negative ratio. Thus, in our superposition, the original circular polarizations interfere, and cancel out, leaving spin-$\sqrt{b/a}$, with a perpendicular (linear) polarization.

2. Later on, we'll model these pictures by twin vectors:

$$\begin{pmatrix} 1 \\ \pm i \end{pmatrix} \in \mathbb{C}^2.$$

3. We'll even superpose them (with coefficients $a, b \in \mathbb{C}$), yielding

$$a \begin{pmatrix} 1 \\ i \end{pmatrix} + b \begin{pmatrix} 1 \\ -i \end{pmatrix}.$$

4. For example, assume that a and b have the same magnitude:

$$a \neq \pm b, \quad \text{and} \quad |a| = |b| > 0.$$

5. This way, they differ in phase only: a makes angle $\arg(a)$ with the real axis, and b angle $\arg(b)$.

6. What does this mean physically? Hint: in our superposition, there is a constant phaseshift: one circular polarization is ahead of the other.
7. In Figure 12.5, for example, spin-up is ahead of spin-down.
8. What is the phaseshift between spin-up and spin-down? Hint: in the complex plane, the phaseshift is the angle between a and b:

$$0 \neq \arg(b/a) = \arg(b) - \arg(a) \neq \pm\pi.$$

9. And what is the half-phaseshift? Hint: this is half of this angle:

$$\arg(\sqrt{b/a}) = \frac{\arg(b) - \arg(a)}{2}.$$

10. In the complex plane, look at two points: 1 and b/a. What do they have in common? Hint: both are on the unit circle.
11. Connect them by a straight line. What is its length? Hint: divide this line into two equal parts. For this, issue a new ray from the origin towards $\sqrt{b/a}$. This ray makes angle

$$\arg(\sqrt{b/a}) = \frac{\arg(b) - \arg(a)}{2}$$

with the real axis. Thanks to this,

$$\frac{|a-b|}{|a|} = \frac{|b-a|}{|a|} = \left|\frac{b}{a} - 1\right| = 2\left|\sin\left(\frac{\arg(b) - \arg(a)}{2}\right)\right|.$$

12. Write our superposition yet more explicitly. Hint:

$$a\begin{pmatrix} 1 \\ i \end{pmatrix} + b\begin{pmatrix} 1 \\ -i \end{pmatrix} = \begin{pmatrix} a+b \\ i(a-b) \end{pmatrix}.$$

13. What is the ratio between the top component (that stands for spin x or y-polarization or E_2) and the bottom component (that stands for spin-y or x-polarization or E_1)? Hint: fortunately, this ratio is real (no imaginary part at all). (This means no phaseshift, and no circular polarization any more.) To see this geometrically, look at Figure 12.5. To see this algebraically, divide:

$$\begin{aligned}
\frac{a+b}{i(a-b)} &= -i\frac{a+b}{a-b} \\
&= -i\frac{(a+b)\left(\bar{a}-\bar{b}\right)}{|a-b|^2} \\
&= \frac{-i}{|a-b|^2}\left(a\bar{a} - a\bar{b} + b\bar{a} - b\bar{b}\right) \\
&= \frac{-i}{|a-b|^2}\left(|a|^2 + 2i\Im\left(b\bar{a}\right) - |b|^2\right) \\
&= \frac{2}{|a-b|^2}\Im\left(b\bar{a}\right) \\
&= \frac{2|a|^2}{|a-b|^2}\Im\left(\frac{b}{a}\right)
\end{aligned}$$

$$
\begin{aligned}
&= \frac{2|a|^2}{|a-b|^2}\sin(\arg(b)-\arg(a)) \\
&= \frac{2\sin\left(\frac{\arg(b)-\arg(a)}{2}\right)\cos\left(\frac{\arg(b)-\arg(a)}{2}\right)}{2\sin^2\left(\frac{\arg(b)-\arg(a)}{2}\right)} \\
&= \operatorname{cotan}\left(\frac{\arg(b)-\arg(a)}{2}\right).
\end{aligned}
$$

14. What is so good about this number? Hint: it is real.
15. What does this mean physically? Hint: E_1 is proportional to E_2, with no phaseshift in between at all.
16. How come? Where is the phaseshift? Hint: the phaseshift is only between the original circular polarizations. But they superpose, interfere, cancel out, and leave an oblique linear polarization.
17. In what direction? Hint: in the complex x-y plane, this is spin-$\sqrt{b/a}$ (in between 1 and b/a), with a perpendicular linear polarization: $i\sqrt{b/a}$-polarization. This way, the spin vector is horizontal (with no z-component at all): the polarization has no circular component.
18. How to see this geometrically? Hint: for simplicity, assume that $\arg(a)$ is bigger than $\arg(b)$ (spin-up is ahead of spin-down). For example, look at Figure 12.5: in the complex x-y plane, $a-b$ points leftwards, so $i(a-b)$ points downwards. In general, $a+b$ and $i(a-b)$ point away from one another, and have a negative ratio, as required.
19. So far, a and b were complex numbers of the same magnitude. But this is going to change: we'll soon pick real a and b (that depend on δ), to design a new quantum-mechanical model.
20. So far, we only looked at the electric field E. We saw that it is horizontal: confined to the x-y plane, and perpendicular to the z-axis. What about the magnetic field? Hint: it is horizontal too: perpendicular to both E and the z-axis.
21. And what is its length? Hint: proportional to $\|E\|$.

12.2.5 Quantum-Mechanical Model

1. So far, we only looked at classical electromagnetics. Next, let's move on to quantum mechanics (which is more accurate). For this, let's go back to our original spin vector, and its tail (Figure 12.3).
2. What is more likely: spin-up (so the tail cycles counterclockwise), or spin-down (so the tail cycles clockwise)? Hint: spin-up is more likely. More precisely, it is $\operatorname{cotan}^2(\delta/2)$ times as likely as spin-down.
3. Can you know δ for sure? Hint: no! It is nondeterministic.
4. Still, how to estimate δ quite well? Hint: in the horizontal x-y plane, place a circular polarizer. Repeat the experiment many times, and gather statistics about many (identical) photons. Count how many went through, and how many didn't. Use this to solve for δ.
5. In the horizontal x-y plane, look at a column vector: the top component stands for spin-x (or y-polarization), and the bottom component for spin-y (or x-polarization) (Section 12.2.4).

6. Model circular polarization by two column vectors: model counterclockwise cycling by $(1, i)^t$, and clockwise cycling by $(1, -i)^t$. Hint: this follows from Section 12.2.3.
7. Are they orthogonal? Hint: calculate their inner product in $\mathbb{C}^2$. Don't forget the complex conjugate (say, on top of the latter vector):

$$\left((1,i)^t, (1,-i)^t\right) = (1,i)(1,-i)^* = (1,i)(1,i)^t = 1 \cdot 1 + i \cdot i = 1 - 1 = 0.$$

8. Normalize them to have norm 1. Hint: divide by $\sqrt{2}$.
9. What do you get? Hint: these two column vectors make an orthonormal basis, useful to span $\mathbb{C}^2$.
10. Use them to superpose our original state (obtained from our original spin vector in Figure 12.3). Hint: sum the following column vectors:

$$\frac{1}{\sqrt{2}} \cos\left(\frac{\delta}{2}\right) \begin{pmatrix} 1 \\ i \end{pmatrix} + \frac{1}{\sqrt{2}} \sin\left(\frac{\delta}{2}\right) \begin{pmatrix} 1 \\ -i \end{pmatrix} = \frac{1}{\sqrt{2}} \begin{pmatrix} \cos\left(\frac{\delta}{2}\right) + \sin\left(\frac{\delta}{2}\right) \\ i\cos\left(\frac{\delta}{2}\right) - i\sin\left(\frac{\delta}{2}\right) \end{pmatrix}.$$

11. What does this mean physically? Hint: interference: constructive in x (the top component), and destructive in y (the bottom component).
12. Look at these two components (the upper, and the lower). What is their relative probability? Hint: square them up (in magnitude), and divide:

$$\frac{\left(\cos\left(\frac{\delta}{2}\right) + \sin\left(\frac{\delta}{2}\right)\right)^2}{\left(\cos\left(\frac{\delta}{2}\right) - \sin\left(\frac{\delta}{2}\right)\right)^2} = \frac{1 + \sin(\delta)}{1 - \sin(\delta)}.$$

13. Thanks to this, we are now ready to estimate linear polarization better. What is more likely: spin-x (so E aligns with y), or spin-y (so E aligns with x)? Hint: the former is more likely. More precisely, it is $(1 + \sin(\delta))/(1 - \sin(\delta))$ times as likely as the latter.
14. Design an experiment to verify this. Hint: in the horizontal x-y plane, place a (linear) y-polarizer. In yet another experiment, place an x-polarizer instead. Repeat this many (independent) times, and gather statistics about many (identical) photons.
15. Which one is more likely to let the photon go through? Hint: y-polarization (the former) is more likely. More precisely, it should be $(1+\sin(\delta))/(1-\sin(\delta))$ times as likely as x-polarization.
16. In particular, test a few simple examples.
17. For $\delta = 0$ (or π), both x- and y-polarization should be as likely. Why? Hint: spin is up (or down), so E cycles in the unit circle in the horizontal x-y plane.
18. Does this agree with our new formula? Hint: yes:

$$1 \pm \sin(0) = 1 \pm \sin(\pi) = 1 \pm 0 = 1.$$

19. For $\delta = \pi/2$, on the other hand, things are completely different: only y-polarization is possible. Why? Hint: this is spin-right, so E aligns with y, all the time.
20. Does this agree with our new formula? Hint: yes:

$$1 \pm \sin(\pi/2) = 1 \pm 1.$$

21. For many more δ's, carry out your experiment, and study the results.
22. Do they agree with our new formula better than with the old (classical) one?
23. Do you believe in quantum mechanics now?
24. In Figure 12.3, our spin vector is in the vertical z-x plane, with no y-component at all. How to extend this, and allow it to have a nonzero y-component as well? Hint: rotate it horizontally (by angle α), around the vertical z-axis (into the page, away from your eyes). With it, rotate the entire x-y plane (by angle α as well). For this, apply the orthogonal matrix

$$O \equiv \begin{pmatrix} \cos(\alpha) & -\sin(\alpha) \\ \sin(\alpha) & \cos(\alpha) \end{pmatrix}.$$

Now, in our superposition, substitute $O(1, \pm i)^t$ for $(1, \pm i)^t$, respectively. Then, superpose these new column vectors (with the same coefficients as before). This will give you the new relative probability: how likely y-polarization is, in comparison with x-polarization (our original coordinates).
25. Write this as a function of α (the angle by which O rotates the entire x-y plane).
26. Can you tell α for sure? Hint: no! It is nondeterministic.
27. Still, how to estimate α quite well? Hint: repeat the same experiment many (independent) times. Gather statistics about many (identical) photons, and estimate the relative probability between y- and x-polarization (the original coordinates). But you already have this algebraically (as a function of α). So, you can now solve for α.
28. Verify this for an example, in which spin-down is as likely as spin-up (so $\delta = \pi/2$). Hint: as before: carry out the experiment many times, and gather statistics about y- and x-polarization (the original coordinates). Their relative probability should be

$$\text{cotan}^2(\alpha) = \frac{\cos^2(\alpha)}{\sin^2(\alpha)}.$$

29. Does this agree with your new formula (once you set $\delta = \pi/2$ in it)?

12.3 Back To The Electron

12.3.1 On The Electron Spin

1. This was a photon. What about an electron? What is its kinetic energy? Hint: its kinetic energy is proportional to the inner product of its spin vector with its momentum vector (in 3-D). Indeed, in Dirac's equation, this is the first term.
2. Can these vectors be measured for certain? Hint: the momentum vector can, because its components commute. But the spin vector can't, because its components don't commute. For this reason, only one spin (in one particular direction) can be measured exactly. For example, spin-up and spin-right don't commute, so only one of them can be measured exactly.
3. Still, in the electron too, how to estimate the spin vector quite well? Hint: repeat the same experiment many times, and gather statistics about a lot of (identical) electrons, and about their spin vector (one component at a time), and its expectation.

4. In quantum mechanics, time mirrors position, and energy mirrors momentum. Dirac's equation extends this even to a very high speed: indeed, in special relativity, momentum-energy mirror one another, and form a relativistic pair,

5. This is how special relativity joins forces with quantum mechanics, to form a complete theory for the electron as well.

12.3.2 Magnetic Energy

1. Consider an electron in a uniform magnetic field B, pointing upwards. What is the Hamiltonian? Hint: the Hamiltonian is a product: the strength of the magnetic field, times the spin matrix:

$$H = \|\mathbf{B}\|\tilde{S}_z = \|\mathbf{B}\|\frac{\bar{h}}{2}\sigma_z = \|\mathbf{B}\|\frac{\bar{h}}{2}\begin{pmatrix} 1 & 0 \\ 0 & -1 \end{pmatrix}.$$

2. What are its eigenvectors? Hint: the standard unit vectors: $(1,0)^t$ and $(0,1)^t$.
3. What are their eigenvalues? Hint: $\pm\|\mathbf{B}\|\bar{h}/2$.
4. What is the physical meaning of this? Hint: a spin-up electron agrees with the magnetic field. This is why it has a positive magnetic energy. A spin-down electron, on the other hand, disagrees. This is why it has a negative magnetic energy, and tends to flip. (Compare this with Figure 12.2.)
5. Thanks to the Schrodinger equation (Section 9.7.2), how should the initial state evolve in time? Hint:

$$v \equiv v(t) \equiv \begin{pmatrix} \cos(\alpha)\exp(-i\phi t) \\ \sin(\alpha)\exp(i\phi t) \end{pmatrix}.$$

6. In this solution, what is the phaseshift between the top and bottom components? Hint: in each component, the frequency is energy over $\bar{h}$: $\phi = \|\mathbf{B}\|/2$. But the bottom component also picks a minus sign. Thus, the phaseshift between them is $2\phi t = \|\mathbf{B}\|t$.
7. Is this dynamic? Does the phaseshift change in time?
8. At what frequency does the phaseshift precess? Hint: $2\phi = \|\mathbf{B}\|$.
9. This is Larmor's frequency.
10. Note: this is only about spin, not about position at all. The electron can keep moving forward, while spinning in this way.
11. In this system, what spin is conserved? Hint: only spin-up, because $\tilde{S}_z$ commutes with H (which is proportional to $\tilde{S}_z$). But not spin-right or spin-in (because $\tilde{S}_x$ and $\tilde{S}_y$ don't commute with H).

Part VI

Quantum Chemistry: Electronic Structure

Part VI: Quantum Chemistry: Electronic Structure

So far, we looked at one electron, and studied its spin. But what about many electrons in the atom? How do they look like? After all, they are too small, and no microscope can help see them!

Thanks to classical physics, we already know how they interact. Better yet, thanks to quantum mechanics, we already know more: if they have the same spin, then

- They are indistinguishable, and are attracted by an exchange force.
- They can't share the same wave function. This is Pauli's exclusion principle.

How to model this mathematically-physically? How to look at the electronic structure in the atom, even with no microscope at all? For this, we have an algebraic tool: the determinant. Thanks to it, we can write the Hartree-Fock system: an eigenvalue problem. Its orthogonal eigenvectors will give us the desired solution.

13

Background: Determinant

To model electrons in the atom, we need an algebraic tool: the determinant. In 2-D, we already illustrated it geometrically (Figure 2.11). But here, we design it from scratch, using group theory. This will help design our quantum-mechanical model later on.

For this, let's look at an interesting group: the permutation group. It will help define the determinant in a new way. This will give us a few attractive properties. Later on, in quantum chemistry, this will help analyze the electronic structure in the atom.

13.1 Permutation

13.1.1 Permutation

How to design a permutation? Consider a set of n natural numbers:

$$\{1,\ 2,\ 3,\ \ldots, n\}$$

(for some natural number n). The braces tell us that this is just a set, with no specific order. (The increasing order used here is just a convention.) Now, what is a permutation? This is a one-to-one mapping from this set onto itself. This means that each number $1 \le i \le n$ is mapped to a distinct number $1 \le p(i) \le n$. The entire mapping is denoted by

$$p\left(\{1,\ 2,\ 3,\ \ldots, n\}\right).$$

So far, we looked at a set, and mapped it onto itself. But what about a list (with a specific order)? To denote it, use triangular brackets. In the beginning, look at our set as a list, with an increasing order:

$$< 1, 2, 3, \ldots, n > .$$

This way, the first number is 1. But, after the permutation, what is the (new) first number? This is the number that was mapped to 1, and took its place. Thus, after the permutation, the numbers reorder:

$$< p^{-1}(1),\ p^{-1}(2),\ p^{-1}(3),\ \ldots,\ p^{-1}(n) > .$$

And how to apply yet another permutation on top? For this, you must first rename the numbers in an increasing order back again:

$$< 1,\ 2,\ 3,\ \ldots,\ n > .$$

Only now are you ready to apply the next permutation, and design the next list, and so on.

13.1.2 Switch

For example, the switch

$$(1 \to 3)$$

switches 1 with 3: at the same time, 1 is mapped to 3, 3 is mapped back to 1, and the rest remain the same: 2 is mapped to 2, 4 to 4, and so on. For this reason, the switch is symmetric, and can also be written as

$$(1 \to 3) = (3 \to 1).$$

And what happens to the list? Before the switch, we assume an increasing order, as always:

$$< 1,\ 2,\ 3,\ \ldots,\ n > .$$

After the switch, on the other hand, the numbers reorder:

$$< 3,\ 2,\ 1,\ 4,\ 5,\ \ldots, n > .$$

We also say that the switch is odd: it picks a minus sign. This is denoted by

$$e((1 \to 3)) = -1.$$

13.1.3 Cycle

A cycle, on the other hand, is more complicated. For example,

$$(1 \to 3 \to 2)$$

maps 1 to 3, 3 to 2, and 2 back to 1 (at the same time). This is why the cycle is indeed cyclic, and can also be written as

$$(1 \to 3 \to 2) = (2 \to 1 \to 3) = (3 \to 2 \to 1).$$

And what happens to the list? Before the cycle, we assume an increasing order, as always:

$$< 1,\ 2,\ 3,\ \ldots,\ n > .$$

After the cycle, on the other hand, the new list is

$$< 2,\ 3,\ 1,\ 4,\ 5,\ \ldots, n > .$$

Why? Because the cycle works like this: it maps 1 to 3. To make room, both 3 and 2 must shift leftwards.

13.2 Decomposition

13.2.1 Composition – Product

Our cycle can also be written as a composition (or product) of two switches:

$$(3 \to 2 \to 1) = (3 \to 2)(2 \to 1).$$

This is read leftwards: first, 1 switches with 2, producing

$$< 2,\ 1,\ 3,\ 4,\ 5,\ \ldots, n > .$$

Then, it also switches with 3, producing

$$< 2,\ 3,\ 1,\ 4,\ 5,\ \ldots, n >,$$

as required.

Why is this decomposition useful? Because it tells us that the cycle is even, not odd. Indeed, each switch picks a minus sign:

$$e\left((3 \to 2 \to 1)\right) = e\left((3 \to 2)\right) e\left((2 \to 1)\right) = (-1)(-1) = 1.$$

13.2.2 3-Cycle

Here is a short notation for our cycle:

$$[3 \to 1] \equiv (3 \to 2 \to 1).$$

Let's use this style to say once again that the cycle is even:

$$e([3 \to 1]) = 1.$$

This is also called 3-cycle. The switch, on the other hand, is also called 2-cycle.

13.2.3 4-Cycle

Likewise, we can also write a yet longer cycle: a 4-cycle:

$$[4 \to 1] \equiv (4 \to 3 \to 2 \to 1).$$

This cycle can be decomposed in three switches:

$$[4 \to 1] = (4 \to 3)(3 \to 2)(2 \to 1).$$

Again, this is read leftwards: 1 switches with 2, then with 3, then with 4. The resulting list is

$$< 2,\ 3,\ 4,\ 1,\ 5,\ 6,\ \ldots,\ n >,$$

as required. This is why this cycle is odd, not even:

$$e([4 \to 1]) = -1,$$

and so on.

13.2.4 How to Decompose a Permutation?

Let's decompose a general permutation p. For this purpose, start from 1. How to map it? Map it to some $1 \le k \le n$. This occupies $k = p(1)$. Next, look at the rest: from 2 to n. How to map them? They must map from

$$\{2, 3, 4, \ldots, n\} \to \{1, 2, 3, \ldots, k-1, k+1, k+2, \ldots, n\}.$$

(The braces tell us that these are only sets: the order is not specified here.) How to do this? In two stages:

- First, mix them (using a smaller permutation q):
$$q : \{2, 3, 4, \ldots, n\} \to \{2, 3, 4, \ldots, n\}.$$
(1 is left unchanged.)
- If $2 \le q(i) \le k$, then subtract 1 from it:
$$q(i) \to q(i) - 1.$$

This way, k is not used, as required. In summary, the original permutation was decomposed as

$$p(\{1,\ 2,\ 3,\ \ldots, n\}) = [k \to 1]q(\{2,\ 3,\ 4,\ \ldots, n\}),$$

with a unique (smaller) permutation q that mirrors p. As a matter of fact, q acts on $\{1, 2, 3, \ldots, n\}$, but leaves 1 unchanged, so this can be disregarded.

13.3 Permutations and their Group

13.3.1 Group of Permutations

Look at all permutations on $\{1, 2, 3, \ldots, n\}$. Place them in a new group:

$$P(\{1,\ 2,\ 3,\ \ldots, n\}).$$

What is a group? It is a set G, closed under product: multiply $a, b \in G$, and obtain $ab \in G$. This must be associative, but not necessarily commutative. Besides, the group must contain one identity (or unit) object: I. Each object must have its own inverse: multiply them, and obtain the identity object back again:

$$aa^{-1} = a^{-1}a = I.$$

In our case, the objects are the permutations. This way, the group is indeed closed: compose two permutations, and obtain a new permutation. This is associative, as required. The identity permutation changes nothing. Finally, each individual permutation has its (unique) inverse.

Thanks to Section 13.2.4, the entire group can now be written as a union of smaller groups:

$$P(\{1,\ 2,\ 3,\ \ldots, n\}) = \cup_{k=1}^{n}[k \to 1]P(\{2,\ 3,\ 4,\ \ldots, n\}).$$

Note that, in this union, the k-cycles $[k \to 1]$ have alternating signs: even, odd, even, odd, and so on:

$$e([k \to 1]) = (-1)^{k-1}.$$

This will be useful later.

13.3.2 How Many Permutations?

Let's denote our group by

$$P \equiv P(\{1, 2, 3, \ldots, n\})$$

for short. How many permutations are there in P? Well, in the above union, k could take n possible values: from 1 to n. Thanks to mathematical induction, P must contain $n!$ different permutations:

$$|P| = n!.$$

Half of them are odd, and half are even. This could be proved by mathematical induction on $n \geq 2$.

13.4 Determinant

13.4.1 Determinant

Let's use the permutation group to define the determinant. For this purpose, consider an $n \times n$ (complex) matrix:

$$A \equiv (a_{i,j})_{1 \leq i,j \leq n}\,.$$

Its determinant is a (complex) number: sum of products of elements. Each product multiplies n elements: one from each row, and one from each column:

$$\det(A) \equiv \sum_{p \in P} e(p) a_{1,p(1)} a_{2,p(2)} a_{3,p(3)} \cdots a_{n,p(n)}.$$

13.4.2 Determinant of the Transpose

Look at the transpose matrix. What is its determinant? To find out, mirror each permutation by its unique inverse. After all, if the permutation is even (odd), then its inverse is even (odd) as well. Indeed,

$$1 = e\left(pp^{-1}\right) = e(p) e\left(p^{-1}\right).$$

Thus, instead of scanning all permutations, scan all their inverses, one by one:

$$\begin{aligned}
\det(A) &\equiv \sum_{p \in P} e(p) a_{1,p(1)} a_{2,p(2)} a_{3,p(3)} \cdots a_{n,p(n)} \\
&= \sum_{p^{-1} \in P} e(p) a_{1,p(1)} a_{2,p(2)} a_{3,p(3)} \cdots a_{n,p(n)} \\
&= \sum_{p^{-1} \in P} e\left(p^{-1}\right) a_{p^{-1}(1),1} a_{p^{-1}(2),2} a_{p^{-1}(3),3} \cdots a_{p^{-1}(n),n} \\
&= \sum_{p \in P} e(p)\, a_{p(1),1} a_{p(2),2} a_{p(3),3} \cdots a_{p(n),n} \\
&= \det\left(A^t\right).
\end{aligned}$$

Thus, the transpose has the same determinant. Let's use this to calculate the determinant of a product of two matrices.

13.4.3 Determinant of a Product

Consider now two (complex) matrices of order n:

$$A \equiv (a_{i,j})_{1\le i,j\le n} \quad \text{and} \quad B \equiv (b_{i,j})_{1\le i,j\le n}.$$

What is the determinant of AB? Thanks to our definition, this is

$$\det(AB) \equiv \sum_{p\in P} e(p)(AB)_{1,p(1)}(AB)_{2,p(2)}\cdots(AB)_{n,p(n)}.$$

Inside the sum, we have a product of n factors of the form

$$(AB)_{i,p(i)} \equiv \left(\sum_{j=1}^{n} a_{i,j}b_{j,p(i)}\right).$$

The above product scans over $i = 1,2,3,\ldots,n$, and multiplies these factors, one by one. Upon opening parentheses, one must pick one particular j from each such factor. Which j to pick? There is no point to pick the same j from two different factors (say the kth and lth factors). After all, the resulting product will soon cancel with a similar product, obtained from a permutation of the form

$$p(k \to l)$$

(which picks a minus sign). So, for this particular j, we get the same 4-product (only with a minus sign), which cancel out:

$$\cdots a_{k,j}b_{j,p(k)}a_{l,j}b_{j,p(l)}\cdots - \cdots a_{k,j}b_{j,p(l)}a_{l,j}b_{j,p(k)}\cdots = 0.$$

So, we better focus on a more interesting option: pick a different j from each factor, say

$$j = q(i),$$

for some permutation $q \in P$. This way,

$$\begin{aligned}
&\det(AB)\\
&\equiv \sum_{p,q\in P} e(p)a_{1,q(1)}b_{q(1),p(1)}a_{2,q(2)}b_{q(2),p(2)}\cdots a_{n,q(n)}b_{q(n),p(n)}\\
&= \sum_{p,q^{-1}\in P} e(p)a_{q^{-1}(1),1}b_{1,pq^{-1}(1)}a_{q^{-1}(2),2}b_{2,pq^{-1}(2)}\cdots a_{q^{-1}(n),n}b_{n,pq^{-1}(n)}\\
&= \sum_{p,q\in P} e(p)a_{q(1),1}b_{1,pq(1)}a_{q(2),2}b_{2,pq(2)}\cdots a_{q(n),n}b_{n,pq(n)}\\
&= \sum_{r,q\in P} e(r)e(q)a_{q(1),1}b_{1,r(1)}a_{q(2),2}b_{2,r(2)}\cdots a_{q(n),n}b_{n,r(n)}\\
&= \det\left(A^t\right)\det(B)\\
&= \det(A)\det(B)
\end{aligned}$$

(where $r = pq$ helps change the summation). In summary, the determinant of the product is the product of determinants:

$$\det(AB) = \det(A)\det(B).$$

Let's use this.

13.4.4 Orthogonal Matrix

Let O be an orthogonal matrix. (In particular, O is real.) What is its determinant? It is ± 1. Indeed,

$$1 = \det\left(O^t O\right) = \det\left(O^t\right)\det(O) = (\det(O))^2 .$$

13.4.5 Hermitian Adjoint

Let A be a general complex matrix. Look at its Hermitian adjoint: A^*. What is its determinant? It is the complex conjugate of the original determinant. Indeed,

$$\det\left(A^*\right) = \det\left(\bar{A}^t\right) = \bar{\det}\left(A^t\right) = \bar{\det}(A).$$

13.4.6 Unitary Matrix

Let U be a unitary matrix. (This way, U can even be complex.) What is its determinant? It is a complex number of absolute value 1. Indeed,

$$1 = \det\left(U^* U\right) = \det\left(U^*\right)\det(U) = |\det(U)|^2.$$

This will be useful later.

13.5 The Characteristic Polynomial

13.5.1 The Characteristic Polynomial

What is the characteristic polynomial? This is a polynomial in the independent variable λ:

$$\det(A - \lambda I).$$

Where does it vanish? At its roots: those λ's that are eigenvalues of A. Let's write it as

$$\det(A - \lambda I) = q_0 + q_1\lambda + q_2\lambda^2 + \cdots + q_{n-1}\lambda^{n-1} + q_n\lambda^n.$$

What are these coefficients? Well, let's start from the leading term: $q_n\lambda^n$. What is the coefficient q_n? Look at $\det(A - \lambda I)$. In it, only one permutation contributes to λ^n: the identity permutation. Indeed, it contributes the product

$$\left(a_{1,1} - \lambda\right)\left(a_{2,2} - \lambda\right)\left(a_{3,3} - \lambda\right)\cdots\left(a_{n,n} - \lambda\right).$$

Upon opening parentheses, pick $-\lambda$ from each factor. This produces the leading term:

$$q_n\lambda^n = (-1)^n\lambda^n.$$

Thus,

$$q_n = (-1)^n.$$

Next, what is q_{n-1}? Again, only the identity permutation contributes to λ^{n-1}. (All others contribute to λ^{n-2} at most.) As discussed above, it produces the product

$$(a_{1,1} - \lambda)(a_{2,2} - \lambda)(a_{3,3} - \lambda) \cdots (a_{n,n} - \lambda).$$

Upon opening parentheses, pick $-\lambda$ from most factors. Only from one factor don't pick $-\lambda$. This can be done in n different ways, leading to

$$q_{n-1}\lambda^{n-1} = (-1)^{n-1}\lambda^{n-1}\sum_{i=1}^{n} a_{i,i}.$$

Thus,

$$q_{n-1} = (-1)^{n-1}\sum_{i=1}^{n} a_{i,i} = (-1)^{n-1}\text{trace}(A),$$

where the trace is the sum of the main-diagonal elements.

By now, we already uncovered two leading coefficients. Next, what is q_0? Again, look at $\det(A - \lambda I)$. This time, look at all permutations. Each one contributes a product of n factors. From these factors, never pick $-\lambda$ any more. After all, we are now interested in the free term, with no power of λ at all. Thus, its coefficient is

$$q_0 = \det(A).$$

Let's use this.

13.5.2 Trace: Sum of Eigenvalues

How to use this? For this purpose, write the characteristic polynomial in a new form: not as a sum, but as a product. Since it has degree n, it has n (complex) roots:

$$\lambda_1,\ \lambda_2,\ \lambda_3,\ \ldots,\ \lambda_n.$$

(Some of these could be the same, but this doesn't matter.) These are the roots: at these λ_i's, the characteristic polynomial vanishes. They are also the eigenvalues of A. Thanks to them, we can now write the characteristic polynomial as

$$\det(A - \lambda I) = (\lambda_1 - \lambda)(\lambda_2 - \lambda)(\lambda_3 - \lambda) \cdots (\lambda_n - \lambda).$$

You can now set $\lambda = \lambda_i$ $(1 \le i \le n)$, and get zero, as required. Furthermore, the leading term is indeed $(-1)^n\lambda^n$, as required.

In the latter form, let's open parentheses, and pick $-\lambda$ from $n-1$ factors only. This way, we obtain $q_{n-1}\lambda^{n-1}$ in a new form:

$$q_{n-1}\lambda^{n-1} = (-1)^{n-1}\lambda^{n-1}\sum_{i=1}^{n}\lambda_i.$$

Thus, the trace is the sum of eigenvalues:

$$\text{trace}(A) = \sum_{i=1}^{n}\lambda_i.$$

13.5.3 Determinant: Product of Eigenvalues

Finally, upon opening parentheses, never pick $-\lambda$ from any factor. This way, we obtain q_0 in a new form:

$$q_0 = \lambda_1 \lambda_2 \lambda_3 \cdots \lambda_n.$$

Thus, the determinant is the product of eigenvalues:

$$\det(A) = \lambda_1 \lambda_2 \lambda_3 \cdots \lambda_n.$$

Let's use this in quantum chemistry.

13.6 Exercises

13.6.1 Permutation: Product of Switches

1. For some $1 \le i < k \le n$, consider a switch of the form $(i \to k)$.
2. What does it do? Hint: at the same time, i goes to k, and k goes back to i.
3. Show that it is symmetric:

$$(i \to k) = (k \to i).$$

4. What is its inverse? Hint: itself.
5. Consider a cycle of the form

$$[k \to i] \equiv (k \to k-1 \to k-2 \to \cdots \to i+1 \to i).$$

6. What does it do? Hint: at the same time, k goes to $k-1$, $k-1$ to $k-2$, ..., $i+1$ to i, and i goes back to k.
7. Is it symmetric? Hint: only if $k = i+1$.
8. Write it as a product of switches:

$$[k \to i] = (k \to k-1)(k-1 \to k-2)\cdots(i+1 \to i).$$

9. What is its inverse? Hint:

$$[k \to i]^{-1} = (i \to i+1)(i+1 \to i+2)\cdots(k-1 \to k) = [i \to k].$$

10. Write our original switch $(i \to k)$ as a composition of two such cycles. Hint:

$$(i \to k) = [i \to k-1] \circ [k \to i].$$

11. Conclude once again that our original switch is odd. Hint:

$$\begin{aligned}
e((i \to k)) &= e\left([i \to k-1] \circ [k \to i]\right) \\
&= e\left([i \to k-1]\right) e\left([k \to i]\right) \\
&= (-1)^{k-1-i+k-i} \\
&= -1.
\end{aligned}$$

12. Consider now a general 3-cycle:

$$(l \to k \to i).$$

13. What does it do? Hint: at the same time, l goes to k, k to i, and i goes back to l.
14. Write it as a product of two switches. Hint:

$$(l \to k \to i) = (l \to k)(k \to i).$$

15. Conclude that it is even.
16. Consider now a general permutation:

$$p \in P.$$

17. Write it as a product of general cycles. Hint: start from 1. It must go to some number, which must go to some other number, and so on, until returning back to 1. This completes one general cycle. The rest is a disjoint (smaller) permutation, which can benefit from an induction hypothesis.
18. Conclude that every permutation can be written as a product of general cycles, each written as a product of switches, each written as a product of two cycles, each written as a product of elementary switches (that switch neighbors).
19. In the language of group theory, the elementary switches (that switch neighbors) *generate* P.

13.6.2 Interchanging Rows/Columns

1. Look at the transpose. What is its determinant? Hint: the same.
2. Look at the Hermitian adjoint. What is its determinant? Hint: the complex conjugate.
3. Look at a Hermitian matrix. What is its determinant? Hint: real.
4. Look at a product of two matrices. What is its determinant? Hint: the product of determinants.
5. Look at an orthogonal matrix. What is its determinant? Hint: ± 1.
6. Look at a unitary matrix. What is its determinant? Hint: a complex number of magnitude 1.
7. Look at a general matrix again. Interchange two rows (or columns). What will happen to the determinant? Hint: it will pick a minus sign. Indeed, all permutations will pick the same switch, with a minus sign.
8. What does this tell you about a matrix of two identical rows (or columns)? Hint: interchanging them would pick ± 1. Thus, the determinant must vanish.

14

The Hartree-Fock System

Look at an individual electron, placed at some point in space. It has an electric charge: -1 (in atomic units, with mass= $\bar{h} = 1$). Why doesn't it repel itself? Thanks to quantum mechanics, we know better: the electron is not really at one individual point, but at many points at the same time (each at some probability density). This is a superposition of many hypothetical positions. Thus, position is nondeterministic: a random variable. Nature is inherently stochastic.

And how about a few electrons in the atom? Each has its own wave function: its orbital. Place them in a matrix: its determinant will then be their joint wave function. In it, they interact: if they have the same spin, then they are indistinguishable, and obey Pauli's exclusion principle, as required.

Still, how to uncover the individual orbitals? Look at the electrostatic and exchange forces, and write their expected energy. This gives the Hartree-Fock system, whose eigenvectors are the orbitals, and whose eigenvalues are their energy levels [1, 12, 18].

14.1 Entanglement vs. Disentanglement

14.1.1 Entangled Particles

In Section 10.1.2, we already looked at a particle in 3-D, and its wave function and probability density. Consider now two particles that may interact (and even collide). In the language of quantum mechanics, they are *entangled*, and have a joint wave function: $w(r_1, r_2)$, where $r_1 = (x_1, y_1, z_1)$ and $r_2 = (x_2, y_2, z_2)$ are their (hypothetical) positions in 3-D. Once normalized, w satisfies

$$\int\int\int\int\int\int |w(r_1, r_2)|^2 dx_1 dy_1 dz_1 dx_2 dy_2 dz_2 = 1$$

(where each integral sign integrates from $-\infty$ to ∞). This way, $|w(r_1, r_2)|^2$ may indeed serve as a legitimate probability density (or distribution), to tell us how likely the particles are to be at r_1 and r_2 at the same time (not quite pointwise, but only weakly: in terms of integration in 6-D)).

14.1.2 Disentangled Particles

But what about three or four or even more? This gets too big to solve! Besides, even with two particles, w tells us nothing about each one! Better assume that they don't interact at all, so w can be factored as

$$w(r_1, r_2) = v^{(1)}(r_1)v^{(2)}(r_2),$$

where $v^{(1)}(r_1)$ and $v^{(2)}(r_2)$ are the wave functions of the individual particles. In the language of quantum mechanics, they are now *disentangled.*

14.2 Electrons in their Orbitals

14.2.1 Atom: Electrons in Orbitals

Consider now an atom, with M electrons. (For now, assume that they are still distinguishable.) Look at the nth electron ($1 \le n \le M$). It has wave function $v^{(n)}$ (its orbital). It also has a (nondeterministic) position:

$$r_n \equiv (x_n, y_n, z_n) \in \mathbb{R}^3.$$

To be at r_n, the electron has probability density $|v^{(n)}(r_n)|^2$. Let's use this.

14.2.2 Potential Energy and its Expectation

What is the potential energy in the atom? This is a random variable too. Better not measure it, or you'd lose your original wave function. Fortunately, you can still tell its expectation. For this purpose, assume that the electrons are entangled again, with a joint wave function:

$$w(r_1, r_2, \ldots, r_M).$$

This way, $|w(r_1, r_2, \ldots, r_M)|^2$ is the probability density to find them at r_1, r_2, ..., r_n at the same time.

The potential energy has a few terms, coming from classical electrostatics. The first comes from attraction to the nucleus (located at the origin). To have its expectation, take the probability density $|w|^2$, multiply it by the potential $1/\|r\|$, sum over all electrons, and integrate in $3M$-D:

$$-\int\int\int\cdots\int\int\int\sum_{n=1}^{M}\frac{|w|^2}{\|r_n\|}dx_1dy_1dz_1dx_2dy_2dz_2\cdots dx_Mdy_Mdz_M.$$

On top of this, there is yet more potential energy, due to electrostatic repulsion of each two electrons from one another:

$$\int\int\int\cdots\int\int\int\sum_{i=1}^{M}\sum_{n=i+1}^{M}\frac{|w|^2}{\|r_i - r_n\|}dx_1dy_1dz_1dx_2dy_2dz_2\cdots dx_Mdy_Mdz_M.$$

This sums all pairs of electrons, indexed by $1 \le i < n \le M$. (This way, each pair appears only once, not twice.) These are the Coulomb integrals. Let's simplify them a little.

14.3 Distinguishable Electrons

14.3.1 Hartree Product

As discussed above, this w is not so useful: it mixes different orbitals. To separate them back again, assume that the electrons are disentangled again, so w can be factored as

$$w(r_1, r_2, \ldots, r_M) = v^{(1)}(r_1)v^{(2)}(r_2)\cdots v^{(M)}(r_M).$$

This is the Hartree product. To have it, the electrons must be not only disentangled but also *distinguishable*. For example, two electrons can be distinguished by spin: one spins up, and the other down. This way, they can't be one and the same. This leaves no ambiguity about who is who: each has its own identity, with no identity crisis. But what about two electrons of the same spin? They are indistinguishable, and must have a more complicated (joint) wave function. Still, for now, assume that the Hartree product is valid.

14.3.2 Potential Energy of The Hartree Product

In the Hartree product, each orbital makes a probability density:

$$\int\int\int |v^{(i)}(r)|^2 dxdydz = 1, \qquad 1 \le i \le M.$$

Thus, the expected potential energy simplifies to read

$$\begin{aligned} &-\sum_{n=1}^{M} \int\int\int \frac{|v^{(n)}|^2}{\|r\|} dxdydz \\ &+\sum_{i=1}^{M} \sum_{n=i+1}^{M} \int\int\int\int\int\int |v^{(i)}(r)|^2 \frac{1}{\|r - \tilde{r}\|} |v^{(n)}(\tilde{r})|^2 dxdydzd\tilde{x}d\tilde{y}d\tilde{z}. \end{aligned}$$

Here, $r \equiv (x, y, z)$ and $\tilde{r} = (\tilde{x}, \tilde{y}, \tilde{z})$ are dummy variables (integrated upon), and don't need index i or n any more.

14.4 Indistinguishable Electrons

14.4.1 Indistinguishable Electrons

And what about electrons of the same spin? In the atom, they are indistinguishable. You can never tell who is who: they have mixed identities, and can be considered as one and the same.

Thus, in the atom, split the electrons into two (disjoint) subsets. For this purpose, let $0 \le L \le M$ be a new integer number. Now, assume that the L former electrons spin up, and the rest spin down.

Let's focus on the L former electrons. What is their joint wave function? It can no longer be a simple Hartree product, which assumes distinguishability.

14.4.2 Slater Determinant

Look at the L former electrons. Thanks to indistinguishability, they can form $L!$ different Hartree products. [After all, $v^{(i)}$ is no longer tied to r_i only.] Sum them up (half with a minus sign):

$$\frac{1}{\sqrt{L!}} \det\left(\left(v^{(n)}(r_i)\right)_{1\le i,n\le L}\right).$$

This will be their new (joint) wave function: the Slater determinant: the determinant of an $L \times L$ matrix.

Why is it good for indistinguishability? Well, interchanging rows (or columns) will only pick a minus sign (Section 13.6.2), with no physical effect at all. Besides, every two electrons must obey the Pauli exclusion principle: they can't share the same orbital, or the determinant would vanish, which is nonphysical.

Thanks to the Slater determinant, the expected potential energy will now simplify yet more. Let's see this.

14.5 Orbitals and their Canonical Form

14.5.1 The Overlap Matrix and its Diagonal Form

Consider two (square-integrable) functions:

$$u, q : \mathbb{R}^3 \to \mathbb{C}.$$

Define their inner product (or overlap) by

$$(u, q) \equiv \int\int\int \bar{u} q dx dy dz.$$

Unlike before, the complex conjugate is now on top of the former, not the latter. Don't worry: this is just a convention. It extends the inner product of vectors (Section 28.2.1). This way, we can now talk about orthogonality, and even orthonormality.

In particular, look at two orbitals. they have their overlap (inner product in 3-D). This makes the new $L \times L$ overlap matrix:

$$O \equiv (O_{i,n})_{1\le i,n\le L} \equiv \left(\int\int\int \bar{v}^{(i)}(r) v^{(n)}(r) dx dy dz\right)_{1\le i,n\le L},$$

where $r = (x, y, z)$ is the dummy variable (integrated upon). Note that O is Hermitian, but not necessarily orthogonal.

A proper orbital should have norm 1: overlap 1 with itself. This way, it makes a legitimate wave function (whose square is a legitimate probability density). Here, however, we don't have this as yet: the main-diagonal elements $O_{i,i}$ may still be different from 1.

Fortunately, O is Hermitian and positive semidefinite (Section 28.5.1). As such, it can be diagonalized by a unitary matrix U:

$$O = U^* D U,$$

where

$$D \equiv \operatorname{diag}\left(D_{1,1}, D_{2,2}, \ldots, D_{L,L}\right)$$

is a real diagonal matrix, with the eigenvalues of O on its main diagonal:

$$D_{n,n} \geq 0, \quad 1 \leq n \leq L.$$

Often, the orbitals are linearly independent: they have no linear combination that vanishes (almost) everywhere. In this case, O is even positive definite, with positive eigenvalues:

$$D_{n,n} > 0, \quad 1 \leq n \leq L.$$

In this case, the overlap matrix has a positive determinant:

$$\det(O) = \det(D) = D_{1,1} D_{2,2} \cdots D_{L,L} > 0.$$

14.5.2 Unitary Transformation

Since U is unitary, its complex conjugate $\bar{U}$ is unitary as well. Let's use it to transform our orbitals. For this purpose, place them in an L-dimensional column vector:

$$\mathbf{v} \equiv \mathbf{v}(r) \equiv \begin{pmatrix} v^{(1)}(r) \\ v^{(2)}(r) \\ \vdots \\ v^{(L)}(r) \end{pmatrix}.$$

Thanks to this, our overlap matrix can now be written as

$$O = \int\int\int \bar{\mathbf{v}}(r) \mathbf{v}^t(r) dx dy dz$$

(where $\mathbf{v}^t$ is a row vector). Thanks to this,

$$\begin{aligned}
D &= \bar{U} O U^* \\
&= U O \bar{U}^t \\
&= U \int\int\int \bar{\mathbf{v}}(r) \mathbf{v}^t(r) dx dy dz \bar{U}^t \\
&= \int\int\int U \bar{\mathbf{v}}(r) \mathbf{v}^t(r) \bar{U}^t dx dy dz \\
&= \int\int\int \overline{\bar{U}\mathbf{v}}(r) \left(\bar{U}\mathbf{v}\right)^t (r) dx dy dz.
\end{aligned}$$

14.5.3 Orthogonal Orbitals

So, we got a new column vector:

$$\bar{U}\mathbf{v},$$

containing L new orbitals. They are easier to work with. Why? Because they have a diagonal overlap matrix. This means that they have zero overlap with one another. In other words, they are orthogonal.

14.5.4 Slater Determinant and Unitary Transformation

And what about the Slater determinant? Does it change at all? Not much. Indeed, in the old orbitals, it was written as

$$\frac{1}{\sqrt{L!}}\det\left(\begin{pmatrix} \mathbf{v}^t(r_1) \\ \mathbf{v}^t(r_2) \\ \vdots \\ \mathbf{v}^t(r_L) \end{pmatrix}\right).$$

In the new orbitals, on the other hand, it now takes the form:

$$\frac{1}{\sqrt{L!}}\det\left(\begin{pmatrix} \mathbf{v}^t(r_1) \\ \mathbf{v}^t(r_2) \\ \vdots \\ \mathbf{v}^t(r_L) \end{pmatrix}\bar{U}^t\right) = \det\left(\bar{U}\right)\frac{1}{\sqrt{L!}}\det\left(\begin{pmatrix} \mathbf{v}^t(r_1) \\ \mathbf{v}^t(r_2) \\ \vdots \\ \mathbf{v}^t(r_L) \end{pmatrix}\right),$$

which is just a phaseshift, with no physical effect at all (Sections 13.4.2–13.4.6).

14.5.5 The Canonical Form

By now, our new orbitals are orthogonal. How to normalize them? Substitute

$$\mathbf{v} \leftarrow D^{-1/2}\bar{U}\mathbf{v}.$$

These new orbitals are orthonormal: their overlap matrix is the $L \times L$ identity matrix. This is the canonical form. No need to calculate it explicitly. Just assume that the original orbitals were in their canonical form all along.

How does this affect the Slater determinant? We already saw that orthogonalization has no effect. Normalization, on the other hand, has a good effect: it normalizes the Slater determinant too, getting it ready to serve as a legitimate wave function. To see this, the canonical form will prove most useful.

14.5.6 Slater Determinant and its Overlap

What is the Slater determinant? Like every determinant, it is a sum of $L!$ different products, using $L!$ different permutations of $\{1, 2, \ldots, L\}$. Let's calculate its overlap with itself. For this purpose, recall that our up-to-date orbitals are already orthonormal. Thanks to this (and Section 13.4.1), we'll have

$$\begin{aligned}
&\frac{1}{L!}\int\int\int\cdots\int\int\int\left|\det\left(\left(v^{(n)}(r_i)\right)_{1\le i,n\le L}\right)\right|^2 dx_1dy_1dz_1\cdots dx_Ldy_Ldz_L \\
&= \frac{1}{L!}\int\int\int\cdots\int\int\int\sum_{p\in P}\left|v^{(p(1))}(r_1)\right|^2\cdots\left|v^{(p(L))}(r_L)\right|^2 dx_1dy_1dz_1\cdots dx_Ldy_Ldz_L \\
&= \int\int\int\left|v^{(1)}(r)\right|^2 dxdydz\int\int\int\left|v^{(2)}(r)\right|^2 dxdydz\cdots\int\int\int\left|v^{(L)}(r)\right|^2 dxdydz \\
&= 1.
\end{aligned}$$

Indeed, what is the integrand here? It is the determinant times its own complex conjugate. In both, pick the same permutation:

$$p \in P \equiv P(\{1, 2, 3, \ldots, L\}).$$

(Otherwise, there would be no contribution at all, thanks to orthogonality.) This makes the same integral for all p's. [To help see this, each dummy variable was renamed as $r \equiv (x, y, z)$, independent of p.] Thus, we get the same integral $L!$ times. In the end, this is divided by $L!$ back again.

14.6 Expected Energy

14.6.1 Coulomb and Exchange Integrals

There is one more Slater determinant, defined for the $M - L$ latter orbitals of the remaining spin-down electrons. Assume that they are in their canonical form as well. So, our up-to-date wave function is a product of two Slater determinants:

$$\begin{aligned} &w(r_1, r_2, \ldots, r_M) \\ &\equiv \frac{1}{\sqrt{L!}} \det\left(\left(v^{(n)}(r_i)\right)_{1 \le i,n \le L}\right) \frac{1}{\sqrt{(M-L)!}} \det\left(\left(v^{(n)}(r_i)\right)_{L < i,n \le M}\right). \end{aligned}$$

With this new wave function, what is the expected potential energy? Fortunately, it is much simpler than the general form in Section 14.2.2. As a matter of fact, it is nearly as simple as in Section 14.3.2: it contains just one more double sum of new integrals: exchange integrals.

In the expected energy, we often integrate on the probability density

$$|w|^2 = \bar{w}\ w.$$

thanks to this, the potential due to attraction to the nucleus will simplify to read

$$\begin{aligned} &-\int\int\int \cdots \int\int\int \sum_{n=1}^{M} \frac{|w|^2}{\|r_n\|} dx_1 dy_1 dz_1 dx_2 dy_2 dz_2 \cdots dx_M dy_M dz_M \\ &= -\int\int\int \cdots \int\int\int \bar{w} \sum_{n=1}^{M} \frac{1}{\|r_n\|} w dx_1 dy_1 dz_1 dx_2 dy_2 dz_2 \cdots dx_M dy_M dz_M \\ &= -\sum_{n=1}^{M} \int\int\int \bar{v}^{(n)} \frac{1}{\|r\|} v^{(n)} dx dy dz. \end{aligned}$$

Why? Well, let's focus on the former Slater determinant in w. Like every determinant, it is the sum of products, each uses a different permutation of $\{1, 2, \ldots, L\}$. Now, to have a nonzero integral, pick the same permutation in $\bar{w}$ as in w. (Otherwise, thanks to orthogonality, the integral would vanish.) These $L!$ permutations will contribute the same integral $L!$ times. [To help see this, the dummy variable was renamed as $r \equiv (x, y, z)$.] In the end, this is divided by $L!$ back again.

In the Coulomb integrals, on the other hand, things get a little more complicated. To have a nonzero contribution, we have two options. In the main option, pick the same permutation in $\bar{w}$ as in w. This will contribute new Coulomb integrals of the form

$$\sum_{i=1}^{M} \sum_{n=i+1}^{M} \int\int\int\int\int\int |v^{(i)}(r)|^2 \frac{1}{\|r-\tilde{r}\|} |v^{(n)}(\tilde{r})|^2 dxdydzd\tilde{x}d\tilde{y}d\tilde{z}.$$

In the secondary option, on the other hand, pick a slightly different permutation in $\bar{w}$: switch i with n beforehand ($1 \le i < n \le L$ or $L < i < n \le M$). For instance, pick p in w, and

$$p(n \to i)$$

in $\bar{w}$. This will pick a minus sign:

$$e\left(p(n \to i)\right) = e(p)e\left((n \to i)\right) = -e(p).$$

This will contribute the so-called exchange integrals:

$$-\sum_{i=1}^{M} \sum_{n>i \text{ same spin}} \int\int\int\int\int\int \bar{v}^{(i)}(r)v^{(i)}(\tilde{r}) \frac{1}{\|r-\tilde{r}\|} \bar{v}^{(n)}(\tilde{r})v^{(n)}(r) dxdydzd\tilde{x}d\tilde{y}d\tilde{z}.$$

Let's see what this means for each individual orbital.

14.7 Effective Energy: One Electron

14.7.1 Expected Potential Energy

Now, let's focus on the nth electron only. What is the effective potential that it feels? Well, it feels attraction to the nucleus:

$$-\int\int\int \bar{v}^{(n)} \frac{1}{\|r\|} v^{(n)} dxdydz.$$

On top of this, it also feels repulsion from all other electrons:

$$+\sum_{i=1}^{M} \int\int\int\int\int\int |v^{(i)}(r)|^2 \frac{1}{\|r-\tilde{r}\|} |v^{(n)}(\tilde{r})|^2 dxdydzd\tilde{x}d\tilde{y}d\tilde{z}.$$

Here, one may ask: does it feel any repulsion from itself? No! There is one fictitious term here: $i = n$. Don't worry: it will soon drop.

On top of this, it also feels an exchange force from all other electrons of the same spin:

$$-\sum_{i, \text{ same spin as } n} \int\int\int\int\int\int \bar{v}^{(i)}(r)v^{(i)}(\tilde{r}) \frac{1}{\|r-\tilde{r}\|} \bar{v}^{(n)}(\tilde{r})v^{(n)}(r) dxdydzd\tilde{x}d\tilde{y}d\tilde{z}.$$

Here, one may ask: does it feel any exchange force from itself? No! There is a new fictitious term here: $i = n$. Fortunately, both cancel out, and drop. This leads to uniformity: the sums go over $i = 1, 2, \ldots, M$, including $i = n$. Thanks to this, the nth orbital will soon solve the same equation as the others.

14.7.2 Expected Kinetic Energy

On top of this, the nth electron also has its own kinetic energy:

$$\frac{1}{2}\int\int\int \nabla^t \bar{v}^{(n)} \cdot \nabla v^{(n)} dxdydz,$$

where ∇ is the gradient: the vector of partial derivatives (and ∇^t is its transpose: the row vector).

14.7.3 Schrodinger's Equation: Integral Form

Together, these terms must sum to the expected total energy of the nth electron:

$$E\int\int\int |v^{(n)}|^2 dxdydz,$$

where E is an energy level: an eigenvalue of the Hamiltonian. After all, in quantum mechanics, energy comes in discrete quanta. Only these energy levels are allowed.

This is Schrodinger's equation for the nth electron: its kinetic and potential energy (including the exchange terms) sum to its entire expected energy.

14.8 The Hartree-Fock System

14.8.1 Basis Functions: Coefficient Matrix

In numerical analysis, we often approximate our orbital, and expand it as a linear combination of basis functions:

$$v^{(n)} \doteq \sum_{j-1}^{K} c_j \psi_j,$$

where the ψ_j's are the basis functions (as in finite elements), and the c_j's are their unknown (complex) coefficients. (As a matter of fact, c_j depends on n too. For simplicity, we disregard this for now, and focus on one particular n.) Plug this into the effective energy in Section 14.7.1, term by term. For this purpose, in each term, replace $\bar{v}^{(n)}$ by $\bar{\psi}_l$, and $v^{(n)}$ by ψ_j. (Although ψ_l is often real, better take its complex conjugate, to be on the safe side.) This will assemble the (l,j)th element in the coefficient matrix A:

$$\begin{aligned}
&a_{l,j} \\
&\equiv -\int\int\int \bar{\psi}_l \frac{1}{\|r\|}\psi_j dxdydz \\
&+ \sum_{i=1}^{M} \int\int\int\int\int\int |v^{(i)}(r)|^2 \frac{1}{\|r-\tilde{r}\|}\bar{\psi}_l(\tilde{r})\psi_j(\tilde{r}) dxdydzd\tilde{x}d\tilde{y}d\tilde{z} \\
&- \sum_{i,\text{ same spin as } n} \int\int\int\int\int\int \bar{v}^{(i)}(r)v^{(i)}(\tilde{r})\frac{1}{\|r-\tilde{r}\|}\bar{\psi}_l(\tilde{r})\psi_j(r) dxdydzd\tilde{x}d\tilde{y}d\tilde{z} \\
&+ \frac{1}{2}\int\int\int \nabla^t \bar{\psi}_l \cdot \nabla \psi_j dxdydz
\end{aligned}$$

$(1 \leq l, j \leq K)$. This way, A is no longer a constant matrix: it depends on the unknown orbitals: the $v^{(i)}$'s in the above sums. Still, for fixed orbitals, A is Hermitian, as required.

14.8.2 The Mass Matrix

How to define also the mass matrix B? Its (l, j)th element is just the overlap of ψ_l with ψ_j:

$$b_{l,j} \equiv \int \int \int \bar{\psi}_l \psi_j dx dy dz$$

$(1 \leq l, j \leq K)$. Now, how to solve for the unknown c_j's? For this purpose, place them in a new K-dimensional column vector:

$$\mathbf{c} \equiv (c_1, \, c_2, \, c_3, \, \ldots, \, c_K)^t \in \mathbb{C}^K.$$

This way, we can now plug our approximation in. The expected effective energy (Sections 14.7.1–14.7.2) will then take a discrete form:

$$\mathbf{c}^* A \mathbf{c} = E \mathbf{c}^* B \mathbf{c},$$

where E is the (unknown) energy level. Thus, this is a nonlinear equation, with two types of unknowns: the vector $\mathbf{c}$, and the scalar E.

14.8.3 The Pseudo-Eigenvalue Problem

What should the energy level E be? It should be minimal. For this purpose, we need to solve a pseudo-eigenvalue problem:

$$A\mathbf{c} = EB\mathbf{c}.$$

The term "pseudo" tells us that this is actually a nonlinear system: A depends on the (unknown) orbitals themselves. Still, look at fixed orbitals: say, the solution. For these, A is Hermitian, as required.

The mass matrix B is Hermitian too. Better yet, it is the same for all orbitals. But A is not: it depends on spin. For $1 \leq n \leq L$ (spin-up), A takes one form. For $L < n \leq M$ (spin-down), on the other hand, it takes another form.

Fortunately, on the right-hand side, we have a good matrix: B: an SPD matrix (symmetric and positive definite). Thus, this is actually a generalized eigenvalue problem. (See exercises at the end of Chapter 28.)

Thus, same-spin orbitals are expanded by vectors that solve the same pseudo-eigenvalue problem, with the same A and B. Are these vectors orthogonal (in the generalized sense)? Well, if they have different eigenvalues, then they sure are. Otherwise, then apply a (generalized) Gram-Schmidt process. Either way, once normalized, they even get orthonormal (in the generalized sense).

14.8.4 Is The Canonical Form Plausible?

All along, we assumed that the orbitals $v^{(n)}$ are in their canonical form. Is this plausible? To see this, plug two eigenvectors in the right-hand side, with B in between. Thanks to (generalized) orthogonality, you'll get zero. Now, multiply their components by corresponding basis functions, and expand two orthogonal orbitals (of zero overlap). Once normalized, they are indeed in their canonical form, as assumed all along.

14.9 Exercises

14.9.1 Electrostatic Potential

1. Consider a ball (made of metal), centered at the origin: $(0, 0, 0)$ (Figure 12.1).
2. Assume that the ball has charge -1 (in atomic units).
3. Outside the ball, consider a point r. What is the electrostatic field at r? Hint: it is radial and outgoing, with magnitude $1/\|r\|^2$. In other words, it is $r/\|r\|^3$.
4. Look at the hyperbolic function: $1/\|r\|$. What is its gradient? Hint:

$$\begin{aligned}\nabla\left(\frac{1}{\|r\|}\right) &= \nabla\left(\left(x^2+y^2+z^2\right)^{-1/2}\right)\\ &= -\frac{1}{2}\left(x^2+y^2+z^2\right)^{-3/2}(2x, 2y, 2z)^t\\ &= -\frac{r}{\|r\|^3}.\end{aligned}$$

5. Is this familiar? Hint: this is the good old electrostatic field (times -1).
6. What is the physical meaning of this? Hint:

$$\text{field} = -\nabla \text{ potential}.$$

7. Look at the radial field r. What is its divergence (sum of partial derivatives, component by component)? Hint:

$$\nabla \cdot r = \nabla \cdot \left((x, y, z)^t\right) = \frac{\partial x}{\partial x} + \frac{\partial y}{\partial y} + \frac{\partial z}{\partial z} = 1 + 1 + 1 = 3.$$

8. Look at the radial unit field: $r/\|r\|$. What is its divergence? Hint:

$$\begin{aligned}\nabla \cdot \left(\frac{r}{\|r\|}\right) &= \frac{\nabla \cdot r}{\|r\|} + r^t \nabla\left(\frac{1}{\|r\|}\right)\\ &= \frac{3}{\|r\|} - r^t \frac{r}{\|r\|^3}\\ &= \frac{3}{\|r\|} - \frac{1}{\|r\|}\\ &= \frac{2}{\|r\|}.\end{aligned}$$

9. Is this familiar? Hint: this is the electrostatic potential (times 2).

Part VII

Introduction to Lie Algebras

Part VII: Introduction to Lie Algebras

How to model physics in 3-D? Use vectors! After all, vectors are linear: added linearly, and even multiplied by a scalar. Moreover, they even enjoy vector product: not commutative, but anti-commutative. This makes an elementary Lie algebra.

Better yet, use matrices. After all, matrices are linear too: added linearly, and even multiplied by a scalar. Better yet, they are even multiplied by one another. In general, they don't commute, but have a (nonzero) commutator. (This is how the Pauli matrices model spin and polarization.) This is the key to many interesting Lie algebras.

In this part, we introduce Lie algebras from scratch. Furthermore, we uncover their inner structure, and decompose them in terms of simple ideals, useful in classical and quantum mechanics alike.

These ideals are Lie algebras in their own right, containing quantum-mechanical observables (times $\sqrt{-1}$). Once exponentiated, they make new Lie groups: the rotation group, and the Lorentz group in special relativity.

15

The Jordan Form of a Matrix and Its Extension to Algebras

To introduce Lie algebras, we need some background in linear (and modern) algebra. For this, we need a fundamental theorem in number theory: the Chinese remainder theorem. This will help design the Jordan decomposition of a matrix. In modern algebra, this will help design a subalgebra, in which each derivation has its own Jordan form. This will give us a good starting point to Lie algebras.

15.1 Nilpotent Matrix and Generalized Eigenvectors

15.1.1 Nilpotent Matrix

Let B be a square (real or complex) matrix. Assume that, if raised to a sufficiently high power, then it would vanish. In other words, for some natural number $m \geq 1$,

$$B^m = (0)$$

(the zero matrix). Then we say that B is nilpotent. Clearly, m is not unique: it could grow, and still work. Later on, we'll see how to pick a good m.

What does this mean for an individual vector? Let u be a nonzero vector. Clearly,

$$B^m u = (0)u = \mathbf{0}$$

(the zero vector). Still, why go to a power as high as m? Maybe there is a (smaller) number

$$1 \leq k \leq m$$

for which

$$B^k u = \mathbf{0}.$$

Unlike m, which depends on B only, k depends on u too. Still, like m, k is not unique: it could grow, and still work. Let's pick a minimal k.

15.1.2 Cycle – Invariant Subspace

Assume that k is minimal. This way,

$$B^{k-1}u \neq \mathbf{0}$$

is a singular eigenvector of B (eigenvalue 0):

$$B\left(B^{k-1}u\right) = B^k u = \mathbf{0}.$$

We can now start from u, and apply B time and again, producing a new cycle of k vectors:

$$u,\ Bu,\ B^2u,\ \ldots,\ B^{k-1}u.$$

These are called generalized eigenvectors. Only the latter is a legitimate eigenvector: the former, on the other hand, are not.

Clearly, the cycle is invariant under B: each vector is mapped to the next one (except for the latter, which is mapped to zero). Thanks to linearity, the cycle spans an invariant subspace.

In particular, if $k = 1$, then the cycle is very short: u only. In this special case, u is a legitimate eigenvector in its own right. Otherwise, u is only a generalized eigenvector, not an eigenvector.

15.1.3 Generalized Eigenvectors – Linear Independence

Let's order our cycles in terms of decreasing length. For example, assume that the longest cycle contains four vectors:

$$u,\ Bu,\ B^2u,\ B^3u.$$

Furthermore, assume that the second longest cycle contains three vectors:

$$v,\ Bv,\ B^2v.$$

Then these come after. These seven vectors are generalized eigenvectors. Are they linearly independent? To make sure, we need one more assumption.

Assume that the eigenvectors B^3u and B^2v are linearly independent of one another. Are all seven vectors linearly independent as well? To check on this, assume that

$$c_0u + c_1Bu + c_2B^2u + c_3B^3u + d_0v + d_1Bv + d_2B^2v = \mathbf{0}.$$

Must all these coefficients vanish:

$$c_0 = c_1 = c_2 = c_3 = d_0 = d_1 = d_2 = 0?$$

To check on this, place them in the top diagram in Figure 15.1: the longer cycle at the top row, and the shorter at the bottom row. Now, apply B to the entire linear combination:

$$B\left(c_0u + c_1Bu + c_2B^2u + c_3B^3u + d_0v + d_1Bv + d_2B^2v\right) = \mathbf{0}.$$

Since $B^4u = B^3v = \mathbf{0}$, drop them:

$$c_0Bu + c_1B^2u + c_2B^3u + d_0Bv + d_1B^2v = \mathbf{0}.$$

This is illustrated geometrically in Figure 15.1: the entire diagram shifts leftwards, and the leftmost cells drop out. Next, apply B once again. This shifts the diagram

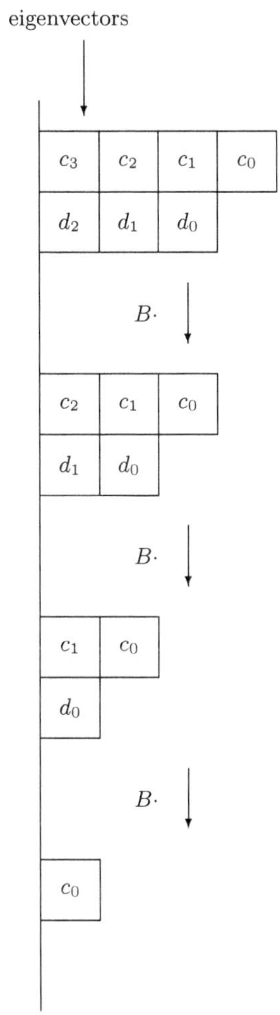

Fig. 15.1. If the eigenvectors B^3u and B^2v are linearly independent, then so are all their generalized eigenvectors. Indeed, apply B to the entire linear combination time and again, shifting the diagram leftwards. Those cells that pass the left edge drop out, leaving the upper-right cell only, which must therefore vanish: $c_0 = 0$.

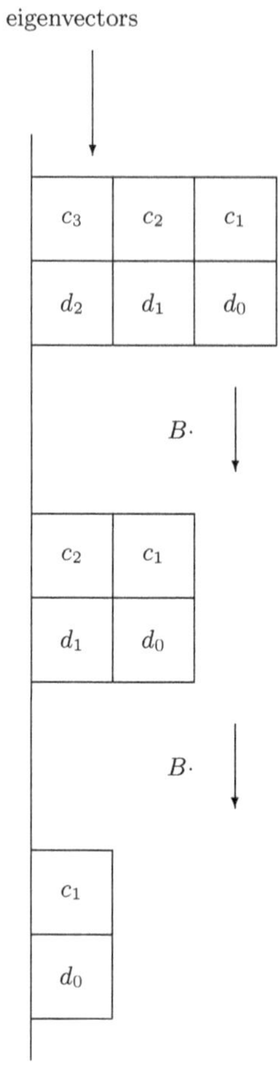

Fig. 15.2. Once c_0 dropped, restart the same process all over again, shifting leftwards as before. This time, only two cells survive. Since the eigenvectors are linearly independent, we must have $d_0 = c_1 = 0$. Drop them, and restart all over again, and so on.

once again, dropping yet another column from the left. Finally, apply B for the last time. This shifts the diagram once again, leaving one last cell on the right: c_0. We can therefore conclude that

$$c_0 = 0,$$

as required. This "shaves" the rightmost cell from the original diagram. The process can now restart all over again (as in Figure 15.2), to conclude that

$$c_1 = d_0 = 0$$

too. This "shaves" the second-right column as well, and so on. In the end, we'll see that all coefficients indeed vanish, as required.

This is not limited to cycles of length three or four only: it could be extended to longer cycles as well, using a nested mathematical induction.

Furthermore, this is not limited to two cycles: it is true for one cycle as well. This gives us our first bonus: the vectors in a cycle are linearly independent of each other. Thus, the length of the cycle can't exceed the dimension of our linear space (the order of B).

Better yet, our process is not limited to two cycles, but could be extended to many more cycles as well. We therefore get our second bonus: if a few eigenvectors are linearly independent, then so are also all their generalized eigenvectors.

15.1.4 Linear Dependence

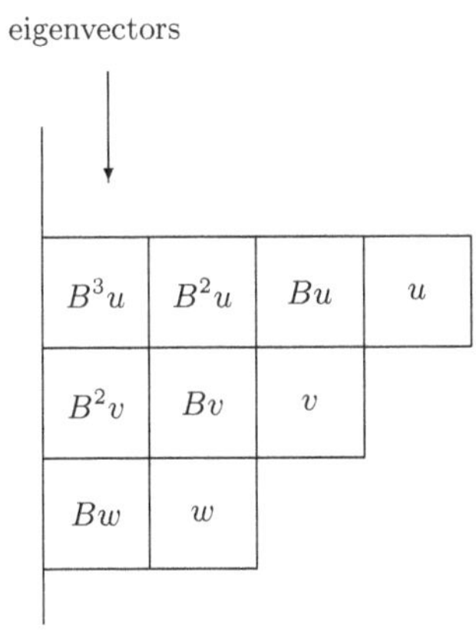

Fig. 15.3. If there is no new eigenvector any more, then there is no new generalized eigenvector either. Indeed, if Bw depends linearly on B^3u and B^2v, then w must depend linearly on their cycles (the vectors in the first and second rows).

Next, let's work the other way around: assume that there is no new eigenvector any more. Is there any new *generalized* eigenvector? To check on this, let w be some vector, with a short cycle of length 2 only:

$$w, \ Bw.$$

This new cycle is short enough: it could be placed at the bottom of our diagram, and still obey our decreasing-length convention (Figure 15.3). Now, assume that Bw is not really a new eigenvector, but depends linearly on the former ones:

$$Bw = \alpha B^3 u + \beta B^2 v$$

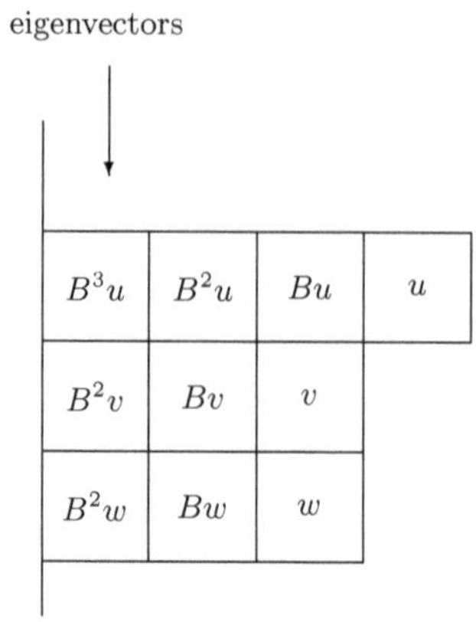

Fig. 15.4. Likewise, if B^2w depends linearly on B^3u and B^2v, then both w and Bw must depend linearly on the two former cycles (the vectors in the first and second rows).

(where α and β are not both zero at the same time). Now, what about w? Does it also depend linearly on the former vectors in the cycles above it?

Well, in the above equation, B maps w to a linear combination of eigenvectors. Still, w is not the only one: there is one more vector mapped in the same way! To design it, just shift the coefficients α and β rightwards, to the column just above w:

$$B\left(\alpha B^2u + \beta Bv\right) = \alpha B^3u + \beta B^2v = Bw.$$

So, there are two vectors that are mapped by B to the same vector. Thus, their difference is mapped to $\mathbf{0}$, and must therefore be an eigenvector in its own right:

$$w - \alpha B^2u - \beta Bv = \gamma B^3u + \delta B^2v.$$

So, w is not really new.

Next, assume that w makes a slightly longer cycle:

$$w, \; Bw, \; B^2w.$$

This is still short enough: it could be placed at the bottom row, without violating our decreasing-length rule (Figure 15.4). Now, assume that the "new" eigenvector B^2w is not really new, but depends linearly on the former eigenvectors:

$$B^2w = \alpha B^3u + \beta B^2v.$$

As discussed above, we can then write Bw as a linear combination as well:

$$Bw = \alpha B^2u + \beta Bv + \gamma B^3u + \delta B^2v.$$

So, B maps w to this linear combination. Still, w is not the only one: there is yet another vector mapped in the same way. To design it, just shift the coefficients α, β, γ, and δ rightwards:

$$B\left(\alpha Bu + \beta v + \gamma B^2 u + \delta Bv\right) = \alpha B^2 u + \beta Bv + \gamma B^3 u + \delta B^2 v = Bw.$$

So, there are two vectors that are mapped by B to the same vector. Thus, their difference is mapped to zero, and is therefore an eigenvector in its own right:

$$w - \alpha Bu - \beta v - \gamma B^2 u - \delta Bv = \zeta B^3 u + \eta B^2 v.$$

So, w is not really new: it depends linearly on the previous generalized eigenvectors.

This is not limited to this small example. In fact, it could be extended to many more rows and columns, using a nested mathematical induction.

Thus, if there is a really new w, then it has a really new eigenvector. Its cycle can't be too long, or it would appear earlier (thanks to our convention). Is this familiar? We already studied this case (Section 15.1.3). Since the eigenvectors are linearly independent, so are also all cycles together. They can even form a new basis: Jordan basis.

15.1.5 Jordan Basis

How to design a Jordan basis? Recall that B is nilpotent:

$$B^m = (0).$$

Recall also that m is not unique: it could grow, and still work. How to pick a good m? In theory, m could be the length of the longest cycle. But we don't know this in advance. So, let m be the order of B. This way, no cycle could be longer:

$$k \le m.$$

(Otherwise, the vectors in the cycle would depend linearly on each other, in violation of Section 15.1.3.) Thus, nothing can stop us: pick some vector w, and design its cycle. Then, pick a (really) new w, and design its cycle as well. If it is longer, then raise it up, to its proper place. (It could even span a previous cycle, and replace it.) Repeat this on and on, designing shorter and shorter cycles, until having a complete basis of m vectors.

15.1.6 Jordan Blocks

Let's order our cycles in a reverse order. In the above example, this gives

$$\left\{\ldots,\ B^2v,\ Bv,\ v,\ B^3u,\ B^2u,\ Bu,\ u\right\}.$$

How does B act on this list? Well, each vector is mapped to the previous one (except for the eigenvectors, which are mapped to $\mathbf{0}$).

Now, in the above list, represent each vector by a standard unit vector:

$$(0,0,0,\ldots,0,1,0,0,0,\ldots,0)^t$$

(only one component is 1, and the rest are 0). For this basis, B looks like

$$B = \begin{pmatrix} \ddots & & & & & & & & \\ & \ddots & & & & & & & \\ & & 0 & 1 & & & & & \\ & & & 0 & 1 & & & & \\ & & & & 0 & & & & \\ & & & & & 0 & 1 & & \\ & & & & & & 0 & 1 & \\ & & & & & & & 0 & 1 \\ & & & & & & & & 0 \end{pmatrix}.$$

This is the Jordan form of B. (As usual, blank spaces stand for zeroes.) Only one diagonal is nonzero: the superdiagonal, just above the main diagonal. In our example, there are two Jordan blocks: at the bottom-right, there is a 4×4 block, with three 1's in the superdiagonal. Just above it, there is a 3×3 block, with two 1's in the superdiagonal.

And how about B^2? How does it look like (in this basis)? Well, just shift the 1's rightwards, to the next higher superdiagonal. (If they exceed the block, then they drop out.) This goes on, power by power. In our example,

$$B^4 = (0),$$

as required.

In a nilpotent matrix, what do we have on the main diagonal? Zero! Why? Because this is the only eigenvalue. And what about a more general matrix (not necessarily nilpotent)? In each Jordan block, on the main diagonal, you can also find a nonzero eigenvalue for a change.

15.2 Matrix and Its Jordan Form

15.2.1 Characteristic Polynomial: Eigenvalue and Its Multiplicity

So far, we looked at a nilpotent matrix: B. Next, let's discuss a more general matrix A (of order n), not necessarily nilpotent. Its Jordan form is slightly different: on its main diagonal, it may also have nonzero. To see this, look at its characteristic polynomial (Section 13.5.1):

$$\det(A - \lambda I) = (\lambda_1 - \lambda)(\lambda_2 - \lambda)(\lambda_3 - \lambda) \cdots (\lambda_n - \lambda),$$

where the λ_i's (the eigenvalues of A) are not necessarily distinct: some of them could be the same. To avoid this, better rewrite

$$\det(A - \lambda I) = (\lambda_1 - \lambda)^{n_1}(\lambda_2 - \lambda)^{n_2}(\lambda_3 - \lambda)^{n_3} \cdots (\lambda_k - \lambda)^{n_k}$$

for a new $k \geq 1$, and new (distinct) λ_i's, each with its own multiplicity: n_i. Clearly, the multiplicities sum to n:

$$n_1 + n_2 + n_3 + \cdots + n_k = n.$$

15.2.2 Invariant Subspace and Its Block

Let's focus on the first eigenvalue: λ_1. Let's subtract λ_1 from the main diagonal in A, and write the result in a block form:

$$A - \lambda_1 I = \begin{pmatrix} C - \lambda_1 I & \\ E & J - \lambda_1 I \end{pmatrix},$$

where I is an identity matrix of a suitable order. Here, what is J? To see this, let

$$V \equiv \{v \mid (A - \lambda_1 I)^{n_1} v = \mathbf{0}\}$$

be the null space of $(A - \lambda_1 I)^{n_1}$. Clearly, V is invariant under $A - \lambda_1 I$. This is how the upper-right block drops from the above decomposition. For this, define J as A, restricted to V:

$$J \equiv A \mid_V .$$

This is why J appears in the lower-right block: it acts on those standard unit vectors that span V.

15.2.3 Block and Its Jordan Form

As a matter of fact, we already know these vectors. After all, $J - \lambda_1 I$ is nilpotent:

$$(J - \lambda_1 I)^{n_1} = (A - \lambda_1 I)^{n_1} \mid_V = (0).$$

Thus, $J - \lambda_1 I$ (and indeed J itself as well) has its own Jordan form (Section 15.1.6):

$$J = \begin{pmatrix} \ddots & & & & & & & & \\ & \ddots & & & & & & & \\ & & \lambda_1 & 1 & & & & & \\ & & & \lambda_1 & 1 & & & & \\ & & & & \lambda_1 & & & & \\ & & & & & \lambda_1 & 1 & & \\ & & & & & & \lambda_1 & 1 & \\ & & & & & & & \lambda_1 & 1 \\ & & & & & & & & \lambda_1 \end{pmatrix}.$$

This is a little more general: on the main diagonal, we now have nonzero.

15.2.4 Block and Its Characteristic Polynomial

How big is J? To see this, decompose the characteristic polynomial of A:

$$\det(A - \lambda I) = \det(C - \lambda I)\det(J - \lambda I),$$

where I is an identity matrix of a suitable order.

So, how could the characteristic polynomial of J look like? Well, it could only contain factors from the characteristic polynomial of A. Could it contain a factor like $\lambda_2 - \lambda$? No! After all, no eigenvector corresponding to λ_2 could belong to V.

Thus, the characteristic polynomial of J must look like

$$\det(J - \lambda I) = (\lambda_1 - \lambda)^l$$

for some natural number l. This also follows from the Jordan form of J, illustrated above.

What could l be? Could

$$l > n_1?$$

No! After all, the characteristic polynomial of J could only contain a factor from the characteristic polynomial of A. On the other hand, could

$$l < n_1?$$

Well, if this were true, then C would have factor $\lambda_1 - \lambda$ in its own characteristic polynomial. This way, C would also have an eigenvector $w \neq \mathbf{0}$, corresponding to λ_1. But then we could design a new eigenvector for A too: just take w, and add a few dummy zeroes:

$$(A - \lambda_1 I)\begin{pmatrix} w \\ \mathbf{0} \end{pmatrix} = \begin{pmatrix} \mathbf{0} \\ Ew \end{pmatrix},$$

so

$$(A - \lambda_1 I)^{l+1}\begin{pmatrix} w \\ \mathbf{0} \end{pmatrix} = (A - \lambda_1)^l \begin{pmatrix} \mathbf{0} \\ Ew \end{pmatrix} = \begin{pmatrix} \mathbf{0} \\ (J - \lambda_1 I)^l \, Ew \end{pmatrix} = \begin{pmatrix} \mathbf{0} \\ (0)Ew \end{pmatrix} = \begin{pmatrix} \mathbf{0} \\ \mathbf{0} \end{pmatrix},$$

so

$$(A - \lambda_1 I)^{n_1}\begin{pmatrix} w \\ \mathbf{0} \end{pmatrix} = \mathbf{0}.$$

Thus, $(w, \mathbf{0})^t$ can't possibly exist: on one hand, it lies outside V (since $w \neq \mathbf{0}$). On the other hand, it lies in the null space of $(A - \lambda_1 I)^{n_1}$, which is nothing but V. To avoid this, we must have

$$l = n_1,$$

meaning that J is an $n_1 \times n_1$ block. Its Jordan form is already illustrated above.

Now, there is nothing special about λ_1: the same could be done for any other λ_i as well. Thus, in its final Jordan form, A will be block-diagonal, with k such blocks, each decomposed in its own Jordan form. How to design this more explicitly? For this, we need some background in number theory.

15.3 Greatest Common Divisor

15.3.1 Divide With Remainder

To have a more explicit Jordan form, we need some background in number theory. For this purpose, let $n > m > 0$ be two natural numbers. What is the ratio n/m? Here, we don't mean the usual fraction n/m, but the maximal integer number that doesn't exceed it:

$$\left\lfloor \frac{n}{m} \right\rfloor = \max\left\{ j \in \mathbb{Z} \mid j \leq \frac{n}{m} \right\},$$

where $\mathbb{Z}$ is the set of integer numbers. This way, we can now divide n by m with remainder. For this, write n as

$$n = km + l,$$

where

$$k = \left\lfloor \frac{n}{m} \right\rfloor,$$

and l is the remainder (or residual) that is too small to be divided by m:

$$0 \leq l < m.$$

This l has a new name: n modulus m:

$$l = (n \bmod m).$$

15.3.2 Congruence: Same Remainder

The above equation could also be written as

$$n \equiv l \bmod m.$$

This means that n and l are *congruent* modulus m (have the same remainder):

$$n = km + l$$
$$l = 0 \cdot m + l.$$

Modulus m, they are indistinguishable: have the same remainder l:

$$(n \bmod m) = l = (l \bmod m).$$

In other words, their difference $n - l$ could be divided by m (evenly, with no remainder at all):

$$m \mid n - l.$$

This is indeed a mathematical equivalence relation.

15.3.3 Common Divisor

Consider now an interesting case: no remainder at all:

$$(n \bmod m) = 0.$$

In this case, we say that m divides n:

$$m \mid n.$$

In general, however, m doesn't divide n. Still, they may share a common divisor (that divides them both). For example, if they are even, then 2 is a common divisor. Still, there could be others. What is the maximal one?

15.3.4 The Euclidean Algorithm

How to calculate the greatest common divisor of n and m? Well, there are two possibilities: if m divides n, then m itself is the greatest common divisor. If not, then

$$\begin{aligned}\mathrm{GCD}(n,m) &= \mathrm{GCD}(km+l,m)\\ &= \mathrm{GCD}(l,m)\\ &= \mathrm{GCD}(m,l)\\ &= \mathrm{GCD}(m,n \bmod m).\end{aligned}$$

So, instead of the original numbers

$$n > m,$$

we can now work with two smaller numbers:

$$m > l = (n \bmod m).$$

This leads to the recursive formula

$$\mathrm{GCD}(n,m) = \begin{cases} m & \text{if } (n \bmod m) = 0\\ \mathrm{GCD}(m, n \bmod m) & \text{if } (n \bmod m) > 0.\end{cases}$$

How to prove this? Use mathematical induction on $n = 2, 3, 4, \ldots$ (see below). This is the Euclidean algorithm.

15.3.5 The Extended Euclidean Algorithm

Mathematical induction even helps write the greatest common divisor explicitly:

$$\mathrm{GCD}(n,m) = an + bm,$$

for some integer numbers a and b (positive or negative or zero). Indeed, if m divides n, then this is easy:

$$\mathrm{GCD}(n,m) = m = 0 \cdot n + 1 \cdot m,$$

as required. If not, then the induction hypothesis helps write

$$\mathrm{GCD}(n,m) = \mathrm{GCD}(m,l) = \tilde{a} \cdot m + \tilde{b} \cdot l$$

(for some new integer numbers $\tilde{a}$ and $\tilde{b}$). Now, in this formula, substitute

$$l = n - km.$$

This gives

$$\begin{aligned}\mathrm{GCD}(n,m) &= \mathrm{GCD}(m,l)\\ &= \tilde{a} \cdot m + \tilde{b} \cdot l\\ &= \tilde{a} \cdot m + \tilde{b}\,(n - km)\\ &= \tilde{b} \cdot n + \left(\tilde{a} - k \cdot \tilde{b}\right) m.\end{aligned}$$

To have the desired linear combination, just define the new coefficients

$$\begin{aligned} a &= \tilde{b}\\ b &= \tilde{a} - k \cdot \tilde{b} = \tilde{a} - \left\lfloor \frac{n}{m} \right\rfloor \tilde{b}.\end{aligned}$$

This is the extended Euclidean algorithm.

15.3.6 The Modular Extended Euclidean Algorithm

Unfortunately, the coefficients a and b could be negative. Worse, they could be too big, and even exceed n. How to avoid this? Well, throughout the above algorithm (including the recursion), introduce one little change: take modulus n. This way, in the induction hypothesis, we can also assume that $\tilde{a}$ and $\tilde{b}$ are between 0 and $n-1$:

$$0 \le \tilde{a}, \tilde{b} < n.$$

Now, rewrite the induction step as

$$\begin{aligned} a &= \tilde{b} \\ s &= \left\lfloor \frac{n}{m} \right\rfloor \tilde{b} \\ b &= \tilde{a} - s. \end{aligned}$$

To make sure that a and b are between 0 and $n-1$ as well, make a little change: take modulus n:

$$\begin{aligned} a &= \tilde{b} \\ s &= \left(\left\lfloor \frac{n}{m} \right\rfloor \tilde{b} \right) \bmod n \\ b &= \begin{cases} \tilde{a} - s & \text{if} \quad \tilde{a} \ge s \\ \tilde{a} + n - s & \text{if} \quad \tilde{a} < s. \end{cases} \end{aligned}$$

This way, thanks to the induction step, a and b are kept between 0 and $n-1$ as well:

$$0 \le a, s, b < n,$$

as required. Furthermore, modulus n, nothing has changed. Thus, we have a new result modulus n:

$$\text{GCD}(n, m) \equiv (an + bm) \equiv bm \bmod n.$$

This will be useful later.

15.4 Modular Arithmetic

15.4.1 Coprime

Let's use the greatest common divisor to define the coprime. For this purpose, consider a natural number n. What is a coprime of n? This is any other number p (prime or not) that shares no common divisor with n:

$$\text{GCD}(n, p) = 1.$$

There may be a few legitimate coprimes. Which one to pick? Better pick a moderate one. For this purpose, start from an initial guess, say the prime number

$$p = 5.$$

Does it divide n? If not, then pick it. And if it does, then keep looking:

$$p \leftarrow \text{ the next prime number},$$

and so on. In at most $(\log_2 n)$ guesses, you'll eventually find a new prime number p that doesn't divide n any more, and can serve as its legitimate coprime.

15.4.2 Modular Multiplication

Here is another interesting task. Let n and j be two natural numbers. How to calculate the product

$$nj \bmod m?$$

Both n and j could be very long, and contain many digits. Their product nj could be even longer, and not easy to store on the computer. How to avoid it?

Fortunately, both n and j could be written in the form in Section 15.3.2:

$$\begin{aligned} n &= \left\lfloor \frac{n}{m} \right\rfloor m + (n \bmod m) \\ j &= \left\lfloor \frac{j}{m} \right\rfloor m + (j \bmod m). \end{aligned}$$

In the right-hand side, only the latter term is interesting. The former, on the other hand, is a multiple of m, which is going to drop anyway. Therefore, instead of nj, better multiply

$$(n \bmod m)(j \bmod m) \bmod m.$$

Here, the modulus operation is carried out not only after but also before starting to multiply. This way, we only multiply moderate numbers that never exceed $m-1$.

15.4.3 Power

And how to calculate a power like

$$n^k \bmod m \;?$$

Here, even for a moderate k, n^k may be too long to store on the computer. How to avoid this? In the algorithm in Section 27.5.4, before using n^2, take modulus m:

$$\text{power}(n,k,m) = \begin{cases} 1 & \text{if } k=0 \\ n \bmod m & \text{if } k=1 \\ n \cdot \text{power}(n^2 \bmod m, (k-1)/2, m) \bmod m & \text{if } k>1 \text{ and } k \text{ is odd} \\ \text{power}(n^2 \bmod m, k/2, m) & \text{if } k>1 \text{ and } k \text{ is even.} \end{cases}$$

Better yet, before doing this, substitute

$$n \leftarrow (n \bmod m).$$

This avoids the long number n^2 from the start. Instead, the recursion is applied to the moderate number

$$(n \bmod m)^2 \le (m-1)^2.$$

This way, throughout the entire calculation, all intermediate products are kept moderate, and never exceed $(m-1)^2$. To prove this, use mathematical induction on k (see exercises below).

Still, this seems a bit too pedant. After all, the modulus operation is expensive too, so why use it too often? Better use it only when absolutely necessary: when the algorithm is about to use a product too long to store.

15.4.4 Modular Inverse

Let p and q be some given coprime numbers (prime or not):

$$\text{GCD}(p, q) = 1.$$

Here is another important task: find the inverse of q modulus p:

$$q^{-1} \bmod p.$$

What is this? This is the unique solution x of the equation

$$qx \equiv 1 \bmod p.$$

Does x exist? Moreover, is x unique? To prove this, consider the set S, containing the integer numbers from 0 to $p - 1$:

$$S = \{0, 1, 2, \ldots, p - 1\}.$$

Define the new mapping

$$M : S \to S$$

by

$$M(s) = (qs \bmod p), \qquad s \in S.$$

Our aim is to show that M maps some number to 1. This number will then serve as our x. To show this, let's look at M, and study its properties. Is M one-to-one? To check on this, consider two numbers: $a, b \in S$ (say, $a \geq b$). Assume that

$$M(a) = M(b).$$

This means that

$$q(a - b) \equiv 0 \bmod p.$$

In other words, p divides $q(a - b)$:

$$p \mid q(a - b).$$

Since p shares no common divisor with q, it must divide $a - b$:

$$a \equiv b \bmod p.$$

Thus, M is indeed one-to-one. As such, it preserves the total number of elements in S:

$$|M(S)| = |S| = p.$$

Thus, M is a one-to-one mapping from S onto S. Thus, it has an inverse mapping:

$$M^{-1} : S \to S.$$

In particular, we can now define

$$x = M^{-1}(1).$$

Thus, the inverse of q modulus p exists uniquely:

$$\left(q^{-1} \bmod p\right) = M^{-1}(1).$$

15.4.5 Using The Extended Euclidean Algorithm

How to solve for x in practice? For this purpose, assume that $q < p$. Otherwise, just substitute

$$q \leftarrow q - p$$

time and again, until q gets small enough. (After all, this makes no difference to x.) Now, in the modular extended Euclidean algorithm (Section 15.3.6), substitute p for n, and q for m:

$$1 = \text{GCD}(p, q) \equiv bq \bmod p,$$

for a new integer number b between 0 and $p-1$, which will now serve as our solution:

$$x = b.$$

Thus, the inverse of q modulus p is now available explicitly:

$$\left(q^{-1} \bmod p\right) = b.$$

This will be useful below.

15.5 The Chinese Remainder Theorem

15.5.1 Modular Equation

We are now ready to prove the Chinese remainder theorem. The proof is not only theoretical but also constructive: it actually designs a practical algorithm.

Consider k natural numbers (prime or not), greater than 1:

$$p_1, \ p_2, \ p_3, \ \ldots, \ p_k > 1.$$

Assume that they are mutually coprime: any two of them share no common divisor. In other words, their greatest common divisor is 1:

$$\text{GCD}(p_i, p_j) = 1 \quad (1 \le i \ne j \le k).$$

Look at the first equation:

$$x \equiv d_1 \bmod p_1,$$

where $0 \le d_1 < p_1$ is a given (integer) number, and x is the desired (integer) solution.

What could x be? Well, this seems too easy: just define

$$x = d_1.$$

Still, this is not what we want. Here, we want x to have a special form.

15.5.2 Using The Coprime

Let P be the product of all p_i's:

$$P = p_1 p_2 p_3 \cdots p_k.$$

Likewise, let P_1 be the product of all p_i's but the first one:

$$P_1 = p_2 p_3 p_4 \cdots p_k = \frac{P}{p_1}.$$

We want x to have the special form

$$x = P_1 x_1.$$

How to design such an x? For this, we must first have x_1. What could x_1 be? Well, we already know what it is. After all, p_2, p_3, p_4, ..., p_k are coprime to p_1. Therefore, their product P_1 is coprime to p_1 as well:

$$\text{GCD}(P_1, p_1) = 1.$$

Let $\tilde{x}_1$ be the inverse of P_1 modulus p_1:

$$P_1 \tilde{x}_1 \equiv 1 \bmod p_1.$$

In fact, $\tilde{x}_1$ is available from the modular extended Euclidean algorithm (Section 15.4.5). Now, define

$$x_1 = d_1 \tilde{x}_1.$$

This way,

$$P_1 x_1 \equiv P_1 d_1 \tilde{x}_1 \equiv d_1 \bmod p_1,$$

as required. So, $P_1 x_1$ is indeed a legitimate solution to the first equation.

But this is not the end of it. To $P_1 x_1$, we could now add a multiple of p_1. For example, let P_2 be the product of all p_i's but p_2:

$$P_2 = p_1 p_3 p_4 p_5 \cdots p_k = \frac{P}{p_2}.$$

This way, to $P_1 x_1$, we could also add P_2, and design a new solution of the form

$$P_1 x_1 + P_2.$$

Indeed, this makes no harm: since P_2 is a multiple of p_1, the first equation is still satisfied.

What is all this good for? After all, we already solved the first equation long ago! Well, let's introduce more equations.

15.5.3 Modular System of Equations

Let's move on to a system of k modular equations:

$$\begin{aligned} x &\equiv d_1 \bmod p_1 \\ x &\equiv d_2 \bmod p_2 \\ x &\equiv d_3 \bmod p_3 \\ &\vdots \\ x &\equiv d_k \bmod p_k, \end{aligned}$$

where

$$\begin{aligned} 0 \le d_1 &< p_1 \\ 0 \le d_2 &< p_2 \\ 0 \le d_3 &< p_3 \\ &\vdots \\ 0 \le d_k &< p_k \end{aligned}$$

are given numbers.

All these equations must be solved simultaneously by the same x. This is not so easy any more. How to find a new x that satisfies all these equations at the same time?

Like before, let's solve each individual equation, and obtain its own solution. This way, we'll have not only x_1 but also x_2, x_3, x_4, ..., x_k. From these, the new x will emerge as a linear combination.

For this purpose, define more products of the form

$$\begin{aligned} P_1 &= \frac{P}{p_1} \\ P_2 &= \frac{P}{p_2} \\ P_3 &= \frac{P}{p_3} \\ &\vdots \\ P_k &= \frac{P}{p_k}. \end{aligned}$$

Each of these products contains all p_i's but one. Let's use them to define our new x:

$$x = P_1x_1 + P_2x_2 + P_3x_3 + \cdots + P_kx_k,$$

where

$$\begin{aligned} P_1x_1 &\equiv d_1 \bmod p_1 \\ P_2x_2 &\equiv d_2 \bmod p_2 \\ P_3x_3 &\equiv d_3 \bmod p_3 \\ &\vdots \\ P_kx_k &\equiv d_k \bmod p_k. \end{aligned}$$

This is the desired x that solves all our equations. For example, it solves the first equation. Indeed, it is not much different from P_1x_1: it just adds a few more multiples of p_1 that make no harm. The same is true for the other equations as well, as required.

15.5.4 Uniqueness

In the above process, x could become too large. To make it look better, let's make it moderate again: subtract P from it. After all, this makes no harm: since P is a product of all p_i's, all equations are still satisfied.

By subtracting P time and again, we can eventually make x as small as

$$0 \le x < P.$$

In this range, x is unique. Indeed, assume that

$$0 \le y < P$$

solves our system too. Without loss of generality, assume that

$$x \ge y.$$

Must x and y be the same? Well, look at their difference:

$$0 \le x - y < P.$$

It solves the homogeneous system

$$\begin{aligned} x - y &\equiv 0 \bmod p_1 \\ x - y &\equiv 0 \bmod p_2 \\ x - y &\equiv 0 \bmod p_3 \\ &\vdots \\ x - y &\equiv 0 \bmod p_k. \end{aligned}$$

In other words, $x - y$ is divided by all p_i's:

$$p_1,\ p_2,\ p_3, \ldots,\ p_k \mid x - y.$$

Thus, their product must divide $x - y$ as well:

$$P \mid x - y.$$

As a result, we must have

$$x - y = 0,$$

or

$$x = y.$$

Thus, in the above range, x is indeed unique. This is the Chinese remainder theorem.

15.6 How to Use The Remainder?

15.6.1 Remainder in Integer Division

What is a remainder? Well, we already used it in a few places. In Section 15.3.1, we looked at two natural numbers $n > m > 0$, and divided them with remainder. In other words, we wrote n in the new form

$$n = lm + r,$$

where l is some (nonnegative) integer number, and the remainder r is smaller than m:

$$0 \le r < m.$$

This was indeed the key to the (extended) Euclidean algorithm (Sections 15.3.4–15.3.5).

15.6.2 Remainder in a Binary Number

Even in its implicit form, the remainder is still useful. Indeed, in Section 27.4.2, we'll write a natural number k in its binary form:

$$k = b_m b_{m-1} b_{m-2} \cdots b_1 b_0,$$

where b_m, b_{m-1}, ..., b_1, b_0 are the binary digits (bits: 0 or 1), ordered rightwards: from the most significant to the least significant one. This is actually a binary polynomial:

$$k = b_0 + b_1 \cdot 2 + b_2 \cdot 2^2 + \cdots + b_m \cdot 2^m = \sum_{j=0}^{m} b_j 2^j = q(2),$$

where q is the polynomial

$$q(x) = \sum_{i=0}^{m} b_i x^i,$$

and x is an (independent) variable.

Given a natural number k, how to uncover its binary form? For this purpose, just divide by 2 with remainder:

$$k = 2l + b_0.$$

What is this remainder? We already know it: the least significant digit b_0. So, we uncovered one bit. What are the rest? Use recursion on l (used in the above formula). Its own least significant digit will serve as our b_1, and so on. Let's extend this to a more general polynomial as well.

15.6.3 Remainder in a Polynomial

Thanks to the remainder, we uncovered the bits, one by one. How? We divided the original number by 2, and uncovered its least meaningful bit: b_0. This could go on and on recursively, uncovering the rest as well.

What are the bits? They are coefficients in a binary polynomial. But why only binary? Why not do the same in just any polynomial of any kind? This is Horner's algorithm.

15.6.4 Remainder in Horner's Algorithm

In Section 27.2.3, we'll consider a polynomial of degree n:

$$p(x) = \sum_{i=0}^{n} a_i x^i,$$

where x is the (independent) variable, and the a_i's are some given coefficients: not necessarily binary any more, but real, or even complex.

Now, the monomial x is a polynomial in its own right (although quite simple). How to divide by it (with remainder)? We already know how: write p in its new form:

$$p(x) = a_0 + x\hat{p}(x),$$

where $\hat{p}(x)$ is a new polynomial:

$$\hat{p}(x) = a_1 + a_2 x + a_3 x^2 + \cdots + a_n x^{n-1} = \sum_{i=0}^{n-1} a_{i+1} x^i.$$

In Section 27.2.3, we'll use this to calculate p at a given point x. Here, on the other hand, we want more: to divide with remainder.

If you haven't noticed, we already got our remainder: the constant a_0. After all, this is a "polynomial" in its own right, albeit degenerate and boring. Furthermore, it is also "smaller" than x. In what sense? In sense of degree: x is of degree 1, and a_0 is of degree 0.

So, polynomials are just like numbers, and can benefit from number theory, and even from the Chinese remainder theorem. This will have a nice application in linear algebra.

15.7 Using The Chinese Remainder Theorem

15.7.1 Characteristic Polynomial: Root and Multiplicity

Let's go back to our $n \times n$ matrix: A. Denote its characteristic polynomial in a slightly different way:

$$P(x) = (x - \lambda_1)^{n_1} (x - \lambda_2)^{n_2} (x - \lambda_3)^{n_3} \cdots (x - \lambda_k)^{n_k},$$

where the λ_i's are the *distinct* eigenvalues of A. (This is not the same k as in Section 15.1.1.) This way, each eigenvalue is still a root of P. Indeed, at $x = \lambda_i$, P vanishes:

$$P(\lambda_i) = 0, \quad 1 \le i \le k.$$

Each λ_i comes with its own multiplicity: n_i. Together, the multiplicities sum to n:

$$n_1 + n_2 + n_3 + \cdots + n_k = n.$$

15.7.2 Multiplicity and Jordan Subspace

The multiplicity n_i has yet another interpretation: it is also the dimension of the null space of $(A - \lambda_i I)^{n_i}$, denoted by V_i:

$$V_i = \{v \mid (A - \lambda_i I)^{n_i} v = \mathbf{0}\} .$$

(See Section 15.2.4.) This is the ith Jordan subspace. It will be useful in the Jordan decomposition below. For this, in the Chinese remainder theorem, plug polynomials instead of numbers. In particular, instead of the p_i's used in the original version, plug new polynomials: factors from the characteristic polynomial P.

15.7.3 The Chinese Remainder Theorem With Polynomials

In the Chinese remainder theorem, we have to solve a system of equations modulus p_i: mutually coprime numbers. But here, we replace numbers with polynomials. After all, we've already established that numbers are a special kind of polynomials. So, consider a new system:

$$\begin{aligned} Q &\equiv \lambda_1 \bmod (x - \lambda_1)^{n_1} \\ Q &\equiv \lambda_2 \bmod (x - \lambda_2)^{n_2} \\ Q &\equiv \lambda_3 \bmod (x - \lambda_3)^{n_3} \\ &\vdots \\ Q &\equiv \lambda_k \bmod (x - \lambda_k)^{n_k} . \end{aligned}$$

To these equations, better add one more equation:

$$Q \equiv 0 \bmod x.$$

Later on, we'll see why. But this is not always necessary. For example, if $\lambda_1 = 0$, then we already have a stronger equation:

$$Q \equiv 0 \bmod x^{n_1} .$$

In this case, there is no need to add any new equation. Instead, just raise the power, to obtain a yet stronger equation:

$$Q \equiv 0 \bmod x^{n_1+1} .$$

(This is done for the sake of uniformity: the degrees of the p_i's sum now to $n + 1$.) This completes our new system.

In this system, what is Q? This is the unknown polynomial: $Q(x)$. In each equation, Q is congruent to a degenerate "polynomial" of degree zero: the constant λ_i. This congruence is modulus a factor from the characteristic polynomial: $(x - \lambda_i)^{n_i}$.

15.8 Matrix and Its Jordan Decomposition

15.8.1 The Solution Polynomial

Thanks to the Chinese remainder theorem, we can solve for the unknown polynomial: $Q(x)$. How does it look like? Well, thanks to our extra equation (Section 15.7.3), Q has factor x in it. Thus, once written as a sum, Q has no constant term at all: it starts from the linear term x^1 onwards. This will be useful later (Chapter 21).

Furthermore, Q is quite special: it is unique modulus $xP(x)$. In other words, Q is the only polynomial of degree n (or less) that solves our system: no other polynomial of degree n (or less) could be added to it.

In $Q(x)$, what does x stand for? Well, x is still unspecified: it could be just anything. Let's give it a concrete meaning.

15.8.2 The Diagonal Part

For this purpose, instead of x, plug the original matrix A. This way, $Q(A)$ is a legitimate $n \times n$ matrix: a linear combination of A and its powers.

What's so good about $Q(A)$? Well, it is diagonal. (More precisely, it is diagonalizable or semisimple, but we're not pedant.) In what sense? Well, thanks to the ith equation, it could be written as a constant, plus an irrelevant term: $(A - \lambda_i I)^{n_i}$.

Why is this irrelevant? Because, in the null space of $(A - \lambda_i I)^{n_i}$, it is always zero. (This is our Jordan subspace V_i.) Thus, in V_i, it is completely invisible, and has no effect at all. In V_i, the only relevant term is the constant $\lambda_i I$. This is true for every $1 \le i \le k$.

15.8.3 The Nilpotent Error

Our new matrix $Q(A)$ approximates A well: the error $A - Q(A)$ is "small." In what sense? Well, thanks to the ith equation, it satisfies

$$A - Q(A) \equiv A - \lambda_i I \mod (A - \lambda_i I)^{n_i}.$$

In V_i, this becomes an equality:

$$(A - Q(A)) \mid_{V_i} = (A - \lambda_i I) \mid_{V_i}.$$

Raise this to power n_i:

$$(A - Q(A))^{n_i} \mid_{V_i} = (A - \lambda_i I)^{n_i} \mid_{V_i} = (0),$$

let alone

$$(A - Q(A))^n \mid_{V_i} = (0).$$

This is true for all $1 \le i \le k$, so

$$(A - Q(A))^n = (0).$$

Thus, the error is nilpotent. In this sense, $Q(A)$ is dominant, and approximates A well.

15.8.4 The Jordan Decomposition

Thanks to the Chinese remainder theorem, A was written as the sum of two parts: its diagonal part, plus a less important part – a nilpotent error:

$$A = Q(A) + (A - Q(A)).$$

Let's use this in a few examples.

15.9 Example: Space of Polynomials

15.9.1 Polynomial and Its Derivatives

Let's see a few examples. Consider a polynomial of degree n (or less):

$$p(x) = a_0 + a_1 x + a_2 x^2 + \cdots + a_n x^n = \sum_{i=0}^{n} a_i x^i.$$

This could be viewed as an $(n+1)$-dimensional vector:

$$(a_0, a_1, a_2, \ldots, a_n)^t.$$

On p, define the differentiation operator:

$$Dp = p'.$$

As a linear operator, D has a Jordan form. As a matter of fact, the Jordan decomposition is just

$$D = (0) + D.$$

Why? Because the "error" D is nilpotent. Indeed, its nth power is the nth derivative of p, which is just a constant:

$$D^n p = p^{(n)} = n! a_n$$

(Section 27.6.3). Thus, the next derivative vanishes:

$$D^{n+1} p = p^{(n+1)} = 0,$$

which means nilpotency.

15.9.2 Jordan Block

In fact, D has just one Jordan block. How to design it? Pick a degenerate "polynomial:" the constant 1. Is this an eigenvector of D? To check on this, just apply D to it:

$$D1 = 1' = 0.$$

Thus, 1 is a singular eigenvector, of eigenvalue 0. How to design a generalized eigenvector? Take the antiderivative. This is the linear monomial: x:

$$Dx = x' = 1.$$

Next, pick the antiderivative of x, and so on:

$$\left\{1, x, \frac{x^2}{2}, \frac{x^3}{3!}, \ldots, \frac{x^n}{n!}\right\}.$$

This basis spans the entire space: all polynomials of degree n (or less). On it, how does D act? Well, D maps each one to the previous one (except for the first one, which is mapped to 0). Thus, in this basis, D takes its Jordan form:

$$D = \begin{pmatrix} 0 & 1 & & & & \\ & \ddots & \ddots & & & \\ & & \ddots & \ddots & & \\ & & & 0 & 1 & \\ & & & & 0 & 1 \\ & & & & & 0 \end{pmatrix}.$$

This is an $(n+1) \times (n+1)$ matrix. On its superdiagonal, it has 1's. (All the rest vanish.) And what are its powers? Shift the 1's rightwards:

$$D^2 = \begin{pmatrix} 0 & & 1 & & & \\ & \ddots & & \ddots & & \\ & & \ddots & & \ddots & \\ & & & 0 & & 1 \\ & & & & 0 & & 1 \\ & & & & & 0 & \\ & & & & & & 0 \end{pmatrix}$$

$$D^3 = \begin{pmatrix} 0 & & & 1 & & & \\ & \ddots & & & \ddots & & \\ & & \ddots & & & \ddots & \\ & & & 0 & & & 1 \\ & & & & 0 & & \\ & & & & & 0 & \\ & & & & & & 0 \end{pmatrix}$$

$$\vdots$$

$$D^n = \begin{pmatrix} 0 & & & & & 1 \\ & \ddots & & & & \\ & & \ddots & & & \\ & & & 0 & & \\ & & & & 0 & \\ & & & & & 0 \\ & & & & & & 0 \end{pmatrix}.$$

The next power vanishes:

$$D^{n+1} = (0),$$

which means nilpotency again.

15.9.3 Polynomials of Any Degree

So far, n was fixed. Still, why stop at n? Why not go on and on indefinitely? For this, consider now the infinite-dimensional space of all polynomials (of any degree). On it, we still have the same Jordan decomposition:

$$D = (0) + D.$$

These are now operators: infinite matrices. The "error" D is now "nilpotent" in a new (nonstandard) sense: a given polynomial p, if differentiated sufficiently many times, gives zero. For instance, if p has degree n, then

$$D^{n+1}p = 0.$$

Still, $n + 1$ is no longer global: it depends on p. This is no longer nilpotency in its standard sense. The Jordan block still looks the same, except infinite: it has no bottom at all:

$$D = \begin{pmatrix} 0 & 1 & & & & & \\ & \ddots & \ddots & & & & \\ & & \ddots & \ddots & & & \\ & & & 0 & 1 & & \\ & & & & 0 & 1 & \\ & & & & & \ddots & \ddots \end{pmatrix}.$$

15.10 Eigenfunctions

15.10.1 The Exponent Eigenfunction

Into our linear space, throw a new function: $\exp(\lambda x)$, where λ is constant (real, or even complex). How to differentiate it? Like a composite function:

$$D\exp(\lambda x) = \lambda \exp'(\lambda x) = \lambda \exp(\lambda x).$$

This is a new eigenfunction of D, with a new eigenvalue: λ. As such, it lies in the null space of $D - \lambda$:

$$(D - \lambda)\exp(\lambda x) = 0.$$

Thus, this exponent belongs to a new Jordan subspace. How to design it, and define the generalized eigenfunctions?

15.10.2 Leibniz Rule

Into our linear space, throw yet another function: exponent times polynomial. How to differentiate it? As in Leibniz rule (the product rule):

$$D(p\exp(\lambda x)) = (Dp)\exp(\lambda x) + pD\exp(\lambda x).$$

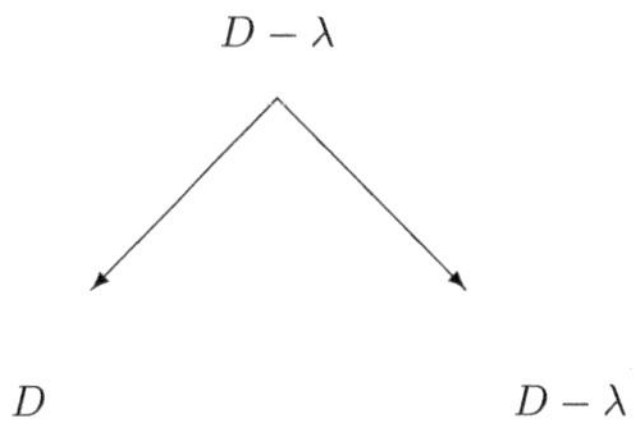

Fig. 15.5. The original operator $D - \lambda$ splits: D on the left, and $D - \lambda$ on the right.

15.10.3 Virtual Binary Tree

Rewrite this as:

$$(D - \lambda)(p \exp(\lambda x)) = (Dp) \exp(\lambda x) + p(D - \lambda) \exp(\lambda x).$$

(The latter term will soon drop, but we leave it for now.) This could be viewed as a new (virtual) binary tree (Figure 15.5). In it, the original operator $D - \lambda$ splits: D on the left, and $D - \lambda$ on the right.

This is a two-level tree. In the next higher power, on the other hand, we'll get a three-level tree.

15.10.4 Three-Level Tree: Pascal's Triangle

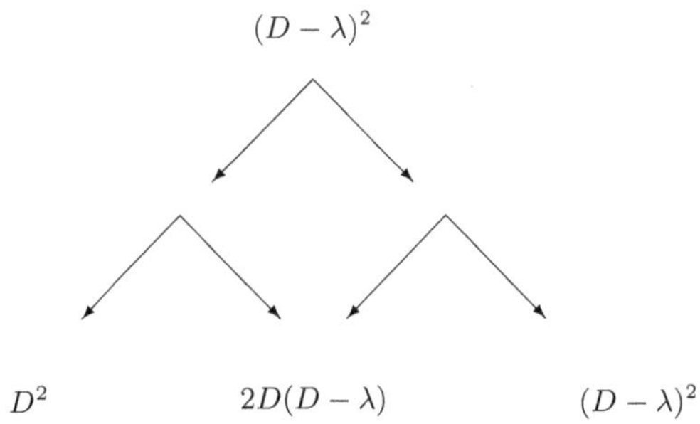

Fig. 15.6. The original operator $(D - \lambda)^2$ splits: D^2 on on the left, $(D - \lambda)^2$ on the right, and $2D(D - \lambda)$ in between (as in Pascal's triangle).

Next, look at a higher power: $(D-\lambda)^2$. How to split it? Use the same trick twice:

$$\begin{aligned}&(D-\lambda)^2(p\exp(\lambda x))\\&=(D-\lambda)(D-\lambda)(p\exp(\lambda x))\\&=(D-\lambda)\left((Dp)\exp(\lambda x)+p(D-\lambda)\exp(\lambda x)\right)\\&=\left(D^2p\right)\exp(\lambda x)+2(Dp)(D-\lambda)\exp(\lambda x)+p(D-\lambda)^2\exp(\lambda x)\end{aligned}$$

(Figure 15.6). Still, this is a rather nonstandard "tree:" at the bottom-middle, both inner branches unite, producing the new coefficient 2. This is as in the second level in Pascal's triangle:

$$1,\ 2,\ 1.$$

15.10.5 How to Mirror The Binomial?

Next, extend this even to a more general power: $(D-\lambda)^k$. This will mirror the binomial formula. In it, most terms will vanish, and only one will survive:

$$\begin{aligned}(D-\lambda)^k(p\exp(\lambda x))&=\sum_{i=0}^{k}\binom{k}{i}D^ip(D-\lambda)^{k-i}\exp(\lambda x)\\&=\exp(\lambda x)D^kp\\&=\exp(\lambda x)p^{(k)}.\end{aligned}$$

(In particular, if k is bigger than the degree of p, then this vanishes.) This is our (infinite-dimensional) Jordan subspace: all polynomials times $\exp(\lambda x)$. This is spanned by its Jordan basis: all monomials times $\exp(\lambda x)$:

$$\operatorname{span}_{k\ge 0}\left\{\frac{x^k}{k!}\exp(\lambda x)\right\}.$$

In this basis, D takes its (infinite) Jordan form:

$$D=\begin{pmatrix}\lambda & 1 & & & & & \\ & \ddots & \ddots & & & & \\ & & \ddots & \ddots & & & \\ & & & \lambda & 1 & & \\ & & & & \lambda & 1 & \\ & & & & & \ddots & \ddots\end{pmatrix}.$$

This is one Jordan block. But the same could be done for all λ's (real, or even complex). This makes infinitely many Jordan blocks, telling us how D acts on a big linear space: all polynomials times all exponents (spanned by its Jordan basis: all monomials times all exponents):

$$\operatorname{span}_{k\ge 0,\ \lambda\in\mathbb{C}}\left\{\frac{x^k}{k!}\exp(\lambda x)\right\}.$$

What is the key to all this? This is our good old exponent. In it, sum is invariant to product:

$$\exp(x)\exp(y)=\exp(x+y).$$

Let's see this principle in a more general framework.

15.11 Algebra and Its Derivation

15.11.1 Leibniz Rule

What is Leibniz rule? So far, it was the same as the product rule, telling us how to differentiate a product of two functions:

$$(fg)' = f'g + fg'.$$

Let's extend to a more general setting, where f and g are not necessarily functions, but members of an algebra: a vector space with one more operation.

15.11.2 Product – Multiplication

In a linear space, you can add vectors, but not multiply. In an algebra, on the other hand, you can: if f and g are in the algebra, then their product fg is in the algebra as well. This way, the algebra is closed under multiplication.

This is vector-times-vector. (Don't confuse it with scalar-times-vector.) How does it work? We don't know yet. We only know that it is distributive (and even bilinear), but not necessarily commutative or even associative (see exercises below).

15.11.3 Derivation

In an algebra, consider a linear operator D: a linear mapping that acts on the algebra, and maps each vector to some vector. Here, D doesn't have to mean differentiation any more: it is much more general.

Still, D could be a derivation. For this, it must behave like one: obey a new Leibniz rule:

$$D(fg) = (Df)g + fDg.$$

This way, in algebraic terms, D indeed behaves like a legitimate derivation: to act on a product, act on the former vector, then on the latter, then add up.

This is more general than before. We can now see that all what we did before (when D was a differentiation) was just an example. Let's extend it to our new setting: an algebra, and its derivation D.

15.12 Product and Its Derivation

15.12.1 Two-Level Tree

Our new Leibniz rule makes a new two-level tree (Figure 15.7, which mirrors Figure 15.5). Better yet, here the picture is even more symmetric. It uses two parameters: τ and λ. Now, to our algebra, apply the new operator $D - \tau - \lambda$. Thanks to our new Leibniz rule,

$$(D - \tau - \lambda)(fg) = ((D - \tau)f)g + f(D - \lambda)g.$$

This is how $D-\tau-\lambda$ splits: $D-\tau$ acts on f, and $D-\lambda$ on g. What's so good about this? Well, if f and g are eigenvectors of D (with eigenvalues τ and λ, respectively), then fg is an eigenvector as well (with eigenvalue $\tau + \lambda$, provided that $fg \neq \mathbf{0}$). Let's raise the power.

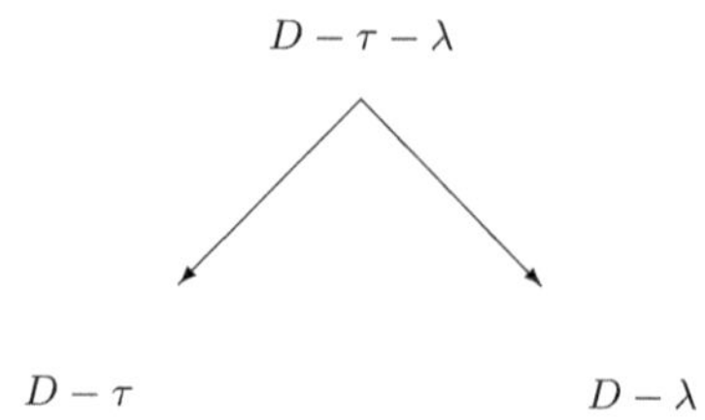

Fig. 15.7. The original operator $D - \tau - \lambda$ splits: $D - \tau$ on the left, and $D - \lambda$ on the right.

15.12.2 Multilevel Tree

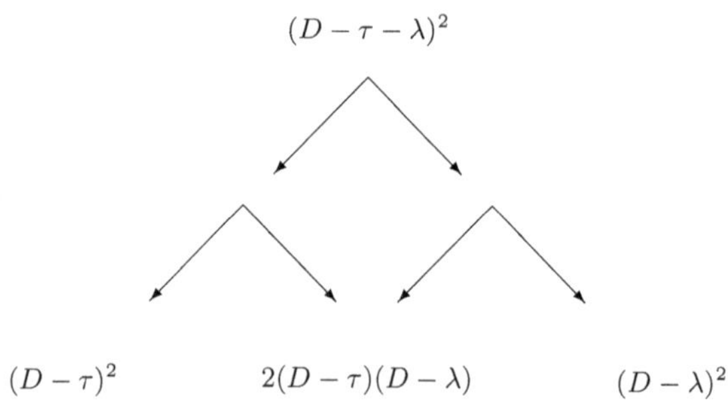

Fig. 15.8. The original operator $(D - \tau - \lambda)^2$ splits: $(D - \tau)^2$ on the left, $(D - \lambda)^2$ on the right, and $2(D - \tau)(D - \lambda)$ in between (as in Pascal's triangle).

And how about $(D - \tau - \lambda)^2$? How does it split? This makes a three-level tree (Figure 15.8, which mirrors Figure 15.6). But this is nonstandard: tt the bottom-middle, two inner branches unite, as in the second level in Pascal's triangle:

$$\begin{aligned}
(D - \tau - \lambda)^2 (fg) &= (D - \tau - \lambda)(D - \tau - \lambda)(fg) \\
&= (D - \tau - \lambda) \left(((D - \tau)f)\, g + f(D - \lambda)g \right) \\
&= \left((D - \tau)^2 f \right) g + 2(D - \tau) f (D - \lambda) g + f (D - \lambda)^2 g.
\end{aligned}$$

15.12.3 Pascal's Triangle – Binomial Formula

And how about the kth power? It splits as in the binomial formula (the kth level in Pascal's triangle):

$$(D-\tau-\lambda)^k(fg)=\sum_{i=0}^{k}\binom{k}{i}(D-\tau)^i f(D-\lambda)^{k-i}g.$$

We can now see that Section 15.10.5 was just a special case of this (with $\tau = 0$).

15.13 Product and Its Jordan Subspace

15.13.1 Two Jordan Subspaces

Does D have a Jordan decomposition? We don't know! After all, the Jordan decomposition was designed only in finite dimension (Section 15.8.4). Our algebra, on the other hand, could be bigger than that: infinite-dimensional.

Still, let's work the other way around: from the bottom up. Start from two individual Jordan subspaces, and merge them into a new (bigger) Jordan subspace, and so on. In the end, you'll get a complete subalgebra, in which D has its Jordan form, as required.

For this purpose, assume that D has (at least) two Jordan subspaces, corresponding to two eigenvalues: τ and λ. Pick f from the former, and g from the latter. What about their product fg? Does it belong to any Jordan subspace?

15.13.2 Product: A New Jordan Subspace

To check on this, look at the sum in Section 15.12.3. In each term in it, there is a high power: either of $D-\tau$ (applied to f), or of $D-\lambda$ (applied to g). Thus, one could pick k so large that all terms vanish. Thus, fg belongs to a new Jordan subspace, corresponding to eigenvalue $\tau+\lambda$.

15.13.3 Example: Polynomials times Exponents

Is this familiar? In our original example, D was differentiation, applied to a product: polynomial times exponent (Section 15.10.5). Each was in its own Jordan subspace: the polynomial was associated with eigenvalue 0, and the exponent with eigenvalue λ. This is why their product was associated with

$$0+\lambda=\lambda.$$

Thanks to our new multilevel tree (Figure 15.8), this is now extended to $\tau \neq 0$ as well. For this purpose, consider the new product

$$q(x)\exp(\tau x)\cdot p(x)\exp(\lambda x).$$

Its former half is associated with eigenvalue τ, and its latter with eigenvalue λ. This is why their product

$$q\exp(\tau x)\cdot p\exp(\lambda x)=(qp)\exp((\tau+\lambda)x)$$

is in a new Jordan subspace, associated with $\tau+\lambda$.

15.14 Subalgebra: Examples

15.14.1 Subalgebra

A general algebra may be too big, and contain vectors that belong to no Jordan subspace at all. How to exclude them? Restrict to a subalgebra, spanned by the generalized eigenvectors of D: complete Jordan subspaces. Why is this a subalgebra? Because it is an algebra in its own right: closed under addition and multiplication alike. Indeed, pick f from one Jordan subspace and g from another, and get fg in yet another.

15.14.2 Subalgebra: Polynomials times Exponents

What was our original example? We started from a big algebra: C^{∞} (all smooth functions). In it, you can differentiate as many times as you like. Then, we focused on a small subalgebra: polynomials times exponents. In it, differentiation has its Jordan form. In it, all Jordan subspaces look alike: one particular exponent times all polynomials.

15.14.3 Finite-Dimensional Algebra

Let's see yet another example: a finite-dimensional algebra. (In the exercises below, we'll see a few examples in physics.) In it, we already saw that every linear mapping has its Jordan form (Section 15.8.4). This applies to our D too (even if it wasn't a derivation): pick f from one Jordan subspace and g from another, and get fg in yet another, as required. Next, let's go back to a general algebra, and see a nice property.

15.15 The Diagonal Part

15.15.1 Is It a Derivation?

Let's go back to a general algebra. In our subalgebra, we already decomposed D in its Jordan decomposition:

$$D = \Lambda + (D - \Lambda),$$

where Λ is diagonal, and the error $D - \Lambda$ is nilpotent. (If our subalgebra is infinite-dimensional, then this is a nonstandard nilpotency: Section 15.9.3.) For this, we assumed that D was a derivation: it obeys its own Leibniz rule. But what about its diagonal part Λ? Is it a derivation in its own right? In other words: does Λ obey its own Leibniz rule? To check on this, pick two vectors from our subalgebra: f and g. Assume that f comes from a Jordan subspace associated with τ. Since Λ is diagonal,

$$\Lambda f = \tau f.$$

Likewise,

$$\Lambda g = \lambda g.$$

And what about fg? It must belong to the Jordan subspace associated with $\tau + \lambda$:

$$\begin{aligned}\Lambda(fg) &= (\tau + \lambda)(fg) \\ &= \tau(fg) + \lambda(fg) \\ &= (\tau f)g + f\lambda g \\ &= (\Lambda f)g + f\Lambda g.\end{aligned}$$

This is a new Leibniz rule for Λ. Thus, Λ is a derivation in its own right.

15.16 Exercises: Hermitian Matrix and Its Eigenbasis

15.16.1 Binary Numbers and Polynomials

1. Consider a natural number n, in its binary representation. What is the rightmost (least significant) binary digit (bit)?
2. Assume that n is even. What is its least significant bit? Hint: 0.
3. Assume that n is odd. What is its least significant bit? Hint: 1.
4. Either way, what is its least significant bit? Hint: $n \bmod 2$.
5. Divide n by 2. What is the remainder? Hint: $n \bmod 2$.
6. How is the remainder related to the least significant bit? Hint: they are the same.
7. What is the next (more significant) bit? Hint: divide n by 2. Disregard the remainder. Look at the result: $\lfloor n/2 \rfloor$. Find its least significant bit.
8. What are the rest? Hint: use recursion (Section 15.6.2).
9. How is this related to Horner's algorithm? Hint: both use the same recursion: divide by x (or 2) with remainder, to shift a list of numbers, time and again.
10. What is the difference? Hint: in the above, we shift the (unknown) bits, to uncover them, one by one. In Horner's algorithm, on the other hand, we work the other way around: shift the (well-known) coefficients, and use them, one by one, to uncover the (unknown) value.
11. Do this in yet another base: base 3 rather than 2.
12. Pick an (arbitrarily long) n at random. Look at its representation in base 3. How likely is it to contain 0's and 2's only, but no 1's at all? Hint: the probability for this is as low as zero.
13. How to add two (arbitrarily long) binary numbers? Hint: write the longer one in the upper line. Write the other one just below it, bit by bit. Start from the rightmost (least significant) bit. Add, bit by bit, all the way leftwards. (As your teacher always said, don't forget the carry!)
14. How to add two polynomials? Hint: likewise: monomial by monomial.
15. How to subtract two binary numbers? Hint: assume that the former is bigger. (Otherwise, interchange their roles, and change sign in the end.) Write it in the upper line. Write the other one just below it, bit by bit. Start from the rightmost bit. Subtract vertically, bit by bit, all the way leftwards.
16. How to subtract two polynomials? Hint: likewise, monomial by monomial.
17. How to multiply two binary numbers? Hint: vertically, leftwards, bit by bit.
18. Is there a better way? Hint: try FFT (Chapter 5 in [47]).
19. How to multiply two polynomials? Hint: likewise, monomial by monomial.
20. Is there a better way?

21. How to divide two binary numbers (with remainder)? Hint: vertically, rightwards, bit by bit (assuming that the latter is nonzero).
22. How to divide two polynomials (with remainder)? Hint: likewise, monomial by monomial.
23. After the division, look at the remainder. What is it? Hint: the former modulus the latter.
24. Use the above to write the Euclidean algorithm for polynomials, to find their greatest common divisor: a new polynomial that divides them both (evenly).
25. Write the extended Euclidean algorithm (Section 15.3.5) for polynomials. This will write their greatest common divisor as their "sum," with some "coefficients," which could be polynomials in their own right.
26. Apply the Chinese remainder theorem, to polynomials.
27. Why is this legitimate? Hint: polynomials mirror numbers, and make a new mathematical ring.
28. Look at two polynomials: $(x - \lambda_1)^{n_1}$, and $(x - \lambda_2)^{n_2}$. Are they coprime? Hint: only if $\lambda_1 \neq \lambda_2$.
29. Look at the solution $Q(x)$. Is it unique? In what sense? Hint: no polynomial of degree n (or less) could be added to it.
30. Show that, if A is an $n \times n$ matrix, then $Q(A)$ is an $n \times n$ matrix as well.
31. Show that $Q(A)$ is diagonal (semisimple).
32. Show that the error $A - Q(A)$ is nilpotent.

15.16.2 Nilpotent Hermitian Matrix

1. Let B be an $l \times l$ nilpotent matrix. Could B have a nonzero eigenvalue? Hint: if it did, then
$$B^i v = \lambda^i v \neq \mathbf{0}, \quad i = 0, 1, 2, 3, \ldots,$$
in violation of nilpotency.
2. Conclude that B has one eigenvalue only: 0.
3. Look at a short cycle of B, of length 2 only:
$$w, \ Bw.$$
What can you say about Bw? Hint: it is a singular eigenvector:
$$B(Bw) = \mathbf{0}.$$
4. Assume also that B is Hermitian. Could it have such a cycle? Hint: no! This would lead to:
$$0 = (\mathbf{0}, w) = (B(Bw), w) = (Bw, Bw) > 0.$$
5. So, B has no cycle of length 2. Still, could it have a cycle of length 3, such as
$$w, \ Bw, \ B^2 w?$$
Hint: no! Since B is Hermitian, this would lead to:
$$0 = (\mathbf{0}, Bw) = \left(B(B^2 w), Bw\right) = \left(B^2 w, B^2 w\right) > 0.$$

6. Could B have a longer cycle? Hint: no! Use the same trick.
7. Conclude that B has only short "cycles," of length 1 each.
8. Conclude that B has no generalized eigenvectors but its eigenvectors.
9. Conclude that B has l linearly-independent eigenvectors.
10. Orthonormalize them. Hint: use a Gram-Schmidt process (Section 9.11.1).
11. What is their joint eigenvalue? Hint: 0 (since B is also nilpotent).
12. Conclude that B is the zero matrix.

15.16.3 Hermitian Matrix: Orthonormal Eigenbasis

1. Let A be a Hermitian matrix (nilpotent or not).
2. Let λ_1 be an eigenvalue of multiplicity n_1.
3. What can you say about λ_1? Hint: it is real (because A is Hermitian).
4. Pick two eigenvectors, corresponding to distinct eigenvalues. What can you say about them? Hint: they are orthogonal to one another (Since A is Hermitian).
5. And what if they correspond to the same eigenvalue? How to orthogonalize them? Hint: use a Gram-Schmidt process.
6. Normalize them.
7. Define the subspace V (Section 15.2.2).
8. What is V? Hint: the null space of $(A - \lambda_1 I)^{n_1}$.
9. Define the matrix J (Section 15.2.2).
10. What is J? Hint: the restriction of A to V.
11. What is the order of J? Hint: n_1.
12. Is J Hermitian? Hint: yes. Indeed, for every two vectors $p, q \in V$, J behaves just like A:
$$(Jp, q) = (Ap, q) = (p, Aq) = (p, Jq).$$
13. Let I be the identity matrix of order n_1. Look at the matrix $J - \lambda_1 I$.
14. Is it Hermitian? Hint: yes (since λ_1 is real).
15. Is it nilpotent? Hint: yes. Indeed, its n_1th power vanishes. Indeed, for every vector $q \in V$,
$$(J - \lambda_1 I)^{n_1} q = (A - \lambda_1 I)^{n_1} q = \mathbf{0}.$$
16. Conclude that $J - \lambda_1 I$ is nilpotent and Hermitian at the same time.
17. Conclude that
$$J - \lambda_1 I = (0)$$
(the zero matrix).
18. Conclude that
$$J = \lambda_1 I.$$
19. Do the same for the other eigenvalues too.
20. Conclude that A has a diagonal Jordan form, corresponding to an orthonormal eigenbasis.

16

Design Your Lie Algebra

By now, we defined an algebra. It is time to define a Lie algebra, and see its properties. In it, each member has an adjoint. Exponentiate it, and have an automorphism that preserves the original Lie algebra, and its algebraic structure: its Lie brackets. To make things more concrete, we'll also design a few specific Lie algebras, used later in quantum and classical mechanics alike.

In this chapter, we use our active style again: questions, followed by their answers. This way, you get to see the theory develop, before your very eyes.

16.1 Derivation and Its Exponent

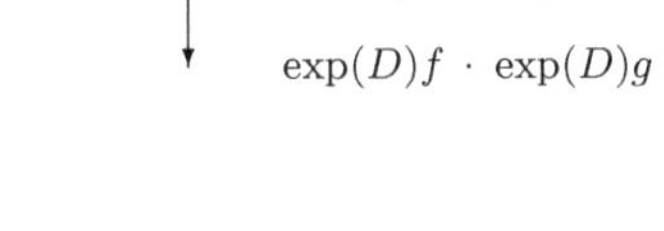

Fig. 16.1. $\exp(D)$ is invariant. In other words, order doesn't matter: you can apply $\exp(D)$ either before or after multiplication.

1. What is an algebra? Hint: a vector space with a new operation: product or multiplication (Section 15.11.1). (This is vector-times-vector, not just scalar-times-vector.)
2. Is it closed under this kind of multiplication? In what sense?

3. What are the properties of this product? Hint: distributive (and even bilinear), but not necessarily commutative or even associative (see below).
4. Let D be a derivation. What does this mean? Hint: it obeys its own Leibniz rule (Section 15.11.3).
5. Write its new Leibniz rule. Hint: pick f and g from the algebra. Then:

$$D(fg) = Df \cdot g + fDg.$$

6. Assume that D is nilpotent. What does this mean? Hint: for some (sufficiently large) m, D^m vanishes (maps every vector to the zero vector).
7. Show that, for every $l \geq m$, D^l vanish as well. Hint: as a linear mapping, D maps the zero vector to itself.
8. Define its exponent:

$$\exp(D) = \sum_{i=0}^{\infty} \frac{D^i}{i!}.$$

9. Is this really infinite? Hint: only m leading terms survive, and the rest vanish. (Later on, in the applications, we'll see an infinite series too.)
10. Is this its Jordan decomposition? Hint: in this series, the first term makes the diagonal part: the identity mapping. The rest is the "tail:"

$$D + \frac{D^2}{2} + \frac{D^3}{3!} + \cdots + \frac{D^{m-1}}{(m-1)!} = \sum_{i=1}^{m-1} \frac{D^i}{i!}.$$

This is nilpotent. Indeed, look at its mth power. It only contains high powers of D that vanish anyway.

11. What is the inverse of $\exp(D)$? Hint: $\exp(-D)$.
12. Write the inverse in terms of the above tail. Hint: design a new (finite) power series. In it, plug the tail and its powers (with alternating signs).
13. How does D (and its powers) act on a product like fg? Hint: mirror the binomial formula:

$$D^n(fg) = \sum_{i=0}^{n} \binom{n}{i} D^i f \cdot D^{n-i} g$$

(Figures 15.7–15.8).

14. In view of this formula, is D^2 a derivation in its own right? Hint: no! It obeys no Leibniz rule:

$$D^2(fg) = D^2 f \cdot g + 2Df \cdot Dg + fD^2 g \neq D^2 f \cdot g + fD^2 g$$

(in general).

15. Likewise, for any $1 < l < m$, is D^l a derivation in its own right? Hint: no! There are intermediate terms.
16. Consider the real exponent function: $\exp(x)$. Is it invariant under sum-product? In what sense? Hint:

$$\exp(x + y) = \exp(x)\exp(y).$$

17. So far, we saw how D (and its powers) act on a product like fg. And how about $\exp(D)$? Hint: mirror the above formulas, and obtain

$$\exp(D)(fg) = \exp(D)f \cdot \exp(D)g.$$

18. By now, what can you say about $\exp(D)$? Hint: it is linear, invertible, and product-preserving (Figure 16.1).
19. Is it a derivation? Hint: no! It obeys no Leibniz rule:

$$\exp(D)(fg) = \exp(D)f \cdot \exp(D)g \neq \exp(D)f \cdot g + f \cdot \exp(D)g$$

(in general).
20. Let E be yet another mapping of the same kind: linear, invertible, and product preserving, but not necessarily a derivation.
21. Denote its inverse by E^{-1}.
22. Look at the composition of E with its inverse:

$$EE^{-1} = E^{-1}E.$$

23. What is this? Hint: the identity mapping that changes nothing.
24. Is it of the same kind? Hint: since it changes nothing, it is indeed linear, invertible, and product preserving (but not a derivation).
25. Recall our original derivation: D. Look at a triple composition:

$$EDE^{-1}.$$

26. How does it act? Hint: leftwards, in three stages: first, apply E^{-1}. Then, apply D. Finally, apply E.
27. Is it a derivation in its own right? Hint: apply it to a product like $EfEg$:

$$\begin{aligned} EDE^{-1}(Ef \cdot Eg) &= ED(fg) \\ &= E(Df \cdot g + fDg) \\ &= EDfEg + EfEDg \\ &= EDE^{-1}EfEg + EfEDE^{-1}Eg. \end{aligned}$$

This is indeed a legitimate Leibniz rule.
28. What are its powers? Hint:

$$\left(EDE^{-1}\right)^l = ED^lE^{-1}$$

($l \geq 0$). Indeed, the inner E and E^{-1} cancel out.
29. Is EDE^{-1} nilpotent?
30. Conclude that it has an exponent.
31. How does it look like? Hint:

$$\exp\left(EDE^{-1}\right) = E\exp(D)E^{-1}.$$

32. Use our original algebra to design a new algebra. Hint: see below.
33. Look at all linear mappings from our original algebra into itself. Is this a legitimate linear space? Hint: verify that a linear combination of two such mappings is still linear, and still maps our algebra into itself.
34. Look at two such mappings: F and G. How to multiply them? Hint: use composition: $F \circ G$.
35. How does $F \circ G$ act? Hint: leftwards. First, apply G. On the result, apply F.

36. Is the new algebra indeed closed under this kind of multiplication? Hint: is $F \circ G$ linear? To check on this, open parentheses in $F \circ G(\alpha f + \beta g)$, where α and β are some scalars.
37. Is this multiplication distributive (and even bilinear)? Hint: let H be yet another linear mapping. Open parentheses in $(\alpha F + \beta G) \circ H$ and $H \circ (\alpha F + \beta G)$.
38. Is this multiplication associative? Hint: yes: composition is.
39. Is it commutative? Hint: no! Composition isn't (in general).
40. Give a few examples of such linear mappings. Hint: D and its powers and $\exp(D)$.
41. In general, mappings don't commute. Still, some do. For example, does D commute with $\exp(D)$? Hint: yes: both are polynomials in D.
42. On our new algebra, define the new operator '$F\circ$' that converts an old mapping G into a new mapping: $F \circ G$.
43. Is it linear? Hint: open parentheses in $F \circ (\alpha G + \beta H)$.
44. Is it a derivation? Hint: no! It obeys no Leibniz rule:

$$F \circ (G \circ H) = (F \circ G) \circ H \neq (F \circ G) \circ H + G \circ (F \circ H)$$

(in general).

16.2 Lie Algebra and Its Adjoint

1. Consider now a special kind of algebra: a Lie algebra [16, 20, 25, 30, 57].
2. It uses a special kind of (bilinear) product, denoted by Lie brackets:

$$\text{Instead of} \quad fg, \quad \text{use now} \quad [f, g].$$

3. What is our Lie algebra? Is it a vector space? Hint: yes.
4. Over what field? Hint: over $\mathbb{C}$, or at least over $\mathbb{R}$ (as in u_n or su_n, defined later).
5. Is it even an algebra? Of what kind? Hint: usually, it is nonassociative: the Lie brackets are bilinear, but often nonassociative.
6. Still, our Lie brackets have a few attractive properties.
7. First of all, they are alternative:

$$[f, f] = \mathbf{0}$$

(the zero vector).

8. Use this (and bilinearity) to prove antisymmetry:

$$[f, g] = -[g, f].$$

Hint:

$$\mathbf{0} = [f + g, f + g] = [f, f] + [f, g] + [g, f] + [g, g] = [f, g] + [g, f].$$

9. Work the other way around: assume antisymmetry, and prove alternativity. Hint: thanks to antisymmetry,

$$[f, f] = -[f, f],$$

so $[f, f]$ must be the zero vector.

10. Besides, a Lie algebra must also satisfy the Jacobi identity (see below).
11. Let h be a fixed member of our Lie algebra. (Later on, in the applications, h will often stand for the Hamiltonian.) Define a special kind of "differentiation:"

$$f' \equiv [f, h].$$

12. To be a legitimate derivation, it must obey a new Leibniz rule. How should it look like? Hint: our product is denoted now by Lie brackets. So, the new Leibniz rule takes the form:

$$[f, g]' = [f', g] + [f, g'].$$

13. Write this more explicitly. Hint: thanks to the definition of f', g', and $[f, g]'$,

$$[[f, g], h] = [[f, h], g] + [f, [g, h]].$$

14. Use antisymmetry to rewrite this as:

$$[[f, g], h] + [[h, f], g] + [[g, h], f] = \mathbf{0}$$

(the zero vector).
15. This is the Jacobi identity. A Lie algebra must satisfy it.
16. Is zero an eigenvalue of the above derivation? What is its eigenvector? Hint: the eigenvector is h itself. Indeed,

$$h' = [h, h] = \mathbf{0}.$$

17. Does our derivation have a Jordan form? Give an example. Hint: if our Lie algebra is finite-dimensional, then every linear mapping has its Jordan form.
18. If our derivation has a Jordan decomposition, what about its diagonal part? Is it a derivation in its own right? In other words, does it obey its own Leibniz rule? Hint: yes (Section 15.15.1).
19. Interpret our derivation the other way around: as a mapping of h:

$$f_{ad}(h) \equiv [f, h].$$

20. This way, f_{ad} maps h.
21. This new mapping is called the adjoint of the original member f:

$$f_{ad} \equiv [f, \cdot].$$

22. How does it act on h? Hint:

$$f_{ad}(h) = [f, h].$$

23. Is it linear? Hint: our Lie brackets are bilinear.
24. What is the adjoint of $[f, g]$? Hint: on each given h,

$$[f, g]_{ad}(h) = [[f, g], h].$$

25. Use this to rewrite the Jacobi identity as:

$$[f, g]_{ad}(h) = -g_{ad}(f_{ad}(h)) + f_{ad}(g_{ad}(h)).$$

16.3 Mappings: New Lie Brackets

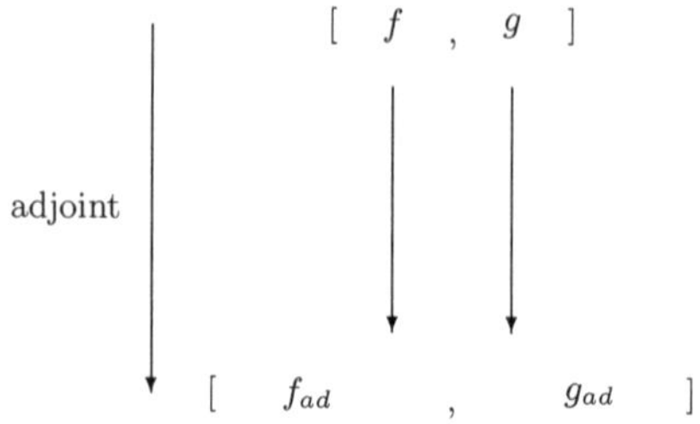

Fig. 16.2. $[f, g]_{ad}$ uses the original Lie brackets, and then takes the adjoint. But you can also work the other way around, and use the new Lie brackets: $[f_{ad}, g_{ad}]$.

1. In the above formula, we act on an individual vector: h. Better yet, drop h altogether, and act as a whole:

$$[f, g]_{ad} = f_{ad} \circ g_{ad} - g_{ad} \circ f_{ad}$$

(where '$\circ$' stands for composition).
2. What is the left-hand side? Hint: it is a mapping: the adjoint of $[f, g]$.
3. Use this formula to explain why the Lie brackets are often called the commutator.
4. How is this formula different from the original Jacobi identity? Hint: it acts on the whole Lie algebra (not just on an individual vector h).
5. Note that this formula actually *defines* new brackets, taking two adjoint mappings f_{ad} and g_{ad}, and producing their commutator:

$$[f_{ad}, g_{ad}] \equiv f_{ad} \circ g_{ad} - g_{ad} \circ f_{ad}$$

(Figure 16.2).
6. Are these new brackets still antisymmetric? Hint: they are inherited from the old ones. Thus,

$$[f_{ad}, g_{ad}] = [f, g]_{ad} = -[g, f]_{ad} = -[g_{ad}, f_{ad}].$$

7. In fact, they make a new representation: each vector in the original Lie algebra is represented by its own adjoint.
8. Prove this more directly. Hint: let f_1 and f_2 be two adjoint mappings. Thanks to the above definition,

$$\begin{aligned}[f_1, f_2] &= f_1 \circ f_2 - f_2 \circ f_1 \\ &= -(f_2 \circ f_1 - f_1 \circ f_2) \\ &= -[f_2 f_1].\end{aligned}$$

9. Do these new brackets still satisfy their own Jacobi identity? Hint: they are inherited from the old ones. Thus, you can throw the "ad" out:

$$[[f_{ad}, g_{ad}], h_{ad}] = [[f, g]_{ad}, h_{ad}] = [[f, g], h]_{ad},$$

and then use the original Jacobi identity.
10. This is how each member in the original Lie algebra is *represented* by its own adjoint.
11. Prove this more directly. Hint: let f_1, f_2, and f_3 be three adjoint mappings. Use the above definition twice:

$$\begin{aligned}[[f_1, f_2], f_3] &= [f_1, f_2] \circ f_3 - f_3 \circ [f_1, f_2] \\ &= (f_1 \circ f_2 - f_2 \circ f_1) \circ f_3 - f_3 \circ (f_1 \circ f_2 - f_2 \circ f_1) \\ &= f_1 \circ f_2 \circ f_3 - f_2 \circ f_1 \circ f_3 - f_3 \circ f_1 \circ f_2 + f_3 \circ f_2 \circ f_1.\end{aligned}$$

Here, the first term mirrors the third: they differ only in sign and cyclic shift:

$$(1, 2, 3) \to (3, 1, 2) \to (2, 3, 1).$$

Likewise, the second term mirrors the fourth. Carry out these three shifts, and sum all 12 terms, and obtain the zero vector, as required.
12. Conclude that these new brackets (on mappings) may indeed serve as legitimate Lie brackets.
13. Conclude that these mappings indeed make a legitimate Lie algebra as well.

16.4 Concrete Case: Adjoint Operator

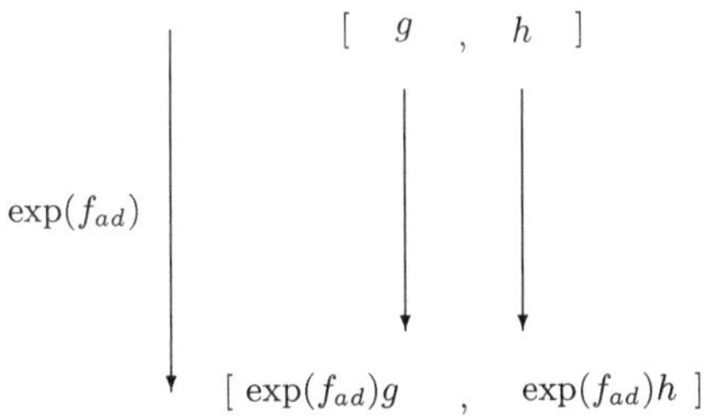

Fig. 16.3. $\exp(f_{ad})$ commutes with Lie brackets. In other words, order doesn't matter: you can apply Lie brackets either before or after applying $\exp(f_{ad})$.

1. So far, f and g were vectors in our original Lie algebra. Next, let's look at a more concrete case: let them be mappings in their own right! This way, the adjoint is

$$f_{ad}(g) \equiv [f, g] = f \circ g - g \circ f$$

(our new Lie brackets).

2. Better yet, drop the g's altogether, and define this as a whole:

$$f_{ad} \equiv f \circ - \circ f.$$

3. Here, what is $f\circ$? Hint: an operator: it takes a mapping g, and composes f with it, to produce a new mapping: $f \circ g$.
4. Is it linear? Hint: see the end of Section 16.1.
5. Is it a derivation? Hint: no! See the end of Section 16.1.
6. On the other hand, what is $\circ f$? Hint: an operator: it takes a mapping g, and composes it with f, to produce a new mapping: $g \circ f$.
7. Is it linear?
8. Is it a derivation?
9. In the above formula, what is the right-hand side? Hint: a new operator: the difference between these two.
10. And what if f and g were no adjoints at all, but just some linear mappings on some vector space? Could you still define f_{ad}? Hint: use the same formula.
11. Let's go more general: what if f and g were some members in an abstract algebra, with some (associative) product '$\circ$' (not necessarily composition)? Could you still define f_{ad}? Hint: use the same formula.
12. Is f_{ad} a legitimate derivation not only with respect to Lie brackets but also with respect to '$\circ$'? Does it obey a Leibniz rule with respect to '$\circ$' as well? Hint: thanks to associativity,

$$\begin{aligned} f_{ad}(g \circ h) &= f \circ g \circ h - g \circ h \circ f \\ &= f \circ g \circ h + (-g \circ f \circ h + g \circ f \circ h) - g \circ h \circ f \\ &= (f \circ g \circ h - g \circ f \circ h) + (g \circ f \circ h - g \circ h \circ f) \\ &= f_{ad}(g) \circ h + g \circ f_{ad}(h). \end{aligned}$$

13. Use this to prove once again that f_{ad} is a legitimate derivation with respect to the new Lie brackets. Hint: write this with g and h interchanged, subtract, and obtain the Jacobi identity back again.
14. Does $f\circ$ commute with $\circ f$? Hint: you can apply $f\circ$ either before or after $\circ f$. It doesn't matter. Indeed, thanks to associativity,

$$(f \circ g) \circ f = f \circ (g \circ f).$$

15. Our new (concrete) operator $f_{ad} = f \circ - \circ f$ is the adjoint of f, which is now a member in our abstract algebra (with some associative product '$\circ$').
16. Is it linear?
17. Assume that $f\circ$ is nilpotent. What does this mean? Hint: for some (sufficiently large) k, $(f\circ)^k$ is the zero operator (on our abstract algebra).
18. Define its exponent:

$$\exp(f\circ) = \sum_{i=0}^{\infty} \frac{(f\circ)^i}{i!}.$$

19. Is this really infinite? Hint: only k leading terms survive, and the rest vanish. (Later on, in the applications, we'll look at an infinite series too.)

20. Show that this is its Jordan decomposition. Hint: the first term is the identity operator. This is the diagonal part. The rest is the tail, which is nilpotent. Indeed, look at its kth power: it contains only high powers of $f\circ$ that vanish anyway.
21. What is the inverse operator? Hint: $\exp(-f\circ)$.
22. Write the inverse in terms of the above tail. Hint: plug the tail (and its powers) in a (finite) power series (with alternating signs).
23. Likewise, look at $\circ f$. Assume that it is nilpotent too (with $\tilde{k}$ rather than k).
24. Define its exponent too.
25. What is its Jordan decomposition?
26. What is its inverse?
27. Now, look at f_{ad}. Raise it to power $n \geq 0$. Hint: since $f\circ$ commutes with $\circ f$,

$$f_{ad}^n = (f\circ - \circ f)^n = \sum_{i=0}^{n}(-1)^{n-i}\binom{n}{i}(f\circ)^i(\circ f)^{n-i}.$$

28. How does this operator act on a given g (a member of our abstract algebra)? Hint: in the right-hand side, plug g in the middle: in between $(f\circ)^i$ and $(\circ f)^{n-i}$.
29. Is f_{ad} nilpotent? Hint: in the above formula, pick n as large as $k+\tilde{k}$. This way, in the right-hand side, all terms vanish.
30. We say now that f is adjoint-nilpotent. Why? Hint: its adjoint is nilpotent.
31. Define the exponent of f_{ad}. Hint:

$$\exp(f_{ad}) \equiv \sum_{i=0}^{\infty}\frac{f_{ad}^i}{i!}.$$

32. Is this really infinite? Hint: no! Only a few terms survive.
33. How many? Hint: at most $k+\tilde{k}$.
34. Show that this exponent is already in its Jordan decomposition.
35. What is its inverse? Hint: $\exp(-f_{ad})$.
36. Show that

$$\exp(f_{ad}) = \exp(f\circ - \circ f) = \exp(f\circ)\exp(-\circ f).$$

37. In the right-hand side, do the exponents commute? Hint: recall that $f\circ$ commutes with $\circ f$.
38. How does this operator act on a given g (a member of our abstract algebra)? Hint: in the right-hand side, plug g in the middle.
39. Use this to have the Jordan decomposition of $\exp(f_{ad})$ in yet another way. Hint: in the right-hand side, each exponent has its own Jordan decomposition. Multiply their diagonals, and obtain the required diagonal.
40. Use this to have the inverse of $\exp(f_{ad})$ in yet another way.
41. Show that $\exp(f_{ad})$ preserves the new Lie brackets (Figure 16.3): for every g and h (members of our abstract algebra),

$$\exp(f_{ad})([g,h]) = [\exp(f_{ad})g, \exp(f_{ad})h],$$

or, more explicitly,

$$\exp(f_{ad})(g\circ h - h\circ g) = \exp(f_{ad})g\circ\exp(f_{ad})h \; - \; \exp(f_{ad})h\circ\exp(f_{ad})g.$$

Hint: in Section 16.1, this was proved not only for f_{ad} but also for any derivation on any algebra.

42. And what if '∘' was nonassociative? Hint: f_{ad} is still an adjoint derivation on a Lie algebra (see below).

16.5 Back To The Original Lie Algebra

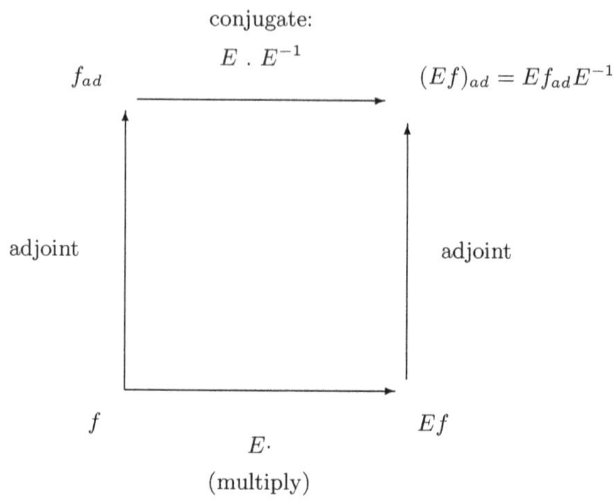

Fig. 16.4. You have two options: either multiply by E and then take the adjoint (rightwards-upwards), or work the other way around: take the adjoint, and then conjugate (plug in between E and E^{-1}): upwards-rightwards.

1. By now, f_{ad} acts on our abstract algebra, and takes the form:

$$f_{ad} = f \circ - \circ f.$$

2. Better yet, go general back again (as in the beginning):

$$f_{ad} \equiv [f, \cdot],$$

   where f is in our original Lie algebra (which may have no '∘' at all).
3. Assume that f_{ad} is nilpotent. Does it have an exponent?
4. Look at $\exp(f_{ad})$. What is its inverse? Hint: $\exp(-f_{ad})$.
5. Does $\exp(f_{ad})$ preserve the original Lie brackets? Hint: in Section 16.1, use our original Lie brackets rather than product (as illustrated in Figure 16.3).
6. Thanks to this, what can you say about $\exp(f_{ad})$? Hint: on our original Lie algebra, it is linear, invertible, and Lie-bracket preserving.
7. Is it a derivation? Hint: no! See Section 16.1.

8. Let E be of the same kind: linear, invertible, and Lie-bracket preserving (but not necessarily a derivation).
9. Look at a triple product of the form:

$$E \circ f_{ad} \circ E^{-1}$$

(where '$\circ$' means composition again).
10. How does it act on a given g (a member of our original Lie algebra)? Hint: leftwards, in three stages: first, apply E^{-1}. Then, apply f_{ad}. Finally, apply E.
11. Is it a derivation in its own right? Hint: yes! See Section 16.1.
12. Is it an adjoint of anything? Hint: yes! It is the adjoint of Ef:

$$(Ef)_{ad} = E \circ f_{ad} \circ E^{-1}.$$

Indeed, look at the right-hand side. Apply it to Eg (where g is a member of our original Lie algebra):

$$\begin{aligned} E \circ f_{ad} \circ E^{-1} Eg &= E \circ f_{ad} g \\ &= E[f, g] \\ &= [Ef, Eg] \\ &= (Ef)_{ad}(Eg). \end{aligned}$$

13. Illustrate it in a commutative diagram. Hint: see Figure 16.4.
14. Is it nilpotent? Hint: to make sure, assume that f_{ad} is nilpotent (on our original Lie algebra).

16.6 Exponentiate the Adjoint

1. Conclude that it has an exponent.
2. Write its exponent as a new triple product:

$$\exp((Ef)_{ad}) = \exp\left(E \circ f_{ad} \circ E^{-1}\right) = E \exp(f_{ad}) E^{-1}.$$

Hint: see Section 16.1.
3. Illustrate this in a commutative diagram. Hint: see Figure 16.5.
4. Work the other way around: conclude that every triple product like $E \exp(f_{ad}) E^{-1}$ can be written as an exponent of some adjoint. Hint:

$$E \exp(f_{ad}) E^{-1} = \exp\left(E \circ f_{ad} \circ E^{-1}\right) = \exp\left((Ef)_{ad}\right).$$

5. This will be useful in Lie groups, later on.

16.7 Examples: Linear Lie Algebras

1. Design a Lie algebra with a geometrical meaning. Hint: see below.
2. Look at $\mathbb{R}^3$ (the 3-D Cartesian space). Is it a Lie algebra?

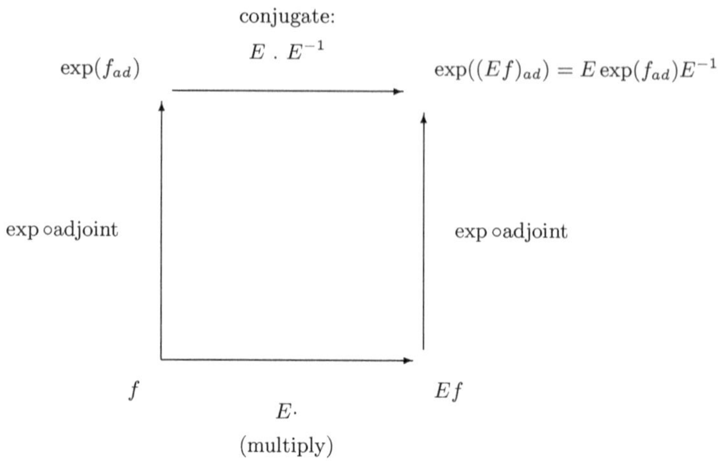

Fig. 16.5. You have two options: either multiply by E and then exponentiate the adjoint (rightwards-upwards), or work the other way around: exponentiate the adjoint, and then conjugate (plug in between E and E^{-1}): upwards-rightwards.

3. Look at vector product (Chapter 2). Use it as Lie brackets:
$$[u, v] \equiv u \times v, \qquad u, v \in \mathbb{R}^3.$$
4. Is it alternative? Hint:
$$[u, u] = u \times u = \mathbf{0} \in \mathbb{R}^3.$$
5. Conclude that it is antisymmetric as well.
6. Does it satisfy the Jacobi identity? Hint: first, check a triplet of standard unit vectors: $\mathbf{i}$, $\mathbf{j}$, and $\mathbf{k}$. Thanks to alternativity,
$$[[\mathbf{i}, \mathbf{j}], \mathbf{k}] = (\mathbf{i} \times \mathbf{j}) \times \mathbf{k} = \mathbf{k} \times \mathbf{k} = \mathbf{0}.$$
And what if two of them are the same, say $\mathbf{i}$, $\mathbf{i}$, and $\mathbf{j}$? Thanks to alternativity and antisymmetry,
$$\begin{aligned} &[[\mathbf{i}, \mathbf{i}], \mathbf{j}] + [[\mathbf{j}, \mathbf{i}], \mathbf{i}] + [[\mathbf{i}, \mathbf{j}], \mathbf{i}] \\ &= \mathbf{0} + [[\mathbf{j}, \mathbf{i}], \mathbf{i}] - [[\mathbf{j}, \mathbf{i}], \mathbf{i}] \\ &= \mathbf{0}. \end{aligned}$$
(There is nothing special about $\mathbf{i}$ and $\mathbf{j}$: the same could be done for any two standard unit vectors.) And what if all three are the same? That's easy:
$$[[\mathbf{i}, \mathbf{i}], \mathbf{i}] = [\mathbf{0}, \mathbf{i}] = \mathbf{0}.$$
Finally, use bilinearity to extend this to any triplet of any three vectors as well.

7. Design yet another Lie algebra, with a physical meaning.
8. For this purpose, look at the $n \times n$ (complex) matrices.
9. Do they make a linear space? Hint: matrix addition is commutative, and distributive with respect to scalar multiplication.
10. Do they make an algebra? Hint: matrix product is bilinear, and even associative.
11. Do they make a legitimate Lie algebra? Hint: the commutator makes good Lie brackets: bilinear, antisymmetric, and satisfying the Jacobi identity (Section 16.3).
12. This is the general linear Lie algebra, denoted by gl_n.
13. More precisely, its full name is

$$gl_n \equiv gl(n, \mathbb{C}).$$

But we mainly work with complex matrices, so we often drop the '$\mathbb{C}$.'
14. In a Lie algebra, what is a subalgebra? Hint: a subspace that is closed under the Lie brackets.
15. Is it a Lie algebra in its own right?
16. A subalgebra of gl_n is called a linear Lie algebra.
17. Give an example.
18. Look at the 3×3 spin matrices (that model spin-one: Section 11.1.1).
19. Multiply them by $i = \sqrt{-1}$. This way, do they span a linear space? Hint: yes: over $\mathbb{R}$.
20. With matrix product, do they span an algebra? Hint: no! they are *not* closed under product.
21. Still, do they span a Lie algebra? Hint: they are closed under the commutator.
22. Is this a linear Lie algebra? Hint: it is a subalgebra of $gl(3, \mathbb{R})$.
23. How is it related to the 3-D Cartesian space, discussed above? Hint: in both, the Lie brackets are cyclic.
24. Conclude that they are isomorphic.
25. Conclude once again that the vector product obeys the Jacobi identity.
26. Look at the 2×2 Pauli matrices (that model spin-one-half: Section 11.2.1).
27. Multiply them by $i = \sqrt{-1}$.
28. This way, do they span a linear space? Hint: yes: over $\mathbb{R}$.
29. With matrix product, do they span an algebra? Hint: no! They are *not* closed under product: their square (the identity matrix) is missing.
30. Still, do they span a Lie algebra? Hint: they are closed under the commutator.
31. Is this a linear Lie algebra? Hint: it is a subalgebra of gl_2.
32. How is it related to the previous examples? Hint: in all three examples, the Lie brackets are cyclic.
33. Conclude that they are isomorphic.
34. Look at gl_n again: the general linear Lie algebra. In particular, this is a vector space. Design a basis for it. Hint: let $1 \le p, q \le n$ be two fixed indices. Define a new $n \times n$ matrix $E^{(p,q)}$, with only one nonzero element (at the pth row and qth column):

$$\left(E^{(p,q)}\right)_{i,j} = \begin{cases} 1 & \text{if } i = p \text{ and } j = q \\ 0 & \text{otherwise.} \end{cases}$$

35. Use the tensor

$$\delta_i^p \equiv \begin{cases} 1 & \text{if} \quad i = p \\ 0 & \text{if} \quad i \neq p \end{cases}$$

to write this more nicely. Hint:

$$\left(E^{(p,q)}\right)_{i,j} = \delta_i^p \delta_j^q.$$

36. How many such matrices are there? Hint: n^2.
37. Are they linearly independent?
38. What is the dimension of gl_n? Hint: n^2.
39. What is the product of two such matrices? Hint: let $1 \leq p, q, r, s \leq n$ be four fixed indices. To have a nonzero element in the product, we must have $q = r$. Indeed, in the following, only the term with $j = q = r$ survives:

$$\begin{aligned}\sum_{j=1}^{n} \left(E^{(p,q)}\right)_{i,j} \left(E^{(r,s)}\right)_{j,k} &= \sum_{j=1}^{n} \delta_i^p \delta_j^q \delta_j^r \delta_k^s \\ &= \delta_q^r \delta_i^p \delta_k^s \\ &= \delta_q^r \left(E^{(p,s)}\right)_{i,k}.\end{aligned}$$

40. Write this more compactly:

$$E^{(p,q)} E^{(r,s)} = \delta_q^r E^{(p,s)}.$$

41. Interchange (p, q) with (r, s):

$$E^{(r,s)} E^{(p,q)} = \delta_p^s E^{(r,q)}.$$

42. Subtract these formulas, and obtain the commutator:

$$\begin{aligned}\left[E^{(p,q)}, E^{(r,s)}\right] &= E^{(p,q)} E^{(r,s)} - E^{(r,s)} E^{(p,q)} \\ &= \delta_q^r E^{(p,s)} - \delta_p^s E^{(r,q)} \\ &= \begin{cases} E^{(p,s)} & \text{if} \quad q = r \quad \text{and} \quad p \neq s \\ -E^{(r,q)} & \text{if} \quad p = s \quad \text{and} \quad q \neq r \\ E^{(p,p)} - E^{(q,q)} & \text{if} \quad q = r \neq p = s \\ (0) & \text{otherwise.} \end{cases}\end{aligned}$$

43. Conclude that the commutator is trace-free: it must have a zero trace.
44. Extend this to any two matrices: their commutator must be trace-free. Hint: write each matrix as a linear combination of these E's, and use bilinearity.

16.8 Trace-Free Matrices

1. Look at the trace-free matrices of order n. Do they make a legitimate Lie algebra? Hint: they are closed under their Lie brackets. Indeed, the commutator is trace-free as well.
2. This is the special linear Lie algebra, denoted by sl_n.

3. Is it indeed linear? Hint: $sl_n \subset gl_n$.
4. Design a basis for it. Hint: take $E^{(p,q)}$ with $p \neq q$, and also $E^{(r,r)} - E^{(r+1,r+1)}$ $(1 \le r < n)$.
5. How many matrices are there in this basis? Hint: $n^2 - 1$.
6. What is the dimension of sl_n? Hint: $n^2 - 1$.
7. Let L be a Lie algebra that lies in between:

$$sl_n \subset L \subset gl_n.$$

8. Is L linear? Hint: $L \subset gl_n$.
9. What is its dimension? Hint: either n^2 or $n^2 - 1$.
10. What is L? Hint: either $L = gl_n$ or $L = sl_n$.
11. Conclude that there is no other Lie algebra in between sl_n and gl_n.
12. sl_n is also called A_l, where $l = n - 1$.
13. In terms of l, what is the dimension of A_l? Hint: $l^2 + 2l$.
14. For example, look at $sl_2 = A_1$.
15. Design a basis for it. Hint: the basis could contain the following matrices:

$$e \equiv \begin{pmatrix} 0 & 1 \\ 0 & 0 \end{pmatrix}, \ f \equiv \begin{pmatrix} 0 & 0 \\ 1 & 0 \end{pmatrix}, \quad \text{and } h \equiv \begin{pmatrix} 1 & 0 \\ 0 & -1 \end{pmatrix}.$$

16. What is the dimension? Hint: 3.
17. Does this agree with our formula? Hint: here, $n = 2$ and $l = 1$, so $n^2 - 1 = l^2 + 2l = 3$, as required.
18. Show that

$$[e, f] = h.$$

19. Does this agree with our formula for the E's (Section 16.7)?
20. Show also that

$$[h, e] = 2e, \quad \text{and} \quad [h, f] = -2f.$$

21. Does this agree with our formula for the E's (Section 16.7)?
22. Look at the adjoint mapping:

$$h_{ad} = [h, \cdot] : sl_2 \to sl_2,$$

defined by

$$v \to [h, v], \qquad v \in sl_2$$

(Section 16.2).
23. What are its eigenvectors? Hint: e, f, and h.
24. What are their eigenvalues? Hint: 2, -2, and 0, respectively.

16.9 Triangular Matrices

1. Next, look at $n \times n$ upper-triangular matrices. Do they make a linear space?
2. What is its basis? Hint: take those $E^{(p,q)}$'s with $p \le q$.

3. From this basis, pick two, and multiply. Look at the product. How does it look like? Hint: we already saw that

$$E^{(p,q)}E^{(r,s)} = \delta_q^r E^{(p,s)}$$

(Section 16.7). Thus, to have a nonzero product, we must have

$$p \le q = r \le s.$$

So, the product is upper-triangular too.

4. Could it be diagonal? Hint: to have $p = s$, we must have

$$p = q = r = s.$$

So, this must be a very special product of the form

$$\left(E^{(q,q)}\right)^2.$$

5. In this case, what is the commutator? Hint: subtract this from itself, and obtain the zero matrix. This is alternativity indeed.
6. So, for two E's picked from our basis, how does their commutator look like? Hint: strictly upper-triangular.
7. Extend this linearly: for any two linear combinations of our $E^{(p,q)}$'s (with $p \le q$), their product is upper triangular, and their commutator is even *strictly* upper triangular. Hint: open parentheses, and use bilinearity.
8. What can you say about our linear space? Is it an algebra? Is it even a Lie algebra? Hint: it is closed under multiplication, and even under the commutator.
9. Denote it by

$$T_0 \equiv \text{span}\left\{E^{(p,q)} \mid q - p \ge 0\right\}.$$

10. Is T_0 linear? Hint: $T_0 \subset gl_n$.
11. What is its dimension? Hint: $n(n+1)/2$.
12. Next, look at the strictly upper-triangular matrices of order n. Do they make a linear space?
13. Is it an algebra? Hint: it is closed under multiplication.
14. Is it a Lie algebra? Hint: it is closed under the commutator.
15. Denote it by

$$T_1 \equiv \text{span}\left\{E^{(p,q)} \mid q - p \ge 1\right\}.$$

16. Is T_1 linear? Hint: $T_1 \subset T_0 \subset gl_n$. Furthermore, $T_1 \subset sl_n \subset gl_n$.
17. What is its dimension? Hint: $n(n-1)/2$.
18. Likewise, for any $k \ge 0$, look at yet stricter matrices (with yet fewer diagonals), and define yet smaller Lie algebras:

$$T_k \equiv \text{span}\left\{E^{(p,q)} \mid q - p \ge k\right\}.$$

19. Pick a basis member from T_k, and another from T_j (for some $j, k \ge 0$). Where is their product? Where is their commutator? Hint: in T_{j+k}.
20. Extend this bilinearly: pick one matrix from T_k, and another from T_j. Where is their product? Where is their commutator? Hint: in $T_{j+k} \subset T_j \cap T_k$.

21. In particular, when $j = k$, the commutator is in $T_{2k} \subset T_k$.
22. Conclude (once again) that T_k is closed under the commutator.
23. Conclude that
$$T_k \subset T_{k-1} \subset T_{k-2} \subset \cdots \subset T_1 \subset T_0$$
are subalgebras of each other.
24. So far, we picked a member from T_j and a member from T_k, and saw that they have their commutator in T_{j+k}. Next, work the other way around: pick an $E^{(p,q)} \in T_{j+k}$. Could it be written as a commutator? (Assume that $j + k > 0$.) Hint: since $q - p \geq j + k > 0$,
$$E^{(p,q)} = E^{(p,p+j)} E^{(p+j,q)} = \left[E^{(p,p+j)}, E^{(p+j,q)}\right].$$
25. In the above exercise, why did we assume that $j + k > 0$? Hint: if $j = k = 0$, then this could fail. For example, $E^{(1,1)} \in T_{j+k} = T_0$ is no commutator: it has trace 1, not 0.
26. Give an example of a strictly upper-triangular matrix. Hint: a nilpotent Jordan block (Section 15.1.6): 1's on the superdiagonal only, and 0 elsewhere.
27. Write it as a sum of our E's.
28. Use this to calculate its powers. Hint: in the multinomial formula, only a few terms survive.
29. Does this agree with Section 15.1.6?
30. Show that its kth power is in T_k $(k \geq 0)$.
31. Conclude that its nth power vanishes.
32. Conclude once again that it is indeed nilpotent.

16.10 Diagonal Matrices

1. Look at $E^{(r,r)}$ and $E^{(p,p)}$ $(1 \leq r, p \leq n)$. Are they strictly upper triangular?
2. Are they upper triangular?
3. Are they diagonal?
4. Do they commute?
5. What is their commutator? Hint: the zero matrix.
6. Is this also true for every two diagonal matrices? Hint: they commute, and are commutator-free.
7. Look at all diagonal matrices of order n. Do they make a Lie algebra? Hint: their commutator is the zero matrix, which is diagonal as well.
8. What is their physical meaning? Hint: in quantum mechanics, a real diagonal matrix could serve as a position matrix (Section 9.4.1).
9. Is this a linear Lie algebra? Hint: it is a subalgebra of $T_0 \subset gl_n$.
10. What is its dimension? Hint: n.

16.11 Orthogonal Lie Algebra: D_l

1. Next, let S be an $n \times n$ symmetric matrix:
$$S = S^t.$$

2. Look at another $n \times n$ matrix, denoted by X.
3. In practice, the convention is that X and S are both real.
4. Still, in theory, they could be even complex. (In what follows, everything remains the same.)
5. Assume that X is S-antisymmetric. What does this mean? Hint: this means that SX is antisymmetric:

$$SX = -(SX)^t = -X^t S^t = -X^t S.$$

6. Look at all such X's. Do they make a Lie algebra? Hint: look at yet another matrix $\tilde{X}$ that is S-antisymmetric as well:

$$S\tilde{X} = -\tilde{X}^t S.$$

   This way, their product satisfies

$$SX\tilde{X} = -X^t S\tilde{X} = X^t \tilde{X}^t S = \left(\tilde{X}X\right)^t S.$$

   Thus, their commutator is S-antisymmetric as well:

$$S[X,\tilde{X}] = S\left(X\tilde{X} - \tilde{X}X\right) = \left(\tilde{X}X - X\tilde{X}\right)^t S = -[X,\tilde{X}]^t S,$$

   as required.
7. This is the orthogonal Lie algebra, denoted by o_n.
8. Is o_n linear? Hint: $o_n \subset gl_n$.
9. If S is also nonsingular, then we say that o_n is nondegenerate.
10. For example, assume that n is even:

$$n = 2l.$$

11. Assume also that S has a block form:

$$S = \begin{pmatrix} & I \\ I & \end{pmatrix},$$

   where I is the $l \times l$ identity matrix.
12. Write X in its block form too:

$$X = \begin{pmatrix} M & N \\ P & Q \end{pmatrix},$$

   where M, N, P, and Q are some $l \times l$ blocks.
13. Write our condition in this block form:

$$\begin{aligned} SX &= \begin{pmatrix} & I \\ I & \end{pmatrix}\begin{pmatrix} M & N \\ P & Q \end{pmatrix} \\ &= \begin{pmatrix} P & Q \\ M & N \end{pmatrix} \\ &= -X^t S \\ &= -\begin{pmatrix} M^t & P^t \\ N^t & Q^t \end{pmatrix}\begin{pmatrix} & I \\ I & \end{pmatrix} \\ &= -\begin{pmatrix} P^t & M^t \\ Q^t & N^t \end{pmatrix}. \end{aligned}$$

14. In this form, each block gives us a new condition. The upper-left block tells us that P must be antisymmetric.
15. Thanks to this, what does P have on its main diagonal? Hint: zero.
16. Thanks to this, how many degrees of freedom are there in P? Hint: $l(l-1)/2$.
17. Likewise, the lower-right block tells us that N must be antisymmetric too.
18. Thanks to this, what does N have on its main diagonal? Hint: zero.
19. Thanks to this, how many degrees of freedom are there in N? Hint: $l(l-1)/2$.
20. Moreover, the upper-right and lower-left blocks tell us that
$$Q = -M^t.$$
Thus, Q and M can't be picked freely. Only one of them can, but the other can't. How many degrees of freedom are there in them? Hint: l^2.
21. In total, how many degrees of freedom are there in X? Hint: $2l^2 - l$.
22. What is the trace of X? Hint: thanks to the latter condition, X must be trace-free.
23. In the above setting, o_n has a special name: D_l.
24. What is its dimension? Hint: $2l^2 - l$.
25. Conclude that $D_l \subset sl_{2l}$.

16.12 Symplectic Lie Algebra: C_l

1. Next, introduce a little change. Assume now that S is not symmetric but *anti-symmetric*:
$$S = -S^t.$$
2. As before, look at an $n \times n$ matrix X that is S-antisymmetric. What does this mean? Hint: this means that SX is symmetric:
$$SX = (SX)^t = X^t S^t = -X^t S.$$
3. Look at all such X's. Do they make a Lie algebra? Hint: look at yet another matrix $\tilde{X}$ that is S-antisymmetric as well:
$$S\tilde{X} = -\tilde{X}^t S.$$
This way, their product still satisfies
$$SX\tilde{X} = -X^t S\tilde{X} = X^t \tilde{X}^t S = \left(\tilde{X}X\right)^t S.$$
Thus, their commutator is S-antisymmetric as well:
$$S[X, \tilde{X}] = S\left(X\tilde{X} - \tilde{X}X\right) = \left(\tilde{X}X - X\tilde{X}\right)^t S = -[X, \tilde{X}]^t S.$$
4. This is the symplectic Lie algebra, denoted by sp_n.
5. Is it linear? Hint: $sp_n \subset gl_n$.
6. If S is also nonsingular, then we say that sp_n is nondegenerate.
7. For example, assume again that n is even:
$$n = 2l.$$

8. Assume also that S can be written in a block form:

$$S = \begin{pmatrix} & I \\ -I & \end{pmatrix}.$$

9. Write X in its block form too:

$$X = \begin{pmatrix} M & N \\ P & Q \end{pmatrix},$$

   where M, N, P, and Q are new $l \times l$ blocks.
10. Write our condition in this block form:

$$\begin{aligned} SX &= \begin{pmatrix} & I \\ -I & \end{pmatrix} \begin{pmatrix} M & N \\ P & Q \end{pmatrix} \\ &= \begin{pmatrix} P & Q \\ -M & -N \end{pmatrix} \\ &= -X^t S \\ &= -\begin{pmatrix} M^t & P^t \\ N^t & Q^t \end{pmatrix} \begin{pmatrix} & I \\ -I & \end{pmatrix} \\ &= \begin{pmatrix} P^t & -M^t \\ Q^t & -N^t \end{pmatrix}. \end{aligned}$$

11. In this form, each block gives us a new condition. The upper-left block tells us that P must be symmetric. How many degrees of freedom are there in P? Hint: $l(l+1)/2$.
12. Likewise, the lower-right block tells us that N must be symmetric too. How many degrees of freedom are there in N? Hint: $l(l+1)/2$.
13. Finally, the upper-right and lower-left blocks tell us that

$$Q = -M^t.$$

   Thus, Q and M can't be picked freely. Only one of them can, but the other can't. How many degrees of freedom are there in them? Hint: l^2.
14. In total, how many degrees of freedom are there in X? Hint: $2l^2 + l$.
15. What is the trace of X? Hint: thanks to the latter condition, X must be trace-free.
16. In the above setting, sp_n has a special name: C_l.
17. What is its dimension? Hint: $2l^2 + l$.
18. Conclude that $C_l \subset sl_{2l}$.

16.13 Orthogonal Lie Algebra: B_l

1. Let's go back to o_n, so S is symmetric again (Section 16.11).
2. Let's make a little change. Assume now that n is not even but odd:

$$n = 2l + 1.$$

3. This leads to a new block form, with one more row and column:

$$S = \begin{pmatrix} 1 & & \\ & & I \\ & I & \end{pmatrix}.$$

4. In X too, in its new block form, there are now one more row and column:

$$X = \begin{pmatrix} d & y & z \\ u & M & N \\ v & P & Q \end{pmatrix},$$

where M, N, P, and Q are new $l \times l$ blocks, y and z are row vectors, u and v are column vectors, and d is a scalar.

5. Write our condition in this block form:

$$\begin{aligned} SX &= \begin{pmatrix} 1 & & \\ & & I \\ & I & \end{pmatrix} \begin{pmatrix} d & y & z \\ u & M & N \\ v & P & Q \end{pmatrix} \\ &= \begin{pmatrix} d & y & z \\ v & P & Q \\ u & M & N \end{pmatrix} \\ &= -X^t S \\ &= -\begin{pmatrix} d & u^t & v^t \\ y^t & M^t & P^t \\ z^t & N^t & Q^t \end{pmatrix} \begin{pmatrix} 1 & & \\ & & I \\ & I & \end{pmatrix} \\ &= -\begin{pmatrix} d & v^t & u^t \\ y^t & P^t & M^t \\ z^t & Q^t & N^t \end{pmatrix}. \end{aligned}$$

6. In this form, the conditions are now:

$$\begin{aligned} P &= -P^t \\ N &= -N^t \\ Q &= -M^t \\ d &= 0 \\ y &= -v^t \\ z &= -u^t. \end{aligned}$$

7. Compare this to D_l (Section 16.11). How many new degrees of freedom are here? Hint: here, there are $2l$ new degrees of freedom (to pick u and v).
8. In total, how many degrees of freedom are there in X? Hint: $2l^2 + l$.
9. What is the trace of X? Hint: thanks to the above conditions, X is trace-free.
10. In this case, o_n has a special name: B_l.
11. What is its dimension? Hint: $2l^2 + l$.
12. Is there a symplectic Lie algebra of the same dimension? Hint: C_l.
13. Conclude that $B_l \subset sl_{2l+1}$.

17

Ideals and Isomorphism Theorems

So far, we saw a few examples of Lie algebras. Later on, we'll use them in physics: quantum mechanics, and special relativity. For this, we need an important subalgebra: an ideal. Thanks to it, we'll see all sorts of isomorphism theorems.

This will mirror group theory: factor subgroups, and isomorphism theorems. Later on, we'll see how to design a Lie group: take your Lie algebra, and exponentiate. This will lead to a few interesting applications.

17.1 Ideals

17.1.1 Two Subalgebras

Let L be a Lie algebra. Let

$$K, M \subset L$$

be two subalgebras. Let's introduce a new notation:

$$[K, M] \equiv \operatorname{span}\{[k, m] \mid k \in K,\ m \in M\}.$$

What is this? This is a new subspace of L, spanned by a few special vectors, obtained from Lie brackets. (By now, we don't know if this is a Lie algebra or not.) Thanks to antisymmetry,

$$[K, M] = [M, K].$$

Still, is this a subalgebra in its own right? In other words, is it closed under Lie brackets? To check on this, pick two vectors of the form $[k, m]$ and $[\tilde{k}, \tilde{m}]$. Plug them in Lie brackets, and check: is

$$[[k, m], [\tilde{k}, \tilde{m}]] \in [K, M]?$$

Not necessarily! This is why $[K, M]$ is only a subspace, but not necessarily a subalgebra.

17.1.2 Subalgebras and Their Sum

Let's continue to "play" with our subalgebras. Let's sum them up:

$$K + M \equiv \{k + m \mid k \in K,\ m \in M\}.$$

Clearly, this is a subspace. Indeed, it is linear: closed under addition and scalar multiplication. Still, is it a subalgebra as well? To check on this, pick two vectors from it, and plug them in Lie brackets:

$$[k + m, \tilde{k} + \tilde{m}] = [k, \tilde{k}] + [k, \tilde{m}] + [m, \tilde{k}] + [m, \tilde{m}].$$

In this sum, $[k, \tilde{m}]$ and $[m, \tilde{k}]$ could lie outside $K + M$. Thus, $K + M$ is only a subspace, but not necessarily a subalgebra.

Nevertheless, we can still look at a special case, in which both $[K, M]$ and $K+M$ are legitimate subalgebras.

17.1.3 Ideal

Let's look at a special subspace:

$$I \subset L.$$

(Don't confuse this with the identity matrix, used often in linear algebra.) To be a legitimate subalgebra, I must be closed under Lie brackets:

$$[I, I] \subset I.$$

Let's require even more:

$$[I, L] \subset I.$$

This way, I is called an ideal of L.

17.1.4 Two Ideals

Let I and J be two ideals of L. What about their sum? Is it a legitimate ideal as well? Well, look at three vectors: $i \in I$, $j \in J$, and $l \in L$. Thanks to bilinearity,

$$[i + j, l] = [i, l] + [j, l] \in [I, L] + [J, L] \subset I + J,$$

as required. Furthermore, this is also true for every linear combination, so

$$[I + J, L] \subset I + J,$$

as required.

Next, what about $[I, J]$? Is it a legitimate ideal as well? Well, look at our three vectors again: $i \in I$, $j \in J$, and $l \in L$. Thanks to the Jacobi identity,

$$[[i, j], l] = -[[l, i], j] - [[j, l], i] \in [[I, L], J] + [[J, L], I] \subset [I, J] + [J, I] = [I, J],$$

as required. Thanks to bilinearity, this also extends to any linear combination of $[i, j]$'s. Thus,

$$[[I, J], L] \subset [I, J],$$

as required.

17.2 Homomorphism and Its Kernel

17.2.1 Homomorphism and Isomorphism

What is a homomorphism between groups? This is a mapping that preserves multiplication. What is a homomorphism between linear spaces? This is a linear mapping.

But here, we need even more: what is a homomorphism between Lie algebras? This is a linear mapping that preserves Lie brackets. If the homomorphism is invertible, then it is also called isomorphism.

17.2.2 Kernel – Null Space

In our homomorphism, what vectors are mapped to $\mathbf{0}$? In group theory, this is called kernel. In linear algebra, this is called null space. Here, in Lie algebras, we use both terms: either kernel or null space.

In particular, if our homomorphism is also an isomorphism, then the kernel is quite small: it contains just one vector: $\mathbf{0}$.

17.2.3 Homomorphism: What Is It?

Let ϕ be a homomorphism on a Lie algebra L. How does ϕ behave? First of all, ϕ is linear. What does this mean? Pick two vectors $k, l \in L$, and a scalar α. Then, ϕ preserves both addition and scalar multiplication:

$$\begin{aligned} \phi(k + l) &= \phi(k) + \phi(l) \\ \phi(\alpha k) &= \alpha\phi(k). \end{aligned}$$

Still, this is not the whole story. After all, L is not only a vector space but also a Lie algebra. Thus, to be a legitimate homomorphism, ϕ must also preserve Lie brackets:

$$\phi([k, l]) = [\phi(k), \phi(l)].$$

Thanks to this, ϕ is indeed a legitimate homomorphism. We then say that our original Lie algebra L is homomorphic to $\pi(L)$ (the image of ϕ).

17.2.4 Kernel: Is It an Ideal?

Assume now that k is in the kernel of ϕ:

$$\phi(k) = \mathbf{0}.$$

This way,

$$\phi([k, l]) = [\phi(k), \phi(l)] = [\mathbf{0}, \phi(l)] = \mathbf{0}.$$

Thus, $[k, l]$ is in the kernel as well. Thus, the kernel is not only a subspace but also an ideal in its own right. This will be useful below.

17.2.5 The Factor (Quotient) Lie Algebra

So far, we started from a given homomorphism, and used it to design a new ideal: its kernel. Next, let's work the other way around: start from a given ideal, and use it to design a new homomorphism.

Let $I \subset L$ be an ideal. In particular, I is a subalgebra. In particular, I is a subspace. In particular, I is a subgroup with respect to addition: '+'. Since addition is commutative, I is actually a normal subgroup. Thanks to this, we can go ahead and design the factor group.

For this purpose, however, we must use the plus sign '+'. This produces a new Lie algebra: the factor Lie algebra:

$$L/I \equiv \{x + I \mid x \in L\}.$$

In it, what is a vector? This:

$$x + I \equiv \{x + i \mid i \in I\}.$$

To design this new vector, $x \in L$ is not unique. Instead, you could also use $x + \tilde{i}$ (for any $\tilde{i} \in I$), and still get the same thing:

$$x + I \equiv \{x + i \mid i \in I\} = \left\{x + \tilde{i} + i \mid i \in I\right\} = x + \tilde{i} + I.$$

Now, pick two such vectors:

$$x + I, \ y + I \in L/I.$$

We already know how to add them:

$$(x + I) + (y + I) \equiv (x + y) + I.$$

By now, L/I is just a group (with respect to addition). How to multiply by a scalar? Like this:

$$\alpha(x + I) \equiv \alpha x + I.$$

Again, this is independent of the particular choice of representative:

$$\alpha(x + \tilde{i} + I) \equiv \alpha(x + \tilde{i}) + I = \alpha x + \alpha\tilde{i} + I = \alpha x + I = \alpha(x + I).$$

By now, L/I is a linear space. Indeed, these operations are distributive in two ways:

$$\begin{aligned} \alpha((x + I) + (y + I)) &= \alpha(x + y + I) \\ &= \alpha(x + y) + I \\ &= \alpha x + \alpha y + I \\ &= \alpha x + I + \alpha y + I \\ &= \alpha(x + I) + \alpha(y + I), \end{aligned}$$

and

$$\begin{aligned}(\alpha+\beta)(x+I) &= (\alpha+\beta)x+I \\ &= \alpha x+\beta x+I \\ &= \alpha x+I+\beta x+I \\ &= \alpha(x+I)+\beta(x+I).\end{aligned}$$

(In the exercises below, we'll prove this in yet another way.) By now, we never used the fact that I was an ideal. It is time to use it. This will help define new Lie brackets (in L/I):

$$[x+I, y+I] \equiv [x,y]+I.$$

Again, this is independent of the particular choice of representative. To see this, pick two vectors $i, \tilde{i} \in I$, and add i to x and $\tilde{i}$ to y. This would change nothing:

$$\begin{aligned}[x+i+I, y+\tilde{i}+I] &\equiv [x+i, y+\tilde{i}]+I \\ &= [x,y]+[i,y]+[x,\tilde{i}]+[i,\tilde{i}]+I \\ &= [x,y]+I\end{aligned}$$

(because I is an ideal). Fortunately, these new Lie brackets are bilinear and alternative, and even satisfy the Jacobi identity (see exercises below). Thus, L/I is indeed a legitimate Lie algebra.

17.2.6 The Canonical Homomorphism

This is the factor Lie algebra: L/I. How to map L onto it? Define the new mapping

$$\pi : L \to L/I$$

by picking the relevant member from the factor Lie algebra:

$$\pi(x) \equiv x+I, \quad x \in L.$$

Is π linear? To check on this, pick two vectors $x, y \in L$, and a scalar α. We can now see that π preserves addition:

$$\pi(x+y) = (x+y)+I = (x+I)+(y+I) = \pi(x)+\pi(y),$$

and scalar multiplication:

$$\pi(\alpha x) = \alpha x+I = \alpha(x+I) = \alpha\pi(x).$$

Thus, π is indeed linear. Does it preserve Lie brackets? Check:

$$\pi([x,y]) \equiv [x,y]+I = [x+I, y+I] = [\pi(x), \pi(y)].$$

Thus, π is indeed a legitimate homomorphism from L onto L/I. Compared to what we did in Section 17.2.4, this is indeed the other way around: given an ideal I, we designed a new homomorphism π whose kernel is I (see below).

Thanks to π, we even get a bonus for free: since π preserves the Lie brackets, we can deduce once again that L/I is indeed a legitimate Lie algebra: its Lie brackets are indeed bilinear and alternative, and even satisfy the Jacobi identity (see exercises below).

17.2.7 The Canonical Homomorphism and Its Kernel

Look at L/I again: what is its zero vector? This is just I. Indeed, pick some member of the form $x + I \in L/I$. Now, add I to it:

$$(x + I) + I = (x + I) + (\mathbf{0} + I) = (x + \mathbf{0}) + I = x + I.$$

Thus, adding I changes nothing. Thus, I is indeed the zero "vector" in L/I.

What is the kernel of π? To find out, pick some vector $x \in L$. To be in the kernel, x must be mapped to I:

$$\pi(x) = x + I = I.$$

To satisfy this, x must belong to I:

$$x \in I.$$

Thus, the kernel of π is exactly I. This leads to isomorphism theorems.

17.3 Isomorphism Theorems

17.3.1 The First Isomorphism Theorem

As in group theory, here too we have a few isomorphism theorems. The first one deals with the following question. Let ϕ be a homomorphism on a Lie algebra L. How to write ϕ as a product? For this purpose, let I be an ideal of L, included in the kernel of ϕ:

$$x \in I \Rightarrow \phi(x) = \mathbf{0}.$$

Let

$$\pi : L \to L/I$$

be the canonical homomorphism. Then, the original homomorphism ϕ could split into two stages:

- First, apply π, to map to some member of L/I.
- To L/I, apply a new homomorphism, named ψ.

In other words, there is a unique homomorphism ψ on L/I, such that

$$\phi = \psi \circ \pi.$$

How to define ψ? Easy:

$$\psi(x + I) \equiv \phi(x), \qquad x \in L.$$

This way, ψ doesn't depend on any particular representative. Indeed, for any fixed $\tilde{i} \in I$,

$$\psi(x + \tilde{i} + I) = \phi(x + \tilde{i}) = \phi(x) + \phi(\tilde{i}) = \phi(x) + \mathbf{0} = \phi(x).$$

Furthermore, ψ is linear on L/I. To see this, pick some $x, y \in L$, and some scalar α. Then,

$$\psi(x + y + I) = \phi(x + y) = \phi(x) + \phi(y) = \psi(x + I) + \psi(y + I),$$

and

$$\psi(\alpha x + I) = \phi(\alpha x) = \alpha\phi(x) = \alpha\psi(x + I).$$

Better yet, ψ even preserves the Lie brackets in L/I:

$$\psi([x + I, y + I]) = \psi([x, y] + I) = \phi([x, y]) = [\phi(x), \phi(y)] = [\psi(x + I), \psi(y + I)]].$$

Thus, ψ is indeed a legitimate homomorphism on L/I, as required. In other words, L/I is homomorphic to $\phi(L)$ (the image of ϕ).

And what if I was the same as the kernel of ϕ? In this case, ψ must be one-to-one. To see this, pick $x, y \in L$ such that $x + I \neq y + I$. This way,

$$\psi(x + I) = \phi(x) \neq \phi(y) = \psi(y + I),$$

as asserted. Thus, ψ is an invertible isomorphism. This way, L/I is not only homomorphic but also *isomorphic* to $\phi(L)$. This mirrors the fundamental theorem of homomorphism in group theory [27].

17.3.2 The Second Isomorphism Theorem

And how to mirror the second isomorphism theorem in group theory? Let L be a Lie algebra. Let

$$I, J \subset L$$

be two ideals. What about their intersection $I \cap J$? Clearly, it is a subspace: closed under addition and scalar multiplication. Still, is it an ideal as well? To check on this, pick two vectors: $x \in I \cap J$, and $l \in L$. Plug them in Lie brackets:

$$[x, l] \in I, \quad \text{and} \quad [x, l] \in J$$

(because both I and J are ideals). In other words,

$$[x, l] \in I \cap J.$$

(Extend this to any linear combination as well.) Thus,

$$[I \cap J, L] \subset I \cap J,$$

as required. We can now define a new homomorphism:

$$\phi : J \to \frac{I + J}{I}$$

by

$$\phi(j) \equiv j + I, \quad j \in J.$$

This mirrors group theory. (In Chapter 5 in [49], substitute I for S, J for T, and ϕ for ξ.) Here, however, we know more: I, J, $I + J$, and $I \cap J$ are not only subgroups (with respect to addition) but also ideals. This way, ϕ is linear: it preserves not only addition but also scalar multiplication. To see this, pick some vector $j \in J$, and some scalar α. Then,

$$\phi(\alpha j) \equiv \alpha j + I = \alpha(j + I) = \alpha\phi(j).$$

Furthermore, ϕ has yet another advantage: it preserves Lie brackets too. To see this, pick two vectors $j, \tilde{j} \in J$. Then,

$$\phi([j,\tilde{j}]) \equiv [j,\tilde{j}] + I = [j + I, \tilde{j} + I] = [\phi(j), \phi(\tilde{j})] .$$

So, ϕ is indeed a legitimate homomorphism. What is its kernel? Thanks to group theory, we already know the answer: $I \cap J$. Moreover, as in group theory, ϕ is not only *into* but also *onto*. Thus,

$$\frac{J}{I \cap J} \simeq \frac{I + J}{I}.$$

(See the end of Section 17.3.1.) Unlike in group theory, this isomorphism tells us much more. It preserves the entire structure of a Lie algebra: not only addition but also scalar multiplication, and even Lie brackets.

17.3.3 The Third Isomorphism Theorem

The next theorem also mirrors group theory. To see this, let L be a Lie algebra. Let I and J be two ideals of L, such that $I \subset J$. In what follows, I and J will play the role of S and T in Chapter 5 in [49]. Likewise, let

$$\pi : L \to L/I$$

and

$$\pi' : L/I \to \frac{L/I}{J/I}$$

play the role of ξ and ξ' there: canonical homomorphisms with the following properties: *onto*, linear, and Lie-bracket preserving. This way, their composition

$$\pi'\pi : L \to \frac{L/I}{J/I}$$

shares the same properties: *onto*, linear, and Lie-bracket preserving. Furthermore, as in group theory, its kernel is exactly J. Indeed,

$$\pi'\pi(l) = \pi'(l + I) = l + I + J/I.$$

When is this zero? Only when $l \in J$. Thus,

$$L/J \simeq \frac{L/I}{J/I}.$$

(See the end of Section 17.3.1.) This is the third isomorphism theorem. Unlike in group theory, here the isomorphism is even stronger: it preserves not only addition but also scalar multiplication and Lie brackets. Why? Because π and π' do.

17.4 Exercises: Canonical Homomorphism

1. Let L be a Lie algebra. Let $I \subset L$ be an ideal. What does this mean? Hint: $[I, L] \subset I$.
2. Look at L/I. Is this a legitimate group? Hint: yes: with respect to addition.
3. What is the group operation? Hint: addition, denoted by '+'.
4. In this group, what is the unit element? Hint: $\mathbf{0} + I = I$.
5. In this group, what is the inverse of a given vector of the form $x + I$? Hint: $-x + I$.
6. Is L/I a linear space as well? Hint: see Section 17.2.5.
7. How many distributive laws does it satisfy? Hint: see Section 17.2.5.
8. Look at the canonical mapping π. Where is it defined? Hint: on L.
9. What is its image? Hint: $\pi(L) = L/I$.
10. Is π linear? Hint: see Section 17.2.6.
11. Use this to prove once again that L/I satisfies two distributive laws. Hint: pick two vectors $x, y \in L$, and two scalars: α and β. Write the distributive laws in L:

$$\begin{aligned} \alpha(x+y) &= \alpha x + \alpha y \\ (\alpha+\beta)x &= \alpha x + \beta x. \end{aligned}$$

Now, since π is linear, it can be safely applied to both sides:

$$\begin{aligned} \alpha(x+y+I) &= \alpha(x+I) + \alpha(y+I) \\ (\alpha+\beta)(x+I) &= \alpha(x+I) + \beta(x+I). \end{aligned}$$

12. Do the same for the associative law for addition. Hint: first, write it in L: for every $x, y, z \in L$,

$$(x+y)+z = x+(y+z).$$

Then, apply π to both sides:

$$(x+y+I)+z+I = x+I+(y+z+I).$$

13. Conclude once again that L/I is indeed a legitimate linear space.
14. Look again at $\pi : L \to L/I$. Does it preserve Lie brackets? Hint: see Section 17.2.6.
15. Use this to prove that L/I is indeed a legitimate Lie algebra: its Lie brackets are indeed bilinear and alternative, and even satisfy the Jacobi identity. Hint: pick three vectors $x, y, z \in L$, and a scalar α. Write the properties in L (bilinearity, alternativity, and Jacobi identity):

$$\begin{aligned} [x+y, z] &= [x, z] + [y, z] \\ [\alpha x, z] &= \alpha[x, z] \\ [x, x] &= \mathbf{0} \\ [[x, y], z] + [[z, x], y] + [[y, z], x] &= \mathbf{0}. \end{aligned}$$

Now, since π is linear and Lie-bracket preserving, it can be safely applied to both sides:

$$\begin{aligned}[x+y+I, z+I] &= [x+I, z+I] + [y+I, z+I] \\ [\alpha x+I, z+I] &= \alpha[x+I, z+I] \\ [x+I, x+I] &= I\end{aligned}$$

$$[[x+I, y+I], z+I] + [[z+I, x+I], y+I] + [[y+I, z+I], x+I] = I.$$

16. This is true not only for π but also for any homomorphism ϕ: its image $\phi(L)$ is a legitimate Lie algebra in its own right. Why? Hint: set I to be the kernel of ϕ.

18

Solvability and Nilpotency

Lie algebras come in two major types:

- A Lie algebra could be solvable, and even nilpotent (shrinks to zero, in some sense). Nilpotency is stronger: if it is nilpotent, then it must be solvable as well. (Don't confuse this with a nilpotent matrix.)
- If not, then it could still be semisimple, and even simple (never shrinks to zero). Simplicity is stronger: if it is simple, then it must be semisimple as well.

These types disagree, and never overlap: if our Lie algebra is solvable, then it can never be semisimple (let alone simple), and vice versa.

This chapter is written in our active style again: questions, followed by their answers. This way, you get to see the theory born in front of your eyes.

18.1 The Derived Algebra

1. Let L be a Lie algebra. Look at $\{\mathbf{0}\}$: the degenerate subspace that contains one vector only: the zero vector. Is this a legitimate subspace? Hint: is it closed under addition and scalar multiplication? Well, $\mathbf{0}+\mathbf{0}=\mathbf{0}$, and $a\mathbf{0}=\mathbf{0}$.
2. Is it a subalgebra? Hint: is it closed under the Lie brackets? Well, thanks to bilinearity, $[\mathbf{0},\mathbf{0}]=\mathbf{0}$.
3. Furthermore, is it an ideal? Hint: thanks to bilinearity, $[\mathbf{0},l]=\mathbf{0}$ (for every $l\in L$).
4. Conclude that this is the smallest ideal.
5. What is the biggest ideal? Hint: see below.
6. Look at L itself. Is this an ideal?
7. Look at $[L,L]$. What is this? Hint: the subspace spanned by all vectors of the form $[l,m]$, for every $l,m\in L$.
8. Is $[L,L]\subset L$? Hint: L is closed under Lie brackets.
9. $[L,L]$ is called the derived algebra.
10. Is $[L,L]$ an ideal? Hint: in Section 17.1.4, set $I=J=L$.
11. On the other hand, is $L\subset[L,L]$? Hint: not always!
12. Give an example of $L=[L,L]$. Hint: let L be the three-dimensional Cartesian space (Section 16.7). Write each standard unit vector as the vector product of the other two.

13. Give yet another example of $L = [L, L]$. Hint: look at the Pauli matrices, with imaginary coefficients (Section 16.7). Write each matrix as the commutator of the other two.
14. Next, look at yet another example: $L = sl_2$. In this case, is $L = [L, L]$? Hint: check on a basis: $e, f, h \in sl_2$ (Section 16.8). Could these be written as a commutator? They sure can:

$$e = \frac{[h,e]}{2}, \; f = \frac{[h,f]}{-2}, \quad \text{and} \quad h = [e,f].$$

15. Extend this to sl_n as well. Hint: scan $i = 1, 2, \ldots, n-1$. For each fixed i, start from $j = i+1$. As before, write $E^{(i,j)}$, $E^{(j,i)}$, and $E^{(i,i)} - E^{(j,j)}$ as a commutator. And what about $j > i+1$? How to fill the ith row (and column), and write as a commutator as well? Like this:

$$E^{(i,j)} = E^{(i,i)}E^{(i,j)} = \left[E^{(i,i)} - E^{(i+1,i+1)}, E^{(i,j)}\right]$$
$$E^{(j,i)} = E^{(j,i)}E^{(i,i)} = \left[E^{(j,i)}, E^{(i,i)} - E^{(i+1,i+1)}\right].$$

16. Conclude that

$$sl_n = [sl_n, sl_n].$$

17. Prove that

$$sl_n = [sl_n, sl_n] = [gl_n, gl_n].$$

Hint: since $sl_n \subset gl_n$,

$$sl_n = [sl_n, sl_n] \subset [gl_n, gl_n].$$

But the commutator is trace-free (Section 16.7), so we also have

$$[gl_n, gl_n] \subset sl_n.$$

18. Conclude that gl_n and sl_n, although different, have the same derived algebra.

18.2 Automorphisms and Their Lie Group

1. What is a homomorphism?
2. What is an isomorphism?
3. An isomorphism is an invertible homomorphism.
4. Give an example.
5. An isomorphism from the Lie algebra onto itself is called an automorphism. Give an example.
6. For example, look at the identity mapping. Is it an automorphism? Why?
7. Write it as an exponent (of some adjoint mapping). Hint: look at $\exp(\mathbf{0}_{ad})$. In its power series, only one term survives: the leading term.
8. For yet another example, look at E (Section 16.5). Is it an automorphism? Why?
9. In our Lie algebra, look at two vectors: f and g.
10. Look at their adjoints:

$$f_{ad} \equiv [f, \cdot], \quad \text{and} \quad g_{ad} \equiv [g, \cdot].$$

11. Assume that they are nilpotent.
12. Conclude that they have an exponent.
13. Look at such an exponent. Is it a legitimate automorphism? Why?
14. It is called an inner automorphism.
15. Moreover, assume that f_{ad} and g_{ad} commute:

$$[f_{ad}, g_{ad}] = [f, g]_{ad} = (0)$$

(the zero mapping).
16. Look at their sum. Raise it to power $n \geq 0$. Hint: thanks to commutativity, the binomial formula works:

$$(f_{ad} + g_{ad})^n = \sum_{i-0}^{n} \binom{n}{i} f_{ad}^i g_{ad}^{n-i}.$$

17. Conclude that their sum is nilpotent as well (and has an exponent too).
18. Conclude that the exponent is invariant (under sum-product). Hint: thanks to the binomial formula,

$$\exp(f_{ad} + g_{ad}) = \exp(f_{ad})\exp(g_{ad}).$$

19. Conclude that these exponents commute:

$$\exp(f_{ad})\exp(g_{ad}) = \exp(f_{ad} + g_{ad}) = \exp(g_{ad} + f_{ad}) = \exp(g_{ad})\exp(f_{ad}).$$

20. What is the inverse of $\exp(f_{ad})$? Hint: $\exp(-f_{ad})$. Indeed, their product is the identity automorphism.
21. Look at all automorphisms. Is this a group?
22. What is the group operation? Hint: composition of automorphisms.
23. What is the unit element? Hint: the identity automorphism.
24. This group is denoted by $aut(L)$.
25. Look at f_{ad}, and all its scalar multiples. Do they commute?
26. Exponentiate them. What do you get? Hint: a new subgroup of inner automorphisms.
27. This is called a one-parameter subgroup of our Lie group.
28. Is it Abelian (commutative)?
29. We also say that this subgroup is *generated* by f.
30. Let's do this in practice. Multiply f_{ad} by a real coefficient t, and exponentiate.
31. Look at the result. Recall that it preserves Lie brackets:

$$\exp(tf_{ad})[g, h] = [\exp(tf_{ad})g, \exp(tf_{ad})h]$$

(where g and h are in our Lie algebra).
32. Differentiate this (with respect to t) at $t = 0$. Hint: differentiating $\exp(tf_{ad})$ (at $t = 0$) gives f_{ad} back again. This gives

$$f_{ad}([g, h]) = [f_{ad}(g), h] + [g, f_{ad}(h)],$$

or

$$[f, [g, h]] = [[f, g], h] + [g, [f, h]].$$

33. Is this familiar? Hint: this is the good old Leibniz rule (the Jacobi identity in our Lie algebra).
34. In this sense, differentiation is the inverse of exponentiation.
35. Better yet, look not only at f_{ad} but also at many other (nilpotent) adjoint derivations that commute with each other (zero commutator).
36. Algebraically, place them in a big matrix, block by block:

$$\begin{pmatrix} \text{first adjoint} & & & \\ & \text{second adjoint} & & \\ & & \text{third adjoint} & \\ & & & \ddots \\ & & & & \ddots \end{pmatrix}.$$

18.3 Entangled Measurables

1. Quantum-mechanically, these commuting adjoints represent measurables (say, position):

$$\begin{pmatrix} X = x\cdot & & \\ & Y = y\cdot & \\ & & Z = z\cdot \end{pmatrix}.$$

2. How does this act on a given wave function $v : \mathbb{R}^3 \to \mathbb{C}$? Hint: store v as an infinite vector (indexed by x, y, and z). This way, X picks x (or multiplies by x), Y by y, and Z by z. Together, this maps

$$v \to XYZv = xyzv(x, y, z)$$

(a new infinite vector, indexed by x, y, and z as well).
3. If v is nonseparable (can't be written as a triple product), then X, Y, and Z are entangled (affect one another).
4. If v is an eigenvector, how does it look like? Hint: it is nonzero only at one particular point: $(x_0, y_0, z_0) \in \mathbb{R}^3$.
5. Is this separable? Hint: yes.
6. In such an eigenvector, are X, Y, and Z entangled? Hint: no.
7. What is its eigenvalue? Hint: under XYZ, it has eigenvalue $x_0 y_0 z_0$.
8. What does this mean physically? Hint: determinism: in this state, the particle is at (x_0, y_0, z_0), not elsewhere.
9. This is an example of disentanglement (in terms of our position measurables). Indeed, at determinism, X, Y, and Z don't affect one another. Still, there are many more examples.
10. So far, we looked at entanglement in terms of position measurables. Next, how about position-momentum, say

$$\begin{pmatrix} X & \\ & P_x \end{pmatrix}?$$

Can you do the same as before? Hint: only in the sense in [50] (since X and P_x don't commute).

11. Do the same as before. What do you get? Hint: a (dual) phase plane, with two possible orientations: either position-momentum, or momentum-position. Each represents a different order of measurement (with different quantum-mechanical results).
12. How to change orientation? Hint: there are two options:
    - Go back-in-time. This way, momentum picks a minus sign (in its definition).
    - Or, instead, let X pick a minus sign.

    Either way, the commutator picks a minus sign, leading to a different orientation: momentum-position (rather than position-momentum).
13. Look again at our original position-momentum plane. In it, draw your joint (position-momentum) wave function (estimated statistically). How to "normalize" it, line by line? Hint: if you measured a fixed momentum first, then normalize horizontally. If, on the other hand, you measured a fixed position first, then normalize vertically.
14. In this (complex) plane, the orientation comes from the dynamics. In particular, the momentum is the same as the position after a quarter of a period.

18.4 Normal Lie Subgroup

1. Let's go algebraic back again. Look at our (commuting) adjoints again.
2. Do they span a new subalgebra? Hint: they are closed under their new Lie brackets (their zero commutator).
3. Design such a subalgebra of maximal dimension. What does "maximal" mean? Hint: it can't grow and remain commutative.
4. This is called a Cartan subalgebra.
5. Look at its members. Look at their exponents. Do they make a new subgroup of inner automorphisms? Hint: thanks to commutativity, the exponent is invariant under sum-product.
6. Is it Abelian (commutative)?
7. Is it maximal? In what sense? Hint: it can't grow and remain Abelian (at least in nontrivial examples).
8. Look at all adjoint derivations. Do they commute? Hint: not necessarily.
9. Do they form a subspace? Hint: their linear combination is an adjoint in its own right.
10. Do they form a subalgebra? Hint: their commutator is an adjoint in its own right.
11. Look at all their exponents (if exist). What are these? Hint: inner automorphisms.
12. Do they commute? Hint: not necessarily.
13. Do they form a subgroup? Hint: not necessarily.
14. Can they *generate* yet more automorphisms? How? Hint: compose them with one another.
15. Look at such a product. Is it inner? Hint: not necessarily.
16. Throw all such products in. Is this a legitimate subgroup of $aut(L)$? Hint: yes: to have the inverse automorphism, compose the inverses reversely.

17. This subgroup is denoted by

$$int(L) \subset aut(L).$$

18. Is it normal? Hint: first, consider a generator (an inner automorphism). Conjugate it with just any $E \in aut(L)$ (plug it in between E and E^{-1}). This way, you get a new inner automorphism (Figure 16.5). Next, conjugate not only one generator but also a product of two.
19. Later on, $int(L)$ will represent a useful Lie group.

18.5 The Adjoint Representation and Its Trace

1. Let L be a Lie algebra. What is a homomorphism on L? Hint: a linear mapping (into some other Lie algebra) that preserves Lie brackets.
2. Let ϕ be a homomorphism. What is its image? Hint:

$$\phi(L) \equiv \{\phi(l) \mid l \in L\}.$$

3. Is it a legitimate Lie algebra in its own right? Hint: is it closed under addition, scalar multiplication, and Lie brackets?
4. How does ϕ act on the derived algebra $[L, L]$? Hint:

$$\phi([l, m]) = [\phi(l), \phi(m)], \quad l, m \in L.$$

   Extend this linearly to every linear combination of $[l, m]$'s.
5. Conclude that

$$\phi([L, L]) = [\phi(L), \phi(L)].$$

6. What is a representation? Hint: a homomorphism into gl_n (for some natural number n).
7. Let ϕ be a representation of L.
8. In the above formula (for some $l, m \in L$), what do we have in the right-hand side? Hint: a commutator of two $n \times n$ matrices.
9. What is its trace? Hint: zero (Section 16.7).
10. Conclude that the image of the derived algebra is trace-free:

$$\phi([L, L]) \subset sl_n.$$

11. Furthermore, assume that L is the same as its derived algebra:

$$L = [L, L].$$

12. Conclude that, in this case, $\phi(L)$ is trace-free:

$$\phi(L) = \phi([L, L]) \subset sl_n.$$

13. For example, L itself could be linear, say:

$$L = sl_2 = [sl_2, sl_2]$$

   (Section 18.1).

14. What is its basis? Hint: e, f, and h (Section 16.8).
15. Write them as standard unit vectors:

$$e \sim \begin{pmatrix} 1 \\ 0 \\ 0 \end{pmatrix}$$
$$h \sim \begin{pmatrix} 0 \\ 1 \\ 0 \end{pmatrix}$$
$$f \sim \begin{pmatrix} 0 \\ 0 \\ 1 \end{pmatrix}.$$

16. On these, how to define ϕ? Hint: ϕ could be the adjoint representation:

$$\phi(e) = [e, \cdot]$$
$$\phi(h) = [h, \cdot]$$
$$\phi(f) = [f, \cdot]$$

(Sections 16.2 and 16.8).
17. For instance, look at $\phi(h) = [h, \cdot]$. What is this? Hint: a mapping of sl_2.
18. What are its eigenvectors? Hint:

$$\phi(h)(e) = [h, e] = 2e$$
$$\phi(h)(h) = [h, h] = 0h$$
$$\phi(h)(f) = [h, f] = -2f.$$

19. In terms of standard unit vectors, how does $\phi(h)$ look like? Hint: it is diagonal:

$$\phi(h) \sim \begin{pmatrix} 2 & & \\ & 0 & \\ & & -2 \end{pmatrix}.$$

This matrix indeed has three eigenvectors: e, h, and f (written as standard unit vectors).
20. And how about $\phi(e)$? How does it look like? Hint: only its superdiagonal is nonzero:

$$\phi(e) \sim \begin{pmatrix} & -2 & \\ & & 1 \\ & & \end{pmatrix}.$$

21. And how about $\phi(f)$? How does it look like? Hint: only its subdiagonal is nonzero:

$$\phi(f) \sim \begin{pmatrix} & & \\ -1 & & \\ & 2 & \end{pmatrix}.$$

22. What is the trace of these matrices? Hint: zero.
23. Why? Hint: because they are in the image of ϕ, which is trace-free.

24. What about any other representation (not necessarily the adjoint)? Is its image still trace-free? Hint: see above.
25. And what if $L = sl_k$ $(k \geq 2$? Is this still true? Hint:

$$\phi(sl_k) = \phi([sl_k, sl_k]) = [\phi(sl_k), \phi(sl_k)] \subset sl_n,$$

where ϕ could be just any representation, say the adjoint:

$$\phi(l) = l_{ad} = [l, \cdot], \qquad l \in sl_k$$

(Section 18.1).

18.6 The Derived Series

1. Define a series of (smaller and smaller) Lie algebras: L, the derived, the derived of the derived, and so on.
2. More precisely, use mathematical induction:

$$\begin{aligned} L^{(0)} &\equiv L \\ L^{(i+1)} &\equiv \left[L^{(i)}, L^{(i)}\right] \qquad (i \geq 0). \end{aligned}$$

3. This is called the derived series.
4. Show that, in this series, each member is included in all previous members. Hint: use mathematical induction.
5. Show that, in this series, each member is an ideal of L. Hint: use mathematical induction, and Section 17.1.4.
6. Look at $L^{(i)}$ (for some fixed $i \geq 0$). What is its own derived series? Hint:

$$\left(L^{(i)}\right)^{(j)} = L^{(i+j)} \qquad (j \geq 0).$$

7. For example, set $L = sl_n$ (for some fixed n). What is its derived series? Hint: a constant series: $L^{(i)} = sl_n$ $(i \geq 0)$.
8. In this case, the derived series never shrinks to $\{(0)\}$ (the zero Lie algebra): L is not solvable.
9. Conclude that sl_n is not solvable.
10. For yet another example, set $L = gl_n$. What is its derived series? Hint: $L^{(i)} = sl_n$ $(i > 0)$.
11. Conclude that gl_n is not solvable either.

18.7 Solvability vs. Simplicity

1. Recall the notations in Section 16.9: T_0 contains upper-triangular matrices, T_1 strictly upper-triangular, T_2 yet stricter, and so on: as k increases, T_k supports fewer diagonals.
2. What is the derived algebra of T_0? Hint: $[T_0, T_0] = T_1$. Indeed, on one hand, $[T_0, T_0] \subset T_1$ (since the commutator is strictly upper triangular). On the other hand, $T_1 = [T_0, T_1] \subset [T_0, T_0]$ (see below).

3. If $j + k > 0$, what is T_{j+k}? Hint: $T_{j+k} = [T_j, T_k]$ (Section 16.9).
4. In particular, what is T_{2k}? Hint: $T_{2k} = [T_k, T_k]$ $(k > 0)$.
5. Set $L \equiv T_0$. What is its derived series? Hint:

$$L^{(0)} \equiv T_0$$
$$L^{(i)} = T_{2^{i-1}} \quad (i > 0).$$

6. Let i increase more and more, until 2^{i-1} is as big as the order of the matrices: $2^{i-1} \geq n$. Show that, for such a big i, $L^{(i)}$ is as small as $\{(0)\}$ (the zero Lie algebra).
7. In such a case, we say that L is solvable.
8. Conclude that T_0 is solvable.
9. Likewise, look at T_k (for some fixed $k \geq 0$). What is its derived series?
10. Conclude that T_k is solvable as well.
11. How to prove this more easily? Hint: without calculating, scan the derived series of T_k, member by member. Each member is included in the corresponding member in the derived series of T_0 (mathematical induction).
12. Let L be a Lie algebra. What is its largest ideal? Hint: L itself (Section 18.1).
13. What is its smallest ideal? Hint: $\{\mathbf{0}\}$ (the zero Lie algebra, which contains one vector only: the zero vector).
14. What happens if the derived algebra is as small as

$$[L, L] = \{\mathbf{0}\}?$$

Hint: in this case, all vectors commute (zero commutator). We then say that L is Abelian (commutative).
15. In this case, is there a nontrivial ideal (different from both L and $\{\mathbf{0}\}$)? Hint: in this case, every subalgebra is an ideal.
16. But what if L isn't Abelian? Could it still have a nontrivial ideal? Hint: yes, it could. If it doesn't, then it is called simple.
17. If L is simple, what is its derived algebra? Hint: since L isn't Abelian, $[L, L] \neq \{\mathbf{0}\}$. Since L has no nontrivial ideal, $[L, L] = L$.
18. What is its derived series? Hint: it is constant: $L^{(i)} = L$ $(i \geq 0)$.
19. Conclude that a simple Lie algebra can never be solvable.
20. Conclude that a solvable Lie algebra can never be simple. Hint: if it were, then it could never be solvable in the first place.
21. Give an example of a simple Lie algebra. Hint: sl_2. Indeed, let $I \subset sl_2$ be an ideal. Could it be nontrivial? Well, what vector could be in I? If $h \in I$, then $[h, e] = 2e \in I$ and $[h, f] = -2f \in I$ as well. Likewise, if e or $f \in I$, then $[e, f] = h \in I$ as well. In either case, $I = sl_2$ (which is not what we try to design). Now, could I contain a more complicated vector like

$$v \equiv c_e e + c_h h + c_f f$$

(for some coefficients c_e, c_h, and c_f)? Well, as an ideal, I must then contain more vectors:

$$[v, e], \quad [[v, e], e], \quad [v, f], \quad [[v, f], f] \quad \in I.$$

So, $c_e = c_h = c_f = 0$ (or $e, h, f \in I$ back again).

22. Could sl_2 be solvable? Hint: never! It is simple: its derived series is constant, and never shrinks to zero.
23. Likewise, look at the three-dimensional Cartesian space, with vector product. Is it a simple Lie algebra? Hint: use a similar proof.
24. Could it be solvable? Hint: no!
25. Could a Lie algebra be neither simple nor solvable? Hint: for example, gl_n is not simple. Indeed, it has a nontrivial ideal: sl_n. Still, gl_n is not solvable either (Section 18.6).

18.8 Solvability and a Homomorphism

1. Let L be a Lie algebra. Let $K \subset L$ be a subalgebra. What can you say about its derived series? Hint: $K^{(i)} \subset L^{(i)}$ (mathematical induction on $i = 0, 1, 2, \ldots$).
2. Conclude that, if L is solvable, then so is K.
3. Furthermore, let ϕ be a homomorphism on L. Look at the homomorphic image: $\phi(L)$. What is its derived algebra? Hint:
$$\phi(L)^{(1)} = [\phi(L), \phi(L)] = \phi([L, L]) = \phi\left(L^{(1)}\right).$$
This pushes the superscript '(1)' into parentheses.
4. How to use this in practice? Assume that we already have the derived algebra of L. How to obtain the derived algebra of $\phi(L)$ for free (with no need to calculate any Lie brackets)? Hint: take $[L, L]$, and apply ϕ to it.
5. Better yet, how to obtain the entire series for free? Hint: push the superscript '(i)' into parentheses:
$$\phi(L)^{(i)} = \phi\left(L^{(i)}\right).$$
6. Prove this by mathematical induction on $i = 0, 1, 2, 3, \ldots$. Hint: for $i = 0$, this is trivial. Assume that we already know this for some fixed $i \geq 0$ (the induction hypothesis). How to push '$(i+1)$' into parentheses? In two stages: first '(i)', then '(1)':
$$\begin{aligned}\phi(L)^{(i+1)} &= \left[\phi(L)^{(i)}, \phi(L)^{(i)}\right] \\ &= \left[\phi\left(L^{(i)}\right), \phi\left(L^{(i)}\right)\right] \\ &= \phi\left(\left[L^{(i)}, L^{(i)}\right]\right) \\ &= \phi\left(L^{(i+1)}\right).\end{aligned}$$
7. How to use this in practice? Hint: given the derived series of L, scan it, member by member, and apply ϕ to each. This gives you the derived series of $\phi(L)$ for free.
8. Conclude that, if L is solvable, then so is $\phi(L)$.
9. Let L be a Lie algebra, and $I \subset L$ an ideal. Let
$$\pi : L \to L/I$$

be the canonical homomorphism (Section 17.4). In the above formula, replace ϕ by π. Hint:

$$(L/I))^{(i)} = (\pi(L))^{(i)} = \pi\left(L^{(i)}\right) = L^{(i)}/I \quad (i \geq 0).$$

10. Assume that we already have the derived series of L. How to obtain the derived series of L/I for free? Hint: scan the derived series of L. Divide each member by I.
11. Conclude that, if L is solvable, then so is L/I. Hint: for a sufficiently large i, $L^{(i)} = \{\mathbf{0}\}$. Thus, the right-hand side gets as small as

$$L^{(i)}/I = \{\mathbf{0} + I\} = \{I\}$$

(where our ideal I is the zero vector in L/I).
12. Prove this in yet another way. Hint: L/I is the homomorphic image of L under π.
13. Work the other way around: assume that L/I is solvable, but we don't know about L. For a sufficiently large i, how does the above formula look like? Hint: for a sufficiently large i, the left-hand side gets as small as $\{I\}$ (where our ideal I is the zero vector in L/I).
14. Conclude that, for such a big i,

$$L^{(i)} \subset I.$$

15. Assume now even more: both L/I and I are solvable (but we still don't know about L). Conclude that, for our big i, $L^{(i)}$ is solvable as well. Hint: it is included in I, which is solvable.
16. Conclude that, if both I and L/I are solvable, then so is L itself.
17. Let L be a Lie algebra. Let $I, J \subset L$ be two ideals. Assume that J is solvable. What about $I \cap J$? Is it solvable as well? Hint: $I \cap J \subset J$.
18. What about $J/(I \cap J)$? Is it solvable as well? Hint: let $\pi : J \to J/(I \cap J)$ be the canonical homomorphism. We already proved that the homomorphic image $\pi(J) = J/(I \cap J)$ is solvable as well.
19. What about $(I + J)/I$? Is it solvable as well? Hint: use the second isomorphism theorem (Section 17.3.2).
20. Assume now that both J and I are solvable. What about $I + J$? Is it solvable as well? Hint: since both I and $(I + J)/I$ are solvable, $I + J$ is solvable as well.

18.9 Radical and Semisimplicity

1. Let L be a Lie algebra. Let $R \subset L$ be the maximal solvable ideal. Then, R is called the radical (if exists).
2. If L is solvable, what is its radical? Hint: in this case, $R = L$.
3. Is the radical unique? Hint: let $I \subset L$ be another solvable ideal. Then, $R + I$ is yet another solvable ideal. But R is maximal, so $R + I \subset R$, so $I \subset R$ as well. Thus, R must include all solvable ideals.
4. If $R = \{\mathbf{0}\}$ is the only solvable ideal, then we say that L is semisimple.

5. If L is simple, must it also be semisimple? Hint: if L is simple, then its only ideals are L (which is not solvable) and $\{\mathbf{0}\}$.
6. If L is solvable, can it still be semisimple? Hint: if L is solvable, then $R = L$, so L can't be semisimple.
7. If L is semisimple, can it still be solvable? Hint: if it were, then it could never be semisimple in the first place.
8. In general, look at L/R. Is it semisimple? Hint: design an ideal in L/R. How could it look like? It must take the form I/R (for some ideal $I \subset L$). Now, if I/R is solvable, then I is solvable as well, so $I \subset R$, so $I/R = \{R\}$ is the zero ideal in L/R.

18.10 The Lower Central Series

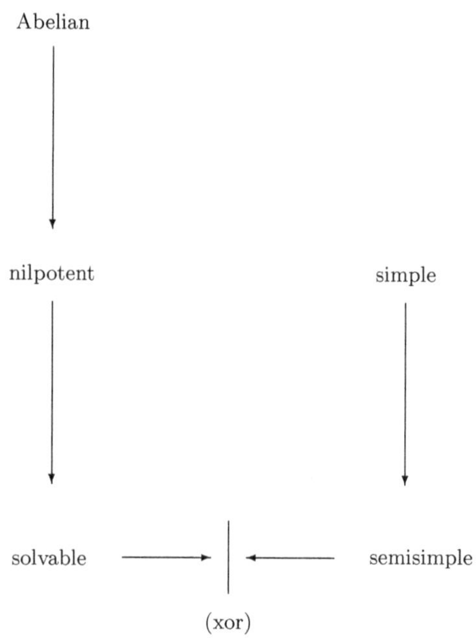

Fig. 18.1. A *xor* relation: a semisimple Lie algebra can't be solvable. A solvable Lie algebra can't be semisimple.

1. Let L be a Lie algebra. Define a new series of (smaller and smaller) Lie algebras. For this, use mathematical induction again:

$$
\begin{aligned}
L^0 &\equiv L \\
L^{i+1} &\equiv \left[L, L^i\right] \quad (i \geq 0).
\end{aligned}
$$

(Here, L^i is just a new notation, not a power.)

2. This is called the lower central series.
3. In this series, what is the first member? Hint: L (the original Lie algebra).
4. And what is the next member? Hint: $[L, L]$ (the derived algebra).
5. Are these familiar? Hint: they are the same as in the derived series:

$$
\begin{aligned}
L^0 &= L^{(0)} = L \\
L^1 &= L^{(1)} = [L, L]
\end{aligned}
$$

(the derived algebra).

6. Show that the lower central series behaves like the derived series:
   - Each member is an ideal of L.
   - Each member is included in all previous members.

   Hint: use mathematical induction.
7. Scan the lower central series, member by member. Show that each member includes the corresponding member from the derived series:

$$
L^{(i)} \subset L^i, \quad i \geq 0.
$$

Hint: use mathematical induction.

8. If the lower central series shrinks to the zero Lie algebra (so $L^i = \{\mathbf{0}\}$ for a sufficiently large i), then L is called nilpotent. (Don't confuse this with a nilpotent matrix.)
9. Conclude that a nilpotent Lie algebra must also be solvable (Figure 18.1).
10. If L is not solvable, could it still be nilpotent? Hint: never!
11. For example, set $L = sl_n$ (for some fixed n). Is it nilpotent? Hint: no! We already saw that sl_n is not solvable.
12. What is its lower central series? Hint: like the derived series, it is constant: $L^i = L^{(i)} = sl_n$ $(i \geq 0)$.
13. For another example, set $L = gl_n$ (for some fixed n). Is it nilpotent? Hint: no! We already saw that gl_n is not solvable.
14. What is its lower central series? Hint: the same as the derived series: $L^i = L^{(i)} = sl_n$ $(i > 0)$.
15. Could a Lie algebra be solvable but not nilpotent? Hint: for example, set $L = T_0$ (the upper-triangular matrices of order n, for some fixed $n > 1$). We already saw that it is solvable. Still, it is not nilpotent. Indeed, thanks to mathematical induction, its lower central series never shrinks:

$$
\begin{aligned}
L^0 &= T_0 \\
L^1 &= [T_0, T_0] = T_1 \\
L^i &= [T_0, T_1] = T_1 \quad (i \geq 2).
\end{aligned}
$$

16. What about $L = T_1$ (the strictly upper-triangular matrices)? Is it nilpotent? Hint: its lower central series is

$$
L^i = T_{i+1} \quad (i \geq 0).
$$

Indeed, thanks to mathematical induction,

$$\begin{aligned} L^0 &= L = T_1 \\ L^{i+1} &= \left[L, L^i\right] = [T_1, T_{i+1}] = T_{i+2} \quad (i \geq 0). \end{aligned}$$

For a sufficiently large i, this shrinks to the zero Lie algebra.

17. Likewise, show that T_k is nilpotent not only for $k = 1$ but also for every $k \geq 1$.

18.11 Nilpotency and a Homomorphism

1. Let L be a Lie algebra. Let $K \subset L$ be a subalgebra. What can you say about its lower central series? Hint: $K^i \subset L^i$ (by mathematical induction).
2. Conclude that, if L is nilpotent, then so is K.
3. Furthermore, let $i \geq 0$ be fixed, and assume now that $K \subset L^i$. What can you say now about the lower central series of K? Hint: for all $j \geq 0$, $K^j \subset L^{i+j}$ (by mathematical induction).
4. For example, set $K = L^i$. What do you get from the above exercise? Hint:

$$(L^i)^j \subset L^{i+j} \quad (j \geq 0).$$

5. Moreover, let ϕ be a homomorphism on L. Look at the homomorphic image: $\phi(L)$. What is its derived algebra? Hint:

$$(\phi(L))^1 = [\phi(L), \phi(L)] = \phi\left([L, L]\right) = \phi\left(L^1\right).$$

This pushes the superscript 1 into parentheses.

6. Better yet, you can even push i into parentheses:

$$(\phi(L))^i = \phi\left(L^i\right) \quad (i \geq 0).$$

Prove this by mathematical induction on $i = 0, 1, 2, \ldots$. Hint: for $i = 0$, this is trivial. Assume that this is true for some fixed $i \geq 0$ (the induction hypothesis). How to push $i + 1$ into parentheses? In two stages: first i, then 1:

$$\begin{aligned} (\phi(L))^{i+1} &= \left[\phi(L), (\phi(L))^i\right] \\ &= \left[\phi(L), \phi(L^i)\right] \\ &= \phi\left(\left[L, L^i\right]\right) \\ &= \phi\left(L^{i+1}\right). \end{aligned}$$

7. How to obtain the lower central series of $\phi(L)$ for free (with no need to calculate any Lie brackets)? Hint: scan the lower central series of L, member by member, and apply ϕ to each.
8. Conclude that, if L is nilpotent, then so is $\phi(L)$.
9. Let L be a Lie algebra, and $I \subset L$ an ideal. Let $\pi : L \to L/I$ be the canonical homomorphism (Section 17.4). In the above formula, replace ϕ by π. Hint:

$$(L/I))^i = (\pi(L))^i = \pi\left(L^i\right) = L^i/I \quad (i \geq 0).$$

10. Assume that we already have the lower central series of L. How to obtain the lower central series of L/I for free? Hint: scan the lower central series of L, member by member. Divide each by I.
11. Conclude that, if L is nilpotent, then so is L/I.
12. Work now the other way around: assume now that L/I is nilpotent, but we don't know about L. What can you say about the lower central series of L? Hint: for a sufficiently large i, $(L/I)^i$ gets as small as $\{I\}$ (the zero ideal in L/I). Thus,
$$L^i \subset I.$$
13. Assume now yet more: both L/I and I are nilpotent. Must L be nilpotent as well? Hint: not necessarily. In fact, this only tells us that $(L^i)^j$ shrinks to zero (as j grows), but L^{i+j} could be bigger!
14. How to make sure that L is nilpotent as well? Hint: assume that I is a special ideal: the center of L, containing those vectors that commute with all vectors in L:
$$I = \{c \in L \mid [c, l] = \mathbf{0},\ l \in L\}.$$
This way,
$$[L, I] = \{\mathbf{0}\}.$$
For our big i, we therefore have
$$L^{i+1} = [L, L^i] \subset [L, I] = \{\mathbf{0}\}.$$
Thus, in this case, L is indeed nilpotent.
15. Let L be a nilpotent Lie algebra. Assume that L is nontrivial: $L \neq \{\mathbf{0}\}$. What about its center? Could it be trivial? Hint: let $k > 0$ be the minimal integer for which $L^k = \{\mathbf{0}\}$. Thus, L^{k-1} must be nontrivial, yet included in the center of L:
$$\left[L, L^{k-1}\right] = L^k = \{\mathbf{0}\}.$$
Thus, the center can't be trivial.
16. Moreover, look at the adjoint representation of L (Section 18.5). Show that, for each individual $x \in L$, $x_{ad} = [x, \cdot]$ is a nilpotent matrix. More precisely, its kth power vanishes:
$$x_{ad}^k y = \mathbf{0}, \quad x, y \in L.$$
Hint: use mathematical induction to show that
$$x_{ad}^i y \in L^i, \quad x, y \in L,\ i = 0, 1, 2, 3, \ldots.$$
17. Conclude that x is adjoint-nilpotent: its adjoint is a nilpotent matrix (Section 16.4).
18. Note that this is true for every $x \in L$. Thus, if L is a nilpotent Lie algebra, then all vectors in it are adjoint-nilpotent.
19. What about the other way around? If all vectors in L are adjoint-nilpotent, must L be a nilpotent Lie algebra? Hint: this will follow from Engel's theorems below.

19

Nilpotency and Engel's Theorems

By now, we already saw a concrete Lie algebra: a linear Lie algebra, containing matrices. Are they nilpotent? Here, we'll study this.

A general Lie algebra, on the other hand, may contain more abstract members, with no notion of nilpotency at all. Still, we can ask: are they adjoint-nilpotent? After all, the adjoint is a linear mapping (or a matrix), which could be nilpotent or not. If it is, then the original member is adjoint-nilpotent. Let's study this.

How to make sure that all members are adjoint-nilpotent? Require your Lie algebra to be nilpotent (shrinking to zero, in some sense). This is easy enough to prove: a global property (nilpotency of the entire Lie algebra) leads to a local property (adjoint-nilpotency of each individual member). But what about the other way around? Suppose that we only know a local property: all members are adjoint-nilpotent. Must our Lie algebra be nilpotent? This is already global: it talks about the entire Lie algebra, as a whole. To prove it, we must first design a common eigenvector.

19.1 Common Eigenvector

19.1.1 Nilpotent Matrices: Local vs. Global Properties

In Chapter 15, we looked at a nilpotent matrix. We saw that it has (at least one) singular eigenvector (of eigenvalue zero). Furthermore, in a suitable basis, it gets (strictly) upper-triangular. But this is quite local: two matrices may disagree: each may have its own eigenvector and basis.

Fortunately, in a linear Lie algebra, we can improve on this, and even get global: if all matrices are nilpotent, then they share the same eigenvector and basis, in which all of them are strictly upper-triangular at the same time. They share a global structure. Why? Because they come from the same Lie algebra.

19.1.2 Nilpotent Matrix and Its Adjoint

In a linear Lie algebra, each member x is a matrix. It has a nice property (Section 16.4): if x is nilpotent, then it also makes two nilpotent operators: $x\circ$ (left multiplication) and $\circ x$ (right multiplication). This way, the adjoint operator

$$x_{ad} \equiv x \circ - \circ x$$

is nilpotent as well (thanks to the binomial formula). Thanks to this, we can move to the adjoint representation, useful in the induction hypothesis below.

19.1.3 Induction on The Dimension of L

Consider a linear Lie algebra:

$$L \subset gl_n$$

(for some natural number n). This way, what are the members of L? They are $n \times n$ matrices. Assume that all of them are nilpotent. As discussed above, they are then adjoint-nilpotent as well. Furthermore, each of them has its own singular eigenvector (of eigenvalue zero). Still, is this global? Do they also share the same singular eigenvector?

To prove this, let's use mathematical induction on the dimension of L. In general, this is *not* n. For example, if $L = sl_n$, then its dimension is $n^2 - 1 \neq n$ (Section 16.8).

For a start, assume that L is one-dimensional. This doesn't mean that $n = 1$: n could be bigger. It only means that L contains just one matrix (up to scalar multiplication). More precisely, for some nonzero matrix $y \neq (0)$,

$$L = \{\alpha y \mid \alpha \text{ is a scalar}\}.$$

How to design an eigenvector of y? We already know how: since y is nilpotent, there is a natural number $l \geq 1$ for which $y^l = (0)$. Pick some nonzero vector $v_0 \neq \mathbf{0}$. Then,

$$y^l v_0 = (0)v_0 = \mathbf{0}.$$

But l is not necessarily minimal. Let $m \leq l$ be the *minimal* natural number for which $y^m v_0 = \mathbf{0}$. This way, the required eigenvector is $y^{m-1}v_0$. Indeed,

$$y\left(y^{m-1}v_0\right) = y^m v_0 = \mathbf{0}.$$

19.1.4 The Induction Step

So far, L was one-dimensional: it contained one matrix only. Assuming that this matrix was nilpotent, we managed to design its (singular) eigenvector. How to extend this to a few (nilpotent) matrices, and design their common eigenvector?

Assume now that L is at least two-dimensional. Assume also the induction hypothesis: every lower-dimensional Lie algebra of nilpotent matrices has a common (singular) eigenvector. How to extend this to L as well?

For this purpose, let $K \subset L$ be a maximal proper subalgebra: K can't be extended (not even by one more matrix) and remain smaller than L. In other words, there is no subalgebra strictly in between K and L. Clearly, K is at least one-dimensional: it contains at least one nonzero matrix (like the above y).

Fortunately, K is lower-dimensional. Thanks to the induction hypothesis, all x's in K share a common eigenvector $v_0 \neq 0$, for which $xv_0 = \mathbf{0}$. Still, this is not good enough: what about those x's outside K? Do they share v_0 too?

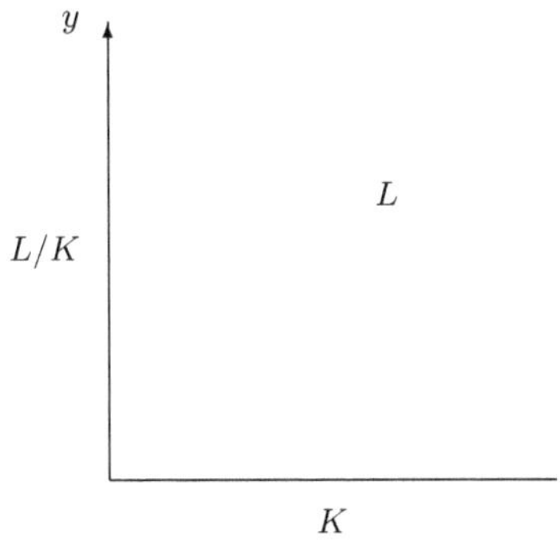

Fig. 19.1. L is spanned by K (horizontal) and L/K (vertical), spanned by one matrix only: y. This way, both K and $\tau(K)$ are lower-dimensional, and can benefit from the induction hypothesis.

By now, we don't know yet that K is an ideal. But don't worry: we can still divide L by K: not as Lie algebras, but only as linear spaces (as in group theory). After all, to divide a linear space, we don't need an ideal. (Indeed, only at the very end of Section 17.2.5 did we use the ideal property, to make sure that the Lie brackets are well-defined.) In our new L/K, we need no Lie brackets: it is still a good linear space, ready for all sorts of linear mappings (Figure 19.1).

Look at some matrix $x \in L$. What is its adjoint? This is the linear mapping

$$x_{ad} = [x, \cdot] : L \to L.$$

Better yet, on our new linear space L/K, define a yet better matrix (of lower order). This is called the factor adjoint:

$$\tau(x) : L/K \to L/K.$$

Thanks to the induction hypothesis, some $\tau(x)$'s will share a common eigenvector of the form $y + K \in L/K$. As an eigenvector, it will be nonzero: $y \notin K$. This will help extend K by one new matrix: y.

Since we don't know yet that K is an ideal, τ can't be defined on the whole of L, but only on K:

$$\tau(k)(l + K) \equiv [k, l] + K, \quad k \in K, \ l \in L.$$

This way, $\tau(k)$ maps L/K uniquely: you can add any $\tilde{k} \in K$ to l, with no harm at all:

$$\tau(k)(l + \tilde{k} + K) \equiv [k, l + \tilde{k}] + K = [k, l] + [k, \tilde{k}] + K = [k, l] + K.$$

If you like, write this as:

$$\tau(x) = \pi \circ x_{ad} \circ \pi^{-1} \quad (x \in K),$$

where $\pi : L \to L/K$ is a canonical projection of linear spaces, and π^{-1} picks a member of L (up to any $\tilde{k} \in K$). Since x_{ad} is nilpotent, so is $\tau(x)$ as well.

Fortunately, both K and $\tau(K)$ are lower-dimensional, and can benefit from the induction hypothesis. Thus, the $\tau(x)$'s share a common eigenvector $y + K \in L/K$, and map it to K (the zero of L/K):

$$\tau(x)(y + K) = [x, y] + K = K, \quad x \in K,$$

or

$$[x, y] \in K, \quad x \in K.$$

So, we can go ahead and extend K to a slightly bigger subalgebra: $K + \text{span}\{y\}$. After all, this is still closed under the Lie brackets. But K is maximal, so

$$K + \text{span}\{y\} = L.$$

We can now see in retrospect that K is an ideal after all. Therefore, τ is actually trivial: for each $x \in K$, $\tau(x)$ is the zero mapping: it maps L/K to its own zero: K. But we don't care about these things any more.

So far, we used the induction hypothesis on $\tau(K)$. Let's use it on K too, to design a common eigenvector $v_0 \neq \mathbf{0}$:

$$xv_0 = \mathbf{0}, \quad x \in K.$$

But what about y? Is $yv_0 = \mathbf{0}$? If so, then we're done: v_0 is our common eigenvector. Otherwise, move on to yv_0. After all, for every $x \in K$, we already know that $[x, y] \in K$, so

$$xyv_0 = xyv_0 - \mathbf{0} = xyv_0 - yxv_0 = (xy - yx)v_0 = [x, y]v_0 = \mathbf{0}.$$

So, yv_0 is a singular eigenvector of every $x \in K$. Does y share it as well? If so, then we're done. Otherwise, move on to y^2v_0, and so on, until $y^{m-1}v_0$, shared by all $x \in L$, as required.

19.2 Strictly Upper-Triangular Form

19.2.1 First Coordinate: The Common Eigenvector

We are now ready to form our new basis. In it, our common eigenvector will take the first coordinate, and span the standard unit vector

$$(1, 0, 0, 0, \ldots, 0)^t.$$

To map it to zero, our (nilpotent) matrix must have a zero column on the left:

$$\begin{pmatrix} 0 & u^t \\ \mathbf{0} & U \end{pmatrix} \begin{pmatrix} 1 \\ 0 \\ \vdots \\ 0 \end{pmatrix} = \mathbf{0}.$$

Here, u^t is an $(n-1)$-dimensional row vector, to make the first row. Below it, U is an $(n-1) \times (n-1)$ block, corresponding to the $n-1$ remaining coordinates (to be specified recursively soon). These new coordinates may be nonperpendicular, but we are not afraid of this. After all, we already saw oblique coordinates before (Figure 8.4).

19.2.2 Invariant Block

How to specify the $n-1$ remaining coordinates? Recursively: mathematical induction on n. After all, blocks like U make a new subalgebra of gl_{n-1}. Indeed, they are invariant under multiplication:

$$\begin{pmatrix} 0 & u^t \\ \mathbf{0} & U \end{pmatrix} \begin{pmatrix} 0 & v^t \\ \mathbf{0} & V \end{pmatrix} = \begin{pmatrix} 0 & v^t V \\ \mathbf{0} & UV \end{pmatrix}.$$

Thus, they are invariant under the commutator too:

$$\begin{aligned} \left[\begin{pmatrix} 0 & u^t \\ \mathbf{0} & U \end{pmatrix}, \begin{pmatrix} 0 & v^t \\ \mathbf{0} & V \end{pmatrix} \right] &= \begin{pmatrix} 0 & u^t \\ \mathbf{0} & U \end{pmatrix} \begin{pmatrix} 0 & v^t \\ \mathbf{0} & V \end{pmatrix} - \begin{pmatrix} 0 & v^t \\ \mathbf{0} & V \end{pmatrix} \begin{pmatrix} 0 & u^t \\ \mathbf{0} & U \end{pmatrix} \\ &- \begin{pmatrix} 0 & u^t V \\ \mathbf{0} & UV \end{pmatrix} - \begin{pmatrix} 0 & v^t U \\ \mathbf{0} & VU \end{pmatrix} \\ &= \begin{pmatrix} 0 & u^t V - v^t U \\ \mathbf{0} & UV - VU \end{pmatrix} \\ &= \begin{pmatrix} 0 & u^t V - v^t U \\ \mathbf{0} & [U, V] \end{pmatrix}. \end{aligned}$$

Likewise, they are invariant under power too:

$$\begin{pmatrix} 0 & u^t \\ \mathbf{0} & U \end{pmatrix}^l = \begin{pmatrix} 0 & v^t U^{l-1} \\ \mathbf{0} & U^l \end{pmatrix}, \quad l \geq 1.$$

Thus, if our original matrix is nilpotent, then U is nilpotent as well.

19.2.3 Mathematical Induction on n

Assume that we already designed $n-1$ new coordinates, for which nilpotent blocks like U get strictly upper triangular. (This is the induction hypothesis.) Thanks to our common (singular) eigenvector, the original (nilpotent) matrices get strictly upper triangular too, as required.

But didn't we forget something? What about the case $n = 1$? This is easy enough: a nilpotent 1×1 "matrix" has just one entry: (0). This is strictly upper triangular, as required.

19.2.4 Nilpotent Linear Lie Algebra

Thus, our Lie algebra is quite special: in our new basis, all matrices get strictly upper triangular:

$$L \subset T_1$$

(Section 16.9). This tells us a new global property: L is a nilpotent Lie algebra.

By now, we only proved this for a linear Lie algebra. But what about a more abstract Lie algebra, containing more abstract members, with no multiplication at all (let alone power or nilpotency)? At best, they could only be adjoint-nilpotent (since the adjoint is a linear mapping, or a matrix). In this case, is our Lie algebra nilpotent (as a whole)? This is Engel's theorem.

19.3 Engel's Theorem

19.3.1 Finite-Dimensional Lie Algebra

This is our original question. Let L be a (finite-dimensional) Lie algebra (not necessarily linear). Assume that all its members are adjoint-nilpotent. In other words, all matrices in $\phi(L)$ are nilpotent, where ϕ is the adjoint representation:

$$\phi(l) = l_{ad} = [l, \cdot], \quad l \in L.$$

This way, we already know that $\phi(L)$ is nilpotent. But what about L itself? Is it nilpotent too?

19.3.2 The Adjoint Mappings and Their Common Eigenvector

Thanks to Section 19.1.4, the $\phi(l)$'s have a common eigenvector $v_0 \in L$:

$$\phi(l)v_0 = [l, v_0] = \mathbf{0}, \quad l \in L.$$

This means that v_0 commutes with all members of L. In other words, v_0 is in Z (the center of L).

19.3.3 Dividing by The Center

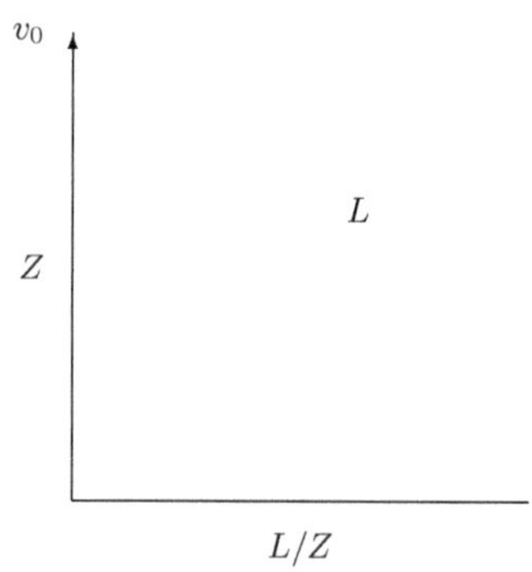

Fig. 19.2. L is spanned by Z (vertically) and L/Z (horizontally), which can benefit from the induction hypothesis.

To prove that L is nilpotent, let's use mathematical induction on its dimension. If L is one-dimensional, then it is Abelian, hence nilpotent, as required.

Assume now that L is at least two-dimensional. We already saw that it has a nontrivial center:

$$Z \neq \{\mathbf{0}\}.$$

Look at L/Z (Figure 19.2). What is its adjoint? It is defined by

$$\tau(l + Z)(m + Z) = [l + Z, m + Z] = [l, m] + Z, \quad l, m \in L.$$

Is this familiar? This is the factor adjoint representation (Section 19.1.4).

19.3.4 Induction on The Dimension of L

For each individual $l \in L$, $\tau(l + Z)$ is a nilpotent mapping from L/Z into itself. Thanks to the induction hypothesis, L/Z is nilpotent. Thus, L is nilpotent as well (Section 18.11). This completes the induction step, to prove Engel's theorem.

20

Weight Space and Lie's Lemma and Theorem

So far, we studied a linear Lie algebra of nilpotent matrices, and designed a common singular eigenvector (of eigenvalue zero). This eigenvector spans a new subspace: the weight space. As a matter of fact, the weight space doesn't have to be one-dimensional: it can even contain a few common eigenvectors.

In general, the eigenvalue doesn't have to be zero. This leads us to the general case, where the matrices don't have to be nilpotent any more. This mirrors linear algebra, where we started from a nilpotent matrix, and advanced to a more general matrix. Here, we improve on linear algebra, and look at many matrices at the same time (each with its own eigenvalue).

For all these matrices, Lie's lemma tells us some good news: the weight space (containing their common eigenvectors) remains invariant: start from a common eigenvector, and apply any matrix to it, and get

- either a new common eigenvector (of the same eigenvalue),
- or zero.

Thanks to this, a (finite-dimensional) solvable Lie algebra has a nice structure:

- It has a chain of nested ideals.
- It has a nilpotent derived algebra.
- If it is also linear, then it is in T_0 (in some basis).
- If it is even nilpotent, then it is even in T_1 (Section 19.2.4).

20.1 Weight Space

20.1.1 Eigenvalue: Linear Functional

Let L be a linear Lie algebra (containing matrices of the same order). Let $I \subset L$ be an ideal. On I, define a linear functional:

$$\lambda : I \to \mathbb{C}.$$

To each matrix $i \in I$, λ assigns a (complex) scalar: $\lambda(i) \in \mathbb{C}$.

Look at a simple example: $I = L$, $\lambda \equiv 0$, and all matrices in L are nilpotent. In this case, we already know that there is a (singular) common eigenvector v_0 (Section 19.1.4). Let v_0 span a new subspace:

$$V_0 \equiv \text{span}\{v_0\}.$$

V_0 is called a weight space. For simplicity, assume that V_0 is one-dimensional. (If there are a few v_0's, then they are indistinguishable in terms of eigenvalue, and can be treated as one and the same.)

20.1.2 More General Common Eigenvector

But this is just an example. In general, I could be smaller than L. Moreover, the eigenvector may be nonsingular: for each matrix $i \in I$, it may have eigenvalue $\lambda(i)$. This makes a new weight space:

$$V \equiv V_\lambda \equiv \{v \mid iv = \lambda(i)v, \ i \in I\}.$$

In V, all vectors are still indistinguishable, and can be treated as one and the same. Indeed, their eigenvalue $\lambda(i)$ depends only on the matrix: $i \in I$.

Our original example was really simple: λ was identically zero, and V_0 contained our singular eigenvector: v_0. In general, λ may assign different eigenvalues. This way, V still contains a common eigenvector, but with no common eigenvalue: each matrix in I has its own eigenvalue.

Given an individual matrix $i \in I$, how to uncover its eigenvalue? Just apply λ to i, and uncover the eigenvalue: $\lambda(i)$. This is better than in linear algebra: in one go, we can now study a lot of matrices at the same time.

20.2 Lie's Lemma

20.2.1 Lie's Lemma: Invariant Weight Space

For all we know, V could be as small as $\{\mathbf{0}\}$. In more interesting cases, it will be (at least) one-dimensional (as in our example). In general, Lie's lemma tells us that it is invariant under our (linear) Lie algebra:

$$LV \subset V.$$

What does this mean? This means that, for each matrix $l \in L$, for each vector $v \in V$, lv is still in the weight space:

$$lv \in V.$$

This way, lv is either

- a common eigenvector of eigenvalue $\lambda(i)$ (shared by every matrix $i \in I$),
- or zero.

In either case,

$$ilv = \lambda(i)lv,$$

or

$$\mathbf{0} = ilv - \lambda(i)lv = ilv - liv = (il - li)v = [i, l]v = i'v = \lambda(i')v$$

(for some other matrix $i' \in I$). So, it is sufficient to prove that

$$\lambda(i') = 0, \quad i' \in [I, L].$$

20.2.2 Invariant Subspace and Its Basis

To prove this, let's design a new basis (for a new subspace): start from $v \neq \mathbf{0}$, and apply l to it, time and again:

$$v,\ lv,\ l^2v,\ l^3v,\ \ldots,\ l^{k-1}v,$$

where l is a fixed matrix in L, and k is the first natural number for which $l^k v$ depends linearly on the previous vectors. In this new basis, v looks like a standard (k-dimensional) unit vector:

$$v \sim \begin{pmatrix} 1 \\ 0 \\ 0 \\ \vdots \\ 0 \end{pmatrix}.$$

Likewise, lv looks like

$$lv \sim \begin{pmatrix} 0 \\ 1 \\ 0 \\ \vdots \\ 0 \end{pmatrix},$$

and so on. (In the end, we'll realize that all these are eigenvectors, and remain in V. But, by now, we don't know this as yet.) Next, let's restrict i, i', and l to our new subspace (spanned by this new basis). As a result, we obtain three $k \times k$ matrices, denoted by B, B', and A (respectively).

How does A look like? In our new basis, A is a Lanczos matrix: 1's on the subdiagonal, some entries on the rightmost column, and 0 elsewhere:

$$A = \begin{pmatrix} & & & & * \\ 1 & & & & * \\ & 1 & & & * \\ & & 1 & & * \\ & & & \ddots & * \\ & & & 1 & * \end{pmatrix}.$$

More importantly, how do B and B' look like? We'll soon see that they are upper-triangular, with a constant main diagonal:

$$B = \begin{pmatrix} \lambda(i) & & & \\ & \lambda(i) & * & \\ & & \ddots & \\ & & & \ddots \\ & & & & \lambda(i) \end{pmatrix},$$

and

$$B' = \begin{pmatrix} \lambda(i') & & & & \\ & \lambda(i') & & * & \\ & & \ddots & & \\ & & & \ddots & \\ & & & & \lambda(i') \end{pmatrix}.$$

To see this, let's focus on B. To design its first column, apply B (or i) to v:

$$Bv = iv = \lambda(i)v.$$

To design the second column, apply B to lv:

$$Blv = ilv = [i, l]v + liv = i'v + \lambda(i)lv = \lambda(i')v + \lambda(i)lv.$$

Thanks to this, we can even design the third column:

$$Bl^2v = il(lv) = [i, l]lv + lilv = i'lv + l(ilv) = i'lv + l\left(\lambda(i')v + \lambda(i)lv\right).$$

In the right-hand side, what is the highest power of l? This is the latter term: $\lambda(i)l^2v$. On the main diagonal, it puts $\lambda(i)$, as required. The rest is unimportant: it goes to the (strictly) upper-triangular part anyway.

20.2.3 Mathematical Induction in The Block

Better yet, carry out mathematical induction on $j = 1, 2, 3, \ldots, k-1$. To design the $(j+1)$st column, look at

$$Bl^jv = il^jv = il(l^{j-1}v) = [i, l]l^{j-1}v + lil^{j-1}v = i'l^{j-1}v + l(il^{j-1}v).$$

This is as required: $\lambda(i)$ goes to the main diagonal, while the rest goes to the (strictly) upper-triangular part (thanks to the induction hypothesis on the previous column, in both B and B' at the same time).

20.2.4 Trace-Free Commutator

Likewise, B' takes the same structure: upper triangular, with a constant main diagonal: $\lambda(i')$. So, our new subspace is invariant under A, B, and B' alike. In it, B' is like i': a trace-free commutator (Section 16.7):

$$0 = \text{trace}(B') = k\lambda(i'),$$

so

$$\lambda(i') = 0.$$

This proves Lie's lemma.

20.3 Lie's Theorem

20.3.1 Linear Solvable Lie Algebra

How to design a nontrivial weight space? How to define its linear functional (on the entire Lie algebra)? Lie's theorem tells us how.

How does our L look like? By now, it is assumed to be linear:

$$L \subset gl_n$$

(for some natural number n). Next, let's assume even more: L is also solvable. How to design a (nontrivial) weight space for the entire L? For this, let's use mathematical induction on the dimension of L (not on n).

20.3.2 Induction on The Dimension of L

If L has dimension zero, then it contains just one matrix: (0). Define $\lambda \equiv 0$ (the zero functional). This way, the weight space gets really big: every n-dimensional vector v is an eigenvector, belonging in V:

$$(0)v = \mathbf{0} = 0v = \lambda v.$$

20.3.3 Dividing by The Derived Algebra

Assume now that L is at least one-dimensional. How to benefit from the induction hypothesis? In L, design a lower-dimensional ideal K. To do this, better apply the canonical homomorphism

$$\pi : L \to L/[L, L],$$

and work underneath, modulus $[L, L]$. In the end, go up, and work in L back again.

Why is this possible? Since L is solvable, its derived algebra is smaller:

$$[L, L] \neq L.$$

Thus, $[L, L]$ is lower-dimensional. So, the factor Lie algebra

$$L/[L, L]$$

is at least one-dimensional. Furthermore, it is Abelian. Indeed, for every $x, y \in L$,

$$[x + [L, L], y + [L, L]] = [x, y] + [L, L] = [L, L]$$

(the zero of $L/[L, L]$).

20.3.4 Abelian Factor Lie Algebra and Its Subspaces

By now, we're already working underneath, modulus $[L, L]$. Since $L/[L, L]$ is Abelian, every subspace of it is an ideal. Let T be a maximal proper subspace (of a slightly lower dimension):

$$\dim(T) = \dim(L/[L, L]) - 1.$$

For example, if $[L, L]$ is nearly the same as L, then T is trivial.

20.3.5 Canonical Homomorphism and Its Kernel

To work underneath, modulus $[L, L]$, we already applied the canonical homomorphism:

$$\pi : L \to L/[L, L]$$

(Section 17.2.7). What is its kernel? This is just $[L, L]$. Its dimension is

$$\dim([L, L]) = \dim(L) - \dim(L/[L, L]).$$

It is time to go up, and rise from T to a new ideal: $K \subset L$ (the origin under π):

$$K \equiv \{l \in L \mid \pi(l) \in T\}.$$

(For example, if T is trivial, then $K = [L, L]$.) Is K an ideal? Let's check:

$$[k, l] \in [L, L] \in T \quad (k \in K,\ l \in L).$$

So, K is solvable as well. Is it lower-dimensional? Well, if we restricted π to K only, then the kernel would remain the same: $[L, L]$. Thus,

$$\begin{aligned}\dim(K) &= \dim([L, L]) + \dim(T)\\ &= \dim([L, L]) + \dim(L/[L, L]) - 1\\ &= \dim(L) - 1.\end{aligned}$$

This is the key to the induction step.

20.3.6 The Induction Step

We can now pick a nonzero vector $y \in L \setminus K$, and decompose:

$$L = K + \text{span}(y).$$

(Compare this with the decomposition in Section 19.1.4, where λ was zero.) Thanks to the induction hypothesis, we can design a new linear functional λ on K, with a new (nontrivial) weight space $V \equiv V_\lambda$. Our only job is to define λ on y as well. (Later on, we'll extend λ linearly to the entire L, as required.)

20.3.7 Using Lie's Lemma

Thanks to Lie's lemma (for L, with its ideal K), V is invariant under y too. Thus, y can be restricted to V, and even take a Jordan form on V, with a new eigenvector $v_1 \in V$, and a new eigenvalue, denoted by $\lambda(y)$:

$$yv_1 = \lambda(y)v_1.$$

Finally, extend λ linearly to the entire L. Our new (extended) λ will then induce a new $V \equiv V_\lambda$, containing at least one eigenvector: v_1, shared by every matrix $l \in L$, each with its own eigenvalue: $\lambda(l)$ (depending on l only). This proves Lie's theorem.

20.4 Upper-Triangular Form

20.4.1 Designing a Basis

Recall that L is a linear solvable Lie algebra. How does it look like? Thanks to Lie's theorem, all matrices in it share a common eigenvector: v_1. Repeat the proof in Sections 19.2.1–19.2.4, with just one change: at the upper-left corner, put $\lambda(l)$ rather than zero (for each matrix $l \in L$). This designs (recursively) a new basis, in which all matrices in L are upper triangular.

20.4.2 Upper-Triangular Form

Thus, in our new basis,

$$L \subset T_0$$

(Section 16.9). This also works the other way around: if

$$L \subset T_0,$$

then L is linear and solvable (Section 18.7). In summary, a Lie algebra is linear and solvable if and only if it is a subalgebra of T_0 (in some basis).

Next, let L be solvable and finite-dimensional (but not necessarily linear any more). How does it look like?

20.5 Solvable m-Dimensional Lie Algebra

20.5.1 The Adjoint Representation

Let L be a solvable m-dimensional Lie algebra (not necessarily linear). Let ϕ be the adjoint representation: for each $l \in L$,

$$\phi(l) = l_{ad} \equiv [l, \cdot]$$

is a linear mapping from L into itself (or an $m \times m$ matrix). Since $\phi(L)$ is solvable and linear, Lie's theorem helps design a new basis, in which

$$\phi(L) \subset T_0 \subset gl_m.$$

20.5.2 Basis for a Solvable Lie Algebra

Since $\phi(L)$ contains linear mappings of L (or $m \times m$ matrices), our new basis contains m (linearly independent) vectors from L:

$$v_1, \ v_2, \ v_3, \ \ldots, \ v_m,$$

useful to span L:

$$L = \text{span}\,(v_1, \ v_2, \ v_3, \ \ldots, \ v_m)\,.$$

Let's group them in nested ideals.

20.5.3 Chain of Nested Ideals

Let's use these vectors to span new ideals:

$$
\begin{aligned}
L_0 &\equiv \{\mathbf{0}\} \\
L_1 &\equiv \operatorname{span}\{v_1\} \\
L_2 &\equiv \operatorname{span}\{v_1,\ v_2\} \\
L_3 &\equiv \operatorname{span}\{v_1,\ v_2,\ v_3\} \\
\dots &\quad \dots \\
L_m &\equiv \operatorname{span}\{v_1,\ v_2,\ v_3,\ \dots,\ v_m\} = L.
\end{aligned}
$$

Are these really ideals? Well, each L_i is invariant under all (upper-triangular) l_{ad}'s, and is therefore a legitimate ideal of L. This decomposes L in a chain of nested ideals:

$$L_0 \subset L_1 \subset L_2 \subset \cdots \subset L_m = L,$$

of higher and higher dimension:

$$\dim(L_i) = i, \qquad 0 \le i \le m.$$

20.5.4 Nilpotent Derived Algebra

Look at the derived algebra: $[L, L]$. Apply the adjoint representation ϕ to it. Thanks to our new basis,

$$\phi([L, L]) = [\phi(L), \phi(L)] \subset [T_0, T_0] = T_1$$

(Sections 16.7 and 18.7). Thus, every member of $[L, L]$ is adjoint-nilpotent. Thanks to Engel's theorem (Section 19.3.4), $[L, L]$ is nilpotent.

This also works the other way around: if $[L, L]$ is nilpotent, then L is solvable. (This follows easily from the definitions: no need to assume that L is finite-dimensional.)

21

Cartan's Criterion for Solvability

Given a Lie algebra L, is it solvable? Assume that L is linear (contains matrices only). Is it solvable? Cartan offers us a simple computational test: scan all matrices $y \in L$, and all matrices $x \in [L, L]$ (the derived algebra). If their product xy is always trace-free, then all x's are nilpotent. As a result, L is indeed solvable.

This is good enough so long as L is linear, so x and y are matrices, with a well-defined trace. But what if L wasn't linear? In this case, work with the adjoint: if $x_{ad}y_{ad}$ is always trace-free, then x is always adjoint-nilpotent. Thanks to Engel's theorem, $[L, L]$ is nilpotent, so L is indeed solvable. This is Cartan's criterion in its general form.

21.1 Matrix: Is It Nilpotent?

21.1.1 Two Subspaces: Good Matrix

Consider a familiar Lie algebra: gl_n, containing all $n \times n$ (complex) matrices. In it, design two new subspaces:

$$A \subset B \subset gl_n.$$

(Later on, we'll specify them.) Let $x \in gl_n$ be a "good" matrix. This means that its adjoint maps B into A:

$$x_{ad}(B) = [x, B] \subset A.$$

This inherits to the square too:

$$x_{ad}^2(B) = x_{ad}\left(x_{ad}(B)\right) \subset x_{ad}(A) \subset x_{ad}(B) \subset A.$$

This inherits to higher powers as well, and even to their linear combination:

$$P\left(x_{ad}\right)(B) \subset A,$$

where P is a polynomial (of complex coefficients), with no constant term.

21.1.2 Criterion for Nilpotency

Here is our criterion: if

$$\text{trace}(xy) = 0$$

for all good y, then x is nilpotent. Let's prove this.

21.1.3 The Jordan Form

How does our x look like? It has a Jordan form (Chapter 15):

$$x = \begin{pmatrix} \lambda_1 & 1 & & & & & & \\ & \lambda_2 & \ddots & & & & & \\ & & \lambda_3 & 1 & & & & \\ & & & \ddots & 0 & & & \\ & & & & \ddots & 1 & & \\ & & & & & \lambda_{n-1} & \ddots & \\ & & & & & & & \lambda_n \end{pmatrix} = \Lambda + (x - \Lambda),$$

where Λ is the diagonal part (that contains the λ_i's), and $x - \Lambda$ is the nilpotent part (that has 1's on the superdiagonal, and 0's elsewhere). Here, we allow duplication: some λ_i's could be the same: we don't care.

How to span our Lie algebra? For this, we already have a basis:

$$gl_n = \text{span}_{1 \le i,j \le n} \left(E^{(i,j)} \right)$$

(Section 16.7). In this basis, Λ takes the form:

$$\Lambda = \sum_{k=1}^{n} \lambda_k E^{(k,k)}.$$

Let's use this.

21.1.4 The Adjoint and Its Jordan Decomposition

Is Λ good? To see this, look at x_{ad}. What is its Jordan decomposition? In particular, what is the diagonal part? Our best candidate is Λ_{ad}. After all, since $x - \Lambda$ is nilpotent, its adjoint

$$(x - \Lambda)_{ad} = x_{ad} - \Lambda_{ad}$$

is nilpotent as well (thanks to the binomial formula: Section 16.4). Still, is Λ_{ad} diagonal? To see this, apply it to a matrix from our basis: $E^{(i,j)}$. This way, in the above sum (over k), only two terms survive: either $k = i$, or $k = j$:

$$\begin{aligned} \Lambda_{ad} E^{(i,j)} &= \left[\Lambda, , E^{(i,j)} \right] \\ &= \left[\sum_{k=1}^{n} \lambda_k E^{(k,k)}, E^{(i,j)} \right] \\ &= \left[\lambda_i E^{(i,i)} + \lambda_j E^{(j,j)}, E^{(i,j)} \right] \\ &= \lambda_i \left[E^{(i,i)}, E^{(i,j)} \right] + \lambda_j \left[E^{(j,j)}, E^{(i,j)} \right] \\ &= \lambda_i E^{(i,j)} - \lambda_j E^{(i,j)} \\ &= (\lambda_i - \lambda_j) E^{(i,j)}. \end{aligned}$$

So, $E^{(i,j)}$ is an eigenvector, of eigenvalue $\lambda_i - \lambda_j$. To write this more clearly, parse $E^{(i,j)}$ as a tall (n^2-dimensional) column vector, and Λ_{ad} as a big $(n^2 \times n^2)$ diagonal matrix:

$$\Lambda_{ad} E^{(i,j)} = \left[\Lambda, E^{(i,j)}\right] = \begin{pmatrix} \ddots & & & & \\ & \ddots & & & \\ & & \lambda_i - \lambda_j & & \\ & & & \ddots & \\ & & & & \ddots \end{pmatrix} \begin{pmatrix} 0 \\ \vdots \\ 0 \\ 1 \\ 0 \\ \vdots \\ 0 \end{pmatrix} \leftarrow (i,j).$$

Thus, in this eigenbasis, Λ_{ad} is indeed the diagonal part in the Jordan decomposition of x_{ad}:

$$x_{ad} = \Lambda_{ad} + (x_{ad} - \Lambda_{ad}).$$

Thus, Λ_{ad} can also be written as

$$\Lambda_{ad} = Q(x_{ad}),$$

where Q is a polynomial with no constant term (Sections 15.8.1–15.8.4). Thus, Λ is indeed good (Section 21.1.1).

21.2 How To Prove Nilpotency?

21.2.1 Rational Linear Functional

Let's treat the λ_k's as "vectors" over $\mathbb{Q}$ (the rational numbers). This makes a new linear space, in which the λ_k's could take rational coefficients, and make all sorts of linear combinations:

$$\left\{ \sum_{k=1}^{n} q_k \lambda_k \mid q_k \in \mathbb{Q},\ 1 \le k \le n \right\}.$$

This could be lower-dimensional (say, three-dimensional): for example,

$$\lambda_1 = 1,\ \lambda_2 = \sqrt{2} + \sqrt{-3},\ \lambda_3 = \sqrt{5} + \sqrt{-7},$$

and the rest are just rational multiples of these three.

On this linear space, define a new linear functional:

$$f : \left\{ \sum_{k=1}^{n} q_k \lambda_k \mid q_k \in \mathbb{Q},\ 1 \le k \le n \right\} \to \mathbb{Q}.$$

For this, define f on a basis of λ_k's, and extend linearly. In our example, you can define three independent functionals:

$$\begin{aligned} f_1(\lambda_1) = 1,\ f_1(\lambda_2) = f_1(\lambda_3) = 0 \\ f_2(\lambda_2) = 1,\ f_2(\lambda_1) = f_2(\lambda_3) = 0 \\ f_3(\lambda_3) = 1,\ f_3(\lambda_1) = f_3(\lambda_2) = 0 \end{aligned}$$

(extended linearly). Let's use them.

21.2.2 Rational Diagonal Matrix

Next, design a new $n \times n$ diagonal matrix:

$$y \equiv \begin{pmatrix} f(\lambda_1) & & & & \\ & f(\lambda_2) & & & \\ & & f(\lambda_3) & & \\ & & & \ddots & \\ & & & & f(\lambda_n) \end{pmatrix} = \sum_{k=1}^{n} f(\lambda_k) E^{(k,k)}.$$

Note that, thanks to linearity,

$$f(\lambda_i) - f(\lambda_j) = f(\lambda_i - \lambda_j), \quad 1 \le i, j \le n.$$

To y, do what we did to Λ (Section 21.1.4): write its adjoint as a big $n^2 \times n^2$ diagonal matrix:

$$y_{ad}E^{(i,j)} = \left[y, E^{(i,j)}\right] = \begin{pmatrix} \ddots & & & \\ & \ddots & & & \\ & & f(\lambda_i - \lambda_j) & & \\ & & & \ddots & \\ & & & & \ddots \end{pmatrix} \begin{pmatrix} 0 \\ \vdots \\ 0 \\ 1 \\ 0 \\ \vdots \\ 0 \end{pmatrix} \leftarrow (i,j).$$

21.2.3 Diagonal Matrices and Polynomials

Using Lagrangian interpolation, we can now design a new polynomial R that agrees with f:

$$R(\lambda_i - \lambda_j) = f(\lambda_i - \lambda_j), \quad 1 \le i, j \le n.$$

By setting $i = j$, we immediately see that

$$R(0) = f(0) = 0,$$

so R has no constant term. We can now use our polynomials on our diagonal matrices:

$$y_{ad} = R(\Lambda_{ad}) = R(Q(x_{ad})) = R \circ Q(x_{ad}).$$

Moreover,

$$R \circ Q(0) = R(Q(0)) = R(0) = 0,$$

so $R \circ Q$ has no constant term either. Thus, y is indeed good (Section 21.1.1).

21.2.4 The Trace-Free Criterion

But we also assumed that xy was trace-free (Section 21.1.2):

$$0 = \text{trace}(xy) = \sum_{k=1}^{n} \lambda_k f(\lambda_k).$$

Apply f to both sides. On the left, thanks to linearity, $f(0) = 0$. On the right, apply f only to the λ_k's (the "vectors"), not to the $f(\lambda_k)$'s (their rational coefficients):

$$0 = \sum_{k=1}^{n} f^2(\lambda_k).$$

Since f is rational (hence real, not complex),

$$f(\lambda_1) = f(\lambda_2) = f(\lambda_3) = \cdots = f(\lambda_n) = 0.$$

Pick a few f's (as in our example), and deduce that

$$\lambda_1 = \lambda_2 = \lambda_3 = \cdots = \lambda_n = 0,$$

so x is indeed nilpotent, as asserted. This proves our criterion for the nilpotency of just one (good) matrix: x (Section 21.1.2).

21.3 Derived Algebra: Is It Nilpotent?

21.3.1 Trace: Order-Invariant

To use our criterion, let's look at the trace, and its properties. Let x, y, and z be three matrices (of the same order). We already know that the commutator is trace-free:

$$\text{trace}([x, y]) = \text{trace}(xy - yx) = 0$$

(Section 16.7), so

$$\text{trace}(xy) = \text{trace}(yx).$$

Thus, the trace is order-blind (or order-invariant): reverse the order of two matrices, and you'd still get the same trace. Let's use this on three matrices:

$$\text{trace}([y, xz]) = \text{trace}(yxz - xzy) = 0,$$

or

$$\text{trace}(yxz) = \text{trace}(xzy).$$

Thus, the trace is even cyclic-shift-blind (or invariant): a shift like

$$yxz \to xzy$$

has no effect on the trace. Thanks to this, we can even look at a mixed triple product, in which two matrices are in Lie brackets:

$$\begin{aligned} \text{trace}([x, y]z) &= \text{trace}(xyz) - \text{trace}(yxz) \\ &= \text{trace}(xyz) - \text{trace}(xzy) \\ &= \text{trace}(x[y, z]). \end{aligned}$$

Thus, the trace is even bracket-shift-blind: shifting the brackets like

$$[x, y]z \to x[y, z]$$

has no effect on the trace. Let's use this.

21.3.2 Derived Algebra: Is It Nilpotent?

Let

$$L \subset gl_n$$

be a linear Lie algebra. Our aim is to check our criterion for all good $x \in [L, L]$ and all good $y \in gl_n$. For this, let's specify our subspaces $A \subset B \subset gl_n$ (Section 21.1.1):

$$B \equiv L, \quad \text{and} \quad A \equiv [L, L].$$

Given a matrix $y \in L$, is it good? Let's check:

$$y_{ad}(B) = y_{ad}(L) = [y, L] \subset [L, L] = A.$$

Thus, $y \in L$ is indeed good, let alone $x \in [L, L] \subset L$. Assume that, for all such x and y, we already checked that our criterion indeed holds: xy is trace-free:

$$\text{trace}(xy) = 0.$$

But what about a good $y \notin L$? No need to check! Indeed, pick some

$$x = [x', x''] \in [L, L]$$

(for some $x', x'' \in L$). Since y is good,

$$y_{ad}(x'') = [y, x''] \in A = [L, L]$$

(even though $y \notin L$). Thus, thanks to Section 21.3.1, xy is automatically trace-free:

$$\text{trace}(xy) = \text{trace}([x', x'']y) = \text{trace}(x'[x'', y]) = \text{trace}([x'', y]x') = 0$$

(thanks to our original assumption, telling us that $[x'', y] \in [L, L]$ times $x' \in L$ is trace-free). Better yet, extend this linearly to every linear combination of such $[x', x'']$'s. Thus, although our original check was limited to $y \in L$, it tells us even more: for every $x \in [L, L]$ (which must be good) and every good $y \in gl_n$ (even not in L), xy is trace-free, as required. Thus, thanks to our criterion, all x's in $[L, L]$ are indeed nilpotent. Let's use this.

21.4 Cartan's Criterion for Solvability

21.4.1 Linear Lie Algebra

Our original assumption was:

$$\text{trace}(xy) = 0, \qquad x \in [L, L], \ y \in L.$$

From this, we deduced even more:

$$\text{trace}(xy) = 0 \qquad (\text{if } x \in [L, L], \ y \in gl_n, \ \text{ and } [y, L] \subset [L, L])$$

(even if $y \notin L$). Thus, all these x's are nilpotent. In other words, $[L, L]$ contains nilpotent matrices only, and must be nilpotent, as a whole (Section 19.2.4). As a result, L is solvable. This is Cartan's little criterion, limited to a *linear* Lie algebra $L \subset gl_n$. But what about a more general Lie algebra, not necessarily linear?

21.4.2 Finite-Dimensional Lie Algebra

Let L be a more general (finite-dimensional) Lie algebra (linear or not). How to write our assumption? After all, L may now contain more abstract members, with no trace at all!

Let ϕ be the adjoint representation of L:

$$\phi(l) \equiv l_{ad} = [l, \cdot] : L \to L, \qquad l \in L.$$

This way, $\phi(L)$ is linear. We can then write our assumption for it: assume that

$$\text{trace}(xy) = 0, \qquad x \in \phi([L, L]) = [\phi(L), \phi(L)], \; y \in \phi(L).$$

Thus, all x's in $\phi([L, L])$ are nilpotent. In other words, all members in $[L, L]$ are adjoint-nilpotent. Thanks to Engel's theorem, $[L, L]$ is nilpotent, as a whole (Section 19.3.4). As a result, L is solvable. This is Cartan's criterion in its general form: relevant to any (finite-dimensional) L (linear or not).

22

The Killing Form and the Simple-Ideal Decomposition

We are now ready to define the Killing form: the trace of the adjoint. This will give us a computational test to check whether a given Lie algebra is semisimple or not. If it is, then it can also be decomposed as the direct sum of its simple ideals. This mirrors linear algebra: a vector space (spanned as the tensor product of its basis vectors), or a Cartesian space (spanned as the tensor product of its axes).

How to study a semisimple Lie algebra? It is sufficient to study its elementary components: simple Lie algebras like A_l, B_l, C_l, and D_l (Chapter 16). Later on, we'll use these in Hamiltonian and quantum mechanics alike.

For example, each three-dimensional ideal could contain three Pauli matrices (times $\sqrt{-1}$), with real coefficients. This way, each ideal models one electron, and its (nondeterministic) spins: spin-up-or-down, right-or-left, and in-or-out. In the same electron, they don't commute, and can't be measured simultaneously. This is their uncertainty principle.

In this example, each three-dimensional ideal models just one electron. Together, they form the entire (semisimple) Lie algebra, to model a lot of electrons. Want to extract an (Abelian) Cartan's subalgebra? From each ideal, pick just one spin. After all, different electrons can be measured simultaneously (for one spin each).

Later on, we'll also use the Killing form in particle physics, to model an interaction between elementary particles.

In this chapter, we go back to our good old style again: questions and answers. This way, you're welcome to rediscover the theory for yourself, step by step.

22.1 Killing Form on an m-Dimensional Lie Algebra

1. Let L be a Lie algebra. Let $I \subset L$ be an ideal. Define the derived series of I. Hint: use mathematical induction (Section 18.6):

$$
\begin{aligned}
I^{(0)} &\equiv I \\
I^{(i+1)} &\equiv \left[I^{(i)}, I^{(i)}\right] \quad (i \geq 0).
\end{aligned}
$$

2. Show that, in this series, each member is included in all previous ones. Hint: use mathematical induction.

3. Show that, in this series, each member is an ideal (not only of I but also of L). Hint: use mathematical induction, and Section 17.1.4.
4. Define the lower central series of I. Hint: use mathematical induction (Section 18.10):

$$I^0 \equiv I$$
$$I^{i+1} \equiv \left[I, I^i\right] \quad (i \geq 0).$$

(Note that this is a new notation, not a power.)
5. Show that, in this series, each member is an ideal (not only of I but also of L). Hint: use mathematical induction, and Section 17.1.4.
6. Conclude that, in this series, each member is included in all previous ones. Hint: use mathematical induction.
7. Assume that L is finite-dimensional. For each pair $x, y \in L$, define the Killing form: take the adjoint of x, multiply it by the adjoint of y, and calculate the trace:

$$\kappa(x, y) \equiv \text{trace}\left(x_{ad} y_{ad}\right).$$

8. Show that κ is bilinear: linear in x, and also in y.
9. Show that κ is symmetric:

$$\kappa(x, y) = \kappa(y, x).$$

Hint: thanks to Section 21.3.1,

$$\kappa(x, y) \equiv \text{trace}(x_{ad} y_{ad}) = \text{trace}(y_{ad} x_{ad}) = \kappa(y, x).$$

10. Pick three vectors: $x, y, z \in L$. Show that κ is invariant under bracket-shift-blind:

$$\kappa([x, y], z) = \kappa(x, [y, z]).$$

Hint: thanks to Section 21.3.1,

$$\begin{aligned}\kappa([x, y], z) &\equiv \text{trace}([x, y]_{ad} z_{ad}) \\ &= \text{trace}([x_{ad}, y_{ad}] z_{ad}) \\ &= \text{trace}(x_{ad} [y_{ad}, z_{ad}]) \\ &= \text{trace}(x_{ad} [y, z]_{ad}) \\ &= \kappa(x, [y, z]).\end{aligned}$$

11. Let S contain those x's that are "orthogonal" to all y's:

$$S \equiv \{x \in L \mid \kappa(x, L) = 0\} \equiv \{x \in L \mid \kappa(x, y) = 0, \ y \in L\}.$$

12. Is S a subspace? Hint: κ is bilinear.
13. Conclude that S is the maximal subspace for which

$$\kappa(S, L) = 0.$$

14. Is S an ideal of L? Hint: pick three vector: $s \in S$, and $y, \tilde{y} \in L$. Is $[s, y] \in S$ as well? Let's check:

$$\kappa([s, y], \tilde{y}) = \kappa(s, [y, \tilde{y}]) = 0.$$

15. Conclude that S is the maximal ideal for which

$$\kappa(S, L) = 0.$$

22.2 Ideal and Its Basis

1. Assume that L is m-dimensional (for some natural number m). Let $0 \le r \le m$ be an integer number. Let $I \subset L$ be an ideal, spanned by r basis vectors:

$$I = \text{span}(x_1, x_2, x_3, \ldots, x_r).$$

2. Introduce $m - r$ more basis vectors, to span the entire Lie algebra:

$$L = \text{span}(x_1, x_2, x_3, \ldots, x_m).$$

3. Pick two vectors: $x \in I$, and $y \in L$.
4. What is the adjoint of y? Hint:

$$y_{ad} \equiv [y, \cdot] : L \to L$$

(a linear mapping, or an $m \times m$ matrix).

5. What is the adjoint of x? Hint:

$$x_{ad} \equiv [x, \cdot] : L \to I$$

(a linear mapping, or an $r \times m$ matrix).

6. Why rectangular matrix? Hint: since I is an ideal, $[x, L] \subset I$.
7. Write both adjoints in a block form:

$$x_{ad} = \begin{pmatrix} x_{ad,rr} & * \\ 0 & 0 \end{pmatrix}$$

and

$$y_{ad} = \begin{pmatrix} y_{ad,rr} & * \\ 0 & * \end{pmatrix}$$

(where 'rr' means an $r \times r$ block, and '$*$' is some block).

8. Use this to multiply:

$$x_{ad}y_{ad} = \begin{pmatrix} x_{ad,rr}y_{ad,rr} & * \\ 0 & 0 \end{pmatrix}.$$

9. Observe that x "wins:" the product takes the same structure as x_{ad}.
10. Prove this in yet another way. Hint: apply $x_{ad}y_{ad}$ in two stages: first y_{ad}, then $x_{ad} : L \to I$.
11. Use this to help calculate the trace:

$$\text{trace}(x_{ad}y_{ad}) = \text{trace}(x_{ad,rr}y_{ad,rr}).$$

12. Conclude that dropping $L \setminus I$ has no effect. Hint: let κ_I be the Killing form of I (rather than L). If $x, y \in I$, then

$$\kappa_I(x, y) = \text{trace}\left(x_{ad,rr}y_{ad,rr}\right) = \text{trace}(x_{ad}y_{ad}) = \kappa(x, y).$$

13. Conclude that, for the ideal S in Section 22.1,

$$\kappa_S(S, S) = \kappa(S, S) = 0.$$

14. In particular, since $[S, S] \subset S$,

$$\kappa_S([S,S],S) = \kappa([S,S],S) = 0.$$

15. Is this familiar? Hint: this is Cartan's criterion, telling us that S is solvable (Section 21.4.2).
16. Conclude that

$$S \subset R,$$

where R is the radical: the maximal solvable ideal of L (Section 18.9).
17. Conclude that, if $S \neq \{\mathbf{0}\}$, then $R \neq \{\mathbf{0}\}$ as well, so L is not semisimple.

22.3 Nilpotent Ideal and Its Basis

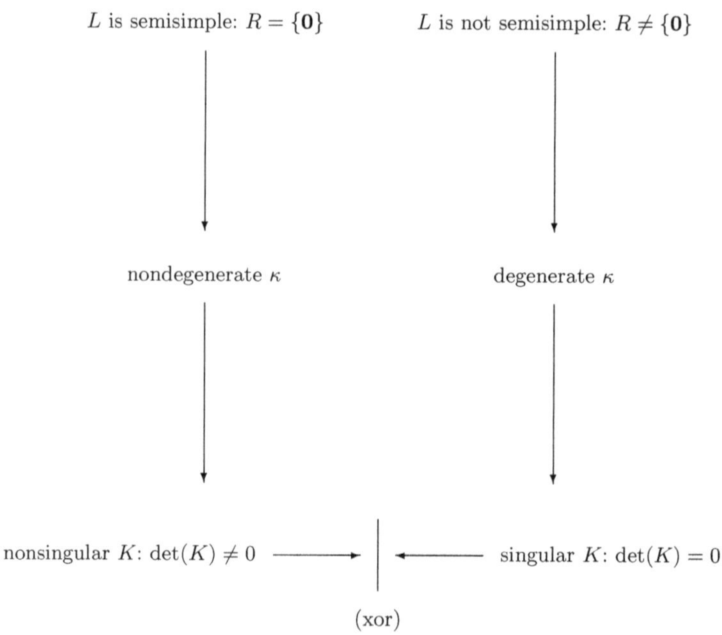

Fig. 22.1. A finite-dimensional Lie algebra L is semisimple (zero radical, and no solvable ideal but zero) if and only if κ is nondegenerate ($\det(K) \neq 0$).

1. Assume now that our ideal I is nilpotent: its lower central series shrinks to zero. Let $n \geq 0$ be the first integer number for which $I^n = \{\mathbf{0}\}$.

2. Design the basis of I cleverly, to span the nested ideals in the lower central series. Say,

$$\begin{aligned}
I^{n-1} &= \operatorname{span}(x_1, x_2), \\
I^{n-2} &= \operatorname{span}(x_1, x_2, x_3, x_4), \\
I^{n-3} &= \operatorname{span}(x_1, x_2, x_3, x_4, x_5, x_6, x_7), \\
I^{n-4} &= \operatorname{span}(x_1, x_2, x_3, x_4, x_5, x_6, x_7, x_8),
\end{aligned}$$

   and so on.
3. Pick some $x \in I$, Look at its adjoint: x_{ad}. Show that it maps like this:

$$x_{ad} : I^i \to I^{i+1}, \quad 0 \le i < n.$$

4. Conclude that $x_{ad,rr}$ has a (strictly) upper-staircase form like this:

$$x_{ad,rr} = \begin{pmatrix}
 & & * & * & * & * & * & * & * & * \\
 & & * & * & * & * & * & * & * & * \\
 & & & & * & * & * & * & * & * \\
 & & & & * & * & * & * & * & * \\
 & & & & & & & * & * & * \\
 & & & & & & & * & * & * \\
 & & & & & & & * & * & * \\
 & & & & & & & & * & * \\
 & & & & & & & & \ddots & * \\
 & & & & & & & & & \ddots
\end{pmatrix}.$$

5. Conclude that $x_{ad,rr}$ is strictly upper triangular, with zero trace:

$$x_{ad,rr} \in T_1 \subset sl_r.$$

6. Pick some $y \in L$. Look at its adjoint: y_{ad}. Show that each I^i is invariant under it:

$$y_{ad} : I^i \to I^i, \quad i \ge 0.$$

   Hint: I^i is an ideal (not only of I but also) of L (Section 22.1).
7. Conclude that $y_{ad,rr}$ has an upper-staircase form (that mirrors $x_{ad,rr}$):

$$y_{ad,rr} = \begin{pmatrix}
* & * & * & * & * & * & * & * & * & * \\
* & * & * & * & * & * & * & * & * & * \\
 & & * & * & * & * & * & * & * & * \\
 & & * & * & * & * & * & * & * & * \\
 & & & & * & * & * & * & * & * \\
 & & & & * & * & * & * & * & * \\
 & & & & * & * & * & * & * & * \\
 & & & & & & & * & * & * \\
 & & & & & & & & \ddots & * \\
 & & & & & & & & & \ddots
\end{pmatrix}.$$

8. Show that x "wins:" $x_{ad,rr}y_{ad,rr}$ is (strictly) upper-staircase (just like $x_{ad,rr}$).
9. Prove this in yet another way. Hint: apply $x_{ad}y_{ad}$ in two stages: first $y_{ad} : I^i \to I^i$, then $x_{ad} : I^i \to I^{i+1}$.
10. Conclude that $x_{ad,rr}y_{ad,rr}$ is strictly upper-triangular, with zero trace:

$$x_{ad,rr}y_{ad,rr} \in T_1 \subset sl_r.$$

11. Conclude that

$$\kappa(x, y) = \text{trace}(x_{ad}y_{ad}) = \text{trace}\left(x_{ad,rr}y_{ad,rr}\right) = 0.$$

12. Conclude that

$$\kappa(x, L) = 0.$$

    Hint: the above is true for every $y \in L$.
13. Conclude that

$$\kappa(I, L) = 0.$$

    Hint: the above is true for every $x \in I$.
14. Conclude that our nilpotent ideal I is included in the ideal S of Section 22.1:

$$I \subset S.$$

15. Must S be nilpotent? Hint: no.
16. In general, must L include a nontrivial nilpotent ideal? Hint: no.
17. Give a condition to make sure that it does.
18. Show that, if L has a nontrivial radical $R \neq \{\mathbf{0}\}$, then R includes a nontrivial nilpotent ideal $I \neq \{\mathbf{0}\}$. Hint: since R is solvable, $[R, R]$ is nilpotent (Section 20.5.4). If $[R, R] \neq \{\mathbf{0}\}$, then you could set $I \equiv [R, R]$. If, on the other hand, $[R, R] = \{\mathbf{0}\}$, then R is Abelian, hence nilpotent, so you could set $I \equiv R$.
19. Conclude that, if $R \neq \{\mathbf{0}\}$, then $S \neq \{\mathbf{0}\}$ either. Hint: S includes a nontrivial nilpotent ideal: I (either R or $[R, R]$).
20. Show that, if $S \neq \{\mathbf{0}\}$, then $R \neq \{\mathbf{0}\}$ either. Hint: $S \subset R$ (Section 22.2).
21. Conclude that $R = \{\mathbf{0}\}$ if and only if $S = \{\mathbf{0}\}$.
22. Recall that, if $R = \{\mathbf{0}\}$, then we say that L is semisimple (has no solvable ideal but zero).
23. S is also called the radical of κ. If it is zero, then we say that κ is nondegenerate.
24. Conclude that L is semisimple if and only if κ is nondegenerate (Figure 22.1).
25. Use our basis to define a new $m \times m$ matrix K:

$$K_{i,j} \equiv \kappa\left(x_i, x_j\right), \quad 1 \le i, j \le m.$$

26. Is K symmetric (same as its own transpose)? Hint:

$$K_{i,j} = \kappa\left(x_i, x_j\right) = \kappa\left(x_j, x_i\right) = K_{j,i},$$

    or

$$K = K^t.$$

27. Is K Hermitian? Hint: only if it is real. Otherwise,

$$K_{i,j} = K_{j,i} \neq \bar{K}_{j,i}.$$

28. Show that S is the same as the null space of $K = K^t$.
29. Conclude that κ is nondegenerate if and only if K is nonsingular, or

$$\det(K) \neq 0$$

Hint: this way, K is invertible, and has no null space at all.
30. This gives you a computational test to check whether L is semisimple or not (Figure 22.1).

22.4 Structure Constants

1. To calculate κ, let's use our basis. Pick two indices $1 \leq i, j \leq m$, and look at $[x_i, x_j]$. This is a new vector in L. Expand it in terms of our basis:

$$[x_i, x_j] = \sum_{l=1}^{m} c_{i,j}^l x_l,$$

for some (complex) coefficients $c_{i,j}^l$.
2. In $c_{i,j}^l$, why is l used as an upper index? Hint: this way, l is both lower and upper, and can be contracted (summed over) even without any '$\sum$' symbol:

$$[x_i, x_j] = c_{i,j}^l x_l.$$

This is Einstein's summation convention.
3. These $c_{i,j}^l$'s are called structure constants. (They depend on our basis.)
4. What law should they obey? Hint: thanks to alternativity, for each fixed i,

$$\mathbf{0} = [x_i, x_i] = c_{i,i}^l x_l.$$

(Here, i is only lower, so it is not summed upon.)
5. But the x_l's are linearly independent, so their coefficients must vanish:

$$c_{i,i}^l = 0, \quad 1 \leq i, l \leq m.$$

6. This is our first law. What is the second one? Hint: thanks to antisymmetry,

$$\mathbf{0} = [x_i, x_j] + [x_j, x_i] = c_{i,j}^l x_l + c_{j,i}^l x_l = \left(c_{i,j}^l + c_{j,i}^l\right) x_l.$$

7. But the x_l's are linearly independent, so their coefficients must vanish:

$$c_{i,j}^l + c_{j,i}^l = 0, \quad 1 \leq i, j, l \leq m.$$

8. In particular, for $i = j$, what is this? Hint: this is our previous law back again.
9. This is our second law.
10. How to write both laws in just one line? Hint: fix $1 \leq l \leq m$, and obtain an antisymmetric matrix:

$$\left(c_{i,j}^l\right)_{1 \leq i,j \leq m}$$

(with two lower indices: i and j).

11. What is the third law? Hint: fix three indices: $1 \le i, j, k \le m$. Then,

$$[[x_i, x_j], x_k] = \left[c^l_{i,j} x_l, x_k\right] = c^l_{i,j} [x_l, x_k] = c^l_{i,j} c^n_{l,k} x_n.$$

Better yet, shift cyclically:

$$(i, j, k) \to (k, i, j) \to (j, k, i),$$

and sum up. Thanks to the Jacobi identity,

$$\begin{aligned} \mathbf{0} &= [[x_i, x_j], x_k] + [[x_k, x_i], x_j] + [[x_j, x_k], x_i] \\ &= \left(c^l_{i,j} c^n_{l,k} + c^l_{k,i} c^n_{l,j} + c^l_{j,k} c^n_{l,i}\right) x_n. \end{aligned}$$

12. But the x_n's are linearly independent, so their coefficients must vanish:

$$c^l_{i,j} c^n_{l,k} + c^l_{k,i} c^n_{l,j} + c^l_{j,k} c^n_{l,i} = 0 \quad (1 \le i < j < k \le m, \quad \text{and} \quad 1 \le n \le m).$$

13. Thanks to our second law, this can be rewritten as

$$c^l_{i,j} c^n_{k,l} + c^l_{k,i} c^n_{j,l} + c^l_{j,k} c^n_{i,l} = 0 \quad (1 \le i < j < k \le m, \quad \text{and} \quad 1 \le n \le m).$$

14. In our third law, we only picked $i < j < k$. Why? Hint: any other pick would give no new law, but only repeat an old law.
15. Conclude that, to be good candidates to serve as legitimate structure constants (for any basis), the $c^n_{i,j}$'s must obey these three laws.
16. So far, we fixed l, and looked at the $m \times m$ (antisymmetric) matrix

$$\left(c^l_{i,j}\right)_{1 \le i,j \le m},$$

with two lower indices: i and j.
17. But what if we fixed i, not l? Hint: this gives yet another $m \times m$ matrix:

$$C^{(i)} \equiv \left(c^l_{i,j}\right)_{1 \le l,j \le m}$$

(where l is the row index, and j the column index).
18. What is its meaning? Hint: see below.
19. Represent our basis in terms of m-dimensional standard unit vectors:

$$x_1 \sim \begin{pmatrix} 1 \\ 0 \\ 0 \\ 0 \\ \vdots \\ \vdots \\ 0 \end{pmatrix}, \ x_2 \sim \begin{pmatrix} 0 \\ 1 \\ 0 \\ 0 \\ \vdots \\ \vdots \\ 0 \end{pmatrix}, \ x_3 \sim \begin{pmatrix} 0 \\ 0 \\ 1 \\ 0 \\ \vdots \\ \vdots \\ 0 \end{pmatrix}, \ \ldots, \ x_j \sim \begin{pmatrix} 0 \\ \vdots \\ 0 \\ 1 \\ 0 \\ \vdots \\ 0 \end{pmatrix} \leftarrow j,$$

and so on.

20. To these vectors, apply the adjoint of x_i: the $m \times m$ matrix $x_{i,ad} \equiv [x_i, \cdot]$:

$$[x_i, x_j] = [x_i, \cdot]\, x_j \sim [x_i, \cdot] \begin{pmatrix} 0 \\ \vdots \\ 0 \\ 1 \\ 0 \\ \vdots \\ 0 \end{pmatrix} = \begin{pmatrix} c^1_{i,j} \\ c^2_{i,j} \\ c^3_{i,j} \\ \vdots \\ \vdots \\ c^m_{i,j} \end{pmatrix}.$$

21. What is this? Hint: the jth column in $x_{i,ad}$, and also in $C^{(i)}$.
22. Do this for all columns (indexed by $1 \le j \le m$). What do you get? Hint: the entire adjoint matrix:

$$x_{i,ad} \equiv [x_i, \cdot] = C^{(i)}, \quad 1 \le i \le m.$$

23. Multiply two such matrices:

$$x_{i,ad} x_{k,ad} = [x_i, \cdot] \cdot [x_k, \cdot] = C^{(i)} C^{(k)}.$$

24. In this product, focus on the (l, n)th element:

$$(x_{i,ad} x_{k,ad})_{l,n} = ([x_i, \cdot] \cdot [x_k, \cdot])_{l,n} = \left(C^{(i)} C^{(k)}\right)_{l,n} = c^l_{i,j} c^j_{k,n}.$$

25. Calculate the trace of this product:

$$\begin{aligned} K_{i,k} &= \kappa\,(x_i, x_k) \\ &= \text{trace}\,(x_{i,ad} x_{k,ad}) \\ &= \text{trace}\,([x_i, \cdot] \cdot [x_k, \cdot]) \\ &= \text{trace}\left(C^{(i)} C^{(k)}\right) \\ &= c^l_{i,j} c^j_{k,l}. \end{aligned}$$

26. Is this symmetric? Hint: interchanging $i \leftrightarrow k$ has no effect.
27. Write an algorithm to find out whether L is semisimple or not. Hint: use the above formula to calculate K. Then, calculate $\det(K)$, and use Figure 22.1.

22.5 Simple-Ideal Decomposition

1. Let L be an m-dimensional semisimple Lie algebra. Show that $m \ge 2$. Hint: if $m = 1$, then L is one-dimensional, hence Abelian, hence nilpotent, hence solvable, not semisimple (Figure 18.1).
2. Conclude that κ is nondegenerate, and K is nonsingular. Hint: see Figure 22.1.
3. Let $I \subset L$ be a nonzero ideal. Could I be solvable? Hint: no (since L is semisimple).

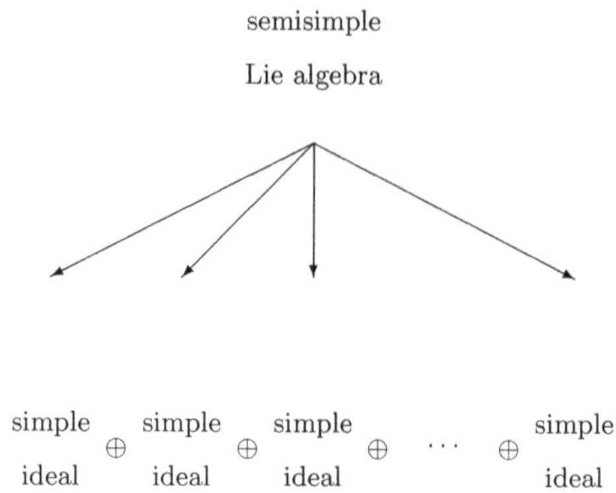

Fig. 22.2. A semisimple (finite-dimensional) Lie algebra is a direct sum of simple ideals.

4. Could I be one-dimensional? Hint: if it were, then it would be Abelian, nilpotent, and solvable, and L could never be semisimple any more.
5. Let J contain those vectors "orthogonal" to I:

$$J \equiv \{l \in L \mid \kappa(I, l) = 0\} \equiv \{l \in L \mid \kappa(i, l) = 0, \ i \in I\}.$$

6. Is J a legitimate subspace of L? Hint: κ is bilinear.
7. Is J a legitimate ideal of L? Hint: pick three vectors: $i \in I$, $j \in J$, and $l \in L$. Is $[l, j] \in J$ as well? In other words, is $[l, j]$ "orthogonal" to i? Well,

$$\kappa(i, [l, j]) = \kappa([i, l], j) = 0$$

(since $[i, l] \in I$).
8. Look at their intersection: $I \cap J$. Is it an ideal as well? Hint: since I is an ideal, $[I \cap J, L] \subset I$. Since J is an ideal, $[I \cap J, L] \subset J$ too. Thus, $[I \cap J, L] \subset I \cap J$, as required.
9. What is the Killing form of $I \cap J$ alone (ignoring all other vectors)? Hint:

$$\kappa_{I \cap J}(I \cap J, I \cap J) = \kappa(I \cap J, I \cap J) = 0$$

(Section 22.2).
10. In particular,

$$\kappa_{I \cap J}([I \cap J, I \cap J], I \cap J) = 0$$

(since $[I \cap J, I \cap J] \subset I \cap J$).
11. Conclude that $I \cap J$ is solvable. Hint: use Cartan's criterion (as in Section 22.2).
12. What could $I \cap J$ be? Hint: it must vanish: L is semisimple, and includes no nonzero solvable ideal.
13. Next, look at $[I, J]$. What is this? Hint: I and J are ideals, so $[I, J] \subset I$ and $[I, J] \subset J$, so $[I, J] \subset I \cap J = \{\mathbf{0}\}$.
14. Pick two vectors: $i \in I$, and $j \in J$. Expand them in our basis. Denote these expansions by i and j as well. What can you say about them? Hint: in our basis, i is orthogonal to $\bar{K}\bar{j}$ (the complex conjugate of K times the complex conjugate of j, expanded in our basis).

15. Look at
$$\bar{K}\bar{J} \equiv \left\{ \bar{K}\bar{j} \mid j \in J \text{ (expanded in our basis)} \right\}.$$
What is this? Hint: $\bar{K}\bar{J}$ is the orthogonal complement of I (in our basis).
16. What is its dimension? Hint: $m - r$ (where $r \geq 2$ is the dimension of I).
17. What is the dimension of J? Hint: $m - r$ too (since K and $\bar{K}$ are nonsingular).
18. Conclude that
$$L = I + J.$$
19. Recall that L is semisimple. Must J be semisimple as well? Hint: let $T \subset J$ be a nonzero ideal of J. Is T an ideal of L too? To check on this, pick two vectors: $t \in T$, and $l \in L$. Write l as $l = i + j$ (for some $i \in I$ and $j \in J$). This way,
$$[t, l] = [t, i + j] = [t, i] + [t, j] = \mathbf{0} + [t, j] = [t, j] \in T$$
(since T is an ideal of J). Thus, T is indeed an ideal (not only of J but also of L). But L is semisimple, so T can't be solvable. This proves that J is semisimple too.
20. Apply recursion: do to J what we did to L before.
21. In the end, what do you get?
22. So far, I wasn't specific. How to pick I? Hint: best pick I simple.
23. How? Hint: see below.
24. Write L (uniquely) as a sum of disjoint simple ideals:
$$L = I_1 + I_2 + I_3 + \cdots + I_n,$$
where
$$I_i \cap I_j = [I_i, I_j] = \{\mathbf{0}\}, \quad 1 \leq i < j \leq n.$$
Hint: by mathematical induction: from the start, pick I a minimal nonzero ideal of L. To be minimal, I must be simple. Define $I_1 \equiv I$. Look at your J: if it is neither zero nor simple, then apply the induction hypothesis to it.
25. Is this unique? Hint: see below.
26. Consider a nonzero ideal $H \subset L$. Pick some $1 \leq j \leq n$. Look at the intersection: $H \cap I_j$. If it is nonzero, how could it look like? Hint: as an ideal of L, it must be the same as I_j.
27. How does H look like? Hint: it is a sum of a few I_j's.
28. If H is also simple, how must it look like? Hint: to be simple (have no inner ideal), H must be a very short "sum:" just one I_j.
29. Conclude that our decomposition of L is indeed unique.
30. It is also called a direct sum, and denoted by
$$L = I_1 \oplus I_2 \oplus I_3 \oplus \cdots \oplus I_n$$
(Figure 22.2). Here, the '$\oplus$' symbol tells us that the I_j's commute, and never interact, as discussed above.
31. How to study semisimple Lie algebras? Hint: it is sufficient to study simple Lie algebras, like A_l, B_l, C_l, and D_l (Chapter 16).
32. Pick some I_j. What is its derived algebra? Hint: clearly, $[I_j, I_j] \subset I_j$. Now, if $[I_j, I_j] = \{\mathbf{0}\}$, then I_j would be Abelian, not simple (Figure 18.1). If, on the other hand, $[I_j, I_j] \neq I_j$, then it would make an inner ideal inside I_j, not letting it be simple. Thus, $[I_j, I_j] = I_j$.

33. What is the derived algebra of L? Hint: since the I_j's commute,

$$\begin{aligned}[L, L] &= [I_1 \oplus I_2 \oplus I_3 \oplus \cdots \oplus I_n, I_1 \oplus I_2 \oplus I_3 \oplus \cdots \oplus I_n] \\ &= [I_1, I_1] \oplus [I_2, I_2] \oplus [I_3, I_3] \oplus \cdots \oplus [I_n, I_n] \\ &= I_1 \oplus I_2 \oplus I_3 \oplus \cdots \oplus I_n \\ &= L.\end{aligned}$$

34. Likewise, show that any ideal of L (like $H \subset L$ above) satisfies

$$[H, H] = H.$$

35. Conclude again that any ideal of L (like $H \subset L$ above) could never be solvable. Hint: its derived series remains constant, and never shrinks to zero.
36. Conclude again that any ideal of L (like $H \subset L$ above) must be semisimple as well. Hint: pick an ideal of H. It must have the same structure: a sum of a few I_j's. As such, it can't be solvable: it has a constant derived series too.

22.6 Electrons and their Spins

1. Look at an electron.
2. It has three (nondeterministic) spins: spin-up-or-down, right-or-left, and in-or-out. (From these three, you can design any other spin, in any other direction in 3-D.)
3. Could they be measured simultaneously? Hint: no! They don't commute.
4. This is their uncertainty principle.
5. Model them by three Pauli matrices (times $\sqrt{-1}$), with real coefficients.
6. What is this? Hint: a simple (three-dimensional) Lie algebra.
7. Next, look at a lot of electrons.
8. Look at their spins.
9. How to model them? Hint: the above Lie algebra (spanned by three Pauli matrices) already models one electron (in terms of spin). Clone it, time and again, in a row: a direct sum of simple (three-dimensional) ideals. This makes a complete (semisimple) Lie algebra.
10. Look at two such ideals. Do they interact? Hint: only if they entangle: have a nonseparable joint wave function, through which they affect one another (in terms of spin).
11. From our big Lie algebra, how to extract an (Abelian) Cartan's subalgebra? Hint: from each simple ideal, pick just one spin.
12. Do they commute? Hint: yes (since they belong to different electrons).
13. Can they be measured simultaneously? Hint: yes (since they commute).
14. For example, look at an electron-positron pair, produced at the same physical process. Look at their spin-up-or-down. It is either up-down or down-up (at equal probability), but neither up-up nor down-down. This is how they entangle, and affect one another (Chapter 25).

Part VIII

Lie Groups/Algebras: Applications

Part VIII: Lie Groups/Algebras: Applications

On top of our Lie algebra, we already built a group of automorphisms. This is a Lie group. Better yet, to design a Lie group, exponentiate. How? From your Lie algebra, pick an anti-Hermitian matrix, and exponentiate it. This will give you a new unitary matrix. Better yet, this will even form a complete Lie group of unitary matrices, ready to use in quantum and classical mechanics alike.

How to use our Lie algebra in practice? Well, it has a few interesting applications. In Hamiltonian mechanics, it helps develop conservation laws. Moreover, in quantum mechanics, it helps model spin. Finally, in particle physics, the Killing form helps model elementary particles, and their interaction.

23

Hamiltonian Mechanics: Energy and Angular Momentum

So far, our Lie algebra contained matrices, or some abstract members. In Hamiltonian mechanics, on the other hand, it will contain smooth functions (defined on a differentiable manifold). Some of these functions will have an important physical meaning: energy, momentum, angular momentum, and more.

In this Lie algebra, we'll have new Lie brackets: the Poisson brackets. These will help write the Hamilton equation, with its new geometrical interpretation [6, 35, 41, 59]. This will help develop conservation laws for a few fundamental quantities: energy, momentum, angular momentum, and more.

Like every Lie brackets, the Poisson brackets are antisymmetric. This will lead to Noether's theorem. How? In the Hamiltonian equation, look at things the other way around. This way, once a new conserved quantity is born, it also introduces a new symmetry line, on which the Hamiltonian is invariant.

For this reason, in its orbit around the sun, the Earth keeps the same rate, energy, and angular momentum. This, however, is not the only conservation law. Our algebraic model leads to a useful algorithm to uncover new conserved quantities, and their symmetry lines.

23.1 Lie Algebra of Smooth Functions

23.1.1 Symplectic Manifold: Darboux Coordinates

Let's start with some geometry. Let n be the dimension (say, $n = 3$ in the Cartesian space). Consider a $2n$-dimensional manifold M, spanned by the (independent) variables $p_1, p_2, \ldots, p_n$ and $q^1, q^2, \ldots, q^n$. (These are just upper indices, not powers.) For $n = 1$, for example, p_1 and $q^1 = x$ span the phase plane (Figure 1.8). More often, we'll use $n = 3$: p_1, p_2, p_3, $q^1 = x$, $q^2 = y$, and $q^3 = z$ will stand for the (unknown) position-momentum in 3-D.

Later on, we'll define the energy: a function of these variables. This will design level sets. In each, we'll have a solution curve: those position-momentum points used during the motion. To find this curve, we'll need to differentiate in M.

Let $C^\infty(M)$ contain smooth functions: differentiable in M, as many times as you like. Let $f \in C^\infty(M)$ be such a function. Its gradient (vector of partial derivatives) can split into two subvectors: first position, then momentum:

$$\nabla f \equiv \begin{pmatrix} \nabla_q f \\ \nabla_p f \end{pmatrix},$$

where ∇_q differentiates with respect to position q^i, and ∇_p with respect to momentum p_i.

But note: we are not in a linear space, but on a differentiable manifold, of a nonstandard (nonlinear) shape. (For example, we could live on the face of the Earth, with $n = 2$.) This is why our functions are called fields. How to differentiate them in M? Fortunately, differentiation is a local process. To carry it out, our coordinates (the q^i's and p_i's) are good enough. They are only local, not global, but this doesn't matter. After all, around each individual point $u \in M$, there is a spot $U \subset M$ (a little $2n$-dimensional domain), where they are well-defined, and can support our Hamilton equation. These are called Darboux coordinates: local, not global: can't be "tied up" into one global coordinate system. To get around this, let's focus on just one spot: $U \subset M$.

23.1.2 Hamiltonian Vector

Thanks to Darboux coordinates, we can now use the $2n \times 2n$ (antisymmetric) matrix

$$S \equiv \begin{pmatrix} & I \\ -I & \end{pmatrix}$$

(where I is the $n \times n$ identity matrix). We already saw it before: it was used to introduce the symplectic Lie algebra (Section 16.12). The only change is in the notation: n here plays the role of l there.

What is the Hamiltonian vector? Look at the gradient, split into two subvectors: ∇_q (position derivatives, followed by ∇_p (momentum derivatives). To this, apply S. This flips the subvectors, and also picks a minus sign: ∇_p, followed by $-\nabla_q$:

$$S\nabla = S\begin{pmatrix} \nabla_q \\ \nabla_p \end{pmatrix} = \begin{pmatrix} \nabla_p \\ -\nabla_q \end{pmatrix}.$$

This is orthogonal to the original gradient:

$$\nabla^t f S \nabla f = \begin{pmatrix} \nabla_q f \\ \nabla_p f \end{pmatrix}^t \begin{pmatrix} \nabla_p f \\ -\nabla_q f \end{pmatrix} = \nabla_q^t f \nabla_p f - \nabla_p^t f \nabla_q f = 0.$$

This looks like alternativity. Why not use this as new Lie brackets?

23.1.3 The Poisson Brackets

Thanks to the Hamiltonian vector, we can now define a special kind of Lie brackets: the Poisson brackets. The convention is to use braces (rather than square brackets): for any two functions $f, u \in C^\infty(M)$,

$$\{f, u\} \equiv \nabla^t u S \nabla f = (S\nabla f)^t \nabla u.$$

And what is the adjoint? Just drop the 'u:'

$$f_{ad} \equiv \{f, \cdot\} \equiv (S\nabla f)^t \nabla.$$

This differentiates at direction $S\nabla f$ (orthogonal to ∇f).

As a differentiation, this satisfies the Leibniz rule, or the Jacobi identity (see exercises below). These could serve as Lie brackets in a new Lie algebra. How to use it geometrically?

23.2 Geometrical Point of View

23.2.1 The Phase Plane

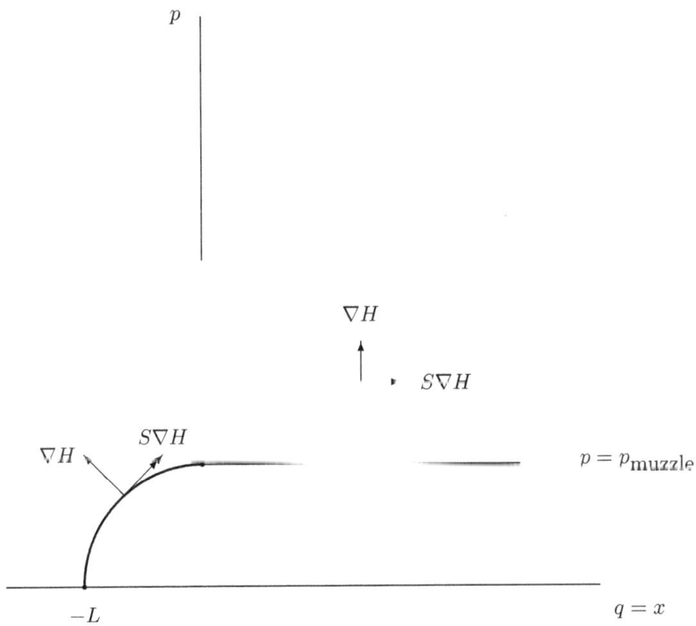

Fig. 23.1. The position-momentum phase plane. Start from $(-L, 0)$: at $-L$, at rest. Then, go rightwards, gaining more and more momentum and speed, until the muzzle at $(0, p_{\text{muzzle}})$. This draws an arc: an energy level set. On it, H is constant, and $S\nabla H$ "pushes" rightwards. (This is mathematical: physically, the particle is on the horizontal x-axis.)

To illustrate this geometrically, look at the 1-D case ($n = 1$): a particle coming from a slingshot. On its way rightwards (until the muzzle), it behaves like a harmonic oscillator (Figure 1.7).

Its 1-D position is denoted by $q = x$ (on the horizontal axis). Its momentum p, on the other hand, is marked on the vertical axis. These are the dynamic variables that span our phase plane (Figures 1.7-1.8).

On the phase plane, define the Hamiltonian: a real function $H \equiv H(q, p)$. This is the total energy: potential plus kinetic. Its level set is a curve, on which it is constant (Figure 23.1). On it, the motion goes on: from $(-L, 0)$ (where the particle is still at rest, deep in the slingshot, with no momentum at all), then rightwards, gaining more and more momentum and speed. This is why the curve curves upwards-rightwards. Physically, however, the particle remains on the horizontal q-axis: 1-D motion.

At the muzzle, the particle takes its maximal momentum: $p = p_{\text{muzzle}}$. From there on, it travels freely rightwards, at the same momentum and speed.

The curve has two important vectors: the normal ∇H, and the tangent $S\nabla H$ (pushing rightwards). This is what powers the motion rightwards, with more and more momentum and speed.

After passing the muzzle, things get calmer: there is no more fuel to power more momentum or speed. ∇H gets vertical, and $S\nabla H$ horizontal. On this horizontal level set, momentum-energy remain constant: conserved forever.

23.2.2 Kinetic Energy in 1-D

For position, we use here a new notation: q. Why not x, as usual? Well, in the phase plane, they look the same: $q = x$. But, in reality, they have different roles: x is limited to a straight motion, but q is more general.

To see this, let $g \equiv g(q)$ be the metric: a positive function defined on the q-axis, assigning a positive weight (or density) to each position q. For simplicity, assume that the energy is only kinetic, not potential. For example, in a flatland, imagine a river, flowing at sea level, at the same altitude, with no ups or downs. This way, the river draws a (curvy) q-axis: q measures length along the river, and $g(q)$ is mass per unit length. Since everything is flat, gravity potential is constant, and has no effect on the motion. But the river can still gain metric: at some q, it may get wider, or more curvy. We, who sail and live in the river, know nothing of all this: from the inside, the river looks as straight and uniform as ever. This is why we need the metric g to give us this new information. This way, as we sail, the Hamiltonian may also change with q:

$$H \equiv \frac{p^2}{2g(q)} = \frac{1}{2}g(q)v^2,$$

where v is the speed of the flow.

23.2.3 Metric Affects Energy

Here, one may ask: how could the metric be in the denominator? After all, a bigger metric should make the energy bigger, not smaller! The answer is that the metric has a mixed effect. On one hand, it affects the denominator linearly. On the other hand, it affects p even quadratically, which affects the numerator (implicitly and indirectly). Thus, in total, g affects H linearly, as required. This is why p and H are also per unit length here.

23.2.4 Energy Affects Momentum

So far, we applied the Hamiltonian vector to a general function: u (Section 23.1.3). Let's apply it to a more concrete function: the momentum $P \equiv p$. Why use a capital letter? Because P is a "new" thing: a physical quantity, not just a variable. As such, it deserves a new name. This way, it can even be differentiated:

$$\frac{\partial P}{\partial q} = 0$$
$$\frac{\partial P}{\partial p} = 1.$$

Thanks to this, we can now see how the kinetic energy acts on the momentum:

$$\begin{aligned}\{H, P\} &= \nabla^t PS\nabla H \\ &= (S\nabla H)^t \nabla P \\ &= \nabla_p^t H \nabla_q P - \nabla_q^t H \nabla_p P \\ &= -\nabla_q^t H \\ &= \frac{p^2 g'(q)}{2g^2(q)}\end{aligned}$$

(since we are in 1-D: $n = 1$). In particular, when g is constant (river with constant curvature and width), this vanishes. This tells us that the momentum isn't "pushed" or raised any more, but remains conserved. Indeed, in the phase plane, H and P share the same (horizontal) level set. On it, they have parallel normals: both ∇P and ∇H point upwards (as in the right part of Figure 23.1).

23.3 Momentum and Its Conservation

23.3.1 Hamilton Equation – General Relativity

What happens when g is not necessarily constant, and may change with q? In this case, the Hamilton equation tells us how P may change in time:

$$\dot{P} = \dot{p} \equiv p'(t) = \{H, P\}.$$

This is the force. (The dot on top stands for time derivative.) This is as in general relativity: even with no potential at all, the metric can still introduce an (implicit) force, leading to a dynamic momentum.

For example, suppose that we live inside a highly nonuniform curve (or river) of nonconstant curvature and density. From the inside, however, the curve seems straight and uniform. Fortunately, g tells us better. For instance, g could increase with q: $g'(q) > 0$. This way, the Hamilton equation is a quadratic differential equation in $p(t)$, telling us how momentum increases dynamically in time:

$$p'(t) = \{H, P\} = \frac{p^2 g'(q)}{2g^2(q)} > 0.$$

This is as in general relativity again: draw force from metric, not from potential. Still, to solve for p, we must also solve for q at the same time:

$$q'(t) = \{H, q\} = \nabla_p^t H = \frac{p}{g(q)}.$$

Once these equations are solved for $q(t)$ and $p(t)$ together, we have the entire dynamics in 1-D.

23.3.2 Inverse Matrix and Its Derivative

So far, g was just a scalar. Next, let g be a (nonsingular) matrix, depending on some parameter. Its inverse is defined by

$$gg^{-1} = I$$

(the identity matrix). Differentiate both sides (with respect to our parameter):

$$g'g^{-1} + g\left(g^{-1}\right)' = 0,$$

or

$$g\left(g^{-1}\right)' = -g'g^{-1},$$

or

$$\left(g^{-1}\right)' = -g^{-1}g'g^{-1}.$$

This will be useful below.

23.3.3 Higher Dimension

Let's move on to the more interesting case of $n > 1$ (say, $n = 3$, or even higher). In this case, it makes sense to introduce two new column vectors:

$$q \equiv \begin{pmatrix} q^1 \\ q^2 \\ q^3 \\ \vdots \\ q^n \end{pmatrix}, \text{ and } p \equiv \begin{pmatrix} p_1 \\ p_2 \\ p_3 \\ \vdots \\ p_n \end{pmatrix}.$$

What happens in higher dimension? The metric g (and its inverse g^{-1}) is no longer a scalar, but an $n \times n$ (symmetric nonsingular) matrix, depending on our new vector: q.

23.3.4 Hamilton System for The Momentum

Assume that our Hamiltonian is still kinetic:

$$H \equiv \frac{1}{2} p^t (g(q))^{-1} p.$$

For a start, let's focus on the kth coordinate. For this, define a "new" constant scalar:

$$P_k \equiv p_k$$

(for some fixed k). This is just a new name for p_k. Thus, its partial derivatives vanish (unless $j = k$):

$$\frac{\partial P_k}{\partial p_j} = \delta_k^j = \begin{cases} 1 & \text{if} \quad j = k \\ 0 & \text{if} \quad j \neq k. \end{cases}$$

We can now write the Hamilton equation for P_k:

$$\begin{aligned} \dot{P}_k &= \{H, P_k\} \\ &= -\nabla_q^t H \nabla_p P_k \\ &= -\frac{\partial H}{\partial q_k} \\ &= \frac{1}{2} p^t \frac{\partial \left(g^{-1}\right)}{\partial q_k} p \\ &= \frac{1}{2} p^t \left(g^{-1} \frac{\partial g}{\partial q_k} g^{-1} \right) p \end{aligned}$$

(Section 23.3.2). (On the right-hand side, the parentheses tell you to calculate first, to save calculations.) Do this for all $1 \leq k \leq n$, and obtain a vector equation (for the unknown vector p):

$$\dot{p} = \frac{1}{2} p^t \left(g^{-1} \ \nabla_q g \ g^{-1} \right) p.$$

This is a system of n coupled (quadratic) differential equations for the unknown momenta $p(t)$, indexed by a lower index. (On the right-hand side, however, q in the denominator is indexed by an upper index, scanned in the end.) But this system is still incomplete: it can't be solved on its own, but only together with yet another system in $q(t)$:

$$\dot{q} - \{H, q\} = \nabla_p H = (g(q))^{-1} p.$$

Once these systems are solved together, the solutions $q(t)$ and $p(t)$ give us the entire dynamics, in higher dimension as well.

23.3.5 Momentum Conservation: Geodesic Equations

How to have the same momentum, all the time? How to have a constant p? For this, our first system must be homogeneous:

$$\dot{p} = \frac{1}{2} p^t \left(g^{-1} \ \nabla_q g \ g^{-1} \right) p = \mathbf{0}.$$

How to uncover p? Just look at the initial p. After all, it never changes. Better yet, use the second system to substitute $\dot{q}$ for $g^{-1}p$, and obtain

$$\dot{q}^t \ \nabla_q g \ \dot{q} = \mathbf{0}.$$

These are the geodesic equations: a system of n equations for the (unknown) "shortest" path $q \equiv q(t)$, along which the kinetic energy remains at its minimum. Why? Because the metric welcomes no force.

23.4 Energy: Symmetry and Dynamics

23.4.1 Invariance on Level set

On a particular level set, energy is blind and invariant. It doesn't care where exactly you are. For all it knows, you could be anywhere: it wouldn't care. After all, energy wouldn't change at all, and wouldn't even know that you moved. Only if you jumped to another level set would energy change.

But this is not only for energy but also for any other physical quantity. On its own level sets, it is blind and invariant as well. For it, this is a symmetry line. On it, you could move back and forth, with no effect. The physical quantity would feel nothing: any experiment would still measure the same result. Only another physical quantity has a chance to help find out where exactly you are on that level set.

So far, we mainly focused on energy, and its level sets. Still, how are they related to those of momentum?

23.4.2 Noether's Theorem: Translation Invariance

Want momentum preserved? Keep the first system homogeneous:

$$\dot{p} = \{H, p\} = -\nabla_q H = \mathbf{0}.$$

This way, $H \equiv H(p)$ depends on p only, not on q. In the right part of Figure 23.1, this makes a horizontal line, independent of $q = x$. In general, both H and p share the same level set. On it, they have the same normal: ∇H and ∇p are parallel. This way, the Hamiltonian vector $S\nabla H$ (tangent to the level set) gets orthogonal to ∇p, and can't affect or "push" it at all. As a result, momentum remains constant too, just like energy.

Next, look at things the other way around: from the perspective of p (rather than H). After all, thanks to antisymmetry,

$$\{p, H\} = -\{H, p\} = \mathbf{0}$$

as well. So, $S\nabla p$ is also tangent to the same level set, hence powerless: can't affect or "push" ∇H. This is what happens at the right part of Figure 23.1: after passing the muzzle, both H and p share the same (horizontal) level set. Its tangent $S\nabla p$ points rightwards, with no effect on ∇H (which keeps pointing upwards). This way, both H and p get translation-blind: invariant under any shift along the q-axis.

This is Noether's theorem: in the q-p phase space, H shares the same level sets with those physical quantities commuting with it. Each level set makes a symmetry line. On it, these physical quantities are just like energy: conserved, and translation-blind: invariant under any shift along the $q(t)$-$p(t)$ solution curve. This is more general than in our simple example: H may contain not only kinetic but also potential energy.

23.4.3 Dynamic Evolution: Liouville Operator

In general, H is total energy: potential plus kinetic. Moreover, a general physical quantity u doesn't commute with H, and has an inhomogeneous Hamiltonian equation:

$$\dot{u} \equiv \frac{du}{dt}(q(t), p(t)) = \{H, u\} \neq \mathbf{0}.$$

This is a dynamic system: the dynamic variables q and p change in time, affecting u indirectly. How? As time goes by, we move to a new state (q, p) in the phase space, with a new u defined there. This change in u can often be observed experimentally in the lab.

How to solve for the unknown $u(q(t), p(t))$? For this, better drop u altogether, and look at the adjoint:

$$H_{ad} \equiv \{H, \cdot\} = \frac{d}{dt}.$$

To integrate, write the Liouville operator:

$$L_H \equiv \exp(tH_{ad}) = \exp(t\{H, \cdot\}).$$

At the initial time $t = 0$, the initial state is $(q(0), p(0))$. This is a point in our phase space. At this point, assume that the initial value is given:

$$u(q(0), p(0)) = u_0(q(0), p(0)),$$

where u_0 is a given function in our phase space (or at least in a little spot around the initial state). To it, apply the Liouville operator, to obtain u later as well:

$$\begin{aligned} u(q(t), p(t)) &= L_H(u_0)(q(0), p(0)) \\ &= \exp(tH_{ad})(u_0)(q(0), p(0)) \\ &= \sum_{i=0}^{\infty} \frac{t^i H_{ad}^i(u_0)}{i!}(q(0), p(0)). \end{aligned}$$

If you like, this is an extended Taylor series around $t = 0$. This will be our u at every time $t \geq 0$. This will give us the entire physical dynamics in terms of the observation u alone, with no need to solve for $q(t)$ or $p(t)$ explicitly any more. You can now measure u in the lab, and compare.

23.5 Angular Momentum and Its Conservation

23.5.1 Potential Energy

So far, the Hamiltonian contained just one term: kinetic energy. This models a river at sea level: constant altitude, all the time. But what about a river with a slope? Thanks to gravity, the potential is no longer constant: it produces more and more momentum, flowing downstream.

How to model this? Look at our position vector: q. Its norm is $\|q\|$ (square root of sum of squares). In the Hamiltonian, introduce a new term: potential energy:

$$H \equiv \frac{1}{2}p^t g^{-1}(q)p + V(q).$$

Let's focus on a practical application.

23.5.2 Kepler-Newton Gravity

Let's focus on 3-D ($n = 3$): motion under Kepler-Newton's gravity laws. For this purpose, assume that our manifold is Riemannian: the metric g is not only symmetric but also positive definite. (Unfortunately, this excludes relativity, where g is indefinite.) This way, g has n orthogonal eigenvectors, with positive eigenvalues. Define its square root: a new matrix $g^{1/2}$, with the same eigenvectors, but with new eigenvalues: the square root of those of g. We can now update our coordinate system, and substitute

$$p \leftarrow (g(q))^{-1/2}p.$$

This way, g is gone, and H simplifies to read

$$H \equiv \frac{\|p\|^2}{2} - \frac{K}{\|q\|},$$

where K is Kepler's constant, and q plays the same role as r in Chapter 2: the difference from the center of mass. We are now ready to prove an important law of nature: conservation of angular momentum.

23.5.3 Conservation of Angular Momentum

In Chapter 2, angular momentum was denoted by $r \times p$. Here, on the other hand, r is replaced by q, so the angular momentum is now $q \times p$. How to apply H to it? First, apply H to q:

$$\{H, q\} = \nabla_p H = p.$$

Next, apply H to p as well:

$$\{H, p\} = -\nabla_q H = -\frac{K}{\|q\|^3}q.$$

Thanks to Leibniz rule, H can now act on the angular momentum too:

$$\{H, q \times p\} = \{H, q\} \times p + q \times \{H, p\} = p \times p - q \times \frac{K}{\|q\|^3}q = \mathbf{0} - \mathbf{0} = \mathbf{0}.$$

23.5.4 Conservation Law: Homogeneous Hamilton System

Thanks to this, the Hamilton equations get homogeneous:

$$\frac{d}{dt}(q(t) \times p(t)) = \{H, q \times p\} = \mathbf{0}.$$

This is conservation of angular momentum: $q \times p$ remains constant, all the time. But this is just one system. To solve for both $q(t)$ and $p(t)$, we need yet another system.

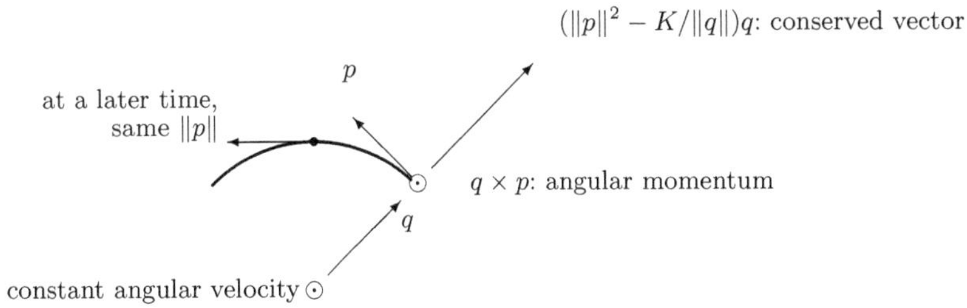

Fig. 23.2. The Earth orbits the sun at constant angular velocity and momentum (pointing upwards, towards your eyes, as indicated by the '⊙'s). On the orbit, they are symmetric: angle-blind. Energy is even more symmetric: invariant on the entire sphere.

23.5.5 Hamiltonian and Its Symmetry

This is why the Earth orbits the sun at the same rate (or angular velocity: Figure 23.2). Thanks to Noether's theorem, angular momentum and energy are both conserved. On the orbit, they are constant: angle-blind (or invariant, or symmetric).

But what about a complete sphere around the sun? On it, energy is still invariant. After all, it depends on $\|p\|$ and $\|q\|$ only. This way, it is still symmetric: doesn't care where exactly you are on the sphere. Angular momentum, on the other hand, is a vector, with a direction (perpendicular to the orbit). If you go to another orbit (even on the same sphere), then this direction changes. This is why angular momentum is not as symmetric as energy: it is symmetric on its own orbit only, but not in the entire sphere around the sun.

23.6 Conserved Vector

23.6.1 Runge-Lenz Conserved Vector

But there is yet another conserved vector: the Runge-Lenz vector:

$$p \times (q \times p) - \frac{K}{\|q\|}q.$$

How to apply H to it? First, apply H to the former term. Thanks to Leibniz rule,

$$\begin{aligned}\{H, p \times (q \times p)\} &= \{H, p\} \times (q \times p) + p \times \{H, q \times p\} \\ &= -\frac{K}{\|q\|^3} q \times (q \times p) + p \times \mathbf{0} \\ &= -\frac{K}{\|q\|^3} q \times (q \times p).\end{aligned}$$

Next, apply H to the latter term (with no minus sign):

$$\begin{aligned}\left\{H, \frac{K}{\|q\|}q\right\} &= \frac{K}{\|q\|}\{H,q\} + Kq\left\{H, \frac{1}{\|q\|}\right\} \\ &= \frac{K}{\|q\|}\nabla_p H + Kq\nabla_p^t H \nabla_q\left(\frac{1}{\|q\|}\right) \\ &= \frac{K}{\|q\|}p + Kq \cdot p^t \nabla_q\left(\frac{1}{\|q\|}\right) \\ &= \frac{K}{\|q\|^3}(q,q)p - Kq \cdot p^t\left(\frac{q}{\|q\|^3}\right) \\ &= \frac{K}{\|q\|^3}\left((q,q)p - (q,p)q\right).\end{aligned}$$

Finally, subtract the latter from the former:

$$\left\{H, p \times (q \times p) - \frac{K}{\|q\|}q\right\} = -\frac{K}{\|q\|^3}q \times (q \times p) - \frac{K}{\|q\|^3}\left((q,q)p - (q,p)q\right) = \mathbf{0}$$

(Section 2.6.6).

23.6.2 Complete System of Conservation Laws

Thanks to this, the Hamilton equations get homogeneous, and make a new conservation law (for the Runge-Lenz vector):

$$\frac{d}{dt}\left(p \times (q \times p) - \frac{K}{\|q\|}q\right) = \left\{H, p \times (q \times p) - \frac{K}{\|q\|}q\right\} = \mathbf{0}.$$

Together with our momentum conservation law (for $q \times p$: Section 23.5.4), we have a complete system for $q(t)$ and $p(t)$: six equations in six unknowns, as required.

23.6.3 Conserved Quantity: Power Series in p

Let's go back to a more general case, with a more general potential V:

$$H = \frac{1}{2}p^t g^{-1} p + V(q).$$

Let O be some conserved quantity (to be specified later). As such, it must obey a homogeneous Hamilton equation:

$$\dot{O} = \{H, O\} = -\{O, H\} = 0.$$

How could O look like? In general, it is a function of both q and p. Better yet, expand it in powers of p (with coefficients depending on q):

$$O \equiv O(q,p) = c(q) + \mathbf{c}^t(q)p + \frac{1}{2}p^t C(q)p.$$

(This is quite short: quadratic in p. Later on, we'll see how to add more terms.) Our (unknown) coefficients are of different types: c is a scalar, $\mathbf{c}$ (bold) is a vector, and C is an $n \times n$ symmetric matrix. Let's design them cleverly, to obey the homogeneous

Hamilton equation for any momentum p that could be introduced in practice later. On our way, we'll also see that our power series must be quite short: only a few terms survive (say, three).

What conditions should our coefficients satisfy? To find out, plug O in the homogeneous Hamilton equation, and make sure that all terms vanish, as required. This way, the equation will indeed be satisfied for any possible momentum. For a start, look at the free term that contains no p at all, and make sure that it indeed vanishes:

$$\begin{aligned}
0 &= \{O, H\} \\
&= \nabla_p^t O \cdot \nabla_q H - \nabla_q^t O \cdot \nabla_p H \\
&= \nabla_p^t O \cdot \nabla_q H - \nabla_p^t H \cdot \nabla_q O \\
&= (\mathbf{c} + Cp)^t \, \nabla_q V - p^t g^{-1} \left(\nabla_q c + \nabla_q \mathbf{c}^t p + \frac{1}{2} p^t \nabla_q C p \right) .
\end{aligned}$$

On the right-hand side, only the first term is free, with no p at all. Thus, it must vanish:

$$\mathbf{c}^t \nabla_q V = 0.$$

In other words, at $p = \mathbf{0}$,

$$\nabla_p^t O \cdot \nabla_q V = 0.$$

What is the physical meaning of this?

23.6.4 Horizontal Angular Momentum

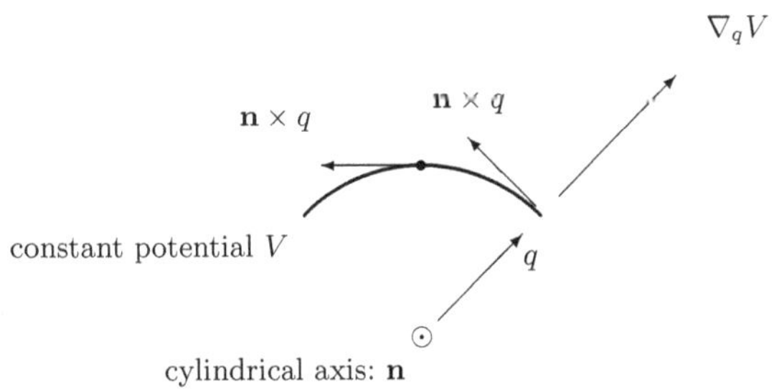

Fig. 23.3. A cylindrical motion. Even before any motion or momentum started, V was already symmetric around $\mathbf{n}$, allowing a constant motion, round and round.

In the Cartesian space, look at our vectors: q and p. Together, they span an (oblique) plane. In it, they draw a parallelogram: q draws one edge, and p another. What is its area? We already know what it is: $\|q\|$ times $\|p\|$ times the sine of the angle in between:

$$\|q \times p\| = \|q\| \cdot \|p\| \sin(q, p).$$

On top of these, add yet another vector: $\mathbf{n}$ (not necessarily perpendicular to q or p). Together, they make a 3-D parallelogram. What is its volume? Take what we did before, and project on $\mathbf{n}$:

$$(\mathbf{n}, q \times p) = \|\mathbf{n}\| \cdot \|q \times p\| \cos(n, q \times p).$$

But these are interchangable. After all, you could turn your 3-D parallelogram round, to sit on its side, and still have the same volume:

$$(\mathbf{n}, q \times p) = (\mathbf{n} \times q, p).$$

Our aim is to model a motion on a theoretical cylinder, with a nonphysical wall (Figure 23.3). How to keep q confined to the wall (not necessarily perpendicular to it)? How to keep p tangent to the wall (not necessarily perpendicular to q)? For this, let

$$\mathbf{n} \equiv (0, 0, 1)^t$$

be a standard (vertical) unit vector. We are only interested in the horizontal angular momentum:

$$O \equiv (\mathbf{n}, q \times p) = (\mathbf{n} \times q, p).$$

Thanks to this definition,

$$\nabla_p O = \mathbf{n} \times q.$$

This is orthogonal to q. How to use this?

23.6.5 Cylindrical Symmetry

How to keep angular momentum conserved (at least horizontally)? More precisely, how to keep our special O constant? One way is to keep V symmetric around $\mathbf{n}$ (and g, say, the identity matrix). After all, in our cylinder, $\mathbf{n}$ is the axis of symmetry. (Vertically, on the other hand, V can change as wildly as you like.) For this, $\nabla_q V$ should be horizontally radial, as in our cylindrical motion:

$$(\nabla_p O, \nabla_q V) = (\mathbf{n} \times q, q) = 0.$$

Is this familiar? This is our formula (end of Section 23.6.3). If you like, we just used it for a special O: linear.

Next, let's go back to our original (quadratic) O. In Section 23.6.3, we looked at the homogeneous Hamilton equation. In it, we focused on the free term, and forced it to vanish. What about the other terms? They should vanish too! This will give us more physical information.

23.6.6 Killing Vector and Tensor

In the homogeneous Hamilton equation for O, collect those terms that are linear in p. In total, they should vanish:

$$(Cp)^t \nabla_q V - p^t g^{-1} \nabla_q c = 0,$$

or

$$p^t \left(C\nabla_q V - g^{-1}\nabla_q c \right) = 0.$$

But this should vanish for any p. To guarantee this, require that

$$gC\nabla_q V = \nabla_q c.$$

In our linear example, $c \equiv 0$ and $C \equiv (0)$, so this indeed holds (Figure 23.3). Next, in the homogeneous Hamilton equation for O, collect quadratic terms (in p). There is just one. It should vanish too:

$$p^t g^{-1} \nabla_q \mathbf{c}^t p = 0.$$

But this should vanish for all p. Thus,

$$g^{-1} \nabla_q \mathbf{c}^t = (0).$$

To simplify this, multiply by g:

$$\nabla_q \mathbf{c}^t = (0).$$

A vector $\mathbf{c}$ that satisfies this is called a Killing vector.

Next, in our homogeneous Hamilton equation for O, look at cubic terms (in p). there is just one. It should vanish too:

$$p^t g^{-1} \nabla_q \left(p^t C p \right) = 0.$$

But this should vanish for all p. To guarantee this, drop the p's altogether, and obtain a new tensor that should vanish. To simplify, multiply by g:

$$\nabla_q C = (0).$$

A matrix C that satisfies this is called a Killing matrix.

Better yet, make O even cubic: to its original power series (Section 23.6.3), add a new (symmetric) tensor of three indices, to store the new coefficients (depending on q, but not on p). As before, it must satisfy a Killing condition: vanish under ∇_q. It is then called a Killing tensor. (Under ∇_p, on the other hand, this may produce a new quadratic term in p, which may change your killing vector back again.) This is Van-Holten's algorithm.

23.7 Exercises: Poisson Brackets as Lie Brackets

23.7.1 Poisson Brackets – Jacobi identity

1. Recall that M is a $2n$-dimensional differentiable manifold, and that S is a $2n \times 2n$ antisymmetric matrix. Hint: see Sections 23.1.1–23.1.2.
2. Recall that, around each individual point $u \in M$, there is a spot $U \subset M$, in which Darboux coordinates guarantee that S is constant.
3. Form the gradient operator, and split it into two subvectors. Hint: place the partial derivatives (with respect to the q^i's and the p_i's) in a tall $2n$-dimensional column vector:

$$\nabla \equiv \begin{pmatrix} \nabla_q \\ \nabla_p \end{pmatrix}.$$

4. Pick a real smooth function $f \in C^{\infty}(M)$.
5. What is ∇f? Hint: the gradient of f: a $2n$-dimensional column vector, containing the partial derivatives of f with respect to the q^i's and the p_i's:

$$\nabla f \equiv \begin{pmatrix} \nabla_q f \\ \nabla_p f \end{pmatrix}.$$

6. What is $\nabla^t f$? Hint: the transpose gradient, containing the partial derivatives, one by one, in a row:

$$\nabla^t f \equiv \left(\nabla_q^t f, \nabla_p^t f\right).$$

7. What is $\nabla\nabla^t f$? Hint: the Hessian of f: a $2n \times 2n$ matrix, containing the second partial derivatives of f.
8. Decompose it into four blocks:

$$\nabla\nabla^t f = \begin{pmatrix} \nabla_q \\ \nabla_p \end{pmatrix} \left(\nabla_q^t f, \nabla_p^t f\right) = \begin{pmatrix} \nabla_q \nabla_q^t f & \nabla_q \nabla_p^t f \\ \nabla_p \nabla_q^t f & \nabla_p \nabla_p^t f \end{pmatrix}.$$

9. In $\nabla\nabla^t f$, what are the off-diagonal elements? Hint: mixed partial derivatives.
10. Is $\nabla\nabla^t f$ symmetric? Hint: just differentiate twice, no matter which one comes first.
11. Conclude that $S\nabla\nabla^t f S$ is symmetric as well.
12. Pick three smooth functions: $f, u, w \in C^{\infty}(M)$.
13. Write the Poisson brackets explicitly:

$$\{u, f\} \equiv \nabla^t f S \nabla u.$$

Hint: see Section 23.1.3.
14. Are the Poisson brackets antisymmetric? Hint: since S is antisymmetric,

$$\{u, f\} \equiv \nabla^t f S \nabla u = -\nabla^t u S \nabla f = -\{f, u\}.$$

15. Conclude that they are also alternative. Hint: $\{f, f\} = -\{f, f\} = 0$.
16. Write the triple Poisson brackets explicitly:

$$\{\{w, u\}, f\} = \nabla^t f S \nabla \left(\nabla^t u S \nabla w\right).$$

17. Use the antisymmetry of S and the Leibniz rule to write this yet more explicitly:

$$\begin{aligned} \{\{w, u\}, f\} &= \nabla^t f S \nabla \left(\nabla^t u S \nabla w\right) \\ &= \nabla^t f S \nabla\nabla^t u S \nabla w - \nabla^t f S \nabla\nabla^t w S \nabla u. \end{aligned}$$

18. Use this to shift cyclically:

$$(w, u, f) \to (f, w, u) \to (u, f, w),$$

and sum up:

$$\begin{aligned} &\{\{w, u\}, f\} + \{\{f, w\}, u\} + \{\{u, f\}, w\} \\ &= \nabla^t f S \nabla\nabla^t u S \nabla w - \nabla^t f S \nabla\nabla^t w S \nabla u \\ &+ \nabla^t u S \nabla\nabla^t w S \nabla f - \nabla^t u S \nabla\nabla^t f S \nabla w \\ &+ \nabla^t w S \nabla\nabla^t f S \nabla u - \nabla^t w S \nabla\nabla^t u S \nabla f \\ &= 0. \end{aligned}$$

Hint: in the middle, you always have a symmetric matrix. For this reason, all six terms cancel out: the first with the last, the second with the third, and the fourth with the fifth.

19. What is this? Hint: this is the Jacobi identity.

24

Lie Groups/Algebras in Quantum Mechanics and Special Relativity

Thanks to Lie algebras, we can now look at quantum mechanics and special relativity from a new angle. In quantum mechanics, we often use the Pauli matrices as observables. But they also have an algebraic face: they span a new Lie algebra: su_2 (over $\mathbb{R}$, not $\mathbb{C}$). For this purpose, multiply them by $i \equiv \sqrt{-1}$, to make them anti-Hermitian. They can then exponentiate, and *generate* a new Lie group: SU_2. This is useful to model two tricky phenomena: the spin of an electron, and the polarization of an electromagnetic wave. Likewise, in special relativity too, use the same methodology as well: exponentiate, and generate a new Lorentz transformation (from one inertial system to another).

Thanks to the adjoint representation, su_2 can also take a new (isomorphic) face: so_3. Like su_2, it is defined over $\mathbb{R}$, not $\mathbb{C}$, and contains antisymmetric matrices only. Once exponentiated, they generate a new Lie group: SO_3. This group has a transparent geometrical interpretation: it rotates the entire Cartesian space (twice as fast). This helps model the boson spin as well. This is how Lie groups and algebras join forces in physics.

In this chapter, we go back to our good old style: questions and answers. This invites you to take an active part in developing physics from scratch. This way, you get to see the theory born in front of your very eyes.

24.1 Matrix Exponent

1. Let A be an $n \times n$ (complex) matrix, with the following Jordan form:

$$A = ODO^{-1},$$

where O contains the generalized eigenvectors of A (column by column), and D is nearly diagonal:

$$D = \text{diag}(\lambda_1,\ \lambda_2,\ \ldots,\ \lambda_n) + \text{superdiagonal of 0's or 1's.}$$

(Some λ_i's could be the same: we don't care.)

2. Show that A has an exponent, of a similar Jordan form. Hint: since A has a finite norm, the following series converges (in norm):

$$\begin{aligned}\exp(A) &\equiv \sum_{j=0}^{\infty} \frac{A^j}{j!} \\ &= \sum_{j=0}^{\infty} \frac{\left(ODO^{-1}\right)^j}{j!} \\ &= \sum_{j=0}^{\infty} \frac{OD^jO^{-1}}{j!} \\ &= O\left(\sum_{j=0}^{\infty} \frac{D^j}{j!}\right)O^{-1} \\ &= O\exp(D)O^{-1}.\end{aligned}$$

3. Conclude that O (and O^{-1}) can be pulled out of parentheses:

$$\exp\left(ODO^{-1}\right) = O\exp(D)O^{-1}.$$

4. In this triple product, write the middle matrix more explicitly. Hint:

$$\begin{aligned}\exp(D) &= \sum_{j=0}^{\infty} \frac{D^j}{j!} \\ &= \sum_{j=0}^{\infty} \frac{(\mathrm{diag}(D))^j}{j!} + \text{strictly upper-triangular part} \\ &= \sum_{j=0}^{\infty} \frac{\mathrm{diag}\left(\lambda_1^j, \lambda_2^j, \ldots, \lambda_n^j\right)}{j!} + \text{strictly upper-triangular part} \\ &= \mathrm{diag}\left(\exp(\lambda_1), \exp(\lambda_2), \ldots, \exp(\lambda_n)\right) + \text{strictly upper-triangular part.}\end{aligned}$$

5. Can you say anything about the (strictly) upper-triangular part? Hint: it remains within each Jordan block. Thus, the eigenvectors remain the same O-columns as before.
6. Conclude that $\exp(A)$ has the same eigenvectors, with exponent eigenvalues.
7. Conclude that the determinant of the exponent is the exponent of the trace. Hint:

$$\begin{aligned}\det(\exp(A)) &= \det(\exp(D)) \\ &= \exp(\lambda_1)\exp(\lambda_2)\cdots\exp(\lambda_n) \\ &= \exp(\mathrm{trace}(D)) \\ &= \exp(\mathrm{trace}(A)).\end{aligned}$$

8. Conclude that a traceless matrix has an exponent of determinant 1.

24.2 Anti-Hermitian Matrix

1. What if A was Hermitian? Can you tell us even more? Hint: in this case, O is unitary, and D and $\exp(D)$ are real and diagonal (no superdiagonal any more). Thus, our Jordan form is

$$A = ODO^*, \quad \text{and} \quad \exp(A) = O\exp(D)O^*.$$

2. And what if A was anti-Hermitian (i times a Hermitian matrix)? What can you tell us now? Hint: in this case, O is unitary, D is diagonal and pure imaginary, and $\exp(D)$ is diagonal and unitary (diagonal elements of magnitude 1). Thus, our Jordan form is

$$A = ODO^*, \quad \text{and} \quad \exp(A) = O\exp(D)O^*$$

   (which is unitary as well).
3. Conclude that, in this case, $\exp(A)$ is unitary too.
4. Prove this in yet another way, using no Jordan form at all. Hint: for an anti-Hermitian A,

$$\begin{aligned}
\exp(A)(\exp(A))^* &= \exp(A)\left(\sum_{j=0}^{\infty}\frac{A^j}{j!}\right)^* \\
&= \exp(A)\sum_{j=0}^{\infty}\frac{(A^*)^j}{j!} \\
&= \exp(A)\sum_{j=0}^{\infty}\frac{(-A)^j}{j!} \\
&= \exp(A)\exp(-A) \\
&= \exp(A - A) \\
&= \exp((0)) \\
&= I.
\end{aligned}$$

5. Conclude again that an anti-Hermitian matrix has a unitary exponent.
6. This is how the Lie algebra u_n exponentiates, and *generates* the unitary Lie group: U_n.
7. Moreover, conclude that a traceless anti-Hermitian matrix has a unitary exponent of determinant 1 (a special unitary exponent).
8. This is how the Lie algebra su_n exponentiates, and *generates* the special unitary Lie group: SU_n.

24.3 Pauli Matrix

1. Recall the 2×2 Pauli matrices:

$$\begin{aligned}
\sigma_z &\equiv \begin{pmatrix} 1 & 0 \\ 0 & -1 \end{pmatrix} \\
\sigma_x &\equiv \begin{pmatrix} 0 & 1 \\ 1 & 0 \end{pmatrix} \\
\sigma_y &\equiv \begin{pmatrix} 0 & -i \\ i & 0 \end{pmatrix}.
\end{aligned}$$

2. Are they Hermitian?

3. Are they trace-free?
4. What is their determinant?
5. Multiply them by a new (complex) parameter: $\theta \in \mathbb{C}$. Then, exponentiate:

$$\begin{aligned}\exp(\theta\sigma_z) &= \exp\left(\begin{pmatrix}\theta & \\ & -\theta\end{pmatrix}\right)\\ &= \sum_{n=0}^{\infty}\frac{1}{n!}\begin{pmatrix}\theta & \\ & -\theta\end{pmatrix}^n\\ &= \sum_{n=0}^{\infty}\frac{1}{n!}\begin{pmatrix}\theta^n & \\ & (-\theta)^n\end{pmatrix}\\ &= \begin{pmatrix}\exp(\theta) & \\ & \exp(-\theta)\end{pmatrix}.\end{aligned}$$

6. What is the determinant of this exponent? Hint: 1.
7. Why? Hint: because $\theta\sigma_z$ is traceless.
8. Is this exponent unitary? Hint: look at its columns. They are orthogonal. Still, are they orthonormal? Only if θ is pure imaginary.
9. To see this, substitute
$$\theta \leftarrow i\eta$$
(for a real parameter $\eta \in \mathbb{R}$).
10. This way, is our exponent unitary?
11. Conclude that this exponent is a special unitary matrix: a unitary matrix of determinant 1.
12. In retrospect, was it a good idea to exponentiate $i\eta\sigma_z$ (rather than $\eta\sigma_z$)? Hint: yes: $i\eta\sigma_z \in su_2$ (traceless and anti-Hermitian), with an exponent in SU_2 (a unitary exponent of determinant 1).
13. For all real η's, look at all these exponents together. What is this? Hint: a one-parameter Lie group.
14. What does it mean physically? Hint: it will mirror the Schrodinger equation (with η meaning time).

24.4 Lorentz Matrix in Special Relativity

1. Next, look at
$$\sigma_x \equiv \begin{pmatrix}0 & 1\\ 1 & 0\end{pmatrix}.$$
2. Is it symmetric?
3. Is it also antisymmetric? In what sense? Hint: see below.
4. To answer this, design an orthogonal Lie algebra: D_1. For this purpose, in Section 16.11, set
$$S \equiv \begin{pmatrix}-1 & \\ & 1\end{pmatrix}.$$
5. Show that σ_x is S-antisymmetric:
$$S\sigma_x = -\sigma_x S = -\sigma_x^t S.$$

6. What is the physical meaning of S? Hint: in special relativity, the first coordinate t picks a minus sign (Figure 8.4).
7. Define the hyperbolic cosine:

$$\cosh(\theta) \equiv \frac{\exp(\theta) + \exp(-\theta)}{2}, \quad \theta \in \mathbb{C}.$$

8. Write it in terms of even powers of θ:

$$\begin{aligned}\cosh(\theta) &\equiv \frac{\exp(\theta) + \exp(-\theta)}{2} \\ &= \frac{1}{2}\left(\sum_{n=0}^{\infty} \frac{\theta^n}{n!} + \sum_{n=0}^{\infty} (-1)^n \frac{\theta^n}{n!}\right) \\ &= \sum_{n=0}^{\infty} \frac{\theta^{2n}}{(2n)!}.\end{aligned}$$

Hint: see exercises at the end of Chapter 26.
9. How is the hyperbolic cosine related to the cosine? Hint: for some $\eta \in \mathbb{C}$, substitute $\theta \leftarrow i\eta$:

$$\cosh(i\eta) \equiv \frac{\exp(i\eta) + \exp(-i\eta)}{2} = \cos(\eta).$$

10. Define also the hyperbolic sine:

$$\sinh(\theta) \equiv \frac{\exp(\theta) - \exp(-\theta)}{2}, \quad \theta \in \mathbb{C}.$$

11. Write it in terms of odd powers of θ:

$$\begin{aligned}\sinh(\theta) &\equiv \frac{\exp(\theta) - \exp(-\theta)}{2} \\ &= \frac{1}{2}\left(\sum_{n=0}^{\infty} \frac{\theta^n}{n!} - \sum_{n=0}^{\infty} (-1)^n \frac{\theta^n}{n!}\right) \\ &= \sum_{n=0}^{\infty} \frac{\theta^{2n+1}}{(2n+1)!}.\end{aligned}$$

12. How is the hyperbolic sine related to the sine? Hint: for some $\eta \in \mathbb{C}$, substitute $\theta \leftarrow i\eta$:

$$\sinh(i\eta) \equiv \frac{\exp(i\eta) - \exp(-i\eta)}{2} = i\sin(\eta).$$

13. Multiply σ_x by $\theta \in \mathbb{C}$. What is the exponent? Hint:

$$\begin{aligned}\exp(\theta\sigma_x) &= \exp\left(\begin{pmatrix} & \theta \\ \theta & \end{pmatrix}\right) \\ &= \sum_{n=0}^{\infty} \frac{1}{n!} \begin{pmatrix} & \theta \\ \theta & \end{pmatrix}^n\end{aligned}$$

$$
\begin{aligned}
&= \sum_{n=0}^{\infty} \frac{1}{(2n)!} \begin{pmatrix} & \theta \\ \theta & \end{pmatrix}^{2n} + \sum_{n=0}^{\infty} \frac{1}{(2n+1)!} \begin{pmatrix} & \theta \\ \theta & \end{pmatrix}^{2n+1} \\
&= \sum_{n=0}^{\infty} \frac{1}{(2n)!} \begin{pmatrix} \theta^{2n} & \\ & \theta^{2n} \end{pmatrix} + \sum_{n=0}^{\infty} \frac{1}{(2n+1)!} \begin{pmatrix} & \theta^{2n+1} \\ \theta^{2n+1} & \end{pmatrix} \\
&= \begin{pmatrix} \cosh(\theta) & \\ & \cosh(\theta) \end{pmatrix} + \begin{pmatrix} & \sinh(\theta) \\ \sinh(\theta) & \end{pmatrix} \\
&= \begin{pmatrix} \cosh(\theta) & \sinh(\theta) \\ \sinh(\theta) & \cosh(\theta) \end{pmatrix}.
\end{aligned}
$$

14. Look at all real θ's together. What is this? Hint: this is the Lorentz group: a one-parameter subgroup of $SL(2, \mathbb{R})$.
15. What is its physical meaning? Hint: with a real θ, it is useful in special relativity (Sections 8.5.2–8.5.3).
16. Here, however, we go more general, and consider even a complex θ.
17. What is the determinant of our exponent? Hint: 1.
18. Is our exponent "orthogonal?" In what sense? Hint: plug it around S (on both sides), and obtain S back again:

$$\exp\left(\theta \sigma_x\right) S \exp\left(\theta \sigma_x\right) = S.$$

19. Design a bilinear form that is blind to our exponent (or invariant under it). Hint:

$$\cdot^t S \cdot \; : \; \mathbb{C}^2 \times \mathbb{C}^2 \to \mathbb{C}.$$

20. How to apply it? Hint: plug two vectors around S: one from the left, and one from the right.
21. What does this mean physically? Hint: proper time is invariant: it remains the same, at all inertial systems.

24.5 More General Setting

1. To see things better, go yet more general.
2. How to generalize this even more?
3. Look at general (complex) matrices S and X, satisfying

$$SX = -X^* S.$$

4. Show that

$$SX^n = (-X^*)^n S = ((-X)^n)^* S, \qquad n \geq 0.$$

Hint: use mathematical induction.

5. Divide by $n!$, and sum up:

$$S \exp(X) = (\exp(-X))^* S = (\exp(X))^{-*} S,$$

or

$$(\exp(X))^* S \exp(X) = S.$$

6. Does this exponent converge? Hint: yes: X has a finite norm.
7. Is $\exp(X)$ "unitary?" In what sense? Hint: it leaves the bilinear form $(\cdot, S\cdot)$ unchanged.
8. What does this mean algebraically? Hint: the bilinear form $(\cdot, S\cdot)$ is invariant under $\exp(X)$.
9. What if S was just the identity matrix? Hint: in this case, X is anti-Hermitian, so $\exp(X)$ is unitary. This confirms what we saw before.
10. From this general setting, how to obtain our original example back again? Hint: substitute $\theta\sigma_x$ for X. This is good enough for a real θ, as in the Lorentz group. For a complex θ, make yet another change: replace the Hermitian adjoint '$*$' by the transpose 't' back again.

24.6 Rotation Matrix

1. Next, look at

$$\sigma_y \equiv \begin{pmatrix} 0 & -i \\ i & 0 \end{pmatrix}.$$

2. Is it Hermitian?
3. Multiply it by $\theta \in \mathbb{C}$. What is the exponent? Hint:

$$\begin{aligned}
\exp(\theta\sigma_y) &= \exp\left(\begin{pmatrix} & -i\theta \\ i\theta & \end{pmatrix}\right) \\
&= \sum_{n=0}^{\infty} \frac{1}{n!} \begin{pmatrix} & -i\theta \\ i\theta & \end{pmatrix}^n \\
&= \sum_{n=0}^{\infty} \frac{1}{(2n)!} \begin{pmatrix} & -i\theta \\ i\theta & \end{pmatrix}^{2n} + \sum_{n=0}^{\infty} \frac{1}{(2n+1)!} \begin{pmatrix} & -i\theta \\ i\theta & \end{pmatrix}^{2n+1} \\
&= \sum_{n=0}^{\infty} \frac{1}{(2n)!} \begin{pmatrix} \theta^{2n} & \\ & \theta^{2n} \end{pmatrix} + \sum_{n=0}^{\infty} \frac{1}{(2n+1)!} \begin{pmatrix} & -i\theta^{2n+1} \\ i\theta^{2n+1} & \end{pmatrix} \\
&= \begin{pmatrix} \cosh(\theta) & \\ & \cosh(\theta) \end{pmatrix} + \begin{pmatrix} & -i\sinh(\theta) \\ i\sinh(\theta) & \end{pmatrix} \\
&= \begin{pmatrix} \cosh(\theta) & -i\sinh(\theta) \\ i\sinh(\theta) & \cosh(\theta) \end{pmatrix} \\
&= \begin{pmatrix} \cos(\eta) & \sin(\eta) \\ -\sin(\eta) & \cos(\eta) \end{pmatrix},
\end{aligned}$$

where $\theta = i\eta$ ($\eta \in \mathbb{C}$).
4. Is this familiar? Hint: for a real η, this is a rotation matrix: it rotates the entire x-y plane by angle η clockwise (Figure 2.12).
5. What is its determinant? Hint: 1.
6. Why? Hint: because $\theta\sigma_y$ is traceless.
7. For all real η's together, what is this? Hint: a one-parameter Lie group: SO_2.
8. In retrospect, was it a good idea to exponentiate $i\eta\sigma_y$ (rather than $\eta\sigma_y$)? Hint: for a real η, $i\eta\sigma_y$ is antisymmetric, with an orthogonal exponent.

24.7 Symmetry in Spacetime

1. Look at σ_y again. What is it physically? Hint: an observable: spin-in.
2. Is it Hermitian indeed?
3. Why is this good? Hint: thanks to this, it has real eigenvalues: possible physical observations.
4. Multiply it by $i = \sqrt{-1}$:

$$i\sigma_y = i\begin{pmatrix} & -i \\ i & \end{pmatrix} = \begin{pmatrix} 0 & 1 \\ -1 & 0 \end{pmatrix}.$$

5. Where did this i come from? Hint: see Chapter 4 in [50].
6. Is it antisymmetric now?
7. Extend it into a new 3×3 matrix. Hint: add five dummy zeroes on the top-left:

$$\begin{pmatrix} & & \\ & & 1 \\ & -1 & \end{pmatrix}.$$

8. Extend it even more, into a new 4×4 matrix. Hint: add seven dummy zeroes on the bottom-right.
9. What do you get? Hint: our original $i\sigma_y$ is now wrapped with zeroes all over. This defines a new 4×4 matrix:

$$Q_{xy} \equiv \begin{pmatrix} & & & \\ & & 1 & \\ & -1 & & \\ & & & \end{pmatrix}.$$

10. Where does it act? Hint: it acts in spacetime (spanned by the t-x-y-z axes).
11. On what plane does it really act? Hint: it really acts on the x-y plane in spacetime. The t-z coordinates, on the other hand, remain idle.
12. In this plane, it will soon exponentiate, and generate a new symmetry: rotation. This will help draw orbits, on which angular momentum is conserved.
13. Likewise, design Q_{yz} and Q_{zx}.
14. What is the commutator of Q_{xy} and Q_{yz}? Hint:

$$[Q_{xy}, Q_{yz}] = -Q_{zx}.$$

15. Conclude that these three are closed under the commutator.
16. Conclude that they span a subalgebra of $sl(4, \mathbb{R})$.
17. Moreover, this subalgebra is isomorphic to $\mathbb{R}^3$, but the other way around: a left-handed vector product, picking a minus sign.
18. Next, look at another Pauli matrix:

$$\sigma_x \equiv \begin{pmatrix} 0 & 1 \\ 1 & 0 \end{pmatrix}.$$

19. Extend it into a new 3×3 matrix. Hint: add five dummy zeroes on the bottom-right.

20. Extend it even more, into a new 4×4 matrix. Hint: add seven dummy zeroes on the bottom-right.
21. What do you get? Hint:

$$Q_{tx} \equiv \begin{pmatrix} & 1 & & \\ 1 & & & \\ & & & \\ & & & \end{pmatrix}.$$

22. Where does it act? Hint: it acts in spacetime.
23. On what plane exactly? Hint: on the t-x plane (leaving the y-z coordinates idle).
24. In this plane, it will soon exponentiate, and generate a new symmetry: a Lorentz transformation to another inertial system (preserving proper time and speed of light).
25. Likewise, define Q_{ty} and Q_{tz} (Section 8.17.2).
26. What is the commutator of Q_{tx} and Q_{xy}? Hint:

$$[Q_{tx}, Q_{xy}] = Q_{ty}.$$

27. What is the commutator of Q_{tx} and Q_{ty}? Hint:

$$[Q_{tx}, Q_{ty}] = Q_{xy}.$$

28. Conclude that these six 4×4 matrices are closed under the commutator.
29. Conclude that they span a (six-dimensional) subalgebra of $sl(4, \mathbb{R})$.

24.8 How To Transform Spacetime?

1. Better yet, extend these matrices even more: into new 5×5 matrices. Hint: add nine dummy zeroes on the top-left.
2. Where do they act? Hint: they act in a new 5-D spacetime, spanned by 1-t-x-y-z (with a dummy 1 on top).
3. Design a new 5×5 matrix, with only one nonzero element (in the leftmost column, at the second-top entry):

$$Q_t \equiv \begin{pmatrix} & & & & \\ 1 & & & & \\ & & & & \\ & & & & \\ & & & & \end{pmatrix}.$$

4. In our 5-D spacetime, how does Q_t act? Hint: it shifts by 1 second (thanks to the dummy 1 in 1-t-x-y-z).
5. Likewise, lower the '1' a little, and design

$$Q_x \equiv \begin{pmatrix} & & & & \\ & & & & \\ 1 & & & & \\ & & & & \\ & & & & \end{pmatrix}.$$

6. In our 5-D spacetime, how does Q_x act? Hint: it shifts by one length unit in x (thanks to the dummy 1 in 1-t-x-y-z).
7. Likewise, define Q_y and Q_z as well.
8. What is the commutator of Q_t and the (extended) Q_{tx}? Hint:

$$[Q_t, Q_{tx}] = -Q_x.$$

9. What is the commutator of Q_x and the (extended) Q_{tx}? Hint:

$$[Q_x, Q_{tx}] = -Q_t.$$

10. What is the commutator of Q_x and the (extended) Q_{xy}? Hint:

$$[Q_x, Q_{xy}] = Q_y.$$

11. What is the commutator of Q_y and the (extended) Q_{xy}? Hint:

$$[Q_y, Q_{xy}] = -Q_x.$$

12. What is the commutator of Q_t and Q_x? Hint:

$$[Q_t, Q_x] = (0).$$

13. Conclude that these ten 5×5 matrices are closed under the commutator.
14. Conclude that they span a (ten-dimensional) subalgebra of $sl(5, \mathbb{R})$.
15. In it, look at Q_t, Q_x, Q_y, and Q_z. They span an Abelian ideal.
16. These ten matrices form a basis (over $\mathbb{R}$, not $\mathbb{C}$).
17. In it, each matrix can be represented by a standard (ten-dimensional) unit vector.
18. Use them to calculate the structure constants (Section 22.4).
19. Use them to design the adjoint matrices.
20. Use them to calculate κ: the Killing form (Section 22.1).
21. Design its matrix: K.
22. Show that $\det(K) = 0$. Hint: its ideal is Abelian, hence nilpotent, hence solvable (Figures 18.1 and 22.1).
23. Look at our ten basis matrices. Multiply them by a real parameter $\eta \in \mathbb{R}$, and exponentiate. What do you get? Hint: ten transformations (of our 5-D spacetime):
    - Three planar rotations (in xy, yz, and zx).
    - Three Lorentz matrices (in tx, ty, and tz). (These are Poincar‘e matrices.)
    - Four translations (in t, x, y, and z):

$$\begin{aligned}\exp(\eta Q_t) &= I + \eta Q_t \\ \exp(\eta Q_x) &= I + \eta Q_x \\ \exp(\eta Q_y) &= I + \eta Q_y \\ \exp(\eta Q_z) &= I + \eta Q_z\end{aligned}$$

    (where I is the 5×5 identity matrix).
24. These are ten symmetries in spacetime:

- Three planar rotations (drawing orbits, on which angular momentum is conserved).
- Three Lorentz transformations (from one inertial system to another, preserving proper time and speed of light).
- Time-translation (during a vertical line in spacetime, on which energy is conserved).
- And x-/y-/z-translation (along a horizontal line, on which momentum is conserved).

25. This is how our new Lie subalgebra generates a new Lie group of symmetry transformations in spacetime.

24.9 The Special Unitary Lie Group

1. Look at the Pauli matrices again (times $i = \sqrt{-1}$): $i\sigma_x$, $i\sigma_y$, and $i\sigma_z$.
2. Are they anti-Hermitian?
3. Are they traceless?
4. What is their commutator? Hint: it is cyclic:

$$\begin{aligned}[i\sigma_x, i\sigma_y] &= -2i\sigma_z \\ [i\sigma_y, i\sigma_z] &= -2i\sigma_x \\ [i\sigma_z, i\sigma_x] &= -2i\sigma_y.\end{aligned}$$

5. With real coefficients (like η above), $i\sigma_x$, $i\sigma_y$, and $i\sigma_z$ span a new Lie algebra: su_2.
6. Consider a matrix in su_2. How does it look like? Hint: it is a linear combination of $i\sigma_x$, $i\sigma_y$, and $i\sigma_z$ (with real coefficients, called angles).
7. Is it traceless?
8. Is it anti-Hermitian? Hint: yes (thanks to the real coefficients).
9. Thus, su_2 contains traceless anti-Hermitian matrices (of order 2).
10. Does it contain all of them? Hint: yes: there are three degrees of freedom.
11. Thanks to these properties, su_2 can now exponentiate, and generate the special unitary Lie group: SU_2 (containing unitary matrices of determinant 1).
12. Our 2×2 matrices act on a state $v \in \mathbb{C}^2$. Still, how to move to 3-D, to see things better? Hint: see below.
13. For this, let's study su_2 algebraically a little.
14. Is it simple? Hint: mirror the proof for sl_2 (Section 18.7).
15. Conclude that su_2 is the same as its derived algebra:

$$su_2 = [su_2, su_2].$$

Hint: the derived algebra is an ideal in its own right. It can't be trivial, or su_2 would be Abelian (which it isn't). So, it must be the same as su_2.

16. Conclude once again that su_2 is trace-free. Hint: a commutator is trace-free (Section 16.7).
17. Look at the above equation. Does it inherit to the adjoint too?
18. Conclude that the adjoint is trace-free as well. Hint: see Section 18.5.
19. Verify this explicitly. Hint: see below.

24.10 The Special Orthogonal Lie Group

1. Use $i\sigma_x$, $i\sigma_y$, and $i\sigma_z$ as a basis (over $\mathbb{R}$, not $\mathbb{C}$). For this, mirror them by standard unit vectors in 3-D:

$$i\sigma_x \sim \begin{pmatrix} 1 \\ 0 \\ 0 \end{pmatrix}$$
$$i\sigma_y \sim \begin{pmatrix} 0 \\ 1 \\ 0 \end{pmatrix}$$
$$i\sigma_z \sim \begin{pmatrix} 0 \\ 0 \\ 1 \end{pmatrix}.$$

2. Thanks to this basis, the adjoint representation can now be interpreted in terms of 3×3 matrices (or vector-product operators):

$$i\sigma_{x,ad} = [i\sigma_x, \cdot] \sim \begin{pmatrix} 0 & 0 & 0 \\ 0 & 0 & 2 \\ 0 & -2 & 0 \end{pmatrix} = -2 \begin{pmatrix} 1 \\ 0 \\ 0 \end{pmatrix} \times$$
$$i\sigma_{y,ad} = [i\sigma_y, \cdot] \sim \begin{pmatrix} 0 & 0 & -2 \\ 0 & 0 & 0 \\ 2 & 0 & 0 \end{pmatrix} = -2 \begin{pmatrix} 0 \\ 1 \\ 0 \end{pmatrix} \times$$
$$i\sigma_{z,ad} = [i\sigma_z, \cdot] \sim \begin{pmatrix} 0 & 2 & 0 \\ -2 & 0 & 0 \\ 0 & 0 & 0 \end{pmatrix} = -2 \begin{pmatrix} 0 \\ 0 \\ 1 \end{pmatrix} \times .$$

3. Are these matrices antisymmetric?
4. Are they traceless?
5. What is the physical meaning of these vector-product operators? Hint: rotation around the x-, y-, or z-axis (Figures 23.2–23.3).
6. Are these matrices familiar? Hint: compare them to Q_{yz}, Q_{zx}, and Q_{xy}, respectively (Section 24.7).
7. Besides, are they also related to the 3×3 spin-one matrices S_x, S_y, and S_z (Figures 11.1–11.2)? Hint:

$$\bar{h} i\sigma_{x,ad} = \bar{h}\,[i\sigma_x, \cdot] = 2iS_x$$
$$\bar{h} i\sigma_{y,ad} = \bar{h}\,[i\sigma_y, \cdot] = 2iS_y$$
$$\bar{h} i\sigma_{z,ad} = \bar{h}\,[i\sigma_z, \cdot] = 2iS_z.$$

8. What does this mean physically? Hint: spin-one is the adjoint of spin-one-half.
9. What tells you that this is a valid physical model? Hint: in spin-one too, iS_x, iS_y, and iS_z are trace-free and antisymmetric too.
10. With real coefficients, they span a new Lie algebra: so_3.
11. It contains traceless antisymmetric matrices (of order 3).
12. Does it contain all of them? Hint: yes: there are three degrees of freedom.

13. Thanks to these properties, so_3 can now exponentiate, and generate the special orthogonal Lie group: SO_3 (containing 3-D rotations: orthogonal matrices of determinant 1).
14. These real coefficients have a geometrical meaning as well: Euler angles. These will help model spin in 3-D.
15. This is how quantum mechanics offers a (discrete) basis for (continuous) classical mechanics.

24.11 Photon and Its Polarizer

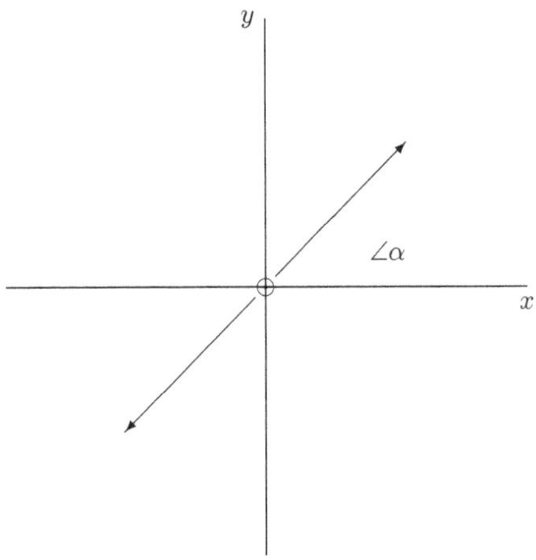

Fig. 24.1. A linearly-polarized electromagnetic wave, propagating toward your eyes (as indicated by '⊙'). At the same time, it also oscillates (obliquely) in the x-y plane, at angle α from the x-axis.

1. How does the boson spin help model the electron spin? Hint: it represents it:

$$\begin{aligned}\bar{h} i\sigma_{x,ad} &= \bar{h}\left[i\sigma_x, \cdot\right] = 2iS_x \\ \bar{h} i\sigma_{y,ad} &= \bar{h}\left[i\sigma_y, \cdot\right] = 2iS_y \\ \bar{h} i\sigma_{z,ad} &= \bar{h}\left[i\sigma_z, \cdot\right] = 2iS_z.\end{aligned}$$

2. On the right-hand side, look at the real coefficient: 2. What is its physical meaning? Hint: in the exponent, the angle doubles (see below).
3. For example, in the original exponent, plug angle η:

$$\begin{aligned}\exp(i\eta\sigma_x) &= \begin{pmatrix}\cosh(i\eta) & \sinh(i\eta)\\ \sinh(i\eta) & \cosh(i\eta)\end{pmatrix}\\ &= \begin{pmatrix}\cos(\eta) & i\sin(\eta)\\ i\sin(\eta) & \cos(\eta)\end{pmatrix}.\end{aligned}$$

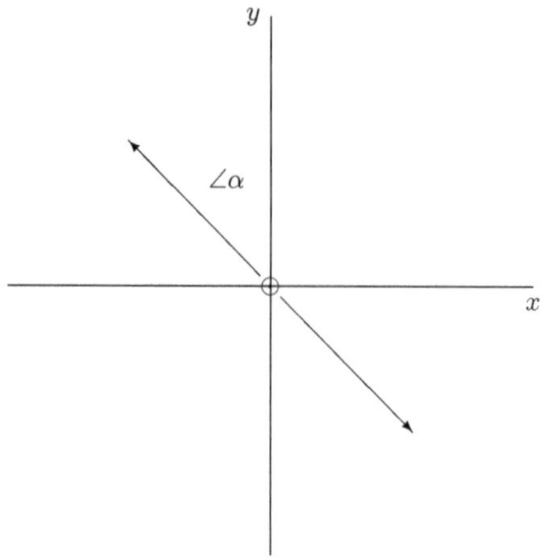

Fig. 24.2. An orthogonal $(\alpha + \pi/2)$-polarization (of eigenvalue -1).

This rotates by angle η in the x-iy plane. But the exponent of the adjoint rotates by angle 2η (in y-z):

$$\begin{aligned}\exp(i\eta\sigma_{x,ad}) &= \exp(2i\eta S_x/\bar{h}) \\ &= \exp\left(\begin{pmatrix} & \\ & 2i\eta\sigma_y \end{pmatrix}\right) \\ &= \begin{pmatrix} 1 & \\ & \exp(2i\eta\sigma_y) \end{pmatrix} \\ &= \begin{pmatrix} 1 & & \\ & \cos(2\eta) & \sin(2\eta) \\ & -\sin(2\eta) & \cos(2\eta) \end{pmatrix}.\end{aligned}$$

4. Use these angles to model (nondeterministic) spin and polarization in quantum mechanics. Hint: see below, and Chapter 12.
5. For this purpose, consider an electromagnetic wave (or a photon), flying upwards (in the positive z-direction) at the speed of light.
6. At the same time, the wave also moves (or oscillates) forward and backward along an oblique line in the x-y plane. We then say that the wave is polarized linearly (Figure 24.1).
7. In this case, the x-oscillation is aligned with the y-oscillation: there is no phase-shift.
8. Now, design a polarizer: a machine that takes a given light ray (polarized at some unknown angle), and tells us how likely it is that this angle was α (where α is known: the angle of the polarizer). This is done by looking at the light that comes out, and measuring how weak it became. In the process, the light changes phase: it becomes polarized at angle α (rather than its original angle).
9. How does the polarizer work mathematically? Hint: see below.
10. Algebraically, the polarizer is an observable: a Hermitian matrix. To design it, use the double angle: 2α:

$$\cos(2\alpha)\sigma_z + \sin(2\alpha)\sigma_x = \begin{pmatrix} \cos(2\alpha) & \sin(2\alpha) \\ \sin(2\alpha) & -\cos(2\alpha) \end{pmatrix}.$$

11. Why use 2α (rather than α)? Hint: this addition actually takes place in 3-D, where our observables are mirrored by

$$\sigma_z \sim \begin{pmatrix} 0 \\ 0 \\ 1 \end{pmatrix}$$
$$\sigma_x \sim \begin{pmatrix} 1 \\ 0 \\ 0 \end{pmatrix}$$
$$\sigma_y \sim \begin{pmatrix} 0 \\ 1 \\ 0 \end{pmatrix}.$$

Now, in 3-D, the angle should be twice as big.

12. Is this a legitimate observable? Hint: is it Hermitian?
13. Does it have an eigenvector of eigenvalue 1? Hint: pick a column vector that models an α-polarization (Figure 24.1):

$$\begin{aligned} &\begin{pmatrix} \cos(2\alpha) & \sin(2\alpha) \\ \sin(2\alpha) & -\cos(2\alpha) \end{pmatrix} \begin{pmatrix} \cos(\alpha) \\ \sin(\alpha) \end{pmatrix} \\ &= \begin{pmatrix} \cos^2(\alpha) - \sin^2(\alpha) & 2\sin(\alpha)\cos(\alpha) \\ 2\sin(\alpha)\cos(\alpha) & \sin^2(\alpha) - \cos^2(\alpha) \end{pmatrix} \begin{pmatrix} \cos(\alpha) \\ \sin(\alpha) \end{pmatrix} \\ &= \begin{pmatrix} \cos(\alpha) \\ \sin(\alpha) \end{pmatrix}. \end{aligned}$$

14. As a Hermitian matrix, this observable must also have an orthogonal eigenvector (Figure 24.2). What is its eigenvalue? Hint: -1:

$$\begin{aligned} &\begin{pmatrix} \cos(2\alpha) & \sin(2\alpha) \\ \sin(2\alpha) & -\cos(2\alpha) \end{pmatrix} \begin{pmatrix} -\sin(\alpha) \\ \cos(\alpha) \end{pmatrix} \\ &= \begin{pmatrix} \cos^2(\alpha) - \sin^2(\alpha) & 2\sin(\alpha)\cos(\alpha) \\ 2\sin(\alpha)\cos(\alpha) & \sin^2(\alpha) - \cos^2(\alpha) \end{pmatrix} \begin{pmatrix} -\sin(\alpha) \\ \cos(\alpha) \end{pmatrix} \\ &= \begin{pmatrix} \sin(\alpha) \\ -\cos(\alpha) \end{pmatrix} \\ &= -\begin{pmatrix} -\sin(\alpha) \\ \cos(\alpha) \end{pmatrix}. \end{aligned}$$

15. What is the physical meaning of this? Hint: α-polarization is a random variable. Our photon is either α-polarized (eigenvalue 1), or $(\alpha + \pi/2)$-polarized (eigenvalue -1), as observed by our polarizer.
16. What is the sum of these eigenvalues? Hint: zero.
17. Why? Hint: because our observable is traceless.
18. Before any experiment was carried out, how likely is the photon to be α-polarized? Hint: this depends on its original state $v \in \mathbb{C}^2$. Normalize it, take its inner product with the former eigenvector, and square this up (in magnitude):

$$\text{probability to be } \alpha\text{-polarized} = \left| \frac{v^t}{\|v\|} \begin{pmatrix} \cos(\alpha) \\ \sin(\alpha) \end{pmatrix} \right|^2 .$$

19. On the other hand, how likely is the photon to be $(\alpha + \pi/2)$-polarized? Hint: do the same with the latter eigenvector:

$$\text{probability to be } (\alpha + \pi/2)\text{-polarized} = \left| \frac{v^t}{\|v\|} \begin{pmatrix} -\sin(\alpha) \\ \cos(\alpha) \end{pmatrix} \right|^2 .$$

20. For example, assume that the photon was originally η-polarized. In this case, what is its original state? Hint: v makes angle η with the positive part of the x-axis:

$$v = \begin{pmatrix} \cos(\eta) \\ \sin(\eta) \end{pmatrix} .$$

21. How likely is the photon to pass the polarizer? Hint: the probability for this is

$$\begin{aligned} \left| \frac{v^t}{\|v\|} \begin{pmatrix} \cos(\alpha) \\ \sin(\alpha) \end{pmatrix} \right|^2 &= \left| (\cos(\eta), \sin(\eta)) \begin{pmatrix} \cos(\alpha) \\ \sin(\alpha) \end{pmatrix} \right|^2 \\ &= |\cos(\eta)\cos(\alpha) + \sin(\eta)\sin(\alpha)|^2 \\ &= \cos^2(\eta - \alpha) \\ &= \cos^2(\alpha - \eta). \end{aligned}$$

22. Design an experiment to observe this probability. Hint: design a Laser beam of η-polarized photons. Let them pass through our polarizer. In the process, light intensity should fall. By how much? By the above probability:

$$\begin{aligned} \text{probability that } \eta \text{ was the same as } \alpha: \quad & \left| (\cos(\eta), \sin(\eta)) \begin{pmatrix} \cos(\alpha) \\ \sin(\alpha) \end{pmatrix} \right|^2 \\ &= \cos^2(\alpha - \eta). \end{aligned}$$

23. Indeed, look at a few special cases: if $\eta = \alpha$, then the entire beam passes (probability 1: Figure 24.1). If, on the other hand, $\eta = \alpha \pm \pi/2$, then nothing passes (probability 0: Figure 24.2). Otherwise, the intensity falls somewhat $(0 <\text{probability}< 1)$.
24. What is the only relevant angle? Hint: $\alpha - \eta$.
25. In what case must the photon pass? Hint: if $\eta = \alpha$ (probability 1).
26. In what case can't the photon pass? Hint: if $\eta = \alpha \pm \pi/2$ (probability 0).
27. Place two polarizers, one by one: an α-polarizer (Figure 24.1), followed by an $(\alpha + \pi/2)$-polarizer (Figure 24.2). What comes out? Hint: nothing.
28. Now, in between, place an $(\alpha + \pi/4)$-polarizer. What comes out? Hint: after passing through the first polarizer, the light gets α-polarized, and its intensity falls somewhat. After passing through the second one, the light gets $(\alpha + \pi/4)$-polarized, and its intensity falls by $\cos^2(\pi/4) = 1/2$. After passing through the third one, the light gets $(\alpha+\pi/2)$-polarized, and its intensity falls by $\cos^2(\pi/4) = 1/2$ once again.

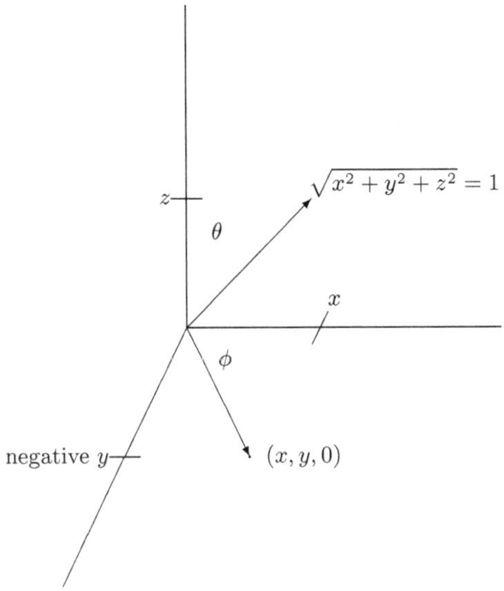

Fig. 24.3. The unit vector (x, y, z) makes angle θ with the vertical z-axis, and projects a shadow $(x, y, 0)$ on the horizontal x-y plane, making angle ϕ with the x-axis.

24.12 General Polarizer: Phaseshift

1. For simplicity, define
$$\theta \equiv 2\alpha.$$
2. Modify your polarizer to detect a slightly different polarization: not quite linear, but also making a narrow ellipse in the horizontal x-y plane. Hint: introduce a small angle ϕ (Figure 24.3), to design a new observable:
$$\cos(\theta)\sigma_z + \sin(\theta)\left(\cos(\phi)\sigma_x + \sin(\phi)\sigma_y\right).$$
3. In this sum, why use the double angle $\theta = 2\alpha$ (rather than α)? Hint: this addition actually takes place in 3-D, where our observables are mirrored by
$$\sigma_z \sim \begin{pmatrix} 0 \\ 0 \\ 1 \end{pmatrix}$$
$$\sigma_x \sim \begin{pmatrix} 1 \\ 0 \\ 0 \end{pmatrix}$$
$$\sigma_y \sim \begin{pmatrix} 0 \\ 1 \\ 0 \end{pmatrix}$$

(Figure 24.3). In 3-D, the angle should be twice as big, as we'll see later in Poincar'e sphere.

4. Is this a legitimate observable? Hint: is it Hermitian?
5. What is the physical meaning of this? Hint: in the new polarizer, there is now a new phaseshift: the y-polarization is now at phase ϕ ahead of the x-polarization.
6. Show this in the eigenvector. Hint: the eigenvector is

$$\begin{pmatrix} \cos(\alpha) \\ \sin(\alpha)\exp(i\phi) \end{pmatrix}.$$

Indeed, for $\phi = 0$ (no phaseshift at all), we already saw this before. Thus, for a nonzero ϕ too,

$$\begin{aligned} &(\cos(\theta)\sigma_z + \sin(\theta)\left(\cos(\phi)\sigma_x + \sin(\phi)\sigma_y\right)) \begin{pmatrix} \cos(\alpha) \\ \sin(\alpha)\exp(i\phi) \end{pmatrix} \\ &= \begin{pmatrix} \cos(\theta) & \sin(\theta)\exp(-i\phi) \\ \sin(\theta)\exp(i\phi) & -\cos(\theta) \end{pmatrix} \begin{pmatrix} \cos(\alpha) \\ \sin(\alpha)\exp(i\phi) \end{pmatrix} \\ &= \begin{pmatrix} \cos(\alpha) \\ \exp(i\phi)\sin(\alpha) \end{pmatrix}. \end{aligned}$$

7. What is the physical meaning of this? Hint: our polarizer detects an elliptic polarization: $\cos(\alpha)$ is the x-polarization, and $\sin(\alpha)$ is the y-polarization (at phase ϕ ahead).
8. While propagating upwards (in the positive z-direction), how does such a wave rotate? Hint: counterclockwise, along a narrow ellipse, in the horizontal x-y plane.
9. What is the orthogonal eigenvector? Hint:

$$\begin{pmatrix} -\sin(\alpha) \\ \cos(\alpha)\exp(i\phi) \end{pmatrix}.$$

Indeed, for $\phi = 0$ (no phaseshift at all), we already saw that it has eigenvalue -1. Thus, for a nonzero ϕ too,

$$\begin{aligned} &(\cos(\theta)\sigma_z + \sin(\theta)\left(\cos(\phi)\sigma_x + \sin(\phi)\sigma_y\right)) \begin{pmatrix} -\sin(\alpha) \\ \cos(\alpha)\exp(i\phi) \end{pmatrix} \\ &= \begin{pmatrix} \cos(\theta) & \sin(\theta)\exp(-i\phi) \\ \sin(\theta)\exp(i\phi) & -\cos(\theta) \end{pmatrix} \begin{pmatrix} -\sin(\alpha) \\ \cos(\alpha)\exp(i\phi) \end{pmatrix} \\ &= -\begin{pmatrix} -\sin(\alpha) \\ \cos(\alpha)\exp(i\phi) \end{pmatrix}. \end{aligned}$$

10. What is the physical meaning of this? Hint: our polarizer wouldn't let this polarization go through at all
11. While propagating upwards (in the positive z-direction), how does such a wave rotate? Hint: counterclockwise, along a new ellipse, in the horizontal x-y plane.
12. How to make it rotate clockwise (rather than counterclockwise)? Hint: replace $\phi \leftarrow -\phi$.
13. Let's see this geometrically.

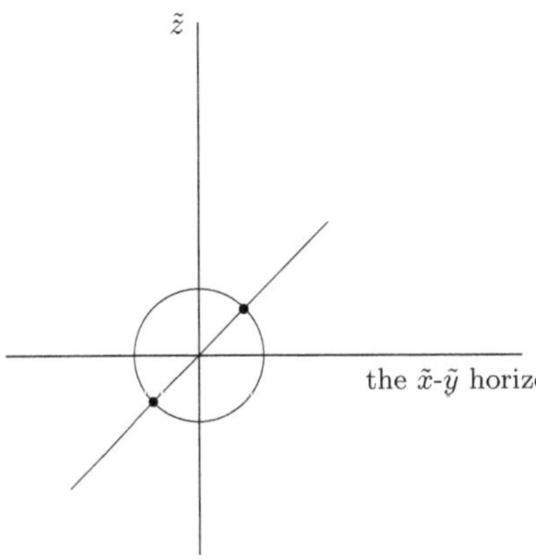

Fig. 24.4. The Poincar'e sphere: a view from the side. It uses new coordinates (a cyclic shift): $\tilde{x} \equiv z$, $\tilde{y} \equiv x$, and $\tilde{z} \equiv y$. For example, a point at the top-right has an antipodal at the bottom-left.

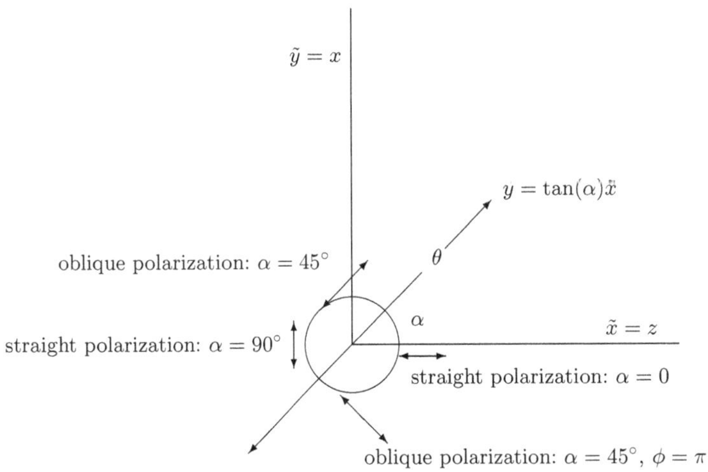

Fig. 24.5. The equator of the Poincar'e sphere: $\tilde{z} = 0$, so $\tilde{x}^2 + \tilde{y}^2 = 1$. (In terms of the old x-y-z coordinates, this is a longitude: $y = 0$, and $z^2 + x^2 = 1$.) No phaseshift at all: linear polarization along the oblique line $\tilde{y} = \tan(\alpha)\tilde{x}$ (if $\phi = 0$) or $\tilde{y} = -\text{cotan}(\alpha)\tilde{x}$ (if $\phi = \pi$).

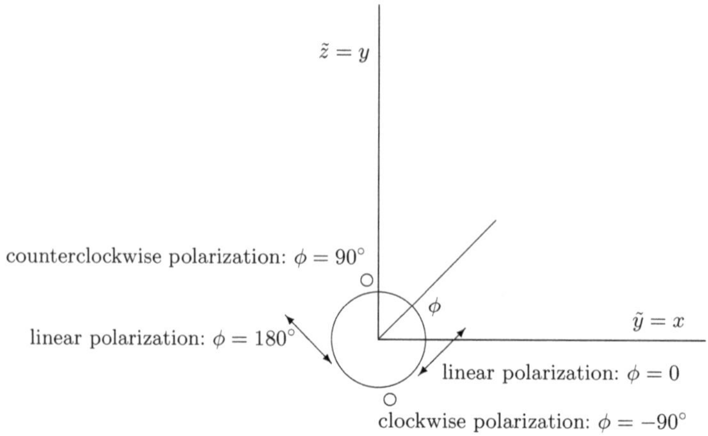

Fig. 24.6. A longitude in the Poincar‘e sphere: $\tilde{x} = 0$, so $\tilde{y}^2 + \tilde{z}^2 = 1$. (In terms of the old x-y-z coordinates, this is the equator: $z = 0$, and $x^2 + y^2 = 1$.) On it, the phaseshift changes from $\phi = 0$ on the right, to $\pi/2$ at the top, to π on the left, to $-\pi/2$ at the bottom.

24.13 Poincar‘e Sphere

1. To see this geometrically, look at the Poincar‘e sphere. This is a sphere of radius 1.
2. Consider a three-dimensional vector of norm 1. How to write it in spherical coordinates? Hint: in Figure 24.3,
$$(x, y, z) = (\sin(\theta)\cos(\phi), \sin(\theta)\sin(\phi), \cos(\theta)) .$$
3. What is the angle between (x, y, z) and the positive part of the z-axis? Hint: θ.
4. What is the range of θ? Hint: $0 \le \theta \le \pi$.
5. What "shadow" does (x, y, z) project on the horizontal x-y plane? Hint: $(x, y, 0)$.
6. What is the length of the shadow? Hint: $\sin(\theta)$.
7. What angle does the shadow make with the positive part of the x-axis? Hint: ϕ.
8. What is the range of ϕ? Hint: $-\pi < \phi \le \pi$.
9. This is 3-D (rotated by SO_3). Next, go back to 2-D (rotated by SU_2). (After all, this is where our state v is defined.) In the process, what happens to the angles? Hint: instead of θ, use just half of it: $\alpha = \theta/2$. The other angle ϕ, on the other hand, remains the same.
10. What is the associated eigenvector? Hint:
$$\begin{pmatrix} \cos(\alpha) \\ \sin(\alpha)\exp(i\phi) \end{pmatrix}$$

(Section 24.12).

11. What is its physical meaning? Hint: the polarization is no longer linear: there is a phaseshift of ϕ. As a result, as the wave propagates, it draws an ellipse (counterclockwise) on the perpendicular plane.
12. On the sphere, our original point has an antipodal. Where? Hint: on the other side of the sphere (Figure 24.4).
13. How to write the antipodal? Hint: each coordinate picks a minus sign.
14. What are its spherical coordinates? Hint: its new angles are $\pi - \theta$ and $\phi \pm \pi$. Indeed,

$$\begin{aligned} &-(\sin(\theta)\cos(\phi), \sin(\theta)\sin(\phi), \cos(\theta)) \\ &= (-\sin(\theta)\cos(\phi), -\sin(\theta)\sin(\phi), -\cos(\theta)) \\ &= (\sin(\pi-\theta)\cos(\phi\pm\pi), \sin(\pi-\theta)\sin(\phi\pm\pi), \cos(\pi-\theta)) . \end{aligned}$$

15. Now, go back to SU_2. What happens to the angles? Hint: instead of $\pi - \theta$, use just half of it:
$$\frac{\pi-\theta}{2} = \frac{\pi}{2} - \frac{\theta}{2} = \frac{\pi}{2} - \alpha.$$
The other angle $\phi \pm \pi$, on the other hand, remains the same.
16. What is the associated eigenvector? Hint:
$$\begin{pmatrix} \cos\left(\frac{\pi}{2}-\alpha\right) \\ \sin\left(\frac{\pi}{2}-\alpha\right)\exp(i(\phi\pm\pi)) \end{pmatrix} = \begin{pmatrix} \sin(\alpha) \\ -\cos(\alpha)\exp(i\phi) \end{pmatrix} .$$
17. Is it familiar? Hint: this is the orthogonal eigenvector (Section 24.12).
18. What is its physical meaning? Hint: elliptic polarization again, but clockwise (rather than counterclockwise).
19. How are these eigenvectors related? Hint: orthogonal.
20. On the Poincar‘e sphere, look at two antipodals (Figure 24.4). What is the relative angle between them? Hint: $\theta + (\pi - \theta) = \pi$.
21. In SU_2, on the other hand, what happens to this angle? Hint: it halves to $\pi/2$. Indeed,
$$\alpha + \left(\frac{\pi}{2} - \alpha\right) = \frac{\pi}{2}.$$
22. In Poincar‘e sphere, look at the equator. Hint: use cyclically-shifted coordinates:
$$\tilde{x} \equiv z, \quad \tilde{y} \equiv x, \quad \text{and} \quad \tilde{z} \equiv y$$
(Figures 24.4-24.5).
23. Look at the new $\tilde{x}$-$\tilde{y}$-$\tilde{z}$ axes. Do they make a right-hand system?
24. How to obtain them from the original x-y-z axes? Hint: use three Euler angles: $\pi/2$, 0, and $\pi/2$ (Section 2.6.3).
25. In the Poincar‘e sphere, look at the equator. What happens physically on it? Hint: on the equator, $\tilde{z} = y = 0$, so $\phi = 0$ or π. This means no phaseshift at all: linear polarization (Figure 24.5).
26. Where is the antipodal? Hint: on the equator too, on the other side, across from the original point.
27. What does this mean physically? Hint: the original point stands for linear polarization at angle α (Figure 24.1). Its antipodal, on the other hand, stands for orthogonal polarization at angle $\alpha + \pi/2$ (Figure 24.2).

28. In particular, look at the rightmost point in Poincar'e sphere: $\tilde{x} = 1$. What happens there? Hint: here, $z = \tilde{x} = 1$, so $\alpha = \theta = 0$: left-right polarization.
29. What happens at the antipodal (the leftmost point on Poincar'e sphere)? Hint: here, $z = \tilde{x} = -1$, so $\theta = \pi$, and $\alpha = \pi/2$: vertical polarization.
30. Next, in Poincar'e sphere, look at the north pole. Hint: set $\tilde{z} = y = 1$, and $z = x = 0$. This way, $\theta = \pi/2$, so $\alpha = \pi/4$. Furthermore, the phaseshift is $\phi = \pi/2$.
31. What does this mean physically? Hint: circular polarization: as the wave propagates, it makes circles counterclockwise in the perpendicular plane.
32. Where is the antipodal? Hint: at the south pole in Poincar'e sphere: $\tilde{z} = y = -1$, and $z = x = 0$. Here too, $\theta = \pi/2$, so $\alpha = \pi/4$. But the phaseshift is now $\phi = -\pi/2$.
33. What does this mean physically? Hint: circular polarization, but the other way around: as the wave propagates, it makes circles clockwise (rather than counterclockwise).
34. What is the relative angle between the north and south pole? Hint: π.
35. But physically, in terms of polarization, what is the physical angle in between? Hint: just half: $\pi/2$.
36. What does this mean physically? Hint: in terms of polarization, they are orthogonal (like every two antipodals on Poincar'e sphere).
37. In Poincar'e sphere, look now at the entire longitude $\tilde{x} = z = 0$ (Figure 24.6). What happens there? Hint: in terms of the old x-y-z coordinates, this is the equator: $x^2 + y^2 = 1$. Here, $\theta = \pi/2$, so $\alpha = \pi/4$. The phaseshift, on the other hand, changes. At $x = 1$, for a start, there is no phaseshift at all (linear polarization, as in Figure 24.1). Next, move along the equator $x^2 + y^2 = 1$. As y grows, the phaseshift grows too: the polarization gets more and more elliptic, making a narrow ellipse in the plane perpendicular to the wave propagation. Upon hitting $y = 1$ and $x = 0$, the ellipse gets as "fat" as a circle. Upon hitting $x = -1$ and $y = 0$, it gets as thin as an oblique line again (as in Figure 24.2). Upon hitting $x = 0$ and $y = -1$, it gets as fat as a circle again, with a circular polarization again (but clockwise rather than counterclockwise).

24.14 Electron and Its Spin

1. Thanks to our algebraic-geometrical framework, we can also model the spin of an electron (or any other fermion).
2. For this purpose, let's start in 3-D. Assume that we have a detector, pointing at a fixed (normalized) direction u:

$$u \equiv \begin{pmatrix} u_1 \\ u_2 \\ u_3 \end{pmatrix} \in \mathbb{R}^3, \qquad \|u\| = 1.$$

3. How likely is a given electron to spin-up in the u-direction? In other words, how likely is it to spin (counterclockwise) around u? Hint: this depends on the angle θ between u and w (the direction around which the electron "really" spins in 3-D).

4. Unfortunately, θ is often unknown.
5. To calculate the above probability, we better go back to SU_2, where the Pauli matrices originally belong. (After all, we have an electron, not a boson.)
6. In the process, θ halves into $\theta/2$.
7. More precisely, the probability to spin (counterclockwise) around u is $\cos^2(\theta/2)$.
8. Still, this is rather theoretical. After all, θ is often unknown. How to get around this? Hint: work in su_2: look at the observable

$$u_1\sigma_x + u_2\sigma_y + u_3\sigma_z.$$

Calculate its eigenvector of eigenvalue 1. Normalize it to have norm 1 in $\mathbb{C}^2$. Take its inner product with the (normalized) state of the electron in $\mathbb{C}^2$. Take the absolute value of this inner product, and square it up. This is the probability to spin (counterclockwise) around u.
9. What is the geometrical meaning of this? Give an example. Hint: look at things in $\mathbb{R}^3$ again (see below).
10. For example, assume that u and w are perpendicular in $\mathbb{R}^3$:

$$\theta = \frac{\pi}{2}.$$

11. What does this mean physically? Hint: the electron is neutral: equally likely to spin clockwise or counterclockwise around u.
12. Confirm this from our formula. Hint: indeed,

$$\cos^2\left(\frac{\theta}{2}\right) = \cos^2\left(\frac{\pi}{4}\right) = \frac{1}{2}.$$

24.15 Electron in a Magnetic Field

1. To see physics, we must go to 3-D back again.
2. In the process, our angle will double: from $\theta/2$ into θ back again.
3. This will give us the full geometrical picture: the electron spin in 3-D.
4. Note: this is only spin, not position. We don't care where the electron is in 3-D, but only how it spins.
5. Moreover, this is only nondeterministic: a random variable. We can only see its expectation.
6. Consider a constant magnetic field, pointing upwards, in the positive z-direction (Figure 24.7):

$$\mathbf{B} \equiv \|\mathbf{B}\| \begin{pmatrix} 0 \\ 0 \\ 1 \end{pmatrix}.$$

7. Let's start from 2-D. Let's guess a dynamic (time-dependent) state for our electron spin:

$$v \equiv v(t) \equiv \begin{pmatrix} \cos\left(\frac{\theta}{2}\right)\exp(-i\phi t) \\ \sin\left(\frac{\theta}{2}\right)\exp(i\phi t) \end{pmatrix}.$$

8. Later on, we'll see that this is much more than just a guess: it actually solves the Schrodinger equation.

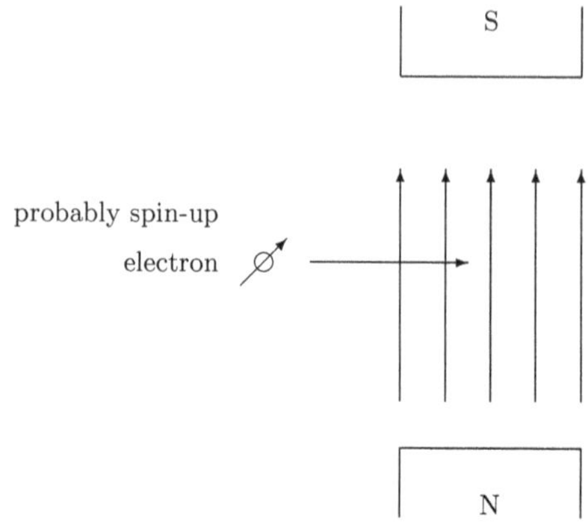

Fig. 24.7. The uniform magnetic field $\mathbf{B} \equiv \|\mathbf{B}\|(0,0,1)^t$, pointing upwards, in the positive z-direction.

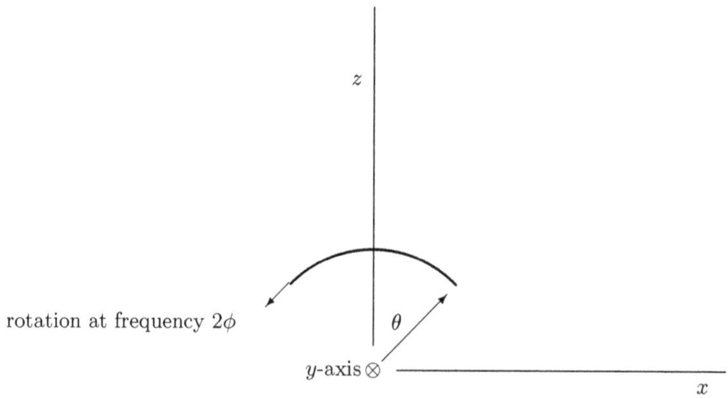

Fig. 24.8. Larmor precession: the nondeterministic spin vector $(\tilde{S}_x, \tilde{S}_y, \tilde{S}_z)$ rotates (in expectation) around the positive part of the z-axis (at frequency 2ϕ). (This is only spin direction, not position.)

9. Here, ϕ is the Larmor frequency. It is proportional to $\|\mathbf{B}\|$.
10. What is the norm of v? Hint: 1, as required.
11. What is the phaseshift between the first and second components of v? Hint: $2\phi t$.
12. At what time is the phaseshift exactly π? Hint: at time $\pi/(2\phi)$.
13. During this time, what happened to v? Hint: it was multiplied by σ_z (from the left):

$$v\left(\frac{\pi}{2\phi}\right) = \sigma_z v(0)$$

(up to an outer phaseshift, with no physical effect).
14. In our magnetic field, how likely is our electron to spin up (counterclockwise, around the z-axis)? Hint: to observe spin-up, the relevant observable is

$$\tilde{S}_z = \frac{\bar{h}}{2}\sigma_z = \frac{\bar{h}}{2}\begin{pmatrix} 1 & \\ & -1 \end{pmatrix}.$$

As a diagonal 2×2 matrix, its relevant eigenvector is $(1,0)^t$. Thus, the probability for this is

$$\left| v^t \begin{pmatrix} 1 \\ 0 \end{pmatrix} \right|^2 = \left| \cos\left(\frac{\theta}{2}\right) \right|^2 = \cos^2\left(\frac{\theta}{2}\right).$$

15. What is the expectation of the spin-up random variable at v? Hint: look at the relevant observable: $\tilde{S}_z = \bar{h}\sigma_z/2$. Its expectation at v follows from Section 9.8.4:

$$\begin{aligned}
&\left(v, \tilde{S}_z v\right) \\
&= \bar{v}^t \tilde{S}_z v \\
&= \frac{\bar{h}}{2}\bar{v}^t \sigma_z v \\
&= \frac{\bar{h}}{2}\left(\cos\left(\frac{\theta}{2}\right)\exp(i\phi t), \sin\left(\frac{\theta}{2}\right)\exp(-i\phi t)\right)\begin{pmatrix} 1 & 0 \\ 0 & -1 \end{pmatrix}\begin{pmatrix} \cos\left(\frac{\theta}{2}\right)\exp(-i\phi t) \\ \sin\left(\frac{\theta}{2}\right)\exp(i\phi t) \end{pmatrix} \\
&= \frac{\bar{h}}{2}\left(\cos^2\left(\frac{\theta}{2}\right) - \sin^2\left(\frac{\theta}{2}\right)\right) \\
&= \frac{\bar{h}}{2}\cos(\theta).
\end{aligned}$$

16. During this calculation, what happened to the angle? Hint: it doubled from $\theta/2$ back to θ, as required in 3-D.
17. Next, what is the expectation of the spin-right random variable (counterclockwise, around the x-axis)? Hint: here, the relevant observable is $\tilde{S}_x = \bar{h}\sigma_x/2$:

$$\begin{aligned}
&\left(v, \tilde{S}_x v\right) \\
&= \bar{v}^t \tilde{S}_x v \\
&= \frac{\bar{h}}{2}\bar{v}^t \sigma_x v \\
&= \frac{\bar{h}}{2}\left(\cos\left(\frac{\theta}{2}\right)\exp(i\phi t), \sin\left(\frac{\theta}{2}\right)\exp(-i\phi t)\right)\begin{pmatrix} 0 & 1 \\ 1 & 0 \end{pmatrix}\begin{pmatrix} \cos\left(\frac{\theta}{2}\right)\exp(-i\phi t) \\ \sin\left(\frac{\theta}{2}\right)\exp(i\phi t) \end{pmatrix}
\end{aligned}$$

$$
\begin{aligned}
&= \frac{\bar{h}}{2}\left(\cos\left(\frac{\theta}{2}\right)\sin\left(\frac{\theta}{2}\right)\exp(2i\phi t) + \sin\left(\frac{\theta}{2}\right)\cos\left(\frac{\theta}{2}\right)\exp(-2i\phi t)\right) \\
&= \frac{\bar{h}}{2}2\sin\left(\frac{\theta}{2}\right)\cos\left(\frac{\theta}{2}\right)\frac{\exp(2i\phi t)+\exp(-2i\phi t)}{2} \\
&= \frac{\bar{h}}{2}\sin(\theta)\cos(2\phi t).
\end{aligned}
$$

18. During this calculation, what happened to the angles? Hint: they doubled: $\theta/2$ to θ, and ϕ to 2ϕ, as required in 3-D.
19. Next, what is the expectation of the spin-in random variable (counterclockwise, around the y-axis)? Hint: here, the relevant observable is $\tilde{S}_y = \bar{h}\sigma_y/2$:

$$
\begin{aligned}
&\left(v, \tilde{S}_y v\right) \\
&= \bar{v}^t \tilde{S}_y v \\
&= \frac{\bar{h}}{2}\bar{v}^t \sigma_y v \\
&= \frac{\bar{h}}{2}\left(\cos\left(\frac{\theta}{2}\right)\exp(i\phi t), \sin\left(\frac{\theta}{2}\right)\exp(-i\phi t)\right)\begin{pmatrix} 0 & -i \\ i & 0 \end{pmatrix}\begin{pmatrix} \cos\left(\frac{\theta}{2}\right)\exp(-i\phi t) \\ \sin\left(\frac{\theta}{2}\right)\exp(i\phi t) \end{pmatrix} \\
&= \frac{\bar{h}}{2}\left(\cos\left(\frac{\theta}{2}\right)\sin\left(\frac{\theta}{2}\right)(-i)\exp(2i\phi t) + \sin\left(\frac{\theta}{2}\right)\cos\left(\frac{\theta}{2}\right)i\exp(-2i\phi t)\right) \\
&= \frac{\bar{h}}{2}2\sin\left(\frac{\theta}{2}\right)\cos\left(\frac{\theta}{2}\right)\frac{\exp(2i\phi t)-\exp(-2i\phi t)}{2i} \\
&= \frac{\bar{h}}{2}\sin(\theta)\sin(2\phi t).
\end{aligned}
$$

20. During this calculation, what happened to the angles? Hint: they doubled: $\theta/2$ to θ, and ϕ to 2ϕ, as required in 3-D.
21. What is the physical meaning of $2\phi t$? Hint: this is the phaseshift.
22. By now, we have three expectations. Place them in one row vector (Figure 24.8):

$$
\frac{\bar{h}}{2}\left(\sin(\theta)\cos(2\phi t), \sin(\theta)\sin(2\phi t), \cos(\theta)\right).
$$

23. How does this look like? Hint: this is a rotation around the z-axis. θ is the angle from the positive z-direction. 2ϕ is the frequency.
24. This is Larmor's precession.
25. In classical electromagnetics, a charged ball spins like this.
26. This is a special case of a more general theorem: the expectations in quantum mechanics obey the same laws as in classical physics.

24.16 Dynamics: Schrodinger's Equation

1. In a quantum-mechanical system, consider a state v. How does v look like? Hint: a complex vector of norm 1. It could even be infinite-dimensional: a wave function.

2. Assume now that $v \equiv v(t)$ is time-dependent. How does it evolve in time? Hint: it must solve the Schrodinger equation:

$$v'(t) \equiv \frac{dv}{dt}(t) = -\frac{i}{\bar{h}} Hv(t),$$

where H is the Hamiltonian of the system.
3. How does $v(t)$ look like? Hint: in terms of the initial state, it takes the form

$$v(t) = \exp\left(-\frac{i}{\bar{h}} Ht\right) v(0).$$

4. What is the evolution operator that produces this kind of dynamics? Hint: $\exp\left(-iHt/\bar{h}\right)$.
5. Is this familiar? Hint: this is actually a Taylor series around $t = 0$. This mirrors the Liouville operator in classical mechanics (Section 23.4.3).
6. Write H in its diagonal form (Section 24.2):

$$H = ODO^*,$$

where O is unitary, and D is real and diagonal.
7. Look at the main-diagonal elements in D. What are they? Hint: the eigenvalues of H.
8. What are they physically? Hint: energy levels.
9. What are the corresponding eigenvectors? Hint: the columns of O.
10. Let

$$u \equiv (1, 0,\ 0,\ 0,\ \ldots,\ 0)^t$$

be a standard unit vector. Look at Ou. What is it? Hint: a column in O: an eigenvector of H (of eigenvalue $D_{1,1}$):

$$HOu = ODO^*Ou = ODu = D_{1,1}Ou.$$

11. What is this physically? Hint: Ou is a state of constant energy level: $D_{1,1}$.

24.17 Evolution Operator

1. How to write the evolution operator more explicitly? Hint: it has a similar diagonal form:

$$\exp\left(-\frac{i}{\bar{h}} Ht\right) = O \exp\left(-\frac{i}{\bar{h}} Dt\right) O^*.$$

2. Conclude that the evolution operator is unitary. Hint: $\exp(-iDt/\bar{h})$ is unitary: diagonal, with elements of magnitude 1.
3. Is this as expected? Hint: the exponent of an anti-Hermitian matrix is indeed unitary (Section 24.1).
4. Is this good? Hint: thanks to this, $v(t)$ keeps norm 1, as required.
5. Apply the evolution operator to the initial state $v(0)$. Hint: this produces the state at time t:

$$v(t) = \exp\left(-\frac{i}{\bar{h}} Ht\right) v(0) = O \exp\left(-\frac{i}{\bar{h}} Dt\right) O^* v(0).$$

6. On the right-hand side, look at the vector $O^*v(0)$. In it, look at the first component (at the top). What is this? Hint: this is the inner product of $v(0)$ with Ou (the first column of O, or the first state of constant energy).
7. How does this individual component evolve in time? Hint: it only precesses (clockwise) at frequency $D_{1,1}/\bar{h}$.
8. Likewise, the other components behave in the same way (only with a different frequency).
9. Algebraically, what happens in this dynamic process? Hint: in the above formula, $v(0)$ is decomposed in terms of columns of O (states of constant energy level). During the evolution, the coefficients are stored in $O^*v(t)$ $(t \geq 0)$. This way, they keep the same sum-of-squares: 1. As time goes by, each coefficient precesses at its own frequency, changing phase, but not magnitude.

24.18 Interference vs. Conservation

1. Consider now a new observable. Could it have a conserved state? How should it look like? Hint: its conserved state must be a column of O as well. Indeed, such a column remains the same, all the time (up to phaseshift). A superposition (sum) of two columns, on the other hand, wouldn't. After all, they precess at different frequencies, thus interfere dynamically: constructively, then destructively, then constructively again, and so on.
2. How should a conserved observable relate to H? Hint: it should have the same eigenvectors: the columns of O. In other words, it should have the same diagonal form (with the same O, but not necessarily the same D). In other words, it should commute with H.
3. Is this familiar? Hint: in classical mechanics, look at the Hamilton equation (Section 23.3.1). There, commutativity makes the right-hand side vanish. This way, the left-hand side vanishes as well. Thus, the new physical quantity is time-independent: conserved.
4. In this sense, classical and quantum mechanics agree.
5. For example, look at the harmonic oscillator (Sections 1.6.1 and 9.9.1). Could momentum be conserved? Hint: only if potential is constant (say, zero). In this case, there is no force, so momentum remains constant too. Indeed, energy is only kinetic, and the Hamiltonian is proportional to momentum squared (which of course commutes with the momentum itself, as required). If, on the other hand, the Hamiltonian also contains a potential term (such as position squared, which doesn't commute with momentum), then momentum is no longer conserved.

24.19 Magnetic Energy

1. For yet another example, consider an electron in a uniform magnetic field (Figure 24.7). What is the Hamiltonian? Hint: the magnetic field $\mathbf{B}$ points in the (positive) z-direction. The magnetic energy is proportional to its strength, and also to the spin-up observable:

$$H = \|\mathbf{B}\|\tilde{S}_z = \|\mathbf{B}\|\frac{\bar{h}}{2}\sigma_z = \|\mathbf{B}\|\frac{\bar{h}}{2}\begin{pmatrix} 1 & \\ & -1 \end{pmatrix}.$$

2. What are its eigenvectors? Hint: $(1,0)^t$ and $(0,1)^t$.
3. What are their eigenvalues? Hint: $\pm\|\mathbf{B}\|\bar{h}/2$.
4. What is the physical meaning of this? Hint: a spin-up electron aligns with the magnetic field. It has eigenvector $(1,0)^t$, and positive magnetic energy. A spin-down electron, on the other hand, has negative magnetic energy. (Compare with Figure 12.2.)
5. In this example, what is O? Hint: the 2×2 identity matrix.
6. Thanks to the Schrodinger equation, how should the initial state evolve in time? Hint: as discussed in Section 24.15,

$$v \equiv v(t) \equiv \begin{pmatrix} \cos\left(\frac{\theta}{2}\right)\exp(-i\phi t) \\ \sin\left(\frac{\theta}{2}\right)\exp(i\phi t) \end{pmatrix}$$

(where θ is the angle between the spin vector and the z-axis).
7. In this solution, what is the phaseshift? Hint: $2\phi t = \|\mathbf{B}\|t$.
8. At what frequency does the phaseshift precess? Hint: $2\phi = \|\mathbf{B}\|$.
9. In this system, what spin is conserved? Hint: only spin-up is conserved, because only $\tilde{S}_z$ commutes with H (which is proportional to $\tilde{S}_z$). Spin-right and spin-in, on the other hand, can't be conserved, because $\tilde{S}_x$ and $\tilde{S}_y$ don't commute with H.

25

Feynman Diagrams and The Killing Form

Quantum mechanics works gradually, step by step: start from statics, and then advance to dynamics. Indeed, the original wave function tells us about static position (in 1-D). This is then extended to 3-D as well, to help model angular momentum and spin. But what about time? Why leave it out? Better throw it in, and treat it geometrically, just like any spatial dimension. This produces spacetime: a new (4-D) differentiable manifold. (To define it, use Zorn's lemma: Chapter 11 in [48].)

This is quantum field theory: integrate not only over a set of static points in space but also over a set of complete paths in spacetime. This is the path integral: paths could superpose and interfere, giving us a complete view over the entire dynamics, on spot, with no need to wait!

In quantum mechanics, we already studied an individual particle, and its dynamics. To see this, we looked at its wave function, and how it evolves in time. This follows from Schrodinger's equation.

Feynman, on the other hand, introduced dynamics in real physical spacetime. He looked at a complete path in spacetime, leading from the initial event to the final event. The particle could follow this path, or any other path connecting these two events (each at its own probability). These paths are then superposed (or summed up, or even integrated upon).

This is the path integral. Thanks to it, we get to see how all paths interfere at the final event in spacetime. We already saw a simple example: the double-slit experiment, with just two paths. (In Figure 9.1, let time flow rightwards, with the particle, and space be vertical.) But, in more interesting cases, there could be infinitely many paths, each with its own Feynman diagram, where virtual particles appear and disappear and apply forces and interactions (provided that energy is conserved).

Not only that. There is more. In quantum mechanics, we mostly focus on just one individual particle. At best, we look at two, and how they entangle. But how is a new particle born? Where did it come from? This is explained in a Feynman diagram. We already saw a simple example: an electron could absorb a photon, and "jump" to the next higher energy level. On the other hand, it could also spit it back, and fall to a lower energy level back again. Here, we'll see higher examples as well: two electrons could even exchange a virtual photon, and repel one another. This is an electrostatic force, in its new quantum-mechanical face.

Feynman's diagrams are geometrical. Still, they also have an algebraic face: the commutator. This will help design the transition matrix. In particle physics, this will help in a few interesting (weak or strong) interactions: bosons, fermions, quarks, gluons, and more. In this context, the Killing form will come handy.

25.1 System: Particle and Anti-Particle

25.1.1 Particle and Its Anti-Particle

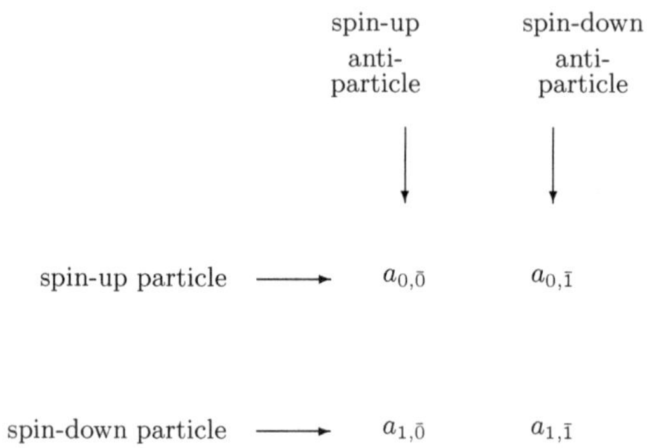

Fig. 25.1. Particle and anti-particle: each could spin up or down, as indicated by their indices: $0 \le i, j < m = 2$. The $i\bar{j}$th option has complex amplitude $a_{i,\bar{j}}$.

In quantum mechanics, consider some particle (say, an electron). Assume that it could be in m hypothetical states (each at some probability), where m is a fixed natural number. For an electron, there are just two states: spin-up-or-down ($m = 2$: Figure 25.1).

25.1.2 Positron: Missing Electron of Negative Energy

Usually, the particle has a positive energy. Still, in theory, it could also have a negative energy. In this case, its anti-particle has positive energy. Later on, we'll see this mathematically.

For example, consider an (imaginary) electron of negative energy. But, it is missing, and only leaves a hole. This hole is called positron (an anti-electron): a missing electron of negative energy. This is why it has positive energy and charge. If an electron meets a positron, then it "falls" into the hole, fills it, and releases the double energy.

The same works for a more general particle (not only an electron). What is its anti-particle? This is its mirror image:

$$\text{anti-particle} = \text{a hole: a missing particle.}$$

This way, it has opposite energy, charge, and spin: all three pick a minus sign.

25.1.3 Complex Amplitude

Together, our particle-anti-particle pair forms a new system, with the following $m \times m$ matrix:

$$A \equiv \left(a_{i,\bar{j}}\right)_{0 \le i,j < m}$$

(Figure 25.1). (The bar reminds us that this is an anti-particle.) What do we have in A? Each entry is called a complex amplitude. Let's focus on our simple example.

25.1.4 Example: Electron-Positron

Our example is quite theoretical: an electron-positron pair. (But later, it will help mirror a more physical process: weak interaction.) In it, $0 \le i, j < m = 2$. The electron could spin either up ($i = 0$) or down ($i = 1$). Likewise, the positron could spin either up ($j = 0$) or down ($j = 1$). So, the system has four possible states: either up-up or up-down or down-up or down-down.

On its own, each option is deterministic, with no ambiguity or uncertainty at all. Still, in A, they are superposed into a more interesting (nondeterministic) state: all options are possible, each at its own complex amplitude and probability. This is our new 2×2 complex matrix:

$$A \equiv \begin{pmatrix} a_{0,\bar{0}} & a_{0,\bar{1}} \\ a_{1,\bar{0}} & a_{1,\bar{1}} \end{pmatrix}$$

(Figure 25.1). In A, each element is a complex amplitude. For example, up-down has complex amplitude $a_{0,\bar{1}}$. This tells us the probability to see up-down: $|a_{0,\bar{1}}|^2$. But these four probabilities must sum to 1, so better assume that A was already normalized:

$$\sum_{i,j=0}^{m-1} |a_{i,\bar{j}}|^2 = 1.$$

25.1.5 Virtual Gauge Boson

This actually makes a boson. To motivate this, imagine a theoretical (nonphysical) situation: the electron "sits" on top of the positron. Why don't they cancel each other? Well, maybe they do! When? We don't know! In fact, thanks to the time-energy uncertainty principle, time is too short to distinguish between existence and nonexistence. The result is a virtual particle that fluctuates ever so fast, to an electron-positron pair and back, and so on. This is a gauge boson: a virtual photon.

In terms of spin, what is the state of our photon? Well, this depends on our original electron-positron. Still, could it be deterministic (say, up-up)? No! It must be nondeterministic: a superposition of up-up *minus* down-down. This means uncertainty: either up-up or down-down (at the same probability). In this virtual photon, there are two processes, illustrated by arrows. At the top arrow, the electron fluctuates from up to down (and back):

$$\begin{array}{l} \text{electron} \; - \\ \text{positron} \; - \end{array} \begin{pmatrix} 0 \\ \bar{0} \end{pmatrix} \rightleftarrows \begin{pmatrix} 1 \\ \bar{1} \end{pmatrix} \begin{array}{l} - \; \text{electron} \\ - \; \text{positron.} \end{array}$$

At the bottom arrow, on the other hand, the positron fluctuates from down to up (and back). But this runs the other way around: back-in-time. Later on, we'll see why.

To get balanced, the top arrow must emit a new $0\bar{1}$-photon. But this also creates a little hole. How to fill it, right from the start? For this, we also have the bottom arrow, which absorbs the $1\bar{0}$-anti-photon (back-in-time). This preserves energy: after all, both photon and anti-photon have the same energy.

Thanks to the frequent fluctuations, we have uncertainty: at a given time, we can't tell whether it is up-up or down-down. After all, both are equally likely. Still, this explanation is a bit too geometrical. How to see this algebraically too? Use Lie groups and algebras.

25.2 Matrix of Complex Amplitudes

25.2.1 Traceless Anti-Hermitian Matrix

So far, we modeled a superposition: up-up minus down-down. Why does down-down pick a minus sign? This way, A takes a familiar form:

$$A = \frac{i}{\sqrt{2}} \sigma_z = \frac{i}{\sqrt{2}} \begin{pmatrix} 1 & 0 \\ 0 & -1 \end{pmatrix}.$$

Thanks to $\sqrt{2}$, A gets normalized, as required. Better yet, thanks to $i = \sqrt{-1}$, A gets anti-Hermitian. This will help make a familiar Lie algebra: u_2, ready to exponentiate, and generate U_2: the unitary Lie group. Better yet, thanks to the minus sign at the bottom-right corner, A even gets traceless. This will make a yet better Lie algebra: su_2, ready to exponentiate, and generate SU_2: the special unitary Lie group (of matrices of determinant 1). For this, we'll also need to set A to

$$A = \frac{i}{\sqrt{2}} \sigma_x = \frac{i}{\sqrt{2}} \begin{pmatrix} 0 & 1 \\ 1 & 0 \end{pmatrix} \quad \text{or} \quad \frac{i}{\sqrt{2}} \sigma_y = \frac{1}{\sqrt{2}} \begin{pmatrix} & 1 \\ -1 & \end{pmatrix}.$$

This excludes any deterministic state. Indeed, thanks to the traceless property, up-up is not allowed on its own. Likewise, thanks to the anti-Hermitian property, up-down is not allowed on its own, and so on.

25.3 Anti-Particle and its Anti-Perspective

25.3.1 Anti-Particle: Opposite Charge and Spin

In the dynamics, exponentiation plays an important role. Why? Because it preserves norm. How? Well, look at an electron (at some energy level). It only precesses (gets multiplied by a Schrodinger exponent):

$$\text{electron} = \exp(i\omega t) = -(-\exp(-i\omega(-t))) = -\text{positron back-in-time.}$$

What do we have here? Four new minus signs: two physical, and two not. In $-t$, the minus sign is nonphysical: back-in-time. (This way, you can now take the anti-perspective, and substitute t for $-t$ back again.) To pay for this, frequency already picked a (physical) minus sign. Besides, in $-\exp$, the minus sign is physical too: the positron is a *missing* electron of negative energy. This will reverse charge and spin alike.

So far, only $-t$ is nonphysical: back-in-time. $-\omega$ and $-\exp$, on the other hand, are physical: anti-particle. And what about the first minus sign (before the whole thing)? It is nonphysical again. Indeed, it only interchanges existence with nonexistence: to appear means to disappear, to absorb means to emit, and vice versa. This gives the anti-particle its new (subjective) coordinates in spacetime: $-x$-$-t$ (rather than x-t). Indeed, while the electron goes from x_0 to x_1, the positron goes the other way around: it arrives at x_0 even before leaving x_1, just in time to fill the hole left by the electron back in the past. This is just a change of perspective: from the anti-perspective (of the anti-particle), our electron looks like

$$\begin{aligned}\text{electron in its arrow} &= \text{positron against the original arrow}\\ &= \text{positron back-in-spacetime.}\end{aligned}$$

25.3.2 Spin vs. Isospin

To see this better, multiply by $\exp(-ipx)$, to have a complete wave. Then, do for position-momentum what we did for time-energy. This way, momentum picks a physical minus sign, and position a nonphysical one. Moreover, in 3-D, all three coordinates pick a minus sign, and become left-handed. This way, spin picks another minus sign. To see this from our own perspective (in the original x-t coordinates), rewind the movie, to see spin go the other way around. This is because you still see your electron. The positron, on the other hand, also has $-\exp$ to offset this. (Want to have an opposite spin? Create an electron-positron pair in the lab.)

Later on, spin will mirror isospin. Project isospin onto momentum, and get energy (Section 12.3.1). Is it positive? If so, then you can even get a weak interaction. This is not so symmetric any more: matter gets a little more stable (and likely) than anti-matter. This is how we got our universe, made of matter, not anti-matter.

25.3.3 Virtual Process

How to see the game better? For this, look at spin again. How to use it (at least in theory)? Look at our virtual process again (Section 25.1.5). The top arrow flips the electron from up to down. The bottom arrow, on the other hand, points the other way around: back-in-spacetime, flipping the positron from down to up back again.

For this, the top arrow spits an up-down gauge boson. The bottom arrow, on the other hand, points in the opposite direction (in spacetime): it absorbs a down-up anti-boson, back-in-spacetime.

25.3.4 Anti-Particle: Complex Amplitude

By now, we presented the positron in terms of an electron (back-in-spacetime). For this, we introduced four new minus signs (Section 25.3.1):

1. Two are nonphysical, and only change perspective: $-t$ reverses time,
2. and the outermost minus sign (at the very beginning) reverses space.
3. Two are physical: $-\omega$ reverses energy,
4. and $-\exp$ tells us that this is a missing particle. These two will really change the complex amplitude.

Consider an $i\bar{j}$-boson, with complex amplitude $a_{i,\bar{j}}$. Now, interchange i and j:

$$\{i \leftrightarrow j\} \Leftrightarrow \{\text{electron} \ \leftrightarrow \ \text{positron}\} \Leftrightarrow \{\text{boson} \ \leftrightarrow \ \text{anti-boson}\}.$$

This should produce its $j\bar{i}$-anti-boson. What is its complex amplitude? Start from $a_{i,\bar{j}}$, and modify it. First, it must pick a complex-conjugate bar on top. Physically, this reverses energy, as required. Besides, there is another physical change: it must also pick a minus sign (since an anti-boson is a missing boson).

What is the geometrical picture? In spacetime, each point is an event. Draw an arrow, pointing at the event. This arrow tells us that something new appeared (was created) at the event. For instance, look at an arrow pointing downwards: it takes the anti-perspective (of the anti-particle). We, in the lab, may disagree: from our perspective, time flows upwards, against the arrow! But don't worry: all this is quite academic: just a matter of orientation, or perspective. From either point of view, the algebraic result agrees with the physics:

$$a_{j\bar{i}} = -\bar{a}_{i\bar{j}},$$

meaning anti-Hermitian.

25.3.5 Dual Perspective

This is why A must be anti-Hermitian. This way, emitting a boson is like absorbing an anti-boson (back-in-spacetime). This is the anti-perspective. You can side with the anti-boson, and take its point of view. Our matrix wouldn't care. After all, A cares for spin-transition only. By adding (or superposing) both perspectives, we get our virtual process back again (Section 25.1.5). In general, let's see this geometrically.

25.4 Transition: Feynman Diagram in Spacetime

25.4.1 Spin Conservation in Scattering

Assume that an i-particle hits an object, bounces off, and changes. This is scattering: it preserves momentum and energy alike. Still, does it preserve spin too? To check on this, take an i-particle, and change it: flip it into a new j-particle:

$$i \to j$$

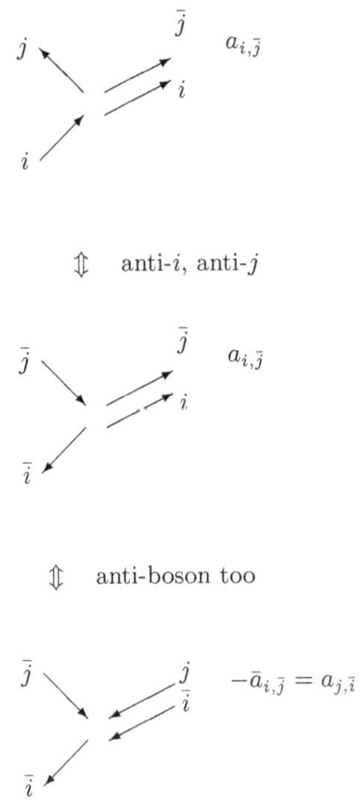

Fig. 25.2. Here, only the top picture takes our perspective: $i \to j$ (emitting an $i\bar{j}$). From the anti-perspective, on the other hand, this is $\bar{i} \leftarrow \bar{j}$ (absorbing $j\bar{i}$, back-in-spacetime).

(Figure 25.2). To preserve spin, an $i\bar{j}$-gauge boson is emitted:

$$i = j + i\bar{j}.$$

Indeed, in terms of spin, j and $\bar{j}$ cancel out. In terms of energy, on the other hand, they add up, and release the extra energy coming from the transition.

25.4.2 Absorbed Anti-Boson: Back-in-Spacetime

In terms of anti-matter, on the other hand, this goes the other way around:

$$\bar{i} \leftarrow \bar{j}$$

(Figure 25.2). For this, absorb $j\bar{i}$-anti-boson (back-in-spacetime). As before, this preserves spin:

$$\bar{i} = \bar{j} + j\bar{i}.$$

This is the anti-perspective, which is as legitimate. In terms of spin, it has complex amplitude

$$a_{j\bar{i}} = -\bar{a}_{i\bar{j}}.$$

This way, our (anti-)boson has spin $j\bar{i}$ at complex amplitude $a_{j\bar{i}}$, as required.

25.4.3 Against the Arrows

Figure 25.2 is quite theoretical. In particular, it tells us about the boson: you can reverse its arrows, and still keep the same matrix. Still, how to make it more practical? For this, look at the bottom picture, and interpret it in our original perspective: against the arrows! This way, the top and bottom pictures are no longer equivalent, but complementary. The particle and anti-particle no longer exchange events in spacetime, but go parallel, side-by-side, hand-in-hand in spacetime. In this interpretation, they model a real physical process: create an i-$\bar{i}$ pair in the lab, transform it into j-$\bar{j}$, and emit an $i\bar{j}$-$j\bar{i}$ boson-anti-boson pair to pay for this.

25.5 Transition: an Algebraic Picture

25.5.1 Transition Matrix

Look at our original transition again. How to draw it better, not only geometrically but also algebraically? Throw the complex amplitude in:

$$i \to ja_{i\bar{j}}.$$

Here, what is i? It is a deterministic state: an m-dimensional (standard unit) row vector:

$$i \sim (0, 0, \ldots, 0, \overset{i}{1}, 0, 0, \ldots, 0).$$

Likewise, j is a row vector as well:

$$j \sim (0, 0, \ldots, 0, \overset{j}{1}, 0, 0, \ldots, 0).$$

On second thought, why flip to j only? Better flip to all possible j's, and superpose over j. This yields the ith row in A:

$$i \to \sum_{j=0}^{m-1} ja_{i\bar{j}} = iA.$$

Better yet, extend this (linearly) once again, and superpose over i as well:

$$v \to vA,$$

where v is a general (m-dimensional) row vector.

25.5.2 State and Its Observable

What is v physically? It is our good old (nondeterministic) state of our particle. What is A physically? Start from an observable: say, σ_z (spin-up) or σ_x (spin-right) or σ_y (spin-in). Multiply by i, and makes our A: anti-Hermitian, ready to exponentiate, and generate a (unitary) Lie group. Better yet, exponentiate the adjoint, and see spin dynamically in 3-D, not only quantum-mechanically but even classically. Still, there is a little change here: A acts from the right, not from the left.

This is quite algebraic. Physically, why must A be anti-Hermitian? We'll soon see why: to help include anti-matter too (from its own anti-perspective). This will agree with Dirac's hypothesis: anti-matter must exist, and back-in-time is as legitimate. Indeed, once the anti-perspective is added as well, the anti-Hermitian property appears back again.

25.5.3 Invariant Anti-Perspective

Next, let's work the other way around (from the anti-perspective):

$$\bar{j} \to -\bar{a}_{i\bar{j}}\bar{i}$$

(Figure 25.2). Here, what is $\bar{j}$? It is a (deterministic) state of the anti-particle: an m-dimensional (standard unit) column vector:

$$\bar{j} \sim (0, 0, \ldots, 0, \overset{j}{1}, 0, 0, \ldots, 0)^t.$$

Likewise, $\bar{i}$ is a column vector as well:

$$\bar{i} \sim (0, 0, \ldots, 0, \overset{i}{1}, 0, 0, \ldots, 0)^t.$$

On second thought, why flip to $\bar{i}$ only? Better flip to all possible $\bar{i}$'s, and superpose over $\bar{i}$. This yields the $\bar{j}$th column in $-\bar{A}$:

$$\bar{j} \to -\sum_{i=0}^{m-1} \bar{a}_{i\bar{j}}\bar{i} = -\bar{A}\bar{j}.$$

Better yet, extend this (linearly) once again, and superpose over $\bar{j}$ as well:

$$v \to -\bar{A}v,$$

where v is now an (m-dimensional) column vector: a new (nondeterministic) state of the anti-particle (going back-in-spacetime). To this, apply the transpose, and interpret v as a row vector (for the anti-particle):

$$v \to -v\bar{A}^t = -vA^* = vA$$

(thanks to the anti-Hermitian property). This looks the same as before: invariant in terms of perspective. Thus, A is perspective-blind: it cares for spin-transition only, regardless of perspective.

25.5.4 Anti-Hermitization

To model real physics, A must be anti-Hermitian. Still, let's consider even a more general A (not necessarily anti-Hermitian or traceless). This will make our $j\bar{i}$-boson more independent: not necessarily an anti-$i\bar{j}$-boson any more. This way, we can even imagine a (nonphysical) deterministic boson, and study it.

Still, if A is *not* anti-Hermitian, how to anti-Hermitize it? Subtract its Hermitian adjoint. This will produce a new anti-Hermitian matrix:

$$A - A^*,$$

as required. Physically, this means superposing both perspectives: ours, plus the anti-perspective. Later on, we'll see that this works for fermions only. For bosons, on the other hand, do something else: add the Hermitian adjoint, and multiply by $i = \sqrt{-1}$. This will produce

$$i\,(A + A^*)\,.$$

This is anti-Hermitian, but not necessarily traceless any more. (To make it traceless too, subtract its trace/m from the main diagonal.) Later on, we'll use this for bosons.

25.5.5 Transition and Its Probability

Look at Figure 25.2 again. At the top, we draw the original transition: $i \to j$ (in our perspective). This could happen at complex amplitude $a_{i,\bar{j}}$, and probability

$$|a_{i,\bar{j}}|^2 \;/\; \sum_{k,l=0}^{m-1} |a_{k,\bar{l}}|^2.$$

At this probability, you'd get to see an $i\bar{j}$-boson. This way, A serves not only as a matrix but also as a (discrete) wave function.

25.6 Two Bosons Interact

25.6.1 Transition of a Boson

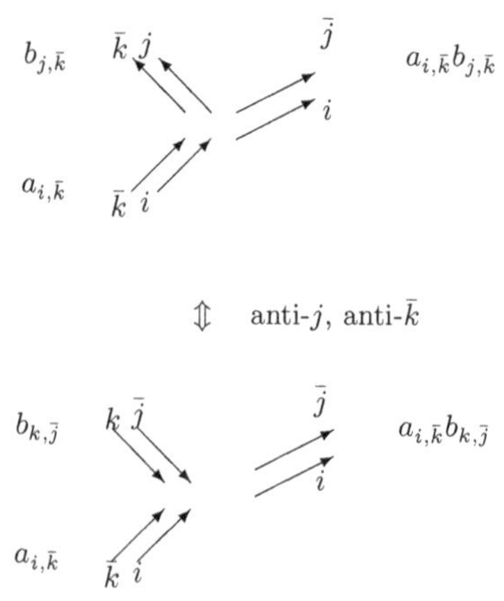

Fig. 25.3. An incoming $i\bar{k}$-boson (of matrix A) flips into an outgoing $j\bar{k}$-boson (of matrix B), emitting a new $i\bar{j}$-gauge boson. Better yet, the outgoing $j\bar{k}$-boson is equivalent to an incoming $k\bar{j}$-anti-boson, back-in-spacetime (same matrix: B).

Look at our original transition again: $i \to j$ (Figure 25.2). Parallel to it, draw a trivial "transition:" $\bar{k} \to \bar{k}$. This makes a new transition (of a complete boson): an incoming $i\bar{k}$-boson (of matrix A) flips into an outgoing $j\bar{k}$-boson (of matrix B). This only changes i to j, leaving $\bar{k}$ idle. But don't worry: later on, in the next step, you'll get to work from the other side, and change $\bar{k}$ too, if you like.

Look at the outgoing boson: $j\bar{k}$ (Figure 25.3). Now, reverse its direction in spacetime. For this, look at its anti-boson: an (incoming) $k\bar{j}$-anti-boson, coming from the upper-left corner, and absorbed (rather than emitted), back-in-spacetime. This is equivalent, and keeps the same matrix: B. On the upper-right, the outgoing boson will then have a new complex amplitude (a new product):

$$a_{i,\bar{k}} b_{k,\bar{j}}.$$

Better yet, do this for all possible k's, and superpose over k. On the upper-right, the outgoing boson will then take a new complex amplitude:

$$\sum_{k=0}^{m-1} a_{i,\bar{k}} b_{k,\bar{j}} = (AB)_{i,\bar{j}}.$$

This will be useful below.

25.6.2 Virtual Gauge Boson

By now, our outgoing boson (on the upper-right) has matrix AB. From this, subtract the anti-perspective (matrix: BA):

$$AB - BA = [A, B].$$

Why does BA pick a minus sign? Well, look at a special case

$$A = B.$$

In this case,

$$[A, B] = [A, A] = (0).$$

This makes sense: a trivial transition like $A \to A$ should indeed emit nothing.

Algebraically, this anti-Hermitizes AB (Section 25.5.4). Physically, this models the virtual gauge boson (emitted on the upper-right). It even gives us its (discrete) wave function. Still, this is a bit too theoretical. How to get more practical?

25.6.3 Two Bosons Emit a New Boson

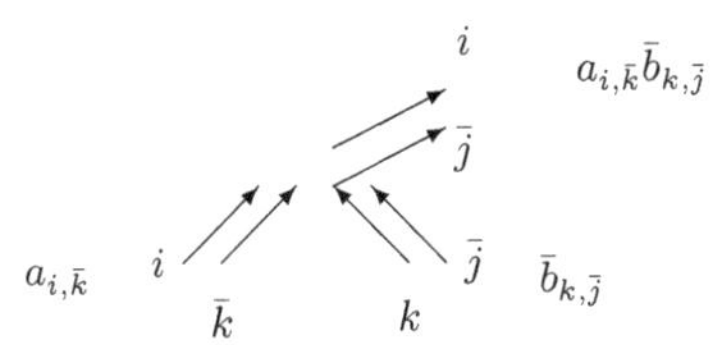

Fig. 25.4. Incoming $i\bar{k}$- and $k\bar{j}$-bosons (of matrices A and $\bar{B}$) interact, and emit a new $i\bar{j}$-boson. Thanks to superposition (over all k's), it has matrix $A\bar{B}$ (anti-Hermitized).

Let's go back to Figure 25.3. By now, we have two incoming bosons: one from below (matrix: A), and one from above, back-in-spacetime (matrix: B). Let's reverse its time. This is no longer equivalent: it has matrix $\bar{B}$ (Figure 25.4). And what about the outgoing boson (on the upper-right)? What is its complex amplitude? It is a new product:

$$a_{i,\bar{k}}\bar{b}_{k,\bar{j}}.$$

Better yet, superpose over all possible k's, and sum up:

$$\sum_{k=0}^{m-1} a_{i,\bar{k}}\bar{b}_{k,\bar{j}} = \left(A\bar{B}\right)_{i,\bar{j}}.$$

Better yet, substitute B for $\bar{B}$ back again. This way, two incoming bosons (of matrices A and B) interact, and emit an outgoing boson of a new matrix: AB? Not quite.

25.6.4 The Commutator

Still, there is a problem. Even if both A and B were anti-Hermitian (and even traceless), AB may not. How to fix it? Anti-Hermitize (Section 25.5.4). Say, subtract its Hermitian adjoint:

$$AB - (AB)^* = AB - B^*A^* = AB - (-B)(-A) = AB - BA = [A, B].$$

This is the commutator: anti-Hermitian (and even traceless). Still, what if

$$A = B?$$

In this case,

$$[A, B] = [A, A] = (0).$$

In Section 25.6.2, this made sense: a trivial transition like $A \to A$ should indeed emit nothing. But here, two physical (identical) incoming bosons can't disappear into thin air. Where did their energy (and spin) go?

25.6.5 Fermions: Pauli's Exclusion Principle

What happens to our incoming particles? Upon scattering (at the center), they have the same position-momentum, and get indistinguishable. (This is why it makes sense to interchange them, and superpose BA too.) If they are fermions, then they also obey Pauli's exclusion principle:

$$A \neq B.$$

This avoids our paradox. Thus, only fermions can use $[A, B]$.

25.6.6 Indistinguishable Bosons

Bosons, on the other hand, obey no exclusion principle at all. They are still indistinguishable, so it still makes sense to interchange them, and superpose BA too (but with no minus sign any more). This is the second kind of anti-Hermitization (Section 25.5.4):

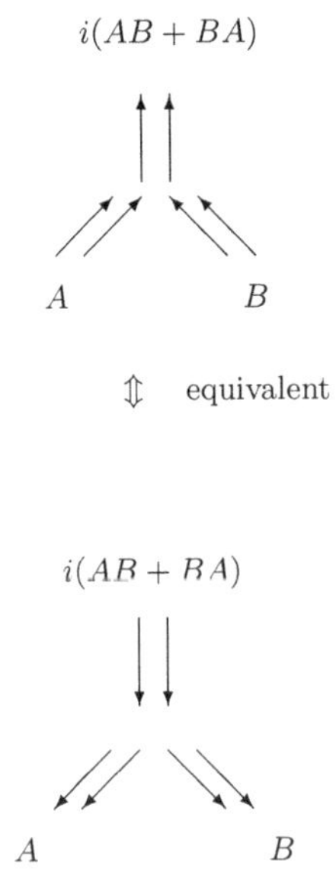

Fig. 25.5. Two incoming bosons (of matrices A and B) unite, and emit a new outgoing boson [of matrix $i(AB + BA)$]. This is also equivalent to the anti-perspective.

- First, Hermitize (add the Hermitian adjoint):

$$AB + (AB)^* = AB + B^*A^* = AB + (-B)(-A) = AB + BA.$$

- Then, multiply by i:

$$i(AB + BA)$$

(Figure 25.5). This is indeed anti-Hermitian, as required (but not necessarily traceless any more). This is suitable to bosons: indistinguishable, but with no exclusion principle: many identical bosons could even "sit" on top of one another.

25.6.7 A Boson Splits Into Two New Bosons

In Figure 25.5, we have two equivalent perspectives. At the top, the process is as before: two incoming bosons unite, and emit an outgoing boson. At the bottom, on the other hand, we have the anti-perspective, which is as legitimate.

In it, let's make a real physical change: reverse time. This is no longer equivalent, but reverse-engineered: the entire picture gets upside-down (Figure 25.6), and the matrices pick a complex-conjugate bar on top. Physically, this is a new process: the incoming boson splits into two outgoing bosons.

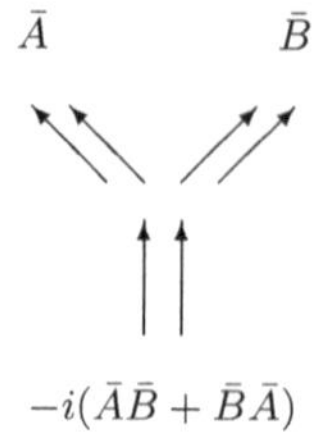

Fig. 25.6. Now, reverse time. This reverse-engineers the above process: the entire picture gets upside-down, and the matrices pick a complex-conjugate bar on top.

25.7 Exercises: Weak and Strong Interaction

25.7.1 Weak Interaction: Quarks and W-Bosons

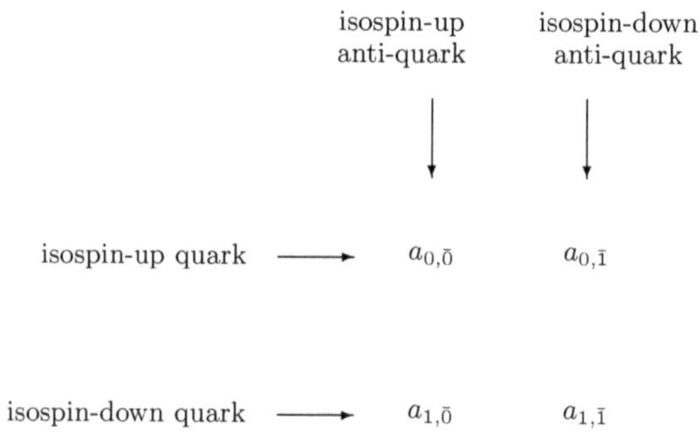

Fig. 25.7. Weak interaction in a quark-anti-quark system. Isospin mirrors spin: an $i\bar{j}$-boson helps flip $i \to j$ (or $\bar{i} \leftarrow \bar{j}$, back-in-spacetime) at complex amplitude $a_{i\bar{j}}$.

1. In the atom, in the nucleus, there are protons and neutrons (which are both fermions).
2. Each proton (or neutron) contains three quarks (which are fermions in their own right).
3. Each quark has a property called isospin (which mirrors spin): either up (0) or down (1).
4. A quark-anti-quark pair forms a new kind of boson: a W-boson. It has an isospin too: up-up ($0\bar{0}$) or up-down ($0\bar{1}$) or down-up ($1\bar{0}$) or down-down ($1\bar{1}$) (Figure 25.7).

5. Physically, what does an $i\bar{j}$-W-boson do? Hint: in our quark-anti-quark system, it helps flip $i \to j$ (or $\bar{i} \leftarrow \bar{j}$, back-in-spacetime).
6. What is its complex amplitude? Hint: $a_{i,\bar{j}}$.
7. At what probability could it be emitted? Hint: $|a_{i,\bar{j}}|^2$ (normalized).
8. In the nucleus, what keeps the protons and neutrons together? Hint: the weak interaction. This is virtual: the proton turns into a neutron and back, all the time.
9. How? Hint: an up-quark from the proton turns into a down-quark, emitting a new $0\bar{1}$-W-boson (a virtual gauge boson, as in Figure 25.2 and Section 25.1.5).
10. As a result, what happens to our proton? Hint: it is not quite a proton any more, but a nondeterministic superposition, equally likely to be a proton or a neutron.
11. Could two W-bosons (of matrices A and B) attract one another? Hint: yes, by emitting a new W-gauge boson (of matrix $[A, B]$, as in Figure 25.3).
12. How? Hint: by flipping an i-quark from boson A into a new j-quark in boson B (or a $\bar{j}$-anti-quark from boson B into an $\bar{i}$-anti-quark in boson A, back-in-spacetime). This way, $i\bar{i}$ could even fluctuate to $j\bar{j}$ and back. (For instance, $0\bar{0}$ to $1\bar{1}$ and back, as in Section 25.1.5.)
13. What is the probability for this? Hint: $|[A, B]_{i,\bar{j}}|^2$ (assuming that both A and B are normalized to have sum-of-squares 1).

25.7.2 Weak Structure Constants

1. Recall the Pauli matrices (Section 11.2.1):

$$\sigma_x \equiv \begin{pmatrix} 0 & 1 \\ 1 & 0 \end{pmatrix}$$
$$\sigma_y \equiv i \begin{pmatrix} 0 & -1 \\ 1 & 0 \end{pmatrix}$$
$$\sigma_z \equiv \begin{pmatrix} 1 & 0 \\ 0 & -1 \end{pmatrix}.$$

2. Are they Hermitian?
3. Are they traceless?
4. Do they anti-commute?
5. Are they cyclic? Hint:

$$\sigma_x \sigma_y = i\sigma_z$$
$$\sigma_y \sigma_z = i\sigma_x$$
$$\sigma_z \sigma_x = i\sigma_y.$$

6. How to anti-Hermitize them? Hint: multiply by i, and define three new matrices:

$$C_1 \equiv i\sigma_x = i \begin{pmatrix} 0 & 1 \\ 1 & 0 \end{pmatrix}$$
$$C_2 \equiv i\sigma_y = \begin{pmatrix} 0 & 1 \\ -1 & 0 \end{pmatrix}$$
$$C_3 \equiv i\sigma_z = i \begin{pmatrix} 1 & 0 \\ 0 & -1 \end{pmatrix}.$$

7. Are they anti-Hermitian?
8. Are they still traceless?
9. With real coefficients, what do they span? Hint: su_2.
10. Do they still anti-commute? Hint: yes: multiplying by a constant doesn't matter.
11. Show that they are anti-cyclic:

$$C_3C_2 = C_1$$
$$C_2C_1 = C_3$$
$$C_1C_3 = C_2.$$

Hint: they are just i times the Pauli matrices.
12. Conclude that they mirror (or represent) vector product in 3-D.
13. Pick A to be either C_1 or C_2 or C_3.
14. Likewise, pick B to be either C_1 or C_2 or C_3. (We'll soon see that some choices are nonphysical.)
15. Expand their commutator as

$$[A, B] = c_1C_1 + c_2C_2 + c_3C_3,$$

for some real coefficients c_1, c_2, and c_3.
16. c_1, c_2, and c_3 are called structure constants (Section 22.4).
17. They depend on A, B, and our basis (C_1, C_2, and C_3):

$$c_1 \equiv c_1(A, B, C_1, C_2, C_3)$$
$$c_2 \equiv c_2(A, B, C_1, C_2, C_3)$$
$$c_3 \equiv c_3(A, B, C_1, C_2, C_3).$$

18. How could W-boson A attract W-boson B? Hint: by flipping an i-quark from A into a new j-quark in B (or an old $\bar{j}$-anti-quark from B into an $\bar{i}$-anti-quark in A, back-in-spacetime).
19. For this, how should $[A, B]$ look like? Hint: it mustn't be diagonal. In other words, it must contain C_1 or C_2 (but not necessarily C_3).
20. What is the probability for this? Hint: $|[A, B]_{i,\bar{j}}|^2$ (assuming that both A and B are normalized to have sum-of-squares 1).
21. What does this tell you about c_1 and c_2? Hint: at least one of them is nonzero. This way, $[A, B]$ has an off-diagonal element, as required.
22. To make this transition come true, what should be emitted? Hint: an $i\bar{j}$-W-boson (mixed: with $i \neq j$). For instance, a $0\bar{1}$-W-boson could convert an isospin-up quark in A to an isospin-down quark in B (as in Section 25.1.5).
23. Explain this transition algebraically, and write its probability in terms of c_1, c_2, and c_3. Hint: the transition is uncertain. Thanks to Figure 25.3, it has its complex amplitude in AB. More precisely, superpose this with $-BA$ (the anti-perspective), and get $[A, B]$ (Section 25.6.2). To it, only C_1 and C_2 contribute an off-diagonal element (to help emit a mixed W-gauge boson). C_3, on the other hand, contributes nothing (since it is diagonal). Thus, the probability for this is

$$|c_1|^2 + |c_2|^2$$

(assuming that both A and B are normalized to have sum-of-squares 1).

24. Look at c_3. It supports no transition at all. What does it do? Hint: c_3 is the complex amplitude of a W-boson that supports no transition, but only fluctuates from $0\bar{0}$ to $1\bar{1}$ and back, all the time:

$$W\text{-boson} \;=\; \begin{pmatrix} 0 \\ \bar{0} \end{pmatrix} \rightleftarrows \begin{pmatrix} 1 \\ \bar{1} \end{pmatrix} \;=\; W\text{-boson}$$

(Section 25.1.5). In a sense, C_3 is too selfish: it doesn't help W-bosons A or B at all. Why? Because it has no off-diagonal entry, required to emit a $0\bar{1}$-W-boson (as in its own top arrow), or absorb a $1\bar{0}$-W-anti-boson (as in its own bottom arrow).

25. So, its own arrows remain inner, and can't help bosons A or B at all. Still, do these arrows preserve spin? Hint: yes. At the top arrow,

$$0 = 1 + 0\bar{1}.$$

At the bottom arrow,

$$\bar{0} = \bar{1} + 1\bar{0}$$

(back-in-spacetime).

25.7.3 Quantum Chromodynamics: Quarks and Gluons

		anti-quark: red ↓	anti-quark: green ↓	anti-quark: blue ↓
quark:	red →	$a_{0,\bar{0}}$	$a_{0,\bar{1}}$	$a_{0,\bar{2}}$
	green →	$a_{1,\bar{0}}$	$a_{1,\bar{1}}$	$a_{1,\bar{2}}$
	blue →	$a_{2,\bar{0}}$	$a_{2,\bar{1}}$	$a_{2,\bar{2}}$

Fig. 25.8. Strong interaction in a quark-anti-quark system, with (nondeterministic) color, indexed by $0 \le i, j < m = 3$. To convert a quark from color i to j, emit an $i\bar{j}$-gluon. This has complex amplitude $a_{i\bar{j}}$.

1. So far, we only used $m = 2$. In this context, a quark has just two options: isospin-up-or-down.
2. Next, let's move on to a more complicated system: $m = 3$. In this context, a quark has three options to be colored: either red (0) or green (1) or blue (2), at some probability each (Figure 25.8).
3. For instance, how likely is the quark to be green, and the anti-quark blue? Hint: this could happen at complex amplitude $a_{1,\bar{2}}$, and probability $|a_{1,\bar{2}}|^2$ (assuming that A is normalized to have sum-of-squares 1).
4. In a proton (or a neutron), there are three quarks. What keeps them together? Hint: this is the strong interaction. For instance, an old red quark flips into a new green quark, emitting a new $0\bar{1}$-gluon (as in Figure 25.2). This attracts the quarks to one another.
5. Could two gluons attract one another as well? Hint: yes (as in Figure 25.3).
6. Instead of the original Pauli matrices, what do we have here? Hint: define eight new (independent) 3×3 matrices:

$$C_1 \equiv \begin{pmatrix} 0 & 1 & 0 \\ -1 & 0 & 0 \\ 0 & 0 & 0 \end{pmatrix}$$

$$C_2 \equiv i \begin{pmatrix} 0 & 1 & 0 \\ 1 & 0 & 0 \\ 0 & 0 & 0 \end{pmatrix}$$

$$C_3 \equiv \begin{pmatrix} 0 & 0 & -1 \\ 0 & 0 & 0 \\ 1 & 0 & 0 \end{pmatrix}$$

$$C_4 \equiv i \begin{pmatrix} 0 & 0 & 1 \\ 0 & 0 & 0 \\ 1 & 0 & 0 \end{pmatrix}$$

$$C_5 \equiv \begin{pmatrix} 0 & 0 & 0 \\ 0 & 0 & 1 \\ 0 & -1 & 0 \end{pmatrix}$$

$$C_6 \equiv i \begin{pmatrix} 0 & 0 & 0 \\ 0 & 0 & 1 \\ 0 & 1 & 0 \end{pmatrix}$$

$$C_7 \equiv i \begin{pmatrix} 1 & 0 & 0 \\ 0 & -1 & 0 \\ 0 & 0 & 0 \end{pmatrix}$$

$$C_8 \equiv i \begin{pmatrix} 0 & 0 & 0 \\ 0 & 1 & 0 \\ 0 & 0 & -1 \end{pmatrix}.$$

7. Are they anti-Hermitian?
8. Are they traceless?
9. With real coefficients, what do they span? Hint: su_3.
10. This is our new basis (over $\mathbb{R}$, not $\mathbb{C}$).
11. Is it unique? Hint: no. If you like, pick any other eight matrices that are 3×3, independent, anti-Hermitian, and traceless.

12. Pick A to be one of the C_i's.
13. Likewise, pick B to be one of the C_i's. (We'll soon see that some choices are nonphysical.)
14. Expand their commutator in terms of our basis:

$$[A, B] = c_1 C_1 + c_2 C_2 + c_3 C_3 + \cdots + c_8 C_8,$$

for some real coefficients $c_1, c_2, c_3, \ldots, c_8$.
15. These coefficients are called structure constants (Section 22.4).
16. They depend on A, B, and our basis $(C_1, C_2, C_3, \ldots, C_8)$.

25.7.4 Strong Structure Constants

1. So far, i and j indexed the quarks (and gave them a color). Next, let $1 \le i, j < 8$ index the matrices in our basis: $C_1, C_2, C_3, \ldots, C_8$.
2. Let A be fixed and specific, as above: one of these eight matrices.
3. B, on the other hand, remains unspecified: either one of these eight matrices.
4. Define a new mapping: $A_{ad} = [A, \cdot]$. What does it do? Hint: it maps B to $[A, B]$:

$$A_{ad} = [A, \cdot] : B \to [A, B].$$

5. How does it look like? Hint: see below.
6. Write it as an explicit 8×8 matrix (in terms of our basis). Hint: to filter out its ith row, apply it to $B \equiv C_i$ (from the right). This will filter out its $(i, \bar{j})$th element:

$$\begin{aligned} A_{ad,i,\bar{j}} = [A, \cdot]_{i,\bar{j}} &= [A, C_i]_j \\ &= c_j \\ &\equiv c_j\,(A, B \equiv C_i, C_1, C_2, C_3, \ldots, C_8) \end{aligned}$$

(depending on A, B, and our basis).
7. For $1 \le i, j \le 8$, these are the strong structure constants.
8. These are the transpose of those in Section 22.4. In what sense? Hint: this matrix works from the right (as in Section 25.5.1).
9. Look at $-A^*_{ad}$. What is this physically? Hint: this is the anti-perspective (applied from the right too, as at the end of Section 25.5.3).
10. Assume that gluon A attracts gluon B (strongly), by converting a red quark in A into a green quark in B, and vice versa (as in Figure 25.3 and Section 25.1.5). How likely is this? Hint: the probability for this is

$$|c_1|^2 + |c_2|^2$$

(assuming that both A and B are normalized to have sum-of-squares 1).
11. And what is the probability to do this by converting red to blue and back? Hint: $|c_3|^2 + |c_4|^2$.
12. And what is the probability to do this by converting green to blue and back? Hint: $|c_5|^2 + |c_6|^2$.

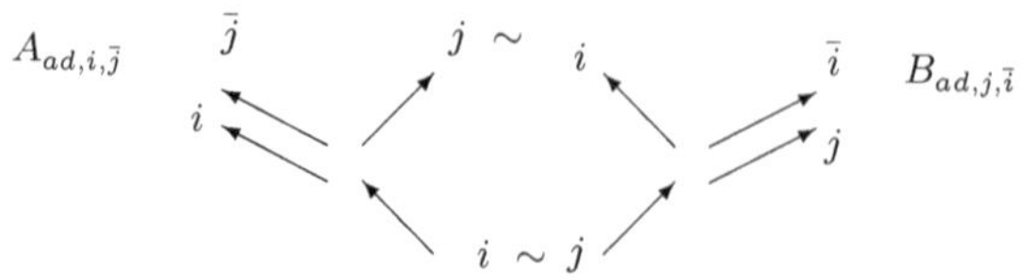

Fig. 25.9. How does gluon A attract gluon B (strongly)? The ith gluon in B interchanges with the jth gluon in A.

25.7.5 Killing Form

1. Look at a strong interaction: gluon A attracts gluon B.
2. For this, the ith gluon in B interchanges with the jth gluon in A (Figure 25.9).
3. What is the complex amplitude? Hint: the product:
$$A_{ad,i,\bar{j}} B_{ad,j,\bar{i}}.$$
4. But i and j are still unspecified, and could be just anything. Superpose over all possible i's and j's, and sum up:
$$\sum_{i,j=1}^{8} A_{ad,i,\bar{j}} B_{ad,j,\bar{i}} = \sum_{i=1}^{8} (A_{ad}B_{ad})_{i,\bar{i}} = \text{trace}\,(A_{ad}B_{ad}) = \kappa(A,B).$$
5. What is this? Hint: this is a number, not a matrix.
6. What is this physically? Hint: this is the complex amplitude that gluon A attracts gluon B (strongly) by exchanging an inner gluon.
7. Better yet, extend this yet more. For this, expand A and B (linearly) in terms of our basis:
$$A = \sum_{i=1}^{8} \alpha_i C_i, \quad \text{and} \quad B = \sum_{j=1}^{8} \beta_j C_j$$
(for some real coefficients α_i's and β_j's).
8. For these new A and B, what is the new complex amplitude? Hint: the new complex amplitude is now
$$\sum_{i,j=1}^{8} \alpha_i K_{i,j} \beta_j,$$
where K is the Killing form:
$$K_{i,j} = \kappa\,(C_i, C_j) = \text{trace}\,(C_{i,ad} C_{j,ad})\,.$$
9. Use our basis for yet another important purpose: to write the commutator. Hint:
$$[A,B] = \sum_{i,j=1}^{8} \alpha_i\,[C_i, C_j]\,\beta_j.$$

10. What is this? Hint: this is a 3×3 matrix.
11. Why? Hint: this is just a linear combination of 3×3 matrices.

Part IX

Appendix: Background in Calculus

Part IX: Appendix: Background in Calculus

On our way, we already saw some background in math: Newton's binomial (Chapter 6), the determinant (Chapter 13), and more. Still, to make the book self-contained, here is some more material in linear algebra and calculus. We start from the very beginning: functions and their derivatives. This leads to the exponent function, and how it converges. This is mirrored in Lie algebras: once exponentiated, they generate Lie groups, useful in quantum and classical mechanics alike.

After introducing differentiation and integration, we move on to partial differentiation in 2- and even 3-D. This material is vital in Newtonian and modern physics alike. Finally, we also discuss matrices and their eigenvalues, with a special focus on three interesting types: Hermitian, unitary, and positive definite. This is particularly useful in quantum physics and chemistry.

26

Functions and Their Derivatives

In this chapter, we introduce elementary calculus from scratch. For this purpose, we start from a new mathematical object: the function.

What is a function? In set theory, it has a narrow face: a mapping. In calculus, on the other hand, it is much more: an analytical-geometrical tool. Indeed, to the original function, we can now apply new analytical operations: differentiation and integration [10, 34]. In particular, we'll see how to differentiate a product of two functions, a composite function, and even an inverse function. Furthermore, we'll use the binomial formula to study a few important functions: the exponent, the logarithm, and the power function. This is mirrored in Lie algebras: once exponentiated, they generate Lie groups, useful in quantum and classical mechanics alike.

26.1 Functions in Physics

26.1.1 Functions

What is a function? It is a mapping: it maps a given set S to some other set. This way, each individual element $x \in S$ is mapped to some $f(x)$:

$$x \to f(x), \quad x \in S.$$

For example, assume that

$$S = \mathbb{R}$$

(the real axis). Assume also that f is a real function:

$$f : \mathbb{R} \to \mathbb{R}.$$

In other words, for each real number $x \in \mathbb{R}$, $f(x)$ is a real number as well. Let's study this kind of function analytically-geometrically. This is indeed calculus.

To analyze, we need to differentiate: measure how fast the function changes. Unfortunately, most functions are nondifferentiable. But don't worry: the functions we really need in practice are differentiable. These are good enough to model physical phenomena like motion, acceleration, and force.

A differentiable function $f(x)$ has a derivative: a new function $f'(x)$ that tells us how fast f grows in a short interval around x. This rate of growth may depend on x: at some x's, f may grow fast. At some other x's, on the other hand, f may remain nearly flat, or even fall. At these x's, the derivative $f'(x)$ may be nearly zero, or even negative.

26.1.2 Time, Position, and Speed

For instance, x could stand for time:

$$x = t.$$

(But this is just an example: our x will be more general.) This way, at time x, $f(x)$ stands for position (or location, or distance): how far the particle got from its start point

$$f(0) = 0.$$

In this case, $f'(x)$ tells us how fast the distance grows: the speed (or velocity) at time $x = t$.

Like the position, the speed is a function of $x = t$ too. After all, it doesn't have to remain constant, but can change with time.

26.1.3 Acceleration

Often, the velocity $f'(x)$ is a differentiable function in its own right. Its derivative $f''(x)$ tells us how fast the speed grows. Is this familiar? This is acceleration.

The acceleration is a function of $x = t$ in its own right. After all, it doesn't have to remain constant, but can change in time.

This is how calculus helps model motion and dynamics, and even introduce more fundamental concepts: force, work, and energy. Thanks to calculus, physics gets easy to study, calculate, and comprehend analytically-geometrically.

26.1.4 Real Function in an Open Domain

Consider the real axis $\mathbb{R}$. Let $\Omega \subset \mathbb{R}$ be an open domain. This means that Ω has no endpoint: each individual number $x \in \Omega$ is surrounded by other numbers in Ω. In fact, x is surrounded by a complete neighborhood: an open interval, with x in the middle:

$$x \in (x - \eta, x + \eta) \subset \Omega \subset \mathbb{R},$$

for some positive parameter $\eta > 0$ (which may depend on x). For instance, Ω could be the entire real axis:

$$\Omega \equiv \mathbb{R}.$$

In this case, η is not necessarily small: it could be as big as you like. For yet another example, Ω could be the open semiaxis:

$$\Omega \equiv (-\infty, 0) \quad \text{or} \quad \Omega \equiv (0, \infty).$$

(Here, the round parentheses tell us that 0 is excluded.) In this case, you must pick η not too big, but as small as $|x|$, say

$$\eta \equiv \eta(x) \equiv \frac{|x|}{2}.$$

For yet another example, Ω could be the union of many open intervals. What can't Ω be? It mustn't be a closed interval, with an endpoint. Indeed, let x be this endpoint: no η is small enough! Thus, a closed interval is *not* an open domain.

Assume now that Ω is indeed an open domain: say, the union of a few open intervals, none of which is closed (or even half-closed). Now, on Ω, define a new real function $f(x)$:

$$f : \Omega \to \mathbb{R}.$$

For each individual $x \in \Omega$, f is defined not only at x but also in a small neighborhood of x. In such a neighborhood, we can now estimate how fast f changes: grows or falls.

26.2 Derivative

26.2.1 Rate of Change: A Geometrical Approach

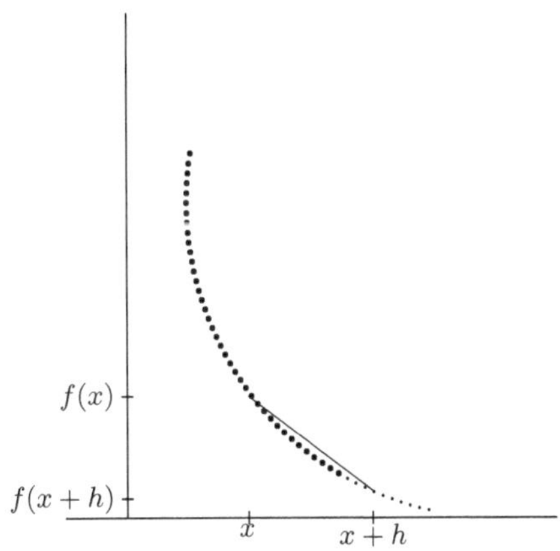

Fig. 26.1. On the graph of f, draw a short string: from $(x, f(x))$ to $(x+h, f(x+h))$. In this picture, it points down-right (since f falls).

Consider some point $x \in \Omega$. How fast does f change at x? To estimate this, pick a small parameter h. Look at two points on the real axis: x and $x+h$ (Figure 26.1). For a sufficiently small $|h|$, both are in Ω:

$$x, \ x+h \in \Omega.$$

Thanks to this, we can now calculate both $f(x+h)$ and $f(x)$, and subtract them from one another:

$$f(x+h) - f(x).$$

Is this big (in magnitude)? If it is, then f changes sharply at x. Otherwise, f is nearly flat at x.

26.2.2 Divided Difference

"Big" relative to what? Relative to h. So, we actually need to look at the divided difference

$$\frac{f(x+h) - f(x)}{h}.$$

This is how f grows per h-unit (from x to $x+h$).

26.2.3 String and Its Slope

What does this mean geometrically? Look at the x-y plane. In it, draw the graph of f (Figure 26.1). On the graph, each point has two coordinates: horizontally, it has coordinate x. Vertically, it has coordinate $f(x)$. Together, these coordinates make our point: $(x, f(x))$ on our graph. Likewise, look at another point: $(x+h, f(x+h))$.

Next, draw the string connecting these points. What is its slope? This is our divided difference, telling us how fast f changes per h-unit (on average, in the short interval $[x, x+h]$).

26.2.4 Tangent and Its Slope

But this is still an average over the (closed) interval $[x, x+h]$. However short, this interval is not short enough. After all, $h > 0$ is fixed and positive, not zero. How to make h infinitesimal? In other words, how fast does f change at x itself?

On our graph, look at the point $(x, f(x))$. Through it, draw the tangent to the graph (Figure 26.2). Its slope is what we want: the derivative of f at x, denoted by $f'(x)$.

Still, this approach is too geometrical. How to calculate this in practice? For this, we must define the derivative not only geometrically but also analytically.

26.2.5 Derivative: Limit of Slopes of Strings

So far, our string got shorter and shorter, approaching the tangent geometrically. This is nice and visual, but not so useful. How to calculate the slope of the tangent? Better work the other way around: for each individual string, calculate its own slope (the divided difference). Then, let h approach zero. This way, the strings will approach the tangent, so their slopes will approach the slope of the tangent, as required. This will be the derivative of f at x:

$$f'(x) \equiv \lim_{h \to 0} \frac{f(x+h) - f(x)}{h}.$$

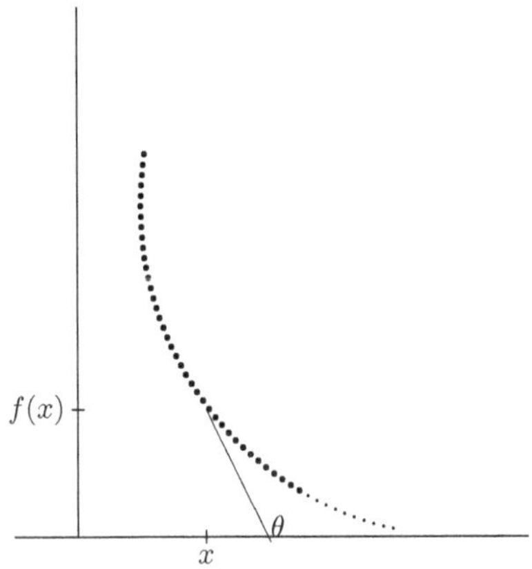

Fig. 26.2. At the point $(x, f(x))$, draw the tangent to the graph of f. Its slope is the derivative of f at x: $f'(x) \equiv \tan(\theta)$. For instance, in this picture, f falls: $f'(x) < 0$.

26.2.6 Limit: Small Error

If this limit exists, then it deserves a new name: $f'(x)$. In this case, we say that f is differentiable at x, and that its derivative at x is $f'(x)$.

How to make sure that $f'(x)$ is indeed a legitimate limit? This is a game between you and me. You pick an $\varepsilon > 0$ (as small as you like), and challenge me to make the error as small as ε. In other words, I have to supply a $\delta > 0$ (as small as I like) for which

$$|h| < \delta \Rightarrow \left| f'(x) - \frac{f(x+h) - f(x)}{h} \right| < \varepsilon.$$

[My δ may depend on your ε: $\delta \equiv \delta(\varepsilon)$.] If we can go on playing the game without fail, then the limit indeed exists, and is denoted by $f'(x)$.

26.2.7 The Second Derivative

If $f'(x)$ exists at every $x \in \Omega$, then the derivative is a real function in its own right:

$$f' : \Omega \to \mathbb{R}.$$

At each individual $x \in \Omega$, we may now ask: is f' differentiable at x? If it is, then it has its own derivative at x:

$$f'' \equiv (f')'.$$

This is the second derivative of f: the derivative of the derivative. Better yet, the nth derivative of f (if exists) is defined by mathematical induction:

$$f^{(0)} \equiv f$$
$$f^{(n+1)} \equiv \left(f^{(n)}\right)' \qquad n \geq 0.$$

(Here, n is the order of the derivative.) Let's see a few examples.

26.3 Examples

26.3.1 The Hyperbolic Function

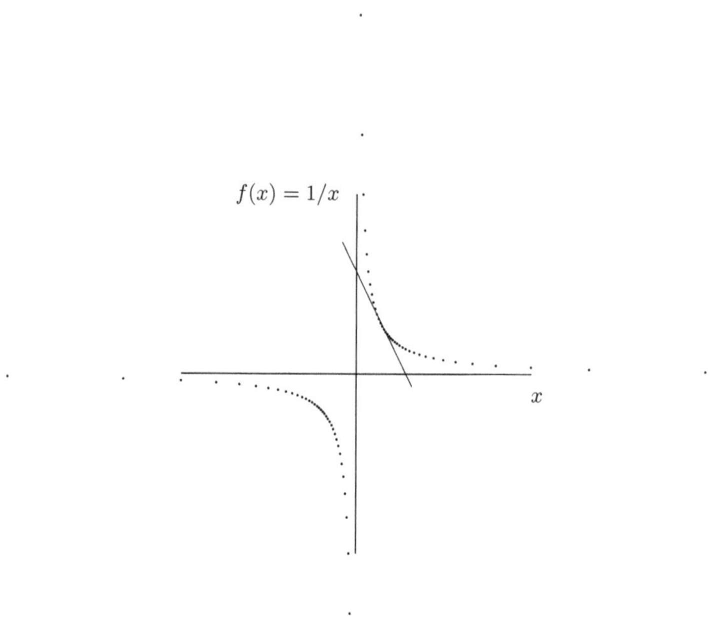

Fig. 26.3. The hyperbolic function $f(x) = 1/x$ is defined at all $x \neq 0$. The graph of this function has a steep tangent: its slope is as sharp as $f'(x) = -1/x^2$.

Consider the following domain:

$$\Omega \equiv \mathbb{R} \setminus \{0\}$$

(the whole real axis, but the origin). It is open: it has no endpoint (since 0 is excluded). In it, define the hyperbolic function:

$$f(x) \equiv \frac{1}{x} \equiv x^{-1}$$

(Figure 26.3). What is its derivative? Well, pick an $x \neq 0$, and a $h > 0$. Look at the divided difference (the slope of the string):

$$\frac{1}{h}\left(\frac{1}{x+h} - \frac{1}{x}\right) = \frac{x-(x+h)}{h(x+h)x} = \frac{-1}{(x+h)x} \to_{h \to 0} \frac{-1}{x^2}.$$

This is the derivative (the slope of the tangent):

$$(x^{-1})' = -x^{-2} \quad (x \neq 0).$$

Next, let's differentiate a few trigonometric functions. For this purpose, we need complex numbers.

26.4 Trigonometric Functions – Complex Numbers

26.4.1 Sine, Cosine, and Polar Decomposition

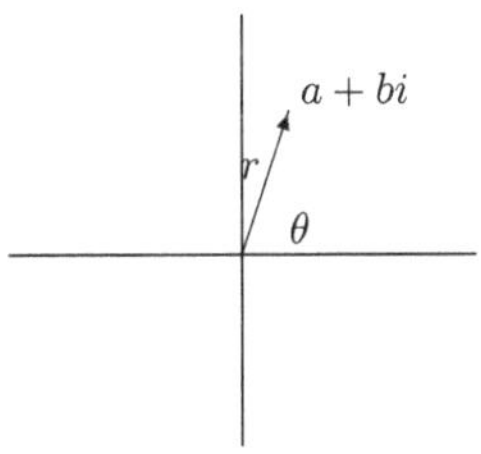

Fig. 26.4. The complex number $a + bi$ has a polar decomposition: $r(\cos(\theta) + \sin(\theta)i)$ (where $r^2 = a^2 + b^2$, and θ is the angle from the real axis)

To differentiate the sine and cosine functions, we need complex numbers. Fortunately, we already introduced them, and their arithmetic operations (Figure 3.17). Let's see their polar decomposition as well (Figure 26.4). This will give us some useful formulas.

In what follows, i stands for the imaginary number:

$$i^2 = -1.$$

In the complex plane, what is the unit circle? Well, it contains those complex numbers of absolute value 1 (or radius 1: distance 1 from the origin):

$$\text{unit circle} = \left\{a + bi \mid a^2 + b^2 = 1\right\}.$$

On the unit circle, consider two points:

- $\cos(\theta) + \sin(\theta)i$, which makes angle θ with the real axis,

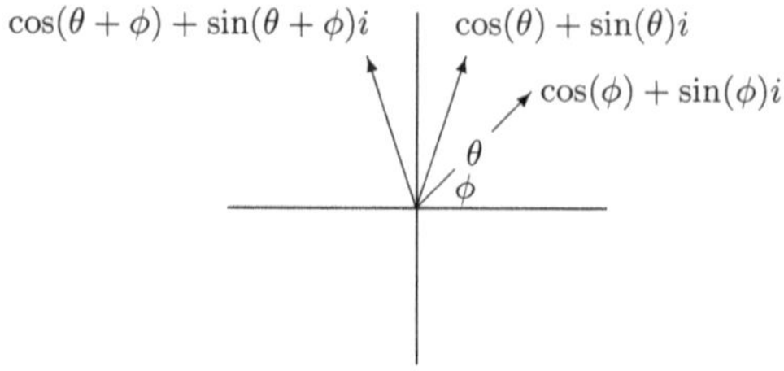

Fig. 26.5. Start from two complex numbers: $\cos(\theta) + \sin(\theta)i$, and $\cos(\phi) + \sin(\phi)i$. Add their angles, and you'd get their product: $\cos(\theta + \phi) + \sin(\theta + \phi)i$.

- and $\cos(\phi) + \sin(\phi)i$, which makes angle ϕ with the real axis

(Figure 26.5). What is their product? It is on the unit circle too, but with a new angle: $\theta+\phi$. Still, how to use this product in practice?, Use linearity: open parentheses, using the distributive and commutative laws:

$$\begin{aligned}&\cos(\theta + \phi) + \sin(\theta + \phi)i\\&= (\cos(\theta) + \sin(\theta)i)(\cos(\phi) + \sin(\phi)i)\\&= (\cos(\theta)\cos(\phi) - \sin(\theta)\sin(\phi)) + (\sin(\theta)\cos(\phi) + \cos(\theta)\sin(\phi))i.\end{aligned}$$

Now, look at the real part alone:

$$\cos(\theta + \phi) = \cos(\theta)\cos(\phi) - \sin(\theta)\sin(\phi).$$

Likewise, look at the imaginary part alone:

$$\sin(\theta + \phi) = \sin(\theta)\cos(\phi) + \cos(\theta)\sin(\phi).$$

Better yet, the real part has another version too: substitute $-\phi$ for ϕ:

$$\cos(\theta - \phi) = \cos(\theta)\cos(-\phi) - \sin(\theta)\sin(-\phi) = \cos(\theta)\cos(\phi) + \sin(\theta)\sin(\phi).$$

Add both versions:

$$\cos(\theta + \phi) + \cos(\theta - \phi) = 2\cos(\theta)\cos(\phi).$$

Likewise, the imaginary part has another version too: substitute $-\phi$ for ϕ:

$$\sin(\theta - \phi) = \sin(\theta)\cos(-\phi) + \cos(\theta)\sin(-\phi) = \sin(\theta)\cos(\phi) - \cos(\theta)\sin(\phi).$$

Add both versions:

$$\sin(\theta + \phi) + \sin(\theta - \phi) = 2\sin(\theta)\cos(\phi).$$

Let's use these formulas.

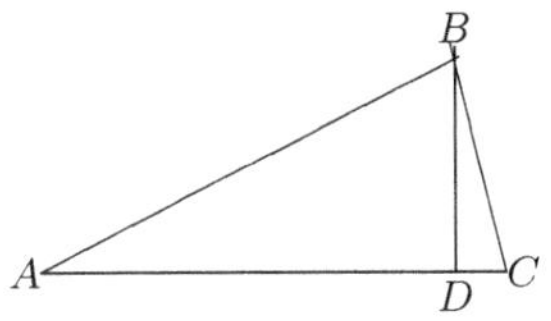

Fig. 26.6. $\triangle ABC$ is isosceles: $AB = AC$. In it, draw a height: BD (vertical to AC): $\angle ADB = 90°$. Now, since $\angle CBD$ is twice as small as $\angle BAC$, it also has a smaller tangent: $\tan(\angle CBD) < \tan(\angle BAC)$.

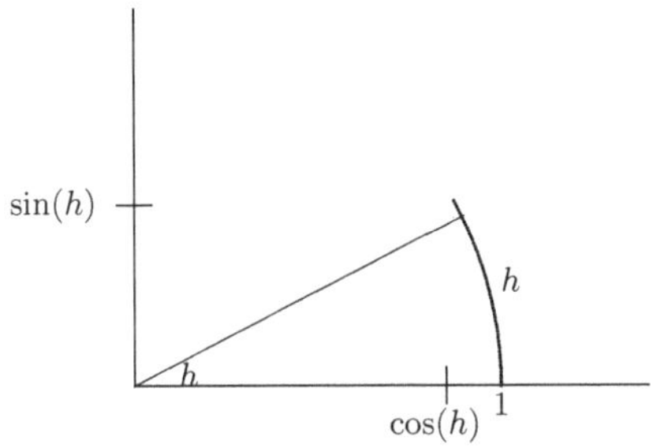

Fig. 26.7. A small angle (or short arc) h. Note that $1-\cos(h) \ll \sin(h)$. Thus, $\sin(h)/h \to 1$.

26.4.2 Trigonometric Functions and Their Derivative

Thanks to these formulas, we can now differentiate the sine and cosine functions. For this purpose, note that, as h approaches 0, $\cos(h)$ approaches 1:

$$\cos(h) \to_{h\to 0} \cos(0) = 1.$$

How fast? Superlinear: even faster than $h \to 0$. Indeed, look at the isosceles triangle $\triangle ABC$ in Figure 26.6:

$$AB = AC \quad \text{and} \quad \angle ABC = \angle ACB.$$

In this triangle, the head angle (vertexed at A) can be written as

$$\angle BAC = 180° - \angle ABC - \angle ACB = 2\,(90° - \angle ACB) = 2\angle CBD > \angle CBD.$$

Since the tangent function is monotonically increasing, it keeps the same inequality:

$$\tan\left(\angle BAC\right) > \tan\left(\angle CBD\right).$$

Now, embed $\triangle ABC$ in the unit circle (Figure 26.7). This way, our inequality takes a new form:

$$\frac{1-\cos(h)}{\sin(h)} = \tan(\angle CBD) < \tan(\angle BAC) = \frac{\sin(h)}{\cos(h)} = \tan(h) \rightarrow_{h\rightarrow 0} 0.$$

Let's use this to prove that $\cos(h)$ approaches 1 superlinearly: faster than $h \rightarrow 0$. For this purpose, we only need to replace $\sin(h)$ by h, or show that both approach zero at the same rate.

Indeed, thanks to the above estimate, we can also compare $\sin(h)$ to h (the length of the arc in Figure 26.7). On one hand, $\sin(h)$ is clearly shorter than this arc:

$$\sin(h) \le h.$$

On the other hand, $\sin(h)$ is also bounded from below. Indeed, as can be seen in Figure 26.7,

$$h \le \sin(h) + (1-\cos(h)) \le \sin(h)\left(1+\tan(h)\right),$$

or

$$\sin(h) \ge \frac{h}{1+\tan(h)}.$$

Upon dividing by h, we obtain

$$1 \ge \frac{\sin(h)}{h} \ge \frac{1}{1+\tan(h)} \rightarrow_{h\rightarrow 0} \frac{1}{1+0} = 1,$$

so

$$\frac{\sin(h)}{h} \rightarrow_{h\rightarrow 0} 1.$$

This is what we wanted:

$$\frac{1-\cos(h)}{h} = \frac{1-\cos(h)}{\sin(h)} \cdot \frac{\sin(h)}{h} \rightarrow_{h\rightarrow 0} 0\cdot 1 = 0,$$

meaning that $\cos(h)$ indeed approaches 1 superlinearly: faster than $h \rightarrow 0$. We are now ready to differentiate the sine function:

$$\begin{aligned}
&\frac{\sin(x+h)-\sin(x)}{h} \\
&= \frac{\sin(x)\cos(h)+\cos(x)\sin(h)-\sin(x)}{h} \\
&= \sin(x)\frac{\cos(h)-1}{h} + \cos(x)\frac{\sin(h)}{h} \\
&\rightarrow_{h\rightarrow 0} \cos(x).
\end{aligned}$$

This gives us the derivative

$$\sin'(x) = \cos(x).$$

Likewise, differentiate the cosine function as well:

$$\begin{aligned}
&\frac{\cos(x+h)-\cos(x)}{h} \\
&= \frac{\cos(x)\cos(h)-\sin(x)\sin(h)-\cos(x)}{h} \\
&= \cos(x)\frac{\cos(h)-1}{h} - \sin(x)\frac{\sin(h)}{h} \\
&\rightarrow_{h\rightarrow 0} -\sin(x).
\end{aligned}$$

This gives us the derivative

$$\cos'(x) = -\sin(x).$$

26.5 Powers and Their Derivative

26.5.1 Using The Binomial Formula

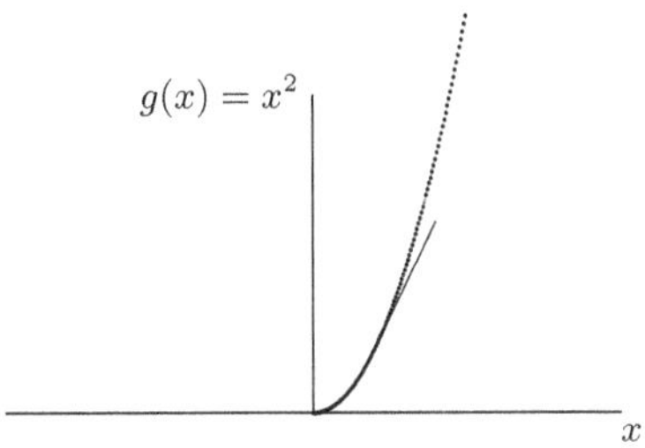

Fig. 26.8. The quadratic function $g(x) = x^2$ has a familiar graph: a parabola. Its tangent has a linear slope: $g'(x) = 2x$.

Assume now that f is a power of x:

$$f(x) \equiv x^n,$$

where $n > 0$ is a fixed natural number. (For $n = 2$, this is a parabola: Figure 26.8.) To differentiate f at x, use Newton's binomial formula (Section 6.5.1):

$$\begin{aligned}
\frac{f(x+h) - f(x)}{h} &= \frac{(x+h)^n - x^n}{h} \\
&= \frac{\sum_{i=0}^{n} \binom{n}{i} x^i h^{n-i} - x^n}{h} \\
&= \frac{\sum_{i=0}^{n-1} \binom{n}{i} x^i h^{n-i}}{h} \\
&= \sum_{i=0}^{n-1} \binom{n}{i} x^i h^{n-i-1} \\
&\to_{h \to 0} \binom{n}{n-1} x^{n-1} \\
&= n x^{n-1}.
\end{aligned}$$

Thus, we have the derivative

$$(x^n)' = n x^{n-1}.$$

Not convinced? Later on, we'll prove this in yet another way.

26.5.2 The Constant Function

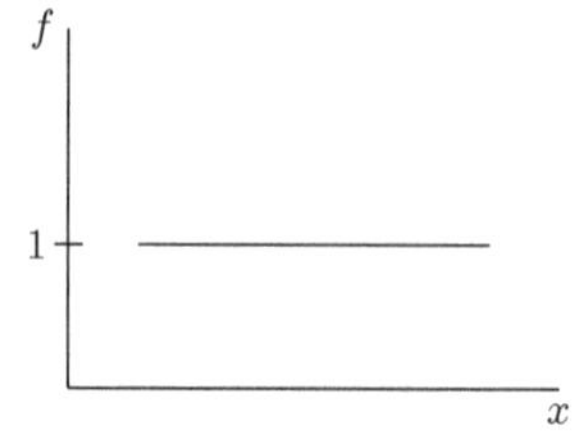

Fig. 26.9. The constant function $f \equiv 1$, and its flat graph. Its tangent has a zero slope: $f' \equiv 0$.

But didn't we forget something? What about the most elementary case: $n = 0$? This is the constant function:

$$f(x) \equiv x^0 \equiv 1.$$

How does it look like? It has a flat graph (Figure 26.9). Its tangent has a zero slope. In other words, it has zero derivative:

$$f' \equiv 0.$$

Indeed,

$$\frac{f(x+h) - f(x)}{h} = \frac{1-1}{h} = \frac{0}{h} = 0.$$

In our formula (end of Section 26.5.1), you can now set $n = 0$ too (provided that $x \neq 0$):

$$\left(x^0\right)' = 0 \cdot x^{0-1} = 0.$$

26.5.3 The Identity Function

Let's check our formula in yet another simple case: $n = 1$. This is the identity function:

$$f(x) = x^1 = x.$$

How does it look like? It has a linear graph (Figure 26.10). Its tangent has a constant slope:

$$f' \equiv 1.$$

Indeed,

$$\frac{f(x+h) - f(x)}{h} = \frac{x+h-x}{h} = \frac{h}{h} = 1.$$

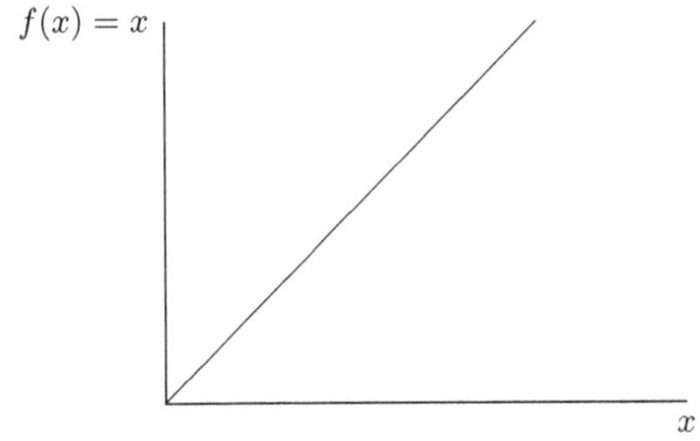

Fig. 26.10. The identity function $f(x) = x$, and its linear graph. Its tangent has a constant slope: $f' \equiv 1$.

This agrees with our formula (with $n = 1$):

$$\left(x^1\right)' = 1 \cdot x^{1-1} = 1 \cdot x^0 = 1.$$

Let's prove our formula in yet another way, using a new principle: Leibniz rule.

26.6 Leibniz (Product) Rule

26.6.1 Product of Two Functions

Let f and g be two given functions. Consider an x for which both $f(x)$ and $g(x)$ are well-defined. The product function fg is defined by

$$(fg)(x) = f(x)g(x).$$

Furthermore, if both f and g are differentiable at x, then fg is differentiable as well, and its derivative is:

$$(fg)' = f'g + fg'.$$

Indeed,

$$\begin{aligned}
&\frac{(fg)(x+h) - (fg)(x)}{h} \\
&= \frac{f(x+h)g(x+h) - f(x)g(x)}{h} \\
&= \frac{f(x+h)g(x+h) - f(x)g(x+h) + f(x)g(x+h) - -f(x)g(x)}{h} \\
&= \frac{f(x+h)g(x+h) - f(x)g(x+h)}{h} + \frac{f(x)g(x+h) - -f(x)g(x)}{h} \\
&= g(x+h)\frac{f(x+h) - f(x)}{h} + f(x)\frac{g(x+h) - -g(x)}{h} \\
&\to_{h\to 0} f'(x)g(x) + f(x)g'(x).
\end{aligned}$$

This way, given f, g, f', and g' at x, you can calculate fg and $(fg)'$.

26.6.2 Leibniz Rule and Mathematical Induction

Thanks to Leibniz rule, we have a new way to prove our formula:

$$(x^n)' = nx^{n-1}, \qquad n \geq 0.$$

(For $n = 0$, this is valid only for $x \neq 0$.) Indeed, by mathematical induction: for $n = 0$, differentiate as in Section 26.5.2:

$$(x^0)' = 1' = \lim_{h \to 0} \frac{1 - 1}{h} = \lim_{h \to 0} \frac{0}{h} = 0.$$

This agrees with our formula. Likewise, for $n = 1$, differentiate as in Section 26.5.3:

$$x' = \lim_{h \to 0} \frac{x + h - x}{h} = \lim_{h \to 0} \frac{h}{h} = \lim_{h \to 0} 1 = 1.$$

This agrees with our formula as well. Next, assume that the formula holds for some $n \geq 1$ (the induction hypothesis). How to prove it for $n + 1$ as well (the induction step)? Just differentiate the product $x^{n+1} = x^n x$:

$$\left(x^{n+1}\right)' = (x^n x)' = (x^n)'x + x^n = (nx^{n-1})x + x^n = nx^n + x^n = (n + 1)x^n,$$

as required. This completes the induction step, and indeed the entire proof.

26.6.3 Differentiation: A Linear Operator

Differentiation (or derivation) is an additive operator: you can open parentheses:

$$(f + g)' = f' + g'.$$

Indeed,

$$\begin{aligned}
(f + g)' &= \lim_{h \to 0} \frac{f(x + h) + g(x + h) - f(x) - g(x)}{h} \\
&= \lim_{h \to 0} \left(\frac{f(x + h) - f(x)}{h} + \frac{g(x + h) - g(x)}{h} \right) \\
&= \lim_{h \to 0} \frac{f(x + h) - f(x)}{h} + \lim_{h \to 0} \frac{g(x + h) - g(x)}{h} \\
&= f' + g'.
\end{aligned}$$

Furthermore, differentiation is also multiplicative: for any constant number a, you can pull a out of parentheses:

$$(ag)' = ag'.$$

Indeed, in Leibniz rule, let f be the constant function $f \equiv a$, which has a zero derivative:

$$(ag)' = a'g + ag' = ag'.$$

Thus, differentiation is a linear operator: additive and multiplicative. Let's use this.

26.6.4 Using The Binomial Coefficients

By now, we saw that differentiation is linear: additive and multiplicative. Thanks to this, we can now calculate the second derivative of x^n:

$$(x^n)'' = (nx^{n-1})' = n(x^{n-1})' = n(n-1)x^{n-2}.$$

This is valid for $n \geq 2$. For $n < 2$, on the other hand, this still makes sense, provided that $x \neq 0$. Thus,

$$(x^n)'' = \begin{cases} n(n-1)x^{n-2} & \text{if } n \geq 2 \\ 0 & \text{if } \ 0 \leq n < 2. \end{cases}$$

Likewise, differentiate more and more, and calculate the kth derivative as well:

$$(x^n)^{(k)} = \begin{cases} \frac{n!}{(n-k)!}x^{n-k} & \text{if } n \geq k \geq 0 \\ 0 & \text{if } \ 0 \leq n < k. \end{cases}$$

Better yet, thanks to the notation in Section 6.6.1, this can also be written as

$$(x^n)^{(k)} = C_{n,k}x^{n-k}.$$

This will be used later.

26.6.5 Mirroring The Binomial Formula

By now, we studied the power function x^n, and calculated its high-order derivatives. For this purpose, we used the binomial coefficients. Next, look again at our functions: f and g. Assume that, at some x, they are available, along with k derivatives. What about fg? What are its derivatives? They mirror the binomial formula:

$$(fg)^{(k)} = \sum_{i=0}^{k} \binom{k}{i} f^{(i)}g^{(k-i)}.$$

After all, to differentiate fg k times, one must differentiate f i times and g $k-i$ times ($0 \leq i \leq k$). Moreover, it doesn't matter whether f is differentiated before or after g (mirroring Section 6.5.1).

26.6.6 Mirroring The Trinomial Formula

Assume now that there is yet another function u, available (along with k derivatives) at the same point: x. Consider now the triple product: fgu. What is its kth derivative at x? For this purpose, mirror the trinomial formula (Section 6.6.3):

$$(fgu)^{(k)} = \sum_{0 \leq i,j,l \leq k, \ i+j+l=k} \frac{k!}{i!j!l!} f^{(i)}g^{(j)}u^{(l)}.$$

In this sum, how many terms are there? Thanks to Section 6.3.5, we already know the answer:

$$\binom{k+3-1}{3-1} = \binom{k+2}{2} = \frac{(k+1)(k+2)}{2}.$$

26.7 Composite Function and Its Derivative

26.7.1 Composite Function

So far, f and g had the same status: they made the product fg. Next, let's compose them on top of each other. For this purpose, let g be a given function, well-defined (and differentiable) at x. Likewise, assume that f is well-defined (and differentiable) at $g(x)$. This way, we can now define the composite function $f \circ g$:

$$(f \circ g)(x) \equiv f(g(x)).$$

What is its derivative at x? To calculate it, define a new limit process:

$$\triangle g \equiv g(x+h) - g(x) \to_{h\to 0} 0.$$

26.7.2 Derivative of The Composite Function

Thanks to this, we can now write the divided difference for $f \circ g$:

$$\begin{aligned}
&\frac{(f \circ g)(x+h) - (f \circ g)(x)}{h} \\
&= \frac{f(g(x+h)) - f(g(x))}{h} \\
&= \frac{f(g(x+h)) - f(g(x))}{g(x+h) - g(x)} \cdot \frac{g(x+h) - g(x)}{h} \\
&= \frac{f(g(x) + \triangle g) - f(g(x))}{\triangle g} \cdot \frac{g(x+h) - g(x)}{h} \\
&\to_{h\to 0} f'(g(x))g'(x).
\end{aligned}$$

Here, g is differentiated at x, and f at $g(x)$, not x. This is why g' is evaluated at x, and f' at $g(x)$, not x. Thus,

$$(f \circ g)'(x) = f'(g(x))g'(x).$$

Let's rewrite this in a more informative way, to remind us of the reasoning behind it.

26.7.3 The Differentiation Operator

The differentiation operator maps the original function to its derivative:

$$f \to f'.$$

In Section 26.6.3, we already saw that this is a linear operator. Let's give it a more informative name: d/dx:

$$\frac{df}{dx} \equiv \frac{d}{dx}(f) \equiv f'.$$

Here, d/dx is neither a fraction nor a ratio, but just a new notation. Why is it better than the prime used so far? Because it reminds us of two important things:

- The differentiation is with respect to x.
- The derivative is the limit of the divided differences.

How to make a high-order derivative? Apply our operator n times:

$$\frac{d^n f}{dx^n} \equiv \left(\frac{d}{dx}\right)^n (f) \equiv f^{(n)}.$$

26.7.4 Differentiating The Composite Function

Let's use our new notation in a composite function. This is a little more complicated: f depends on x indirectly: f depends on g, which depends on x. Thus, to differentiate f with respect to x, one should actually differentiate $f \circ g$:

$$\frac{df}{dx} \equiv \frac{d(f \circ g)}{dx} = \frac{df}{dg} \cdot \frac{dg}{dx},$$

where dg/dx is evaluated at x, and df/dg at $g(x)$.

Don't forget: this is differentiation, not division. Still, it came from divided differences, as in the original proof in Section 26.7.2. This is why this formula is easier to remember: the dg in the numerator offsets the dg in the denominator (mirroring $\triangle g$ in the original proof).

26.8 The Inverse Function and Its Derivative

26.8.1 The Inverse Function

Assume that g is a monotone function, so g' never vanishes. (In Figure 26.11, for example, g' is always negative.) Now, in our composite function pick

$$f = g^{-1}.$$

This is *not* the reciprocal, but the inverse mapping: it undoes g, and maps back to x:

$$(f \circ g)(x) = (g^{-1} \circ g)(x) = g^{-1}(g(x)) = x.$$

This way, the composition $g^{-1} \circ g$ is just the identity function that does nothing.

How to analyze g^{-1} without calculating it explicitly? For this purpose, look at Figure 26.12. It mirrors Figure 26.11, and looks at things the other way around. Indeed, the axes have new roles: the vertical axis stands now for the independent variable y, and the horizontal axis for the dependent variable: $g^{-1}(y)$. Later on, we'll use this geometrically: reciprocal slope. Here, however, we still work algebraically.

26.8.2 Differentiating The Inverse Function

In the above equation. differentiate both sides. The right-hand side is easy enough to differentiate: the identity function has derivative 1 (Figure 26.10). To differentiate the left-hand side, on the other hand, we need to work a little more, and use our formula (Section 26.7.2):

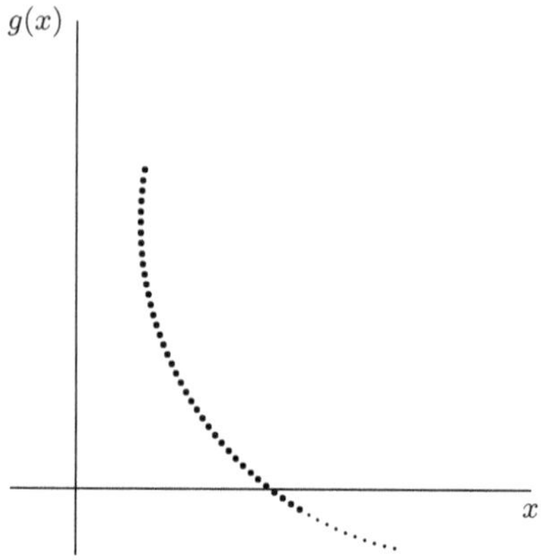

Fig. 26.11. A monotone function g: g' never vanishes.

$$\left(g^{-1}\right)' g' = 1.$$

But remember: these derivatives are *not* at the same point! g' is at x (as usual), but $(g^{-1})'$ is at $g(x)$, not x. Let's make this clear:

$$\left(g^{-1}\right)' (g(x)) = \frac{1}{g'(x)}.$$

Let's see this geometrically too.

26.8.3 Geometrical Point of View

How to see this geometrically? Draw the graph of g^{-1}. How to do this easily? Look at the original graph of g (Figure 26.11). Rotate it by 90° (counterclockwise). This way, the x-axis gets vertical, and the y-axis horizontal (pointing leftwards, not rightwards).

In this rotation, what happens to the slope of the tangent? It gets reciprocal! This agrees with our formula.

26.8.4 Example: Square Root

For example, consider the quadratic function:

$$g(x) = x^2, \quad x > 0$$

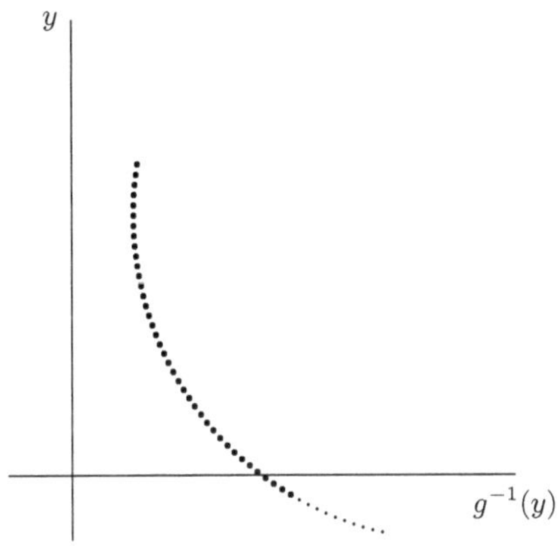

Fig. 26.12. The inverse function g^{-1}. (Don't confuse it with the reciprocal). How to draw it? In the original graph of g, interchange the dependent and independent variables.

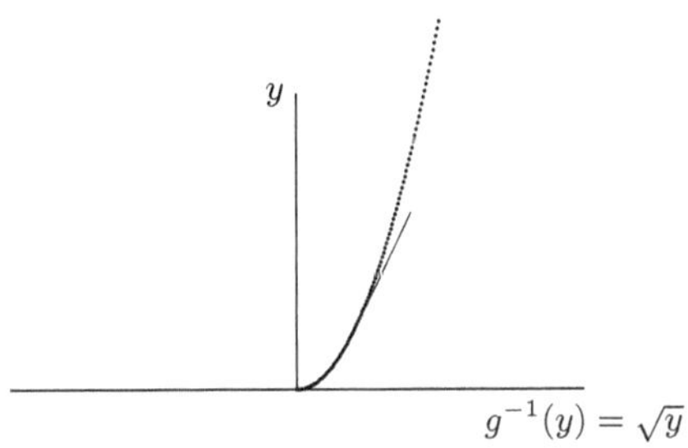

Fig. 26.13. Look at the quadratic function: $y = g(x) = x^2$. Interchange the dependent and independent variables. This is a new function: the square root. The slope of the string is now $h/\triangle g$: the reciprocal of $\triangle g/h$, used in the original derivative.

(Figure 26.8). Its inverse is the square root:

$$g^{-1}(y) = \sqrt{y}, \quad y > 0.$$

No need to draw its graph. In Figure 26.13, we got it for free, by looking at the axes the other way around: y serves now as the independent variable, and $x = \sqrt{y}$ as the dependent variable. The slope is now the cotangent: the reciprocal of the original slope in Figure 26.8. Indeed, in each string, the original slope was $\triangle g/h$, but the new slope is $h/\triangle g$.

26.8.5 The Tangent Point

In Figure 26.8, the tangent point is at $(x, g(x))$. In Figure 26.13, the tangent point is still there, but has a new name:

$$\left(g^{-1}(y), y\right) = (x, g(x)).$$

Our independent variable is now y, not x. At what y should the derivative be evaluated? At $y = g(x)$, as in our original formula.

26.8.6 Differentiating The Square Root

In our example, g is the quadratic function. Its derivative is

$$\left(x^2\right)' = 2x.$$

And what is the inverse function? This is the square root. What is its derivative? Well, at $y = x^2$, it is the reciprocal:

$$(\sqrt{y})' = \frac{1}{2x} = \frac{1}{2\sqrt{y}} = \frac{1}{2}y^{-1/2}.$$

This agrees with our formula (Section 26.5.1, with $n = 1/2$). Let's extend this even more.

26.9 Application: Power Functions

26.9.1 Inverse Powers

So far, how did we use the composite function $f \circ g$? Well, we picked $f = g^{-1}$, to help study the inverse function. Next, let's pick something else:

$$g(x) = x^n, \quad \text{and} \quad f(x) = \frac{1}{x}$$

(for some natural number n, and a nonzero $x \neq 0$). This way,

$$(f \circ g)(x) = f(g(x)) = \frac{1}{x^n} = x^{-n}.$$

Thanks to Sections 26.3.1 and 26.7.2, we now have

$$\left(x^{-n}\right)' = f'(g(x))g'(x) = \frac{-1}{g^2(x)}g'(x) = \frac{-1}{x^{2n}}nx^{n-1} = -nx^{-n-1}.$$

To simplify this, substitute n for $-n$:

$$(x^n)' = nx^{n-1}$$

(where n is now negative). This extends our formula to a negative n too. Still, must n be an integer?

26.9.2 Roots

By now, we already saw that $n = 1/2$ is also allowed. Let's extend this to any other root as well. How? For any natural number n (and a positive $x > 0$), set now

$$g(x) = x^n.$$

This is a monotone function. What is its inverse? This is the nth root:

$$g^{-1}(y) = +y^{1/n}.$$

(In Figure 26.13, $n = 2$ was used.) Fortunately, we already know how to differentiate the inverse function:

$$\left(g^{-1}\right)'(g(x)) = \frac{1}{g'(x)}.$$

In our case, $y = g(x) = x^n$, so

$$\left(y^{1/n}\right)' = \frac{1}{nx^{n-1}} = \frac{1}{n}x^{1-n} = \frac{1}{n}y^{(1-n)/n} = \frac{1}{n}y^{1/n-1}.$$

To simplify this, substitute x for y, and α for $1/n$:

$$(x^\alpha)' = \alpha x^{\alpha-1}$$

(where α is now even more general than before: either an integer number, or even the reciprocal of a natural number). Let's extend this to any rational α as well.

26.9.3 Rational Powers

By now α is a rational number: of numerator 1. How to extend this to any rational α, with any numerator? For instance, look at two natural numbers: n and m. Now, for $x > 0$, define

$$g(x) = x^{1/m}, \quad \text{and} \quad f(x) = x^n.$$

What is their composition? It is a rational power:

$$(f \circ g)(x) = f(g(x)) = \left(x^{1/m}\right)^n = x^{n/m}.$$

Fortunately, we already know how to differentiate a composite function:

$$\begin{aligned}(f \circ g)'(x) &= f'(g(x))g'(x) \\ &= n(x^{1/m})^{n-1} \cdot \frac{1}{m}x^{1/m-1} \\ &= \frac{n}{m}x^{n/m-1/m+1/m-1} \\ &= \frac{n}{m}x^{n/m-1}.\end{aligned}$$

To simplify this, substitute α for n/m:

$$(x^\alpha)' = \alpha x^{\alpha-1}$$

(for every positive rational α). To extend this to a negative α as well, use the same trick as in Section 26.9.1. Could α be even irrational? To prove this, we need the exponent function.

26.10 The Exponent Function

26.10.1 Infinite Series

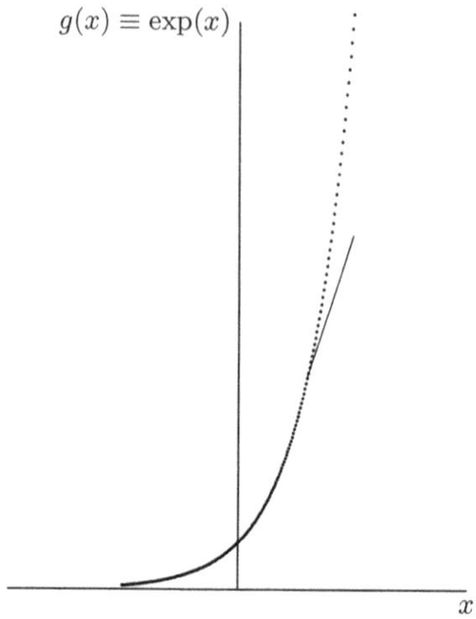

Fig. 26.14. The exponent function: $g(x) \equiv \exp(x) \equiv e^x$. What is the slope of its tangent? This is the same function: $g'(x) = \exp(x)$ as well.

A rational number can be written as a ratio of two natural numbers: n/m. An irrational number, on the other hand, can't. Here are a few examples: $\sqrt{2}$, $\sqrt{3}$, π, and e (the natural exponent). How to define e? There are two (equivalent) ways, each with its own advantages and applications. To see this, better look at e in a wider context: the exponent function $\exp(x)$ or e^x (Figure 26.14). To define it, use an infinite series:

$$\exp(x) \equiv \sum_{i=0}^{\infty} \frac{x^i}{i!}.$$

In this series, let's look at the individual terms, and see how fast they decay.

26.10.2 Individual Terms: Fast Decay

In this series, let x be fixed, and look at the individual terms (as a discrete function of i). As i grows, how fast do they decay (in magnitude)? Very fast: as fast as any (geometrical) power series of the form $\sum q^i$ (even with $q > 0$ as small as you like: Figure 26.15). To see this, write the ith term in terms of the previous term:

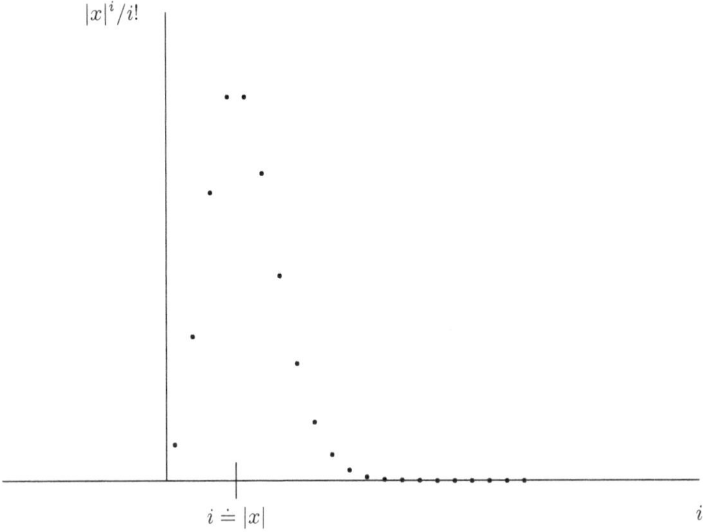

Fig. 26.15. The exponent function is an infinite series. For a fixed x, here are 21 leading terms (in magnitude): $1,\ |x|,\ |x|^2/2,\ \ldots,\ |x|^{20}/20!$. As i grows, they decay.

$$\frac{x^i}{i!} = \frac{x}{i} \cdot \frac{x^{i-1}}{(i-1)!}, \qquad i \geq 1.$$

In other words, the ith term is x/i times the previous term. Now, even for a very big $|x|$, you can still pick a k as big as

$$k \geq \frac{|x|}{q}.$$

From this k onwards, the individual terms use $i \geq k$, so

$$\frac{|x|}{i} \leq \frac{|x|}{k} \leq q.$$

Thus, from this k onward, each term is as small as q times the previous term. Thus, our terms approach zero quite fast (in magnitude):

$$\left|\frac{x^i}{i!}\right| = \frac{|x|^i}{i!} \to_{i\to\infty} 0.$$

26.10.3 The Tail: Fast Decay

Better yet, look at the entire tail (that starts from the kth term onward). It is nearly as small:

$$
\begin{aligned}
\left|\sum_{i=k}^{\infty} \frac{x^i}{i!}\right| &\leq \sum_{i=k}^{\infty} \frac{|x|^i}{i!} \\
&\leq \frac{|x|^k}{k!} \sum_{i=0}^{\infty} \left(\frac{|x|}{k}\right)^i \\
&\leq \frac{|x|^k}{k!} \sum_{i=0}^{\infty} q^i \\
&= \frac{|x|^k}{k!} \cdot \frac{1}{1-q}.
\end{aligned}
$$

For instance, pick

$$q \equiv \frac{1}{2}.$$

Then, for each given x, pick k as big as

$$k \equiv k(|x|) \geq 2|x|.$$

This way, the entire tail is as small as twice the kth term alone (in magnitude). Thus, our series indeed converges, and our definition is legitimate.

26.10.4 New Tail: How Small Is It?

Assume now that we're given an arbitrarily small number $\varepsilon > 0$. Pick a new k, even bigger than before: so big that the individual term gets as small as

$$\frac{|x|^k}{k!} \leq \frac{\varepsilon}{6}.$$

(Since the terms decay, this is indeed possible.) Still, k must now depend not only on $|x|$ but also on ε:

$$k \equiv k(|x|, \varepsilon).$$

Thanks to our estimate, our new tail is now as small as

$$\left|\sum_{i=k}^{\infty} \frac{x^i}{i!}\right| \leq 2\frac{|x|^k}{k!} \leq \frac{\varepsilon}{3}.$$

Why do we want the tail to be so small? Because, later on, we'll split the error into three parts, each as small as $\varepsilon/3$. For this purpose, let's pick an even bigger k: so big that

$$2k^{-2} \exp(|x|) \leq \frac{\varepsilon}{3}.$$

As before, this new k may depend on both $|x|$ and ε:

$$k \equiv k(|x|, \varepsilon).$$

Fortunately, picking a bigger k doesn't spoil what we did so far. This will be useful later.

26.11 Exponent: New Definition

26.11.1 Sequence and Its Limit

Our exponent is an infinite sum: finite sums, getting longer and longer:

$$\exp(x) \equiv \sum_{i=0}^{\infty} \frac{x^i}{i!} \equiv \lim_{n\to\infty} \sum_{i=0}^{n} \frac{x^i}{i!}.$$

But this is not the only sequence. Here is another one:

$$\exp(x) \equiv \lim_{n\to\infty} \left(1 + \frac{x}{n}\right)^n.$$

26.11.2 Bank Interest

What is this? Suppose that you have some spare money, and want to invest it. Your bank offers you an annual interest of

$$x = 0.01 = 1\%.$$

Is this a good deal? Unfortunately, the interest is paid monthly: at the end of each month. So, in one year, your deposit would multiply by

$$\left(1 + \frac{x}{12}\right)^{12}.$$

Is this a good deal? Well, across the street, there is another bank, which offers the same annual interest, paid daily: at the end of each day. With this bank, after one year, your deposit would multiply by

$$\left(1 + \frac{x}{365}\right)^{365}.$$

This is more attractive, but still not enough. After all, this is your money, and it should work for you! After each second, the interest should be paid on spot, with no delay! So, if the bank were honest, then they'd offer a yet better deal: after one year, your original deposit should multiply by

$$\exp(x) \equiv \lim_{n\to\infty} \left(1 + \frac{x}{n}\right)^n.$$

26.11.3 Using The Binomial Formula

By now, we defined the exponent in two ways. Are they the same? To see this, use the binomial formula to write the latter sequence in terms of (longer and longer) sums:

$$\begin{aligned}
\left(1 + \frac{x}{n}\right)^n &= \sum_{i=0}^{n} \binom{n}{i} \frac{x^i}{n^i} \\
&= \sum_{i=0}^{n} \frac{n!}{(n-i)!i!} \cdot \frac{x^i}{n^i} \\
&= \sum_{i=0}^{n} \frac{n!}{(n-i)!n^i} \cdot \frac{x^i}{i!}.
\end{aligned}$$

Is this familiar? This is quite similar to our original series. But here, the ith term contains a new coefficient:

$$\begin{aligned}\frac{n!}{(n-i)!n^i} &= \frac{n(n-1)(n-2)\cdots(n-i+1)}{n^i}\\ &= \left(1-\frac{1}{n}\right)\left(1-\frac{2}{n}\right)\cdots\left(1-\frac{i-1}{n}\right).\end{aligned}$$

26.11.4 The Coefficient: Smaller Than 1

This coefficient is smaller than 1. Thus, if $n \geq k$, then the tail is as small as the original tail:

$$\left|\sum_{i=k}^{n}\binom{n}{i}\frac{x^i}{n^i}\right| \leq \sum_{i=k}^{n}\binom{n}{i}\frac{|x|^i}{n^i} \leq \sum_{i=k}^{\infty}\frac{|x|^i}{i!} \leq \leq \frac{\varepsilon}{3}.$$

26.11.5 Leading Terms

So, we can focus on the k leading terms, for which i is as small as

$$i < k.$$

For these, let's show that the coefficient is very close to 1.

By now, k is already picked and fixed. It is now time to pick n as well. Let's pick n as big as

$$n \geq k^4.$$

26.11.6 The Coefficient: Is It Nearly 1?

This way, the coefficient is nearly 1. How to see this? Open parentheses: from each factor, pick either 1 or a fraction: $-1/n$, or $-2/n$, or $-3/n$, etc. This way, the coefficient is now written as a new sum of products:

$$\begin{aligned}&\left(1-\frac{1}{n}\right)\left(1-\frac{2}{n}\right)\cdots\left(1-\frac{i-1}{n}\right)\\ &= 1-\left(\sum_{j=1}^{i-1}\frac{j}{n}\right)+\left(\sum_{j<l=1}^{i-1}\frac{jl}{n^2}\right)-\left(\sum_{j<l<m=1}^{i-1}\frac{jlm}{n^3}\right)+\cdots.\end{aligned}$$

How far is this from 1? Well, after 1, we have a sum that contains $i-1$ terms, each as small as

$$\frac{i}{n} < \frac{k}{n} \leq k^{-3}.$$

So, in total, this sum is as small as

$$\left|\sum_{j=1}^{i-1}\frac{j}{n}\right| < i\cdot k^{-3} < k\cdot k^{-3} = k^{-2}.$$

Next, we have a longer sum, containing as many as $(i-1)(i-2)/2$ terms, each as small as

$$\frac{i^2}{n^2} < \frac{k^2}{n^2} \le k^{-6}.$$

So, in total, this second sum is as small as

$$\left|\sum_{j<l=1}^{i-1} \frac{jl}{n^2}\right| < i^2 k^{-6} < k^2 k^{-6} = k^{-4},$$

and so on. Together, these sums are as small as a new power series:

$$\begin{aligned}
&\left|\left(\sum_{j=1}^{i-1} \frac{j}{n}\right) - \left(\sum_{j<l=1}^{i-1} \frac{jl}{n^2}\right) + \left(\sum_{j<l<m=1}^{i-1} \frac{jlm}{n^3}\right) - \cdots\right| \\
&\le \left(\sum_{j=1}^{i-1} \frac{j}{n}\right) + \left(\sum_{j<l=1}^{i-1} \frac{jl}{n^2}\right) + \left(\sum_{j<l<m=1}^{i-1} \frac{jlm}{n^3}\right) + \cdots \\
&\le \sum_{i=1}^{\infty} \left(k^{-2}\right)^i \\
&= k^{-2} \frac{1}{1-k^{-2}} \\
&\le 2k^{-2}.
\end{aligned}$$

26.11.7 Leading Terms: Small Error

Thus, in our new definition, in each of the k leading terms (for which $i < k$), the error is as small as

$$\left|\frac{n!}{(n-i)!n^i} \cdot \frac{x^i}{i!} - \frac{x^i}{i!}\right| \le 2k^{-2} \frac{|x|^i}{i!}.$$

Together, these errors are as small as

$$2k^{-2} \exp(|x|) \le \frac{\varepsilon}{3}.$$

26.11.8 Equivalent Definitions

To see that both definitions are the same, we need to show that the error is as small as you like. Well, we just did! In fact, we split the error into three parts, each as small as $\varepsilon/3$:

- The tail in the new definition.
- The total error in the k leading terms.
- The tail in the old definition.

Thus, the total error is as small as

$$\begin{aligned}
&\left|\left(1+\frac{x}{n}\right)^n - \sum_{i=0}^{\infty}\frac{x^i}{i!}\right| \\
&= \left|\sum_{i=k}^{n}\binom{n}{i}\frac{x^i}{n^i} + \sum_{i=0}^{k-1}\binom{n}{i}\frac{x^i}{n^i} - \sum_{i=0}^{k-1}\frac{x^i}{i!} - \sum_{i=k}^{\infty}\frac{x^i}{i!}\right| \\
&= \left|\sum_{i=k}^{n}\binom{n}{i}\frac{x^i}{n^i} + \sum_{i=0}^{k-1}\left(\binom{n}{i}\frac{x^i}{n^i} - \frac{x^i}{i!}\right) - \sum_{i=k}^{\infty}\frac{x^i}{i!}\right| \\
&\le \left|\sum_{i=k}^{n}\binom{n}{i}\frac{x^i}{n^i}\right| + \sum_{i=0}^{k-1}\left|\binom{n}{i}\frac{x^i}{n^i} - \frac{x^i}{i!}\right| + \left|\sum_{i=k}^{\infty}\frac{x^i}{i!}\right| \\
&\le \frac{\varepsilon}{3} + \frac{\varepsilon}{3} + \frac{\varepsilon}{3} \\
&= \varepsilon.
\end{aligned}$$

This is true for those n's that are as big as

$$n \ge k^4$$

(where k is defined as in Section 26.10.4, and depends on both $|x|$ and ε). But this could be done for every $\varepsilon > 0$ (as small as you like). So, the error must vanish, and both definitions must agree.

26.11.9 The Natural Exponent

As a special case, set $x = 1$. This is the natural exponent, denoted by $e = e^1 = \exp(1)$:

$$e \equiv \lim_{n\to\infty}\left(1+\frac{1}{n}\right)^n = \sum_{i=0}^{\infty}\frac{1}{i!}.$$

26.12 Exponent and Its Derivative

26.12.1 Uniform Convergence

Consider an $\varepsilon > 0$, as small as you like. Consider also a given x. Assume that we already worked hard to find a suitable

$$k \equiv k(|x|, \varepsilon)$$

(as in Section 26.10.4). Now, what about a smaller number y, for which

$$|y| \le |x|?$$

Can we use the same k? Fortunately, we can. After all, k satisfies the same requirements not only for the original x but also for every $y \in [-|x|, |x|]$:

•

$$k \ge 2|x| \ge 2|y|,$$

- $$\frac{|y|^k}{k!} \le \frac{|x|^k}{k!} \le \frac{\varepsilon}{6},$$

- and
$$2k^{-2}\exp(|y|) \le 2k^{-2}\exp(|x|) \le \frac{\varepsilon}{3}.$$

Thus, the same k could be picked for the entire interval $[-|x|, |x|]$. We then say that the convergence is uniform in $[-|x|, |x|]$: it has the same rate or "speed" for all y's in this interval. Better yet, pick a bigger k, suitable even for the bigger interval $[-2|x|-1, 2|x|+1]$:

$$k \equiv k(2|x|+1, \varepsilon).$$

This new k is suitable not only for our original x but also for a complete neighborhood around x. This will allow differentiation, term by term.

26.12.2 Differentiation Term by Term

In Section 26.6.3, we already saw that differentiation is additive:

$$(f+g)' = f' + g'$$

(where f and g are two differentiable functions). Still, what about three? What about four, or even more? Easy: use mathematical induction to prove additivity for as many functions as you like. Still, what about infinitely many? For example, look at the exponent function (in its original definition). This is an infinite sum of powers of x. Can it be differentiated term by term? If it can, then

$$\begin{aligned}
\exp'(x) &= \sum_{i=0}^{\infty} \left(\frac{x^i}{i!}\right)' \\
&= \sum_{i=1}^{\infty} \frac{ix^{i-1}}{i!} \\
&= \sum_{i=1}^{\infty} \frac{x^{i-1}}{(i-1)!} \\
&= \sum_{i=0}^{\infty} \frac{x^i}{i!} \\
&= \exp(x).
\end{aligned}$$

Let's prove this.

26.12.3 The Error in The Derivative

Let $\varepsilon > 0$ be fixed (as small as you like). Now, in the derivative of the exponent, look at the error (slope of string minus slope of tangent):

$$\left|\frac{\exp(x+h)-\exp(x)}{h}-\sum_{i=0}^{\infty}\frac{x^i}{i!}\right|$$
$$=\left|\sum_{i=1}^{\infty}\frac{(x+h)^i-x^i}{i!h}-\sum_{i=0}^{\infty}\frac{x^i}{i!}\right|.$$

How to make this as small as ε? In other words, how to pick h small enough? Well, this error is a difference between two infinite sums. The latter is already familiar: the exponent function itself (in its original definition). As a matter of fact, we already picked a k for which its tail is as small as $\varepsilon/3$.

26.12.4 Divided Difference: Slope of String

And what is the former sum in the error? This is the divided difference of the exponent, term by term. What about its tail? Is it also as small?

Fortunately, k was already picked big enough for the entire interval $[-2|x|-1, 2|x|+1]$, including many new numbers, denoted by

$$y_i \in [-2|x|-1, 2|x|+1].$$

In fact, each term could even use its own y_i, and the tail is still small:

$$\left|\sum_{i=k}^{\infty}\frac{y_i^i}{i!}\right| \le \sum_{i=k}^{\infty}\frac{|y_i|^i}{i!} \le \sum_{i=k}^{\infty}\frac{(2|x|+1)^i}{i!} \le \frac{\varepsilon}{3}.$$

This is uniform convergence, in its strong form. In particular, our new y_i's could be in a small neighborhood around x. Let's use them to estimate the divided differences (slope of strings).

26.12.5 String and Its Maximal Slope

Look at the ith term: a power of x. What is the slope of its string? Well, as can be seen in Figure 26.16, it is less than the slope of the tangent at $|x|+h$:

$$\begin{aligned}\left|\frac{(x+h)^i-x^i}{i!h}\right| &\le \max_{y\in[x-h,x+h]}\left|\frac{iy^{i-1}}{i!}\right| \\ &= \max_{y\in[x-h,x+h]}\left|\frac{y^{i-1}}{(i-1)!}\right| \\ &\le \frac{(|x|+h)^{i-1}}{(i-1)!}.\end{aligned}$$

26.12.6 Slope: Small Tail

Let's sum up (over $i=k+1, k+2, \ldots$). Thus, the slope has a small tail:

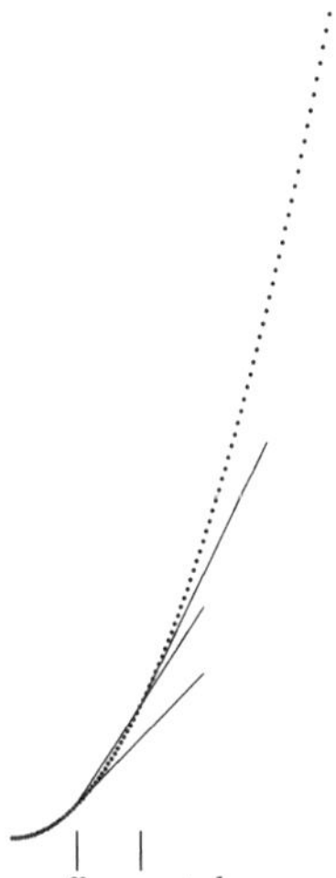

Fig. 26.16. What is the slope of the string leading from $(x, f(x))$ to $(x+h, f(x+h))$? Well, it is in between the minimal and maximal derivative in $[x, x+h]$.

$$\begin{aligned}\left|\sum_{i=k+1}^{\infty} \frac{(x+h)^i - x^i}{i!h}\right| &\le \sum_{i=k+1}^{\infty} \left|\frac{(x+h)^i - x^i}{i!h}\right| \\ &\le \sum_{i=k+1}^{\infty} \left|\frac{(|x|+h)^{i-1}}{(i-1)!}\right| \\ &= \sum_{i=k}^{\infty} \left|\frac{(|x|+h)^i}{i!}\right| \\ &\le \frac{\varepsilon}{3}.\end{aligned}$$

26.12.7 Slope: Leading Terms

So, we can now focus on the k leading terms: $1 \le i \le k$. Each is a power of x, easy to differentiate: just let the string approach the tangent at x. In other words, in the divided differences, pick h_i's so small that

$$\left|\frac{(x+h_i)^i - x^i}{i!h_i} - \frac{x^{i-1}}{(i-1)!}\right| \le \frac{\varepsilon}{3k}, \qquad 1 \le i \le k.$$

After all, thanks to uniform convergence, k was already picked and fixed. Finally, define h as the minimal h_i:

$$h \equiv \min_{1 \le i \le k} h_i.$$

26.12.8 Total Error in The Derivative

This way, in the derivative of the exponent, the total error splits into three parts (each as small as $\varepsilon/3$):

$$\begin{aligned}
&\left|\frac{\exp(x+h)-\exp(x)}{h}-\sum_{i=0}^{\infty}\frac{x^i}{i!}\right| \\
&=\left|\sum_{i=1}^{\infty}\frac{(x+h)^i-x^i}{i!h}-\sum_{i=0}^{\infty}\frac{x^i}{i!}\right| \\
&=\left|\sum_{i=k+1}^{\infty}\frac{(x+h)^i-x^i}{i!h}+\sum_{i=1}^{k}\frac{(x+h)^i-x^i}{i!h}-\sum_{i=0}^{k-1}\frac{x^i}{i!}-\sum_{i=k}^{\infty}\frac{x^i}{i!}\right| \\
&=\left|\sum_{i=k+1}^{\infty}\frac{(x+h)^i-x^i}{i!h}+\sum_{i=1}^{k}\left(\frac{(x+h)^i-x^i}{i!h}-\frac{x^{i-1}}{(i-1)!}\right)-\sum_{i=k}^{\infty}\frac{x^i}{i!}\right| \\
&\le\left|\sum_{i=k+1}^{\infty}\frac{(x+h)^i-x^i}{i!h}\right|+\sum_{i=1}^{k}\left|\frac{(x+h)^i-x^i}{i!h}-\frac{x^{i-1}}{(i-1)!}\right|+\left|\sum_{i=k}^{\infty}\frac{x^i}{i!}\right| \\
&\le\frac{\varepsilon}{3}+k\frac{\varepsilon}{3k}+\frac{\varepsilon}{3} \\
&=\varepsilon.
\end{aligned}$$

This could be done for every $\varepsilon > 0$ (as small as you like). For each such ε, a new k must be picked and fixed, good for the entire interval $[-2|x|-1, 2|x|+1]$. Thanks to uniform convergence, this is indeed possible. We can now play our ε-h game, without fail. Thus, our infinite sum

$$\exp(x)=\sum_{i=0}^{\infty}\frac{x^i}{i!}$$

is indeed the correct derivative: it was indeed a good idea to differentiate term by term in the first place. As a result, the exponent function is quite special and unique: invariant under differentiation. This is quite useful in physics.

26.12.9 Natural Logarithm

In the exercises below, we'll see that the exponent function is positive, and monotonically increasing. As such, it has an inverse function: the natural logarithm $\log(y)$. For $y > 0$, it is defined implicitly by

$$\log(\exp(x)) = x$$

(Figure 26.17). Fortunately, we already know how to differentiate an inverse function (Section 26.8.2):

$$\log'(\exp(x))=\frac{1}{\exp'(x)}=\frac{1}{\exp(x)}.$$

To simplify this, substitute x for $\exp(x)$:

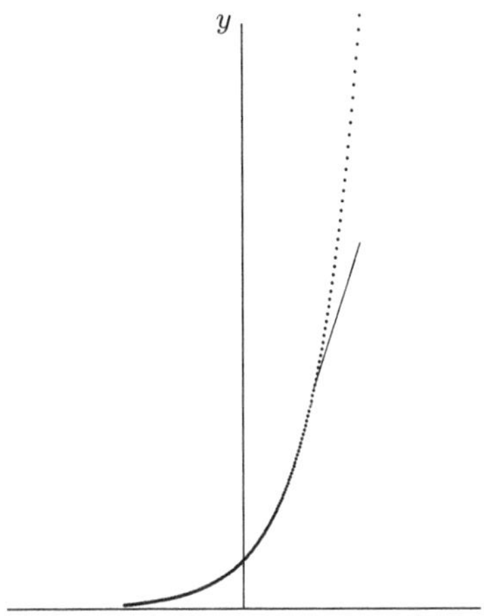

Fig. 26.17. In the graph of the exponent function, interchange dependent and independent variables. This way, the divided difference is now $h/\triangle g$: the reciprocal of the original divided difference $\triangle g/h$.

$$\log'(x) = \frac{1}{x}$$

for every positive number $x > 0$.

By now, we're quite experienced in differentiation. Let's use this in a new task: integration.

26.13 Integration

26.13.1 How to Calculate an Area?

Look at the graph of the function f. For simplicity, assume that the entire graph is positive (as in Figure 26.18). What is the area under the graph? More precisely, recall that f is defined in the open domain

$$\Omega \subset \mathbb{R}.$$

Consider a closed interval

$$[a, x] \subset \Omega.$$

Now, from the x-axis, issue two parallel verticals upwards: one from $(a, 0)$, hitting the graph at $(a, f(a))$, and another from $(x, 0)$, hitting the graph at $(x, f(x))$.

Now, look at the area in between, under the graph. It is bounded by three straight lines: a vertical from the left, another vertical from the right, and the

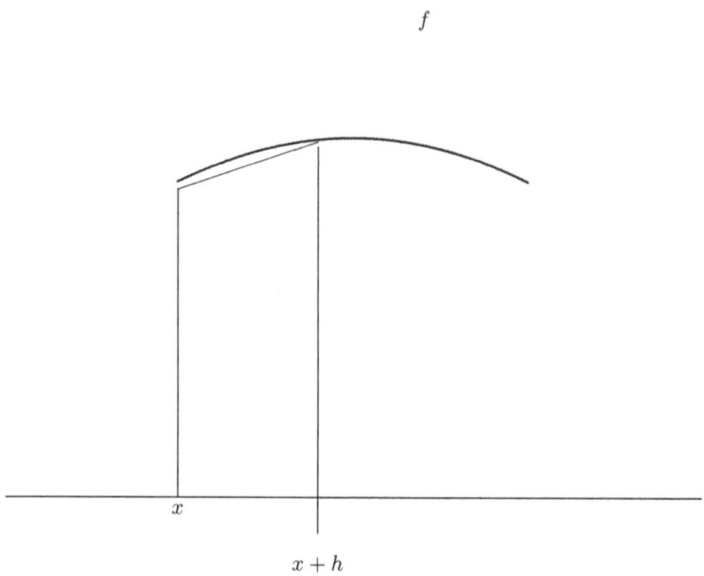

Fig. 26.18. The fundamental theorem of calculus: $f(x)$ is the derivative of $F(x) \equiv \int_a^x f(s)ds$.

horizontal x-axis from below. On top of these, it is also bounded by the graph of f from above. In particular, what is the right edge? It is the vertical leading from $(x, 0)$ to $(x, f(x))$. Thus, as x grows, the area gets bigger and bigger. How fast? To calculate this, write the area as a new function:

$$F(x) \equiv \int_a^x f(s)ds.$$

More explicitly, $F(x)$ is the limit of the area under the strings in Figure 26.18:

$$F(x) \equiv \int_a^x f(s)ds \equiv \lim_{n\to\infty} \sum_{i=0}^{n-1} h \frac{f(a+ih) + f(a+(i+1)h)}{2},$$

where

$$h \equiv \frac{x-a}{n}$$

is the width of each strip in Figure 26.18.

In our case, f is positive, so the area is positive too. In more complicated cases, the graph could go underneath the x-axis, contributing a negative area. Next, let's study the new function F, not only geometrically but also analytically.

26.13.2 The Fundamental Theorem of Calculus

As x grow, how fast does F grow? To find out, we need to differentiate F. Fortunately, the fundamental theorem of calculus tells us that

$$F'(x) = f(x).$$

Indeed, consider one strip in Figure 26.18. From above, it is bounded by the graph of f. Approximate it by a straight string. (After all, the error is as small as h^2.) This gives the divided difference, as required:

$$\begin{aligned} F'(x) &\equiv \lim_{h\to 0} \frac{F(x+h)-F(x)}{h} \\ &= \lim_{h\to 0} \frac{\int_x^{x+h} f(s)ds}{h} \\ &= \lim_{h\to 0} \frac{f(x)+f(x+h)}{2h} h \\ &= \lim_{h\to 0} \frac{f(x)+f(x+h)}{2} \\ &= f(x). \end{aligned}$$

This completes the proof. In practice, however, F is often unknown. How to uncover it?

26.13.3 Primitive Function (Antiderivative)

This is an inverse problem: to reverse-engineer. So far, we were given a function, and calculated its derivative. Now, we work the other way around: we're given the derivative, and seek the original function (which is still unknown). This is why the original function is called the primitive function (or antiderivative): an unknown function F, whose derivative is f. There may be many good candidates: we only need one.

This will help solve our practical problem: for a given function f, how to calculate the area under its graph? Easy: find one primitive function G, satisfying

$$G' = f.$$

How different is G from F? Not much:

$$(G-F)' = G' - F' = f - f = 0,$$

so

$$G - F = K$$

(a constant, as in Figure 26.9). How to find K? Set $x = a$:

$$G(a) = K + F(a) = K + \int_a^a f(s)ds = K + 0 = K.$$

This uncovers F, as required:

$$F(x) = G(x) - K = G(x) - G(a).$$

26.13.4 Definite vs. Indefinite Integral

A primitive function like G is also called an *indefinite* integral of f. Such a function is often denoted by

$$\int^x f(x)dx.$$

This is just a notation: it stands for no area as yet. To calculate an area, we must also specify a. This is why the area is denoted by

$$\int_a^x f(s)ds$$

(Section 26.13.1). This is something else: the *definite* integral of f (over the interval $[a, x]$). So, how to calculate the *definite* integral (or the area)? Just find some *indefinite* integral G. Thanks to it, the area is simply

$$\int_a^x f(s)ds = G(x) - G(a).$$

This result is often denoted by

$$\int_a^x f(s)ds = G\mid_a^x .$$

This is useful in physics.

26.14 Exercises: Taylor Series

26.14.1 Lipschitz Continuity

1. Consider two converging sequences:

$$a_n \to_{n\to\infty} a$$
$$b_n \to_{n\to\infty} b.$$

Look at the products: $a_n b_n$. What is its limit? Hint:

$$a_n b_n \to_{n\to\infty} ab.$$

To see this, let $\varepsilon > 0$ be a positive parameter (as small as you like). Pick n so big that both errors are very small:

$$|a_n - a| \le \frac{\varepsilon}{2|b| + 1}$$
$$|b_n - b| \le \frac{\varepsilon}{2|a| + 2}$$
$$|a_n| \le |a| + 1.$$

This way, the total error splits into two parts, each as small as $\varepsilon/2$:

$$\begin{aligned}
|a_nb_n - ab| &= |a_nb_n - a_nb + a_nb - ab| \\
&\le |a_nb_n - a_nb| + |a_nb - ab| \\
&= |a_n| \cdot |b_n - b| + |b| \cdot |a_n - a| \\
&\le |a_n|\frac{\varepsilon}{2|a|+2} + |b|\frac{\varepsilon}{2|b|+1} \\
&\le \frac{\varepsilon}{2} + \frac{\varepsilon}{2} \\
&= \varepsilon.
\end{aligned}$$

2. Let f be a given real function, defined in an open domain $\Omega \subset \mathbb{R}$. Assume also that f is differentiable in the closed interval

$$[a, b] \subset \Omega.$$

Show (geometrically) that, for every two points

$$x, x + h \in [a, b],$$

the divided difference is as small (in magnitude) as the maximal derivative:

$$\left|\frac{f(x+h) - f(x)}{h}\right| \le \max_{y \in [a,b]} |f'(y)|.$$

Hint: in Figure 26.16, estimate the slope of the string.
3. Multiply this by $|h|$, to obtain

$$|f(x+h) - f(x)| \le \max_{y \in [a,b]} |f'(y)| \cdot |h|.$$

4. We then say that f is Lipschitz-continuous in $[a, b]$. This property will be useful below.
5. Prove this once again, using integrals. Hint:

$$\begin{aligned}
|f(x+h) - f(x)| &= \left|\int_x^{x+h} f'(s)ds\right| \\
&\le \left|\int_x^{x+h} |f'(s)|ds\right| \\
&\le \max_{y \in [a,b]} |f'(y)| \cdot \left|\int_x^{x+h} ds\right| \\
&= \max_{y \in [a,b]} |f'(y)| \cdot |h|.
\end{aligned}$$

26.14.2 Exponent and Its Properties

1. Look at the exponent function, in its second definition:

$$\exp(x) \equiv \lim_{n \to \infty} \left(1 + \frac{x}{n}\right)^n$$

(Section 26.11.1).

2. Use it to obtain a few properties.
3. Show that
$$\exp(x) \geq 0, \qquad -\infty < x < \infty.$$
Hint: even for a negative x, pick n so big that
$$1 + \frac{x}{n} > 0.$$
4. Show that
$$\exp(0) = 1.$$
5. Show that, for any x,
$$0 \leq \exp(-|x|) \leq 1 \leq \exp(|x|).$$
6. Show that $\exp(x)$ is monotonically increasing (at least weakly). Hint: its derivative is either positive or zero:
$$\exp'(x) = \exp(x) \geq 0, \qquad -\infty < x < \infty.$$
(Later on, we'll even show that this is positive.)
7. Show that
$$\exp(x + y) = \exp(x)\exp(y),$$
where x and y are any two real (or even complex) numbers. Hint: see below.
8. Let x be a fixed number. For $n = 1, 2, 3, \ldots$, show (in two different ways) that the sequence
$$\left(1 + \frac{|x|}{n}\right)^n$$
increases monotonically with n (even strictly). In other words,
$$1 + |x| < \left(1 + \frac{|x|}{2}\right)^2 < \left(1 + \frac{|x|}{3}\right)^3 < \cdots \leq \exp(|x|).$$
Hint: use either the bank interest in Section 26.11.2, or the representation in Section 26.11.3, obtained from the binomial formula.
9. Let n be a fixed natural number. Let x be a given number. Use the binomial formula to show that
$$\left(1 + \frac{x}{n}\right)^{n-1} = \sum_{i=0}^{n-1} \binom{n-1}{i} \frac{x^i}{n^i}.$$
10. Conclude that
$$\left|\left(1 + \frac{x}{n}\right)^{n-1}\right| \leq \left(1 + \frac{|x|}{n}\right)^{n-1}.$$
11. What is its primitive function? Hint: see below.

12. Differentiate the composite function
$$\left(1+\frac{x}{n}\right)^n.$$
Hint: the inner fraction has derivative
$$\left(\frac{x}{n}\right)' = \frac{1}{n}.$$
Thus, the composite function has derivative
$$\left(\left(1+\frac{x}{n}\right)^n\right)' = \frac{1}{n}n\left(1+\frac{x}{n}\right)^{n-1} = \left(1+\frac{x}{n}\right)^{n-1}.$$

13. Conclude that this function is Lipschitz-continuous: for every h satisfying $|h| < 1$,
$$\begin{aligned}\left|\left(1+\frac{x+h}{n}\right)^n - \left(1+\frac{x}{n}\right)^n\right| &\le \max_{y\in[x-1,x+1]}\left|\left(1+\frac{y}{n}\right)^{n-1}\right|\cdot|h| \\ &\le \max_{y\in[x-1,x+1]}\left(1+\frac{|y|}{n}\right)^{n-1}|h| \\ &\le \max_{y\in[x-1,x+1]}\left(1+\frac{|y|}{n}\right)^{n}|h| \\ &\le \max_{y\in[x-1,x+1]}\exp(|y|)|h| \\ &\le \exp(|x|+1)|h|.\end{aligned}$$

14. Now, let x and y be two given numbers. In the above inequality, substitute $x+y$ for x, to read
$$\left|\left(1+\frac{x+y+h}{n}\right)^n - \left(1+\frac{x+y}{n}\right)^n\right| \le \exp(|x+y|+1)|h|.$$

15. Now, in the above inequality, set
$$h \equiv \frac{xy}{n}.$$
This way, for a sufficiently large n, $|h| < 1$, as required. Conclude that
$$\left|\left(1+\frac{x+y}{n}+\frac{xy}{n^2}\right)^n - \left(1+\frac{x+y}{n}\right)^n\right| \le \exp(|x+y|+1)\left|\frac{xy}{n}\right| \to_{n\to\infty} 0.$$

16. Conclude that
$$\begin{aligned}\exp(x)\exp(y) &= \left(\lim_{n\to\infty}\left(1+\frac{x}{n}\right)^n\right)\left(\lim_{n\to\infty}\left(1+\frac{y}{n}\right)^n\right) \\ &= \lim_{n\to\infty}\left(1+\frac{x}{n}\right)^n\left(1+\frac{y}{n}\right)^n \\ &= \lim_{n\to\infty}\left(\left(1+\frac{x}{n}\right)\left(1+\frac{y}{n}\right)\right)^n \\ &= \lim_{n\to\infty}\left(1+\frac{x}{n}+\frac{y}{n}+\frac{xy}{n^2}\right)^n \\ &= \lim_{n\to\infty}\left(1+\frac{x+y}{n}\right)^n \\ &= \exp(x+y).\end{aligned}$$

17. Conclude that

$$\exp(x)\exp(-x) = \exp(x - x) = \exp(0) = 1.$$

18. Conclude that the reciprocal always exists:

$$\exp(-x) = \frac{1}{\exp(x)}.$$

19. Conclude that

$$\exp(x) > 0.$$

20. Conclude that $\exp(x)$ is monotonically increasing (even strictly). Hint: it has a positive derivative:

$$\exp'(x) = \exp(x) > 0.$$

21. Conclude that the exponent function indeed has an inverse function: the natural logarithm.

26.14.3 Ratio and Its Derivative

1. Let $g(x)$ be a nonzero differentiable function. This way, its reciprocal $1/g$ is well-defined. Differentiate it. Hint: for $x \neq 0$, look at the hyperbolic function $1/x$, and differentiate it:

$$\left(\frac{1}{x}\right)' = -\frac{1}{x^2}$$

(Section 26.3.1). Now, define $f(x) \equiv 1/x$, and look at the composite function $f \circ g$, and differentiate it (Section 26.7.2):

$$(f \circ g)' = f'(g(x))g'(x) = \frac{-g'(x)}{g^2(x)}.$$

2. Let $\alpha > 0$ be a positive rational number:

$$\alpha \equiv \frac{n}{m} > 0,$$

for some natural numbers n and m. In Section 26.9.3, we already saw that

$$(x^\alpha)' = \alpha x^{\alpha-1}, \quad x > 0.$$

Show that

$$(x^{-\alpha})' = -\alpha x^{-\alpha-1}, \quad x > 0.$$

Hint: in the above, set $g(x) = x^\alpha$.

3. In our formula, substitute $-\alpha$ for α, and obtain an extended formula:

$$(x^\alpha)' = \alpha x^{\alpha-1},$$

where α is now any rational number, positive or negative:

$$\alpha \equiv \pm\frac{n}{m}.$$

4. Let $f(x)$ and $g(x) \neq 0$ be two differentiable functions. Look at their ratio: f/g. Differentiate it. Hint: in Leibniz rule (Section 26.6.1), substitute $1/g$ for g:

$$\begin{aligned}\left(\frac{f(x)}{g(x)}\right)' &= \left(f(x)\frac{1}{g(x)}\right)' \\ &= f'(x)\frac{1}{g(x)} + f(x)\left(\frac{1}{g(x)}\right)' \\ &= \frac{f'(x)}{g(x)} + f(x)\frac{-g'(x)}{g^2(x)} \\ &= \frac{f'(x)g(x) - f(x)g'(x)}{g^2(x)}.\end{aligned}$$

5. Write this formula more concisely:

$$\left(\frac{f}{g}\right)' = \frac{f'g - fg'}{g^2}.$$

6. Show that, for $x \neq -1$, the function

$$\frac{1+2x}{1+x}$$

is monotonically increasing (even strictly). Hint: thanks to the above formula,

$$\begin{aligned}\left(\frac{1+2x}{1+x}\right)' &= \frac{(1+2x)'(1+x) - (1+2x)(1+x)'}{(1+x)^2} \\ &= \frac{2(1+x) - (1+2x)}{(1+x)^2} \\ &= \frac{2+2x-1-2x}{(1+x)^2} \\ &= \frac{1}{(1+x)^2} \\ &> 0.\end{aligned}$$

7. Conclude that the nonlinear process in Section 5.2.1 is unstable: in each step, the relative error multiplies by a monotonically-increasing factor, much bigger than 1.

26.14.4 Power Series and Its Derivative

1. Let x be a real (or even complex) number. Show that

$$(1-x)\sum_{i=0}^{n} x^i = 1 - x^{n+1}.$$

Hint: open parentheses, and cancel telescopical terms.

2. For $x \neq 1$, divide this by $1-x$:

$$\sum_{i=0}^{n} x^i = \frac{1-x^{n+1}}{1-x}.$$

3. For $|x| < 1$, show that

$$|x|^n \to_{n\to\infty} 0.$$

4. Conclude that, for $|x| < 1$, the infinite geometrical power series converges:

$$\sum_{i=0}^{\infty} x^i = \frac{1}{1-x}.$$

5. For any natural number k, multiply this by x^k, and obtain the tail:

$$\sum_{i=k}^{\infty} x^i = x^k \sum_{i=0}^{\infty} x^i = \frac{x^k}{1-x}.$$

6. For $x \neq 1$, show that

$$\left(\frac{1}{1-x}\right)' = \frac{1}{(1-x)^2}.$$

Hint: differentiate the composite function $f \circ g$, where $g(x) = 1 - x$, and $f(x) = 1/x$.
7. Prove this in yet another way. Hint: differentiate the ratio f/g, where $g(x) = 1 - x$, and $f \equiv 1$.
8. Assume again that $|x| < 1$. Show that

$$\sum_{i=0}^{\infty} (-1)^i x^i = \frac{1}{1+x}.$$

Hint: in our infinite series, substitute $-x$ for x.
9. In the formula

$$\left(\frac{1}{1-x}\right)' = \frac{1}{(1-x)^2},$$

is it legitimate to substitute $-x$ for x, and obtain

$$\left(\frac{1}{1+x}\right)' = \frac{1}{(1+x)^2}?$$

Hint: no! The differentiation is with respect to x, not $-x$.
10. Differentiate more carefully:

$$\left(\frac{1}{1+x}\right)' = \frac{-1}{(1+x)^2}.$$

26.14.5 Uniform Convergence

1. For $n = 1, 2, 3, \ldots$, consider the infinite sequence

$$\frac{n+1}{n} = 1 + \frac{1}{n}.$$

Show that it has the following properties:
- It decreases monotonically with n.

- 1 is a lower bound:
$$\frac{n+1}{n} = 1 + \frac{1}{n} \geq 1.$$
- 1 is the maximal lower bound.
- 1 is also the limit:
$$\frac{n+1}{n} = 1 + \frac{1}{n} \to_{n\to\infty} 1.$$

  Hint: let $\varepsilon > 0$ be a given positive parameter (as small as you like). Pick an n so big that
$$\frac{1}{n} \leq \varepsilon.$$

2. Let p be a fixed natural number. For $n = 1, 2, 3, \ldots$, consider the new infinite sequence
$$\left(\frac{n+1}{n}\right)^p = \frac{(n+1)^p}{n^p}.$$
3. Use the binomial formula to estimate it:
$$\begin{aligned}\frac{(n+1)^p}{n^p} &= \frac{\sum_{i=0}^{p}\binom{p}{i} n^{p-i}}{n^p} \\ &= \sum_{i=0}^{p}\binom{p}{i} n^{-i} \\ &= 1 + \sum_{i=1}^{p}\binom{p}{i} n^{-i} \\ &\leq 1 + n^{-1}\sum_{i=1}^{p}\binom{p}{i} \\ &= 1 + \frac{2^p - 1}{n}.\end{aligned}$$
4. Show that it has the same properties:
   - It decreases monotonically with n.
   - 1 is a lower bound:
$$\frac{(n+1)^p}{n^p} \geq 1.$$
   - 1 is the maximal lower bound.
   - 1 is also the limit:
$$\frac{(n+1)^p}{n^p} \to_{n\to\infty} 1.$$
5. Assume that
$$|x| \leq L < M < 1.$$
   As k grows, who wins: k^p or L^k?
6. Show that L^k wins:
$$\left|k^p x^k\right| = k^p |x|^k \leq k^p L^k < M^k.$$
   Hint: as in Section 26.10.1, estimate the ratio between $(k+1)^p L^{k+1}$ and $k^p L^k$:
$$\frac{(k+1)^p L^{k+1}}{k^p L^k} = \frac{(k+1)^p}{k^p} L \to_{k\to\infty} L < M < 1.$$

7. Consider the polynomial-geometrical series

$$\sum_{i=0}^{\infty} i^p x^i.$$

8. Conclude that, for a sufficiently large k, it has a small tail:

$$\begin{aligned}\left|\sum_{i=k}^{\infty} i^p x^i\right| &\le \sum_{i=k}^{\infty} i^p |x|^i \\ &\le \sum_{i=k}^{\infty} i^p L^i \\ &< \sum_{i=k}^{\infty} M^i \\ &= \frac{M^k}{1-M}.\end{aligned}$$

9. Here is yet another polynomial-geometrical series:

$$\sum_{i=0}^{\infty} (i+1)(i+2)\cdots(i+p)x^i.$$

10. Show that, for a sufficiently large k (perhaps bigger than before), it has a small tail too:

$$\begin{aligned}\left|\sum_{i=k}^{\infty} (i+1)(i+2)\cdots(i+p)x^i\right| &\le \sum_{i=k}^{\infty} (i+1)(i+2)\cdots(i+p)|x|^i \\ &\le \sum_{i=k}^{\infty} (i+1)(i+2)\cdots(i+p)L^i \\ &< \sum_{i=k}^{\infty} M^i \\ &= \frac{M^k}{1-M}.\end{aligned}$$

Hint: here too, the ratio is now as small as

$$\frac{(i+2)(i+3)\cdots(i+p+1)L^{i+1}}{(i+1)(i+2)\cdots(i+p)L^i} = \frac{i+p+1}{i+1}L \to_{i\to\infty} L < M < 1.$$

11. Conclude that the convergence is uniform in the entire interval $[-L, L]$: the same (big) k could be picked to have an arbitrarily small tail, no matter which x is used in each individual term.

26.14.6 Differentiation Term by Term

1. Look again at our power series:

$$\frac{1}{1-x} = \sum_{i=0}^{\infty} x^i, \qquad |x| < 1.$$

Use the same technique as in Sections 26.12.2–26.12.8 to differentiate it term by term, not only once but also p times:

$$\begin{aligned}
\frac{p!}{(1-x)^{p+1}} &= \left(\frac{1}{1-x}\right)^{(p)} \\
&= \left(\sum_{i=0}^{\infty} x^i\right)^{(p)} \\
&= \sum_{i=0}^{\infty} \left(x^i\right)^{(p)} \\
&= \sum_{i=p}^{\infty} \left(x^i\right)^{(p)} \\
&= \sum_{i=0}^{\infty} \left(x^{i+p}\right)^{(p)} \\
&= \sum_{i=0}^{\infty} (i+1)(i+2)\cdots(i+p)x^i \\
&= \sum_{i=0}^{\infty} C_{i+p,p} x^i,
\end{aligned}$$

where $|x| < 1$, and $C_{i+p,p}$ is defined in Section 6.6.1.

2. Divide this by $p!$:

$$\begin{aligned}
(1-x)^{-p-1} &= \sum_{i=0}^{\infty} \frac{C_{i+p,p}}{p!} x^i \\
&= \sum_{i=0}^{\infty} \frac{(i+p)!}{i!p!} x^i \\
&= \sum_{i=0}^{\infty} \binom{i+p}{p} x^i,
\end{aligned}$$

with coefficient

$$\frac{C_{i+p,p}}{p!} = \binom{i+p}{p}.$$

3. What is this coefficient geometrically? Hint: this is the total number of points in the discrete p-dimensional simplex of size i (Section 6.3.4).
4. Estimate it. Hint: it is contained in a p-dimensional hypercube of size i:

$$\binom{i+p}{p} \le (i+1)^p.$$

26.14.7 The Extended Binomial Formula

1. Extend the definition of $C_{a,i}$ to apply to a negative a as well. Hint: see Section 6.6.1.
2. Use this to obtain

$$\begin{aligned} C_{-p-1,i} &\equiv (-p-1)(-p-2)\cdots(-p-i) \\ &= (-1)^i(p+1)(p+2)\cdots(p+i) \\ &= (-1)^i(p+i)(p+i-1)\cdots(p+1) \\ &= (-1)^i C_{p+i,i}. \end{aligned}$$

3. Substitute this in our series:

$$\begin{aligned} (1-x)^{-p-1} &= \sum_{i=0}^{\infty} \frac{C_{i+p,i}}{i!} x^i \\ &= \sum_{i=0}^{\infty} (-1)^i \frac{C_{-p-1,i}}{i!} x^i. \end{aligned}$$

4. Extend the binomial coefficient too:

$$\binom{n}{i} \equiv \frac{C_{n,i}}{i!},$$

where n is now not only positive but also negative.
5. With this new definition, what happens when $i > n \geq 0$? Hint: the binomial coefficient still vanishes, as before.
6. Substitute this new definition in our series:

$$(1-x)^{-p-1} = \sum_{i=0}^{\infty} (-1)^i \binom{-p-1}{i} x^i.$$

7. Substitute n for $-p-1$, and x for $-x$. This gives the extended binomial formula:

$$(1+x)^n = \sum_{i=0}^{\infty} \binom{n}{i} x^i,$$

where n is now not only positive but also negative.
8. What happens at $n \geq 0$? Hint: most of these terms vanish, yielding the good old version back again.

26.14.8 Taylor Expansion

1. Let f be a given function, defined in an open domain Ω, containing the origin:

$$0 \in \Omega \subset \mathbb{R}.$$

2. Assume that, at 0, f is differentiable as many times as you like. In other words, at 0, f has infinitely many derivatives:

$$f(0),\ f'(0),\ f''(0),\ f'''(0),\ f''''(0),\ \ldots.$$

3. Divide them by suitable factorials:

$$f(0),\ f'(0),\ \frac{f''(0)}{2},\ \frac{f'''(0)}{3!},\ \frac{f''''(0)}{4!},\ \ldots.$$

4. Use these as coefficients in a new power series, defined in a neighborhood of 0. For this purpose, assume that there is an $\eta > 0$, for which

$$f(x) = \sum_{i=0}^{\infty} \frac{f^{(i)}(0)}{i!} x^i, \qquad |x| < \eta.$$

5. This is the Taylor expansion of f around 0. It tells us how f looks like at

$$x \in (-\eta, \eta),$$

but not elsewhere.

6. Show that the (extended) binomial formula is actually a Taylor series. Hint: to see this, set

$$f(x) \equiv (1+x)^n, \quad |x| < 1,$$

where n is a fixed integer number (positive or negative). In this case, the coefficients are

$$\frac{f^{(i)}(0)}{i!} = \frac{C_{n,i}}{i!}(1+0)^{n-i} = \binom{n}{i}.$$

7. Let

$$\alpha \in \mathbb{Q}$$

be some rational number. Define the new function

$$f(x) \equiv (1+x)^\alpha, \qquad x > 0.$$

Hint: see Section 26.9.3.

8. What is its Taylor series? Hint: in this case, the coefficients are

$$\frac{f^{(i)}(0)}{i!} = \frac{C_{\alpha,i}}{i!}(1+0)^{\alpha-i} = \binom{\alpha}{i}$$

(Section 6.6.2). This way, the Taylor series makes a new binomial formula:

$$f(x) \equiv (1+x)^\alpha = \sum_{i=0}^{\infty} \binom{\alpha}{i} x^i, \quad |x| < 1.$$

26.14.9 Power Function: General Definition

1. So far, α was rational. What about an irrational α? How to define

$$(1+x)^\alpha?$$

2. Define:

$$(1+x)^\alpha \equiv \sum_{i=0}^{\infty} \binom{\alpha}{i} x^i, \quad |x| < 1.$$

In this definition, α is any real (or even complex) number.

3. How to differentiate this? Hint: term by term.
4. Is this legitimate? Hint: use the same technique as in Sections 26.12.2–26.12.8.
5. Conclude that the derivative is

$$\begin{aligned}
((1+x)^\alpha)' &= \sum_{i=0}^{\infty} \binom{\alpha}{i} (x^i)' \\
&= \sum_{i=1}^{\infty} \binom{\alpha}{i} i x^{i-1} \\
&= \sum_{i=1}^{\infty} \frac{C_{\alpha,i}}{i!} i x^{i-1} \\
&= \sum_{i=1}^{\infty} \frac{C_{\alpha,i}}{(i-1)!} x^{i-1} \\
&= \sum_{i=1}^{\infty} \frac{\alpha C_{\alpha-1,i-1}}{(i-1)!} x^{i-1} \\
&= \alpha \sum_{i=1}^{\infty} \frac{C_{\alpha-1,i-1}}{(i-1)!} x^{i-1} \\
&= \alpha \sum_{i=0}^{\infty} \frac{C_{\alpha-1,i}}{i!} x^{i} \\
&= \alpha \sum_{i=0}^{\infty} \binom{\alpha-1}{i} x^{i} \\
&= \alpha (1+x)^{\alpha-1}.
\end{aligned}$$

6. This extends the formula

$$((1+x)^\alpha)' = \alpha(1+x)^{\alpha-1}$$

to any real (or even complex) α (for $|x| < 1$).
7. Repeat this i times (with i as big as you like):

$$((1+x)^\alpha)^{(i)} = C_{\alpha,i}(1+x)^{\alpha-i}, \quad |x| < 1, \quad 0 \le i < \infty.$$

8. Evaluate this derivative at $x = 0$, and divide by $i!$, to obtain the coefficient

$$\frac{C_{\alpha,i}}{i!} = \binom{\alpha}{i}, \quad 0 \le i < \infty.$$

9. Use these coefficients to reproduce the Taylor series (with $\eta \equiv 1$):

$$(1+x)^\alpha = \sum_{i=0}^{\infty} \binom{\alpha}{i} x^i, \quad |x| < 1.$$

10. Is this familiar? Hint: this is the original definition: a most general binomial formula.

26.14.10 Complex Numbers: Polar Decomposition

1. Look at the exponent function:

$$\exp(x) \equiv \sum_{i=0}^{\infty} \frac{x^i}{i!}, \quad |x| < \infty.$$

What is its Taylor series (with $\eta \equiv \infty$)? Hint: the coefficients are

$$\frac{\exp^{(i)}(0)}{i!} = \frac{\exp 0)}{i!} = \frac{1}{i!}.$$

So, the Taylor series is the same as the original definition.
2. What is the Taylor expansion of $\sin(x)$ around 0?
3. Show that it contains odd powers only:

$$\sin(x) = \sum_{i=0}^{\infty} \frac{(-1)^i x^{2i+1}}{(2i+1)!}, \quad |x| < \infty.$$

Hint: use the formula

$$\sin''(x) = \cos'(x) = -\sin(x)$$

(Section 26.4.2).
4. What is the Taylor expansion of $\cos(x)$ around 0?
5. Show that it contains even powers only:

$$\cos(x) = \sum_{i=0}^{\infty} \frac{(-1)^i x^{2i}}{(2i)!}, \quad |x| < \infty.$$

Hint: use the formula

$$\cos''(x) = -\sin'(x) = -\cos(x)$$

(Section 26.4.2).
6. Differentiate term by term.
7. Why is this legitimate? Hint: as in the exponent function.
8. Does this give the correct derivative?
9. Multiply the expansion of $\sin(x)$ by $i \equiv \sqrt{-1}$, and add to the expansion of $\cos(x)$.
10. What is this?
11. Show that this is

$$\cos(x) + i\sin(x) = \exp(ix).$$

Hint: check this, term by term.
12. What is the absolute value of this? Hint:

$$|\exp(ix)| = \cos^2(x) + \sin^2(x) = 1.$$

13. In the complex plane, where is $\exp(ix)$? Hint: on the unit circle (Figures 26.5 and 3.17).
14. Let $0 \leq \theta < 2\pi$ be a fixed angle. In the complex plane, look at the arrow from the origin to $\exp(i\theta)$. Show that it makes angle θ with the real axis. Hint: this is the polar decomposition (Section 26.4.1).

27

Polynomials and Partial Derivatives

So far, we focused on functions of one independent variable: x. Next, consider functions of two (or even three) independent variables: x, y, and even z. For simplicity, let's focus on polynomials. (The same works for more general functions as well.)

Geometrically, we are now in 2-D ($\mathbb{R}^2$) or even 3-D ($\mathbb{R}^3$). Here, there are many possible directions to differentiate. This leads to partial (and even directional) derivatives, useful in physics.

27.1 Polynomials and Their Arithmetic Operations

27.1.1 Polynomial of One Variable

What is a polynomial? It is a real function

$$p : \mathbb{R} \to \mathbb{R},$$

defined by

$$p(x) \equiv a_0 + a_1 x + a_2 x^2 + \cdots + a_n x^n = \sum_{i=0}^{n} a_i x^i,$$

where n is a nonnegative integer number (the degree of the polynomial), and $a_0, a_1, a_2, \ldots, a_n$ are fixed real numbers (the coefficients of the polynomial). Usually, it is assumed that $a_n \neq 0$. (Otherwise, just drop it.)

How to define a concrete polynomial of degree n? Just specify its coefficients: $a_0, a_1, a_2, \ldots, a_n$. Thus, the polynomial is mirrored by the $(n+1)$-dimensional vector

$$(a_0, a_1, a_2, \ldots, a_n) \in \mathbb{R}^{n+1}.$$

27.1.2 Real vs. Complex Polynomial

This is a real polynomial. What is a complex polynomial? It is different in one thing: the coefficients $a_0, a_1, a_2, \ldots, a_n$ (and the independent variable x) are now complex. This makes the polynomial a complex function:

$$p : \mathbb{C} \to \mathbb{C}.$$

Let's focus on real polynomials. (The same works with Complex polynomials as well.) For a start, let's look at our polynomials as algebraic objects, that enjoy addition, multiplication, and even composition. Later on, we'll also treat them as analytic objects, and differentiate them.

27.1.3 Addition

By now, our polynomial $p(x)$ is mirrored by an $(n+1)$-dimensional vector. This helps add polynomials. To see this, let $q(x)$ be yet another polynomial (of degree m):

$$q(x) \equiv \sum_{i=0}^{m} b_i x^i.$$

Without loss of generality, assume that $m \leq n$. (Otherwise, just interchange the roles of p and q.) Now, there are two options: if $m = n$, then we are ready to add. If, on the other hand, $m < n$, then introduce $n - m$ dummy zeroes:

$$b_{m+1} \equiv b_{m+2} \equiv \cdots \equiv b_n \equiv 0,$$

to make q as long as p. We're now ready to add, term by term. For this, use our vectors: add them, coefficient by coefficient. This will produce a new polynomial: $p + q$:

$$(p+q)(x) \equiv p(x) + q(x) = \sum_{i=0}^{n} (a_i + b_i) x^i.$$

For a given x, this gives us a new value. But, algebraically, this is much more: a list of new coefficients, mirrored by our new vector (the sum of the original vectors). This works in subtraction as well:

$$(p-q)(x) \equiv p(x) - q(x) = \sum_{i=0}^{n} (a_i - b_i) x^i.$$

This is a new polynomial: $p - q$, with a new vector of new coefficients.

27.1.4 Scalar Multiplication

How to multiply by a scalar c? Term by term, of course! Again, this is mirrored by our vector: multiply it by c, coefficient by coefficient. This produces the new polynomial cp:

$$(cp)(x) \equiv c \cdot p(x) = c \sum_{i=0}^{n} a_i x^i = \sum_{i=0}^{n} (c a_i) x^i.$$

For a given x, this gives us a new value. But, algebraically, this is much more: a new list of coefficients.

27.1.5 Multiplying Polynomials: Convolution

This is linear algebra. Next, consider two polynomials: p and q. How to multiply them by one another? First, define a new function: pq:

$$(pq)(x) \equiv p(x)q(x).$$

Still, is pq a *polynomial* in its own right? For this, we must design its vector of coefficients. Fortunately, the product of

$$p(x) \equiv \sum_{i=0}^{n} a_i x^i \quad \text{and} \quad q(x) \equiv \sum_{j=0}^{m} b_j x^j$$

can be written as

$$(pq)(x) \equiv p(x)q(x) = \sum_{i=0}^{n} a_i x^i \sum_{j=0}^{m} b_j x^j = \sum_{i=0}^{n} \sum_{j=0}^{m} a_i b_j x^{i+j}.$$

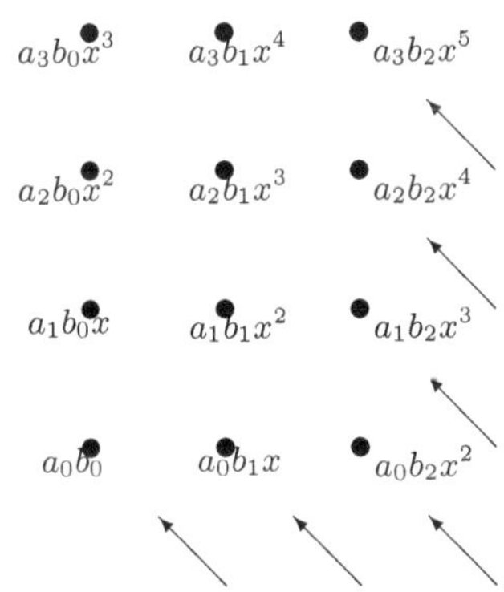

Fig. 27.1. Multiply two polynomials: $p(x) = a_0 + a_1x + a_2x^2 + a_3x^3$ times $q(x) = b_0 + b_1x + b_2x^2$. Work diagonal by diagonal (where the kth diagonal contains x^k).

This is a double sum, which scans the $(n+1) \times (m+1)$ grid

$$\{(i,j) \mid 0 \le i \le n,\ 0 \le j \le m\}.$$

In this rectangle, (i,j) contributes $a_ib_jx^{i+j}$. Better scan diagonal by diagonal (Figure 27.1). After all, on the kth diagonal, (i,j) satisfies $i + j = k$, and contributes $a_ib_jx^{i+j} = a_ib_{k-i}x^k$.

So, our diagonals are indexed by $k = i + j = 0, 1, 2, \ldots, n + m$. This designs a new (outer) sum over k. How to scan the kth diagonal? This is our inner sum. It uses our row-index i, which mustn't exceed the original grid:

$$(pq)(x) = \sum_{i=0}^{n} \sum_{j=0}^{m} a_i b_j x^{i+j} = \sum_{k=0}^{n+m} \sum_{i=\max(0,k-m)}^{\min(k,n)} a_i b_{k-i} x^k.$$

This makes pq a legitimate polynomial, with its own vector of coefficients:

$$(c_0, c_1, c_2, c_3, \ldots, c_{n+m}) \equiv (c_k)_{k=0}^{n+m},$$

where the new coefficient c_k is defined in the inner sum:

$$c_k \equiv \sum_{i=\max(0,k-m)}^{\min(k,n)} a_i b_{k-i}.$$

This is also called convolution (see exercises below).

27.1.6 Example: Scalar Multiplication

Look at a simple example, with a trivial q (of degree $m = 0$):

$$q \equiv b_0.$$

In this (degenerate) case, c_k is just

$$c_k = \sum_{i=k}^{k} a_i b_{k-i} = a_k b_0$$

($0 \leq k \leq n$). Thus, in this case,

$$(pq)(x) = \sum_{k=0}^{n} c_k x^k = \sum_{k=0}^{n} b_0 a_k x^k = b_0 \sum_{i=0}^{n} a_i x^i = b_0 \cdot p(x).$$

This is as in scalar multiplication (Section 27.1.4).

27.2 Polynomial and Its Value

27.2.1 Value at a Given Point

Let x be a given point. How to calculate $p(x)$? Here is a naive algorithm:

- Calculate the individual powers of x:

  $$x^2, x^3, x^4, \ldots, x^n.$$

  Fortunately, this can be done recursively:

  $$x^i = x \cdot x^{i-1} \quad (i = 2, 3, 4, \ldots, n).$$

  Is this expensive? Not too much: just $n - 1$ scalar-times-scalar operations.

- Next, multiply by the corresponding coefficient:

$$a_1x, a_2x^2, a_3x^3, \ldots, a_nx^n.$$

  This costs n more scalar-times-scalar operations.
- Finally, sum up:

$$a_0 + a_1x + a_2x^2 + \cdots + a_nx^n = p(x).$$

  This costs n scalar-plus-scalar additions.

Thus, the total cost (complexity) is $2n - 1$ multiplications and n additions. How to improve on this? Introduce parentheses, and take a common factor out of them.

27.2.2 Using the Distributive Law

Consider a simple problem: let a, b, and c be some given numbers. Calculate

$$ab + ac.$$

The naive approach uses three arithmetic operations:

- Calculate ab.
- Calculate ac.
- Add up: $ab + ac$.

How to improve on this? Use the distributive law: introduce parentheses, and take a out of them:

$$ab + ac = a(b + c).$$

This is easier: it requires just two arithmetic operations:

- Add $b + c$.
- Multiply by a: $a(b + c)$.

Let's use this in our original problem.

27.2.3 Recursion: Horner's Algorithm

Let's calculate

$$p(x) = \sum_{i=0}^{n} a_i x^i$$

(where x is a given argument). How to take x out of parentheses? For this purpose, write p in a new form:

$$p(x) = a_0 + xp_1(x),$$

where $p_1(x)$ is a new polynomial:

$$p_1(x) = a_1 + a_2x + a_3x^2 + \cdots + a_nx^{n-1} = \sum_{i=0}^{n-1} a_{i+1}x^i.$$

Fortunately, p_1 has a lower degree: $n - 1$. Thus, $p_1(x)$ can be calculated recursively. This is Horner's algorithm.

27.2.4 Complexity: Mathematical Induction

This is cheaper: it requires just n scalar-times-scalar and n scalar-plus-scalar operations. To prove this, use mathematical induction on n. For $n = 0$, there is nothing to calculate:

$$p \equiv a_0.$$

Now, for $n = 1, 2, 3, \ldots$, assume that $p_1(x)$ can be calculated in $n-1$ multiplications and $n-1$ additions (the induction hypothesis). On top of this, to calculate

$$p(x) = a_0 + xp_1(x),$$

we also need to multiply by x, and to add a_0. The total cost is n multiplications and n additions, as asserted. This completes the induction step, and indeed the entire proof.

27.3 Composition

27.3.1 Mathematical Induction

So far, we used Horner's algorithm to calculate the value of a polynomial. Let's use it for yet another purpose: to compose two polynomials, p and q:

$$(p \circ q)(x) \equiv p(q(x)).$$

This is how functions are composed (Section 26.7.1). This tells us how to calculate $p \circ q$ (at any given argument x):

- Calculate $q(x)$.
- Apply p, to calculate $p(q(x))$.

This defines $p \circ q$ as a new *function*. Still, is it a legitimate *polynomial* as well? What is its vector of coefficients?

To design this, use mathematical induction on the degree of p: n. For $n = 0$, p is constant: $p \equiv a_0$. So, the composition is

$$(p \circ q)(x) = p(q(x)) = a_0$$

as well. So, the vector of coefficients is

$$(a_0).$$

27.3.2 The Induction Step

Now, for $n = 1, 2, 3, \ldots$, assume that the induction hypothesis holds: for every polynomial p_1 of degree $n-1$ or less (as in Section 27.2.3), we already have the entire vector of coefficients of $p_1 \circ q$. Multiply this by q (Section 27.1.5):

$$q \cdot (p_1 \circ q).$$

To this vector of coefficients, add a_0 at the first entry:

$$p \circ q = a_0 + q \cdot (p_1 \circ q).$$

This is not only theoretical but also constructive: in practice, call it recursively.

27.4 Natural Number as a Polynomial

27.4.1 Decimal Polynomial

Why are polynomials so attractive? Because every natural number k is actually a polynomial. After all, k must lie somewhere in between

$$10^n \le k < 10^{n+1}$$

(for some integer $n \ge 0$). Thus, k is a decimal number of $n+1$ digits:

$$a_n a_{n-1} a_{n-2} \cdots a_1 a_0.$$

This is just a common notation for a polynomial:

$$k = a_0 + a_1 \cdot 10 + a_2 \cdot 10^2 + \cdots + a_n \cdot 10^n = \sum_{i=0}^{n} a_i \cdot 10^i = p(10).$$

This is a decimal polynomial, whose coefficients are digits between 0 and 9. Now, work the other way around: to obtain k, evaluate p at 10.

27.4.2 Binary Polynomial

On second thought, why use base 10? Better use base 2. After all, our k must lie somewhere in between

$$2^m \le k < 2^{m+1}$$

(for some integer $m \ge 0$). Thus, k has $m+1$ bits (binary digits: 0 or 1):

$$b_m b_{m-1} b_{m-2} \cdots b_1 b_0.$$

This is just a common notation for a binary polynomial:

$$k = b_0 + b_1 \cdot 2 + b_2 \cdot 2^2 + \cdots + b_m \cdot 2^m = \sum_{j=0}^{m} b_j 2^j = q(2).$$

Now, work the other way around: to obtain k, evaluate q at 2. In practice, store q bit by bit. In cryptography, this helps store very long numbers efficiently.

27.5 Monomial and Its Value

27.5.1 Monomial

Suppose that we have a very simple polynomial, with just one monomial: x^k. How to calculate its value? In this case, Horner's algorithm is not so attractive any more: it gets sequential (and recursive): for $i = 2, 3, 4, \ldots, k$, calculate

$$x^i = x \cdot x^{i-1}.$$

This is a bit expensive: $k-1$ multiplications. This is worthwhile if you need many monomials: x^2, x^3, x^4, …, x^k. But what if you only need x^k? Is there a more efficient way?

27.5.2 Horner's Algorithm: Implicit Form

Let's apply Horner's algorithm (implicitly) to q:

$$k = q(2) = b_0 + 2q_1(2)$$

(where q_1 is a binary polynomial of degree $m - 1$ or less). No need to specify q or q_1 explicitly: it is good enough to know that they exist.

27.5.3 Mathematical Induction

We can now calculate x^k in just $2m$ multiplications! Indeed, by mathematical induction on m: for $m = 0$, there is no work at all: $k \equiv b_0$, so either $x^k = x^0 = 1$, or $x^k = x^1 = x$.

27.5.4 The Induction Step

Now, assume that the induction hypothesis holds: we already have a good algorithm for polynomials of degree $m-1$ or less (like our q_1). This is relevant to any argument: not only x but also x^2. Thus, if x^2 was available, then we could calculate

$$\left(x^2\right)^{q_1(2)}$$

in just $2(m - 1)$ multiplications. (This is the induction hypothesis.) Let's use this to calculate our original monomial:

$$x^k = x^{q(2)} = x^{b_0+2q_1(2)} = x^{b_0}x^{2q_1(2)} = x^{b_0}\left(x^2\right)^{q_1(2)} = \begin{cases} x \cdot \left(x^2\right)^{q_1(2)} & \text{if } b_0 = 1 \\ \left(x^2\right)^{q_1(2)} & \text{if } b_0 = 0. \end{cases}$$

27.5.5 Complexity: Total Cost

To calculate this right-hand side, we may need two more multiplications:

- Calculate $x^2 = x \cdot x$. Later on, this will help calculate $(x^2)^{q_1(2)}$ recursively.
- Multiply $(x^2)^{q_1(2)}$ by x (if $b_0 = 1$).

Thus, the total cost is at most

$$2 + 2(m - 1) = 2m$$

multiplications. This completes the induction step.

27.5.6 Recursion

Better yet, let's get rid of q and q_1 altogether. Recall that

$$k = q(2) = b_0 + 2q_1(2).$$

Thus, if k is even, then

$$b_0 = 0 \quad \text{and} \quad q_1(2) = \frac{k}{2}.$$

If, on the other hand, k is odd, then

$$b_0 = 1 \quad \text{and} \quad q_1(2) = \frac{k-1}{2}.$$

Thus, our recursion gets simpler:

$$x^k = \begin{cases} x \cdot (x^2)^{(k-1)/2} & \text{if } k \text{ is odd} \\ (x^2)^{k/2} & \text{if } k \text{ is even.} \end{cases}$$

This will be useful later.

27.6 Differentiation

27.6.1 Derivative of a Polynomial

Look again at our original polynomial:

$$p(x) = \sum_{i=0}^{n} a_i x^i.$$

What is its derivative? If $n = 0$, then the derivative is zero. If, on the other hand, $n > 0$, then the derivative is a polynomial of degree $n - 1$:

$$p'(x) = \sum_{i=1}^{n} a_i i x^{i-1} = \sum_{i=0}^{n-1} (i+1) a_{i+1} x^i.$$

27.6.2 Second Derivative

In either case, this is a new polynomial of degree $\max(0, n-1)$. Differentiate it, and obtain the second derivative of p (Section 26.2.7):

$$p''(x) = \frac{d^2 p}{dx^2}(x) = (p'(x))'.$$

27.6.3 Higher Derivatives

This is a new polynomial of degree $\max(0, n-2)$. Differentiate it, and obtain p''', and so on. In general, for $i = 0, 1, 2, \ldots$, the ith derivative of p is defined recursively by

$$p^{(i)} \equiv \begin{cases} p & \text{if } i = 0 \\ \left(p^{(i-1)}\right)' & \text{if } i > 0. \end{cases}$$

In this notation, the zeroth derivative is just the function itself:

$$p^{(0)} \equiv p,$$

and its derivative is

$$p^{(1)} \equiv p'.$$

Furthermore, the nth derivative of p is just a constant, or a polynomial of degree 0:

$$p^{(n)} = a_n n!.$$

For this reason, every higher derivative must vanish:

$$p^{(i)} = 0, \quad i > n.$$

27.7 Integration

27.7.1 Indefinite Integral

Look at our original polynomial again:

$$p(x) = \sum_{i=0}^{n} a_i x^i.$$

As discussed in Section 26.13.2, it has a primitive function (or antiderivative, or indefinite integral): a new polynomial of degree $n+1$:

$$P(x) \equiv \sum_{i=0}^{n} \frac{a_i}{i+1} x^{i+1} = \sum_{i=1}^{n+1} \frac{a_{i-1}}{i} x^i.$$

Indeed, the derivative of P is just the original polynomial p:

$$P'(x) = p(x),$$

as required.

27.7.2 Definite Integral Over an Interval

How to calculate the area underneath the graph of p? This is the definite integral:

$$\int_a^b p(x)dx = P(b) - P(a)$$

(Sections 26.13.2–26.13.4). Let's see a few examples.

27.7.3 Examples

Here are a few examples. If $p(x) = x$ (a linear polynomial), then $P(x) = x^2/2$, so

$$\int_a^b x dx = \frac{b^2 - a^2}{2} = (b-a)\frac{a+b}{2}.$$

This is the length of the interval, multiplied by the value of p (at the midpoint). This is the trapezoidal (or trapezoid, or trapezium) rule.

If, on the other hand, $p(x) = x^2$ (a quadratic polynomial), then $P(x) = x^3/3$, so

$$\int_a^b x^2 dx = \frac{b^3 - a^3}{3}.$$

Finally, if $p \equiv 1$, then $P(x) = x$, so

$$\int_a^b dx = b - a.$$

This is the length of our interval: $[a, b]$.

27.7.4 Definite Integral Over The Unit Interval

Consider our general polynomial again: p. Now, look at the unit interval: $a \equiv 0$, and $b \equiv 1$. What is the definite integral over it? Thanks to our formula,

$$\int_0^1 p(x)dx = P(1) - P(0) = P(1) = \sum_{i=1}^{n+1} \frac{a_{i-1}}{i}.$$

27.8 Sparse Polynomials

27.8.1 Sparse Polynomials

So far, we considered a dense polynomial, with many nonzero terms. For it, Horner's algorithm makes sense (Section 27.2.3). But what about a sparse polynomial, with only a few nonzero terms? For it, Horner's algorithm must be modified.

How could a sparse polynomial look like? Well, it could contain just one nonzero term. This is just a monomial:

$$p(x) = x^k.$$

For a given x, how to calculate this? Well, Horner's algorithm is no good: it reduces to the naive algorithm in Section 27.5.1, which requires as many as $k-1$ multiplications. By now, we already know better:

$$x^k = \begin{cases} 1 & \text{if } k = 0 \\ x & \text{if } k = 1 \\ x \cdot (x^2)^{(k-1)/2} & \text{if } k > 1 \text{ and } k \text{ is odd} \\ (x^2)^{k/2} & \text{if } k > 1 \text{ and } k \text{ is even.} \end{cases}$$

How to make Horner's algorithm agree with this?

27.8.2 How to Write The Sparse Polynomial?

For this purpose, write the sparse polynomial as follows:

$$p(x) = a_0 + a_k x^k + a_l x^l + \cdots + a_n x^n.$$

In this writing style, we specify three terms only:

- The first term a_0 is either zero or nonzero.
- The next nonzero term is $a_k x^k$, where $a_k \neq 0$ is the next nonzero coefficient ($k \geq 1$).
- The next nonzero term is $a_l x^l$, where $a_l \neq 0$ is the next nonzero coefficient ($l > k$), and so on.

27.8.3 The Improved Horner Algorithm

In the original Horner algorithm, we pulled x out of parentheses. Here, on the other hand, we pull even x^k:

$$p(x) = a_0 + x^k p_1(x),$$

where the new polynomial

$$p_1(x) \equiv a_k + a_l x^{l-k} + \cdots + a_n x^{n-k}$$

has degree $n - k$ only. This leads to the improved Horner algorithm:

$$p(x) = \begin{cases} a_0 + x^k p_1(x) & \text{if } a_0 \neq 0 \\ x^k p_1(x) & \text{if } a_0 = 0, \end{cases}$$

where $p_1(x)$ is calculated recursively by the same algorithm itself, and x^k is calculated efficiently, as before:

$$x^k = \begin{cases} x & \text{if } k = 1 \\ x \cdot (x^2)^{(k-1)/2} & \text{if } k > 1 \text{ and } k \text{ is odd} \\ (x^2)^{k/2} & \text{if } k > 1 \text{ and } k \text{ is even.} \end{cases}$$

Clearly, if p is dense, then this is our good old version back again. Let's mirror this in yet another task: composition.

27.8.4 Polynomial and Its Power

So far, our task was easy: to calculate the value of our polynomial. Next, let's move on to a more complicated task: to design the entire vector of coefficients.

Let q be a given polynomial (dense or sparse). Assume that we already have its coefficients. Now, how about q^k? How to design its coefficients as well? Easy: just mirror x^k:

$$q^k = \begin{cases} 1 & \text{if } k = 0 \\ q & \text{if } k = 1 \\ q \cdot (q^2)^{(k-1)/2} & \text{if } k > 1 \text{ and } k \text{ is odd} \\ (q^2)^{k/2} & \text{if } k > 1 \text{ and } k \text{ is even.} \end{cases}$$

In this recursion, we use q^2. This is the vector of coefficients of q^2, calculated as in Section 27.1.5.

27.8.5 Composition

Assume that p is still sparse, and still has the same form as before:

$$p(x) = a_0 + x^k p_1(x).$$

How to compose $p \circ q$, and have all coefficients at the same time? Just mirror Horner's algorithm (in its improved version):

$$p \circ q = a_0 + q^k \cdot (p_1 \circ q).$$

Fortunately, p_1 is of degree $n - k < n$. Thus, $p_1 \circ q$ can be obtained recursively (including all its coefficients). Multiply this by q^k, and obtain $q^k \cdot (p_1 \circ q)$ (all coefficients). Finally, if $a_0 \neq 0$, then add a_0 to the first coefficient.

27.9 Two Independent Variables

27.9.1 Polynomial of Two Variables

So far, we looked at a polynomial of one independent variable: x. Let's move on to a polynomial of two independent variables: x and y. This is a real function of the form

$$p : \mathbb{R}^2 \to \mathbb{R},$$

defined by

$$p(x, y) = \sum_{i=0}^{n} a_i(x) y^i,$$

where $a_i(x)$ is a polynomial in one variable: x. This way, $p(x, y)$ is a real polynomial. A complex polynomial, on the other hand, is a little different: a complex function of the form

$$p : \mathbb{C}^2 \to \mathbb{C}.$$

It has the same structure as before, except that x, y, and $a_i(x)$ can be even complex. In what follows, we'll focus on real polynomials. Complex polynomials can be treated in the same way.

27.9.2 Arithmetic Operations

Arithmetic operations are carried out as before (Sections 27.1.3–27.1.5). The only difference is that the coefficients are no longer constant but polynomials in x. Let's see how this works.

27.9.3 Addition

Let

$$p(x, y) = \sum_{i=0}^{n} a_i(x) y^i \quad \text{and} \quad q(x, y) = \sum_{j=0}^{m} b_j(x) y^j$$

(for some natural numbers $m \leq n$). How to add them? Introduce dummy zeroes:

$$b_{m+1} \equiv b_{m+2} \equiv \cdots \equiv b_n \equiv 0.$$

We're now ready to add, term by term:

$$(p + q)(x, y) = p(x, y) + q(x, y) = \sum_{i=0}^{n} (a_i + b_i)(x) y^i.$$

After all, we already know how to add polynomials of one variable:

$$(a_i + b_i)(x), \qquad 0 \leq i \leq n.$$

These are complete vectors of coefficients, as required.

27.9.4 Multiplication

And how to multiply them? Like this:

$$\begin{aligned}(pq)(x,y) &= p(x,y)q(x,y)\\ &= \left(\sum_{i=0}^{n} a_i(x)y^i\right)\left(\sum_{j=0}^{m} b_j(x)y^j\right)\\ &= \sum_{i=0}^{n}\left(\sum_{j=0}^{m}(a_ib_j)(x)y^{i+j}\right).\end{aligned}$$

This is a sum of $n+1$ polynomials of two variables, which we already know how to do. In it, a_ib_j has a complete vector of coefficients, which we already know how to calculate.

27.10 Partial Differentiation

27.10.1 Partial Derivatives

So far, our polynomial

$$p(x,y) = \sum_{i=0}^{n} a_i(x)y^i$$

was viewed as an algebraic object. Next, let's view it as an analytic object, with a new operation: partial differentiation. For this purpose, consider a particular point

$$(x,y) \in \mathbb{R}^2.$$

At this point, how fast does p grow in the x-direction? To answer this, view y as a fixed (or frozen) parameter, and differentiate p as a function of x alone. This is the partial derivative of p with respect to x:

$$p_x(x,y) \equiv \sum_{i=0}^{n} a_i'(x)y^i.$$

Next, work the other way around. How fast does p grow in the y-direction? To find out, view x as a fixed parameter, and differentiate p as a function of y alone. This is the partial derivative of p with respect to y. If $n=0$, then it is zero. Otherwise,

$$p_y(x,y) \equiv \sum_{i=1}^{n} a_i(x)iy^{i-1} = \sum_{i=0}^{n-1}(i+1)a_{i+1}(x)y^i.$$

What are the partial derivatives? They are polynomials of two variables in their own right. So, they depend on our point (x,y). (After all, the rate of growth may change from point to point.) Thus, the partial derivatives can be differentiated as well (with respect to x or y). This will produce the second partial derivatives of p.

27.10.2 Second Partial Derivatives

For instance, look at p_x, and differentiate it with respect to y. This produces the mixed partial derivative of p:

$$p_{xy}(x,y) \equiv (p_x(x,y))_y .$$

Still, we could also work the other way around: differentiate with respect to y, then with respect to x:

$$p_{yx}(x,y) = \sum_{i=0}^{n-1}(i+1)a'_{i+1}(x)y^i = p_{xy}(x,y).$$

This is the $(1,1)$st partial derivative of p. After all, $x^1 = x$ and $y^1 = y$, so we could also write

$$p_{xy}(x,y) = p_{x^1y^1}(x,y).$$

With this notation, the $(0,0)$th partial derivative is nothing but p itself:

$$p_{x^0y^0}(x,y) = p(x,y).$$

This will be useful below.

27.10.3 High-Order Partial Derivatives

Let's continue this process. For instance, the $(2,1)$st partial derivative of p is

$$p_{x^2y^1}(x,y) \equiv p_{xxy}(x,y) \equiv (p_{xx}(x,y))_y .$$

How to extend this yet more? Recursively, of course! For this purpose, look at the rectangular grid (Figure 27.2). In it, draw parallel diagonals. The mth diagonal contains the (i,j)th partial derivatives, with

$$i+j=m.$$

To design them, start from the origin, and march diagonal by diagonal. After all, the mth diagonal depends on he $(m-1)$st diagonal only. This is nothing but mathematical induction on the order $m = 1,2,3,\ldots$:

$$p_{x^iy^j} \equiv \begin{cases} p & \text{if } i=j=0 \\ \left(p_{x^{i-1}y^j}\right)_x & \text{if } \quad i>0 \\ \left(p_{x^iy^{j-1}}\right)_y & \text{if } \quad j>0. \end{cases}$$

Fortunately, there is no ambiguity: if both $i>0$ and $j>0$, then both definitions agree. Indeed, prove this in the same inductive process:

$$\left(p_{x^iy^{j-1}}\right)_y = \left(p_{x^{i-1}y^{j-1}}\right)_{xy} = \left(p_{x^{i-1}y^{j-1}}\right)_{yx} = \left(p_{x^{i-1}y^j}\right)_x .$$

This completes the induction step, and indeed the entire mathematical induction.

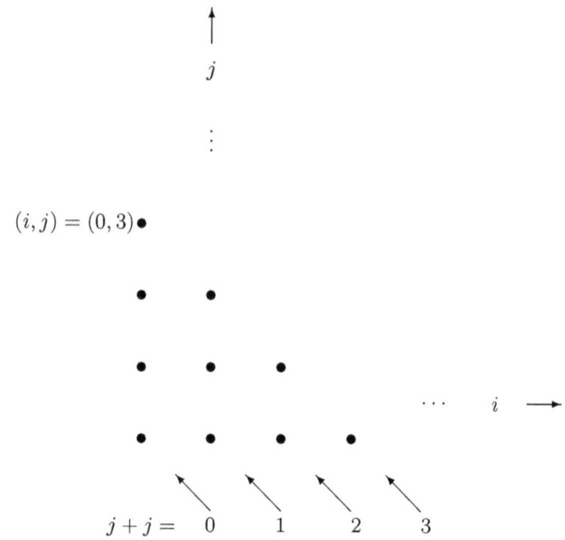

Fig. 27.2. Partial derivatives of order $m = i + j = 0, 1, 2, 3, \ldots$. To design them, march diagonal by diagonal.

27.10.4 Using a Discrete Triangle

How many partial derivatives of order m or less are there? Let's mirror them by a discrete triangle of size m (Figure 6.5). The total number of points in it is

$$\binom{m+2}{2} = \frac{(m+2)!}{2! \cdot m!} = \frac{(m+1)(m+2)}{2}.$$

And how many partial derivatives of order m exactly are there? As in Section 6.3.5, the answer is

$$\binom{m+2-1}{2-1} = \binom{m+1}{1} = m+1.$$

Here they are:

$$p_{x^0y^m}, \; p_{x^1y^{m-1}}, \; p_{x^2y^{m-2}}, \; \ldots, \; p_{x^my^0}.$$

27.10.5 Degree

In a polynomial of one variable, the degree is n: the highest power. In a polynomial of two variables like

$$p(x, y) = \sum_{i=0}^{n} a_i(x) y^i,$$

on the other hand, the degree is often higher. To calculate it, write $a_i(x)$ more explicitly:

$$a_i(x) = \sum_j a_{i,j}x^j$$

(where the $a_{i,j}$'s are the coefficients). This way,

$$p(x,y) = \sum_i a_i(x)y^i = \sum_i \sum_j a_{i,j}x^j y^i.$$

The degree of p is the maximal sum $i + j$ for which $a_{i,j} \neq 0$. Geometrically, this is the maximal m in Figure 27.2 (the index of the last diagonal). This is why we often denote the degree by m (which is often higher than n).

27.10.6 Degree and Discrete Triangle

Look at a polynomial $p(x,y)$ of degree m. What is the maximal number of terms in it? To find out, mirror it by a discrete triangle of size m (Figure 6.5). The total number of points in it is

$$\binom{m+2}{2} = \frac{(m+2)!}{m! \cdot 2!} = \frac{(m+1)(m+2)}{2}.$$

27.11 Three Independent Variables

27.11.1 Polynomial of Three Variables

Next, let's look at a polynomial of three independent variables:

$$p(x,y,z) \equiv \sum_{i=0}^{n} a_i(x,y)z^i.$$

In the arithmetic operations, there is just one change: the a_i's are now polynomials of two variables. Fortunately, we already know how to "play" with them algebraically.

27.12 Partial Differentiation

27.12.1 Partial Derivatives

To have the x-partial derivative, view both y and z as fixed parameters, and differentiate $p(x,y,z)$ as a function of x alone:

$$p_x(x,y,z) \equiv \sum_{i=0}^{n} (a_i)_x(x,y)z^i.$$

In this sum, the a_i's are differentiated with respect to x, as required. Fortunately, we already know how to do this.

Likewise, to have the y-partial derivative, view both x and z as fixed parameters, and differentiate $p(x, y, z)$ as a function of y alone:

$$p_y(x, y, z) \equiv \sum_{i=0}^{n} (a_i)_y(x, y)z^i.$$

Finally, to have the z-partial derivative, view both x and y as fixed parameters, and differentiate $p(x, y, z)$ as a function of z alone. If $n = 0$, then this is zero. Otherwise,

$$p_z(x, y, z) \equiv \sum_{i=1}^{n} a_i(x, y)iz^{i-1} = \sum_{i=0}^{n-1} (i + 1)a_{i+1}(x, y)z^i.$$

27.12.2 Directional Derivative

So far, we differentiated in a Cartesian direction: x or y or z. How about an oblique direction? How fast does p grow in such a direction? To find out, let $\mathbf{n}$ be a fixed 3-D vector:

$$\mathbf{n} \equiv \begin{pmatrix} n_1 \\ n_2 \\ n_3 \end{pmatrix} \in \mathbb{R}^3.$$

Assume that $\mathbf{n}$ is a unit vector:

$$\|\mathbf{n}\|^2 \equiv n_1^2 + n_2^2 + n_3^2 = 1.$$

How to differentiate p in direction $\mathbf{n}$? This is the directional derivative of p in direction $\mathbf{n}$:

$$p_n(x, y, z) \equiv n_1 p_x(x, y, z) + n_2 p_y(x, y, z) + n_3 p_z(x, y, z).$$

This depends on our point: (x, y, z). (After all, the rate of growth may change from point to point.) Still, this goes without saying, so we can simply write

$$p_n \equiv n_1 p_x + n_2 p_y + n_3 p_z.$$

Let's write this in a better style.

27.12.3 Differential Operators

Define three differential operators:

$$\begin{aligned} \frac{\partial}{\partial x} &: p \to p_x \\ \frac{\partial}{\partial y} &: p \to p_y \\ \frac{\partial}{\partial z} &: p \to p_z. \end{aligned}$$

Thanks to these, our directional differentiation takes the form:

$$\frac{\partial}{\partial n} \equiv n_1 \frac{\partial}{\partial x} + n_2 \frac{\partial}{\partial y} + n_3 \frac{\partial}{\partial z}.$$

How to have the ith directional derivative? Apply this i times. To open parentheses, use the trinomial formula.

27.12.4 Using The Binomial Formula

For instance, set

$$\mathbf{n} \equiv \frac{1}{\sqrt{2}}(1,1,0).$$

In this case,

$$\frac{\partial}{\partial n} = \frac{1}{\sqrt{2}}\left(\frac{\partial}{\partial x} + \frac{\partial}{\partial y}\right).$$

Apply this i times to a function like $x^a y^b$. Thanks to the binomial formula,

$$\begin{aligned}
\frac{\partial^i}{\partial n^i}(x^a y^b) &= \left(\frac{1}{\sqrt{2}}\left(\frac{\partial}{\partial x} + \frac{\partial}{\partial y}\right)\right)^i (x^a y^b) \\
&= \frac{1}{2^{i/2}} \sum_{k=0}^{i} \binom{i}{k} \left(\frac{\partial}{\partial x}\right)^k \left(\frac{\partial}{\partial y}\right)^{i-k} (x^a y^b) \\
&= \frac{1}{2^{i/2}} \sum_{k=0}^{i} \binom{i}{k} C_{a,k} C_{b,i-k} x^{a-k} y^{b-(i-k)}
\end{aligned}$$

(where $C_{a,k}$ is as in Section 6.6.1). At those points at which $x = y$, say

$$(x, y, z) = (x, x, 0),$$

our result simplifies yet more:

$$\frac{x^{a+b-i} C_{a+b,i}}{2^{i/2}}$$

(Section 6.6.1). You can even work the other way around: use this to prove the formula in Section 6.6.1 in the first place.

27.12.5 Using The Trinomial Formula

Look at another example:

$$\mathbf{n} \equiv \frac{1}{\sqrt{3}}(1,1,1),$$

so

$$\frac{\partial}{\partial n} = \frac{1}{\sqrt{3}}\left(\frac{\partial}{\partial x} + \frac{\partial}{\partial y} + \frac{\partial}{\partial z}\right).$$

Thanks to the trinomial formula (Section 6.6.3), we can now apply this i times to a function like $x^a y^b z^c$:

$$\begin{aligned}
\frac{\partial^i}{\partial n^i}(x^a y^b z^c) &= \left(\frac{1}{\sqrt{3}}\left(\frac{\partial}{\partial x} + \frac{\partial}{\partial y} + \frac{\partial}{\partial z}\right)\right)^i (x^a y^b z^c) \\
&= \frac{1}{3^{i/2}} \sum_{0 \le l,j,m \le i,\ l+j+m=i} \frac{i!}{l!j!m!} \left(\frac{\partial}{\partial x}\right)^l \left(\frac{\partial}{\partial y}\right)^j \left(\frac{\partial}{\partial z}\right)^m (x^a y^b z^c) \\
&= \frac{1}{3^{i/2}} \sum_{0 \le l,j,m \le i,\ l+j+m=i} \frac{i!}{l!j!m!} C_{a,l} C_{b,j} C_{c,m} x^{a-l} y^{b-j} z^{c-m}.
\end{aligned}$$

At those points at which $x = y = z$, or

$$(x, y, z) = (x, x, x),$$

our result simplifies yet more:

$$\frac{x^{a+b+c-i} C_{a+b+c,i}}{3^{i/2}}$$

(Section 6.6.3).

27.12.6 High-Order Partial Derivatives

How to design a mixed partial derivative? For instance, look at p_x, and differentiate it with respect to z:

$$p_{xz}(x, y, z) \equiv (p_x(x, y, z))_z.$$

If you like, work the other way around: differentiate p with respect to z, and then with respect to x:

$$p_{zx}(x, y, z) = \sum_{i=0}^{n-1} (i+1) \left(a_{i+1}\right)_x (x, y) z^i = p_{xz}(x, y, z).$$

In general, the (i, j, k)th partial derivative is obtained from a differential operator:

$$p_{x^i y^j z^k}(x, y, z) = \left(\left(\frac{\partial}{\partial x} \right)^i \left(\frac{\partial}{\partial y} \right)^j \left(\frac{\partial}{\partial z} \right)^k p \right) (x, y, z).$$

For instance, the $(2, 1, 0)$th partial derivative of p is

$$p_{x^2 y^1 z^0}(x, y, z) = p_{xxy}(x, y, z).$$

In particular, the $(0, 0, 0)$th partial derivative of p is p itself:

$$p_{x^0 y^0 z^0}(x, y, z) \equiv p(x, y, z).$$

27.12.7 Mathematical Induction: Triangle by Triangle

And how about high-order partial derivatives? In a cubic grid, (i, j, k) mirrors the (i, j, k)th partial derivative (Figure 27.3). Its order is the sum: $m = i + j + k$. For a fixed m, this draws an oblique triangle: partial derivatives of order m. To define them, use mathematical induction on $m = 1, 2, 3, \ldots$:

$$p_{x^i y^j z^k} \equiv \begin{cases} p & \text{if } i = j = k = 0 \\ \left(p_{x^{i-1} y^j z^k}\right)_x & \text{if } \quad i > 0 \\ \left(p_{x^i y^{j-1} z^k}\right)_y & \text{if } \quad j > 0 \\ \left(p_{x^i y^j z^{k-1}}\right)_z & \text{if } \quad k > 0. \end{cases}$$

As discussed in Section 27.10.3, there is no ambiguity here. This can be proved in the same inductive process.

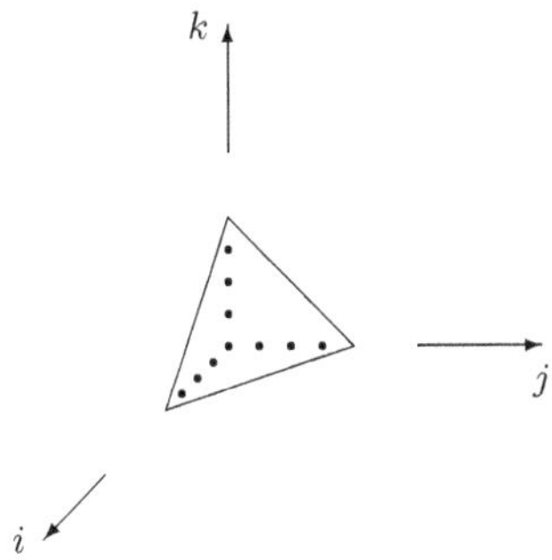

Fig. 27.3. Partial derivatives of order $m = i + j + k = 1, 2, 3, \ldots$. March, triangle by triangle.

27.12.8 Using a Discrete Tetrahedron

How many partial derivatives of order m or less are there? Well, they are already mirrored by a discrete tetrahedron of size m (Figures 27.3 and 6.6). In it, the total number of points is

$$\binom{m+3}{3} = \frac{(m+3)!}{3! \cdot m!} = \frac{(m+1)(m+2)(m+3)}{6}.$$

And how many partial derivatives of order m exactly are there? Thanks to Section 6.3.5, the answer is

$$\binom{m+3-1}{3-1} = \binom{m+2}{2} = \frac{(m+2)!}{2! \cdot m!} = \frac{(m+1)(m+2)}{2}.$$

27.12.9 Degree and Discrete Tetrahedron

Let's write our $a_i(x, y)$'s more explicitly:

$$a_i(x, y) = \sum_j \sum_k a_{i,j,k} x^k y^j$$

(where the $a_{i,j,k}$'s are some coefficients). This helps write p more explicitly too:

$$p(x, y, z) = \sum_{i=0}^{n} \sum_j \sum_k a_{i,j,k} x^k y^j z^i.$$

Thanks to this sum, we have the degree: the maximal $i + j + k$ for which $a_{i,j,k} \neq 0$. This is often denoted by m (which could be higher than n).

Look at a polynomial $p(x, y, z)$ of degree m. How many terms could it contain? Fortunately, they are mirrored by a discrete tetrahedron of size m (Figures 27.3 and 6.6). In it, the total number of points is

$$\binom{m+3}{3} = \frac{(m+3)!}{m! \cdot 3!} = \frac{(m+1)(m+2)(m+3)}{6}.$$

27.13 Exercises: Convolution

27.13.1 Convolution of Vectors

1. Let
$$u \equiv (u_0, u_1, u_2, \ldots, u_n) \equiv (u_i)_{0 \le i \le n}$$
be an $(n+1)$-dimensional vector.
2. Likewise, let
$$v \equiv (v_0, v_1, v_2, \ldots, v_m) \equiv (v_i)_{0 \le i \le m}$$
be an $(m+1)$-dimensional vector.
3. Add dummy zeroes, and complete both u and v into longer $(n+m+1)$-dimensional vectors:
$$u_{n+1} \equiv u_{n+2} \equiv \cdots \equiv u_{n+m} \equiv 0,$$
and
$$v_{m+1} \equiv v_{m+2} \equiv \cdots \equiv v_{n+m} \equiv 0.$$
4. The convolution of u and v (denoted by $u * v$) is a new $(n+m+1)$-dimensional vector, with the following components:
$$(u * v)_k \equiv \sum_{i=0}^{k} u_i v_{k-i}, \qquad 0 \le k \le n+m.$$
5. Show that it is commutative:
$$u * v = v * u.$$
6. Show that it is also associative.
7. Show that it is also distributive.
8. Use u to supply coefficients, to help define a new polynomial p (of degree n):
$$p(x) \equiv \sum_{i=0}^{n} u_i x^i = \sum_{i=0}^{n+m} u_i x^i.$$
9. Likewise, use v to supply coefficients, to help define a new polynomial q (of degree m):
$$q(x) \equiv \sum_{i=0}^{m} v_i x^i = \sum_{i=0}^{n+m} v_i x^i.$$
10. Look at the product: pq. It is a new function. But is it a legitimate polynomial in its own right?
11. Design its vector of coefficients. Hint: this is the convolution vector: $u*v$. Indeed,
$$(pq)(x) = \sum_{k=0}^{n+m} (u * v)_k x^k.$$
12. These are isomorphic algebras: polynomials (with their product) mirror vectors (with their convolution).
13. Thus, both algebras share the same properties.
14. In particular, both are commutative, associative, and distributive.
15. For polynomials, these properties are easy to prove.
16. Use this isomorphism to prove these properties for vector convolution as well.

27.13.2 Polar Decomposition

1. Recall the Taylor series:

$$\exp(x) = 1 + x + \frac{x^2}{2} + \frac{x^3}{3!} + \frac{x^4}{4!} + \cdots = \sum_{n=0}^{\infty} \frac{x^n}{n!}.$$

   (See exercises at the end of Chapter 26.)
2. Truncate this series after $k+1$ terms. This approximates $\exp(x)$ by a new polynomial of degree k:

$$\exp(x) \doteq \sum_{n=0}^{k} \frac{x^n}{n!}.$$

3. For a moderate $|x|$, this is a good approximation.
4. For a big $|x|$, increase k more and more, until the tail is small enough.
5. For a given x, how to calculate this polynomial? Hint: use Horner's algorithm in a new version that avoids division by $n!$. The solution can be found in Chapter 1 in [47].
6. Likewise, do the same to the sine function, and to its Taylor series:

$$\sin(x) = x - \frac{x^3}{3!} + \frac{x^5}{5!} - \frac{x^7}{7!} + \cdots = \sum_{n=0}^{\infty} (-1)^n \frac{x^{2n+1}}{(2n+1)!}.$$

7. Likewise, do the same to the cosine function, and to its Taylor series:

$$\cos(x) = 1 - \frac{x^2}{2} + \frac{x^4}{4!} - \frac{x^6}{6!} + \cdots = \sum_{n=0}^{\infty} (-1)^n \frac{x^{2n}}{(2n)!}.$$

8. Conclude that

$$\exp(i\theta) = \cos(\theta) + i\sin(\theta),$$

   where $i \equiv \sqrt{-1}$, and $0 \leq \theta < 2\pi$ is the angle with the real axis (Figures 26.4–26.5). Hint: add, term by term. (See exercises at the end of Chapter 26.)

28

Matrix and Its Eigenvalues

To model physics, we often use linear algebra: matrices and their eigenvalues. In special relativity, for example, we often use a 2×2 Lorentz matrix (Section 8.5.3). In quantum mechanics, on the other hand, we even use an infinite matrix: an operator. This helps model position, momentum, and even energy. This is where math meets physics at the core: eigenvalues take their physical meaning as energy levels.

It is time to introduce eigenvalues (and eigenvectors) from scratch, especially in Hermitian, orthogonal, unitary, and positive definite matrices. These are most useful in classical and quantum mechanics alike.

28.1 Complex Matrix and Its Hermitian Adjoint

28.1.1 The Hermitian Adjoint

A real matrix has real elements (entries). Here, on the other hand, we consider a more general case: an $m \times n$ (rectangular) complex matrix, with complex elements in $\mathbb{C}$. These use complex arithmetic (Section 3.9). Thanks to it, they are still linear: distributive and associative. Instead of the transpose, we now have a better notion: the Hermitian adjoint.

Let $i \equiv \sqrt{-1}$ be the imaginary number. Recall that the complex number

$$c \equiv a + bi$$

has the complex conjugate

$$\bar{c} \equiv a - bi$$

(Figure 3.18). This helps define the Hermitian adjoint of a given complex matrix. This could be done in two stages:

- First, take the transpose.
- Then, for each individual element, take the complex conjugate.

This way, the original $m \times n$ matrix

$$A \equiv (a_{i,j})_{1 \le i \le m,\ 1 \le j \le n}$$

makes a new $n \times m$ matrix, denoted by A^* (or $A^\dagger$):

$$A^*_{j,i} \equiv \bar{a}_{i,j}, \quad 1 \le i \le m,\ 1 \le j \le n.$$

In particular, if A is real, then

$$A^* = A^t.$$

Thus, the Hermitian adjoint actually extends the transpose to the wider class of complex matrices. This is how it also inherits a few nice properties:

$$(A^*)^* = A,$$

and

$$(BA)^* = A^*B^*,$$

where the number of columns in B is the same as the number of rows in A.

28.1.2 Hermitian (Self-Adjoint) Matrix

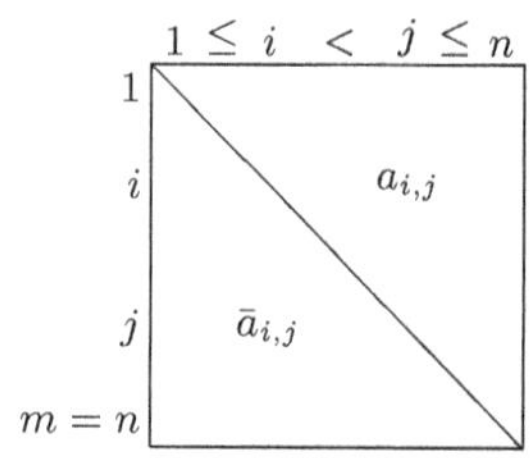

Fig. 28.1. In a Hermitian matrix, the lower triangular part mirrors the upper triangular part, with a bar on top.

Consider now a square matrix A, of order $m = n$. We say that A is Hermitian (or self-adjoint) if it is the same as its Hermitian adjoint:

$$A = A^*.$$

In this case, the main diagonal acts like a mirror (Figure 28.1). In fact, the lower triangular part mirrors the upper triangular part, with a bar on top:

$$a_{j,i} = (A^*)_{j,i} = \bar{a}_{i,j}, \quad 1 \le i, j \le n.$$

This is why the main diagonal must be real:

$$a_{i,i} = \bar{a}_{i,i}, \quad 1 \le i \le n.$$

28.2 Inner Product and Norm

28.2.1 Inner (Dot) Product

Let

$$u \equiv \begin{pmatrix} u_1 \\ u_2 \\ \vdots \\ u_n \end{pmatrix} \quad \text{and} \quad v \equiv \begin{pmatrix} v_1 \\ v_2 \\ \vdots \\ v_n \end{pmatrix}$$

be two column vectors in $\mathbb{C}^n$. They can also be viewed as narrow $n \times 1$ "matrices," with just one column. This way, they also have their own Hermitian adjoint:

$$u^* = (\bar{u}_1, \bar{u}_2, \ldots, \bar{u}_n) \quad \text{and} \quad v^* = (\bar{v}_1, \bar{v}_2, \ldots, \bar{v}_n).$$

These are $1 \times n$ "matrices," or n-dimensional row vectors.

Thus, u^* has n "columns," and v has n "rows." What is u^* times v? This is a new 1×1 "matrix," or just a new (complex) number:

$$(u, v) \equiv u^* v = \sum_{j=1}^{n} \bar{u}_j v_j.$$

This is the inner (or scalar, or dot) product of u and v.

28.2.2 Bilinearity

Our inner product is bilinear. In what sense? Well, let $c \in \mathbb{C}$ be some complex number. Then, from our inner product, pull c out (possibly with a bar on top):

$$(cu, v) = \bar{c}(u, v), \quad \text{and} \quad (u, cv) = c(u, v).$$

Furthermore, let w be yet another vector. Then, we also have a distributive law:

$$(u + w, v) = (u, v) + (w, v) \quad \text{and} \quad (u, v + w) = (u, v) + (u, w).$$

In this sense, our inner product is bilinear.

28.2.3 Skew-Symmetry

Moreover, our inner product is also skew-symmetric: interchanging u and v just picks a bar on top. Indeed,

$$(v, u) = \sum_{j=1}^{n} \bar{v}_j u_j$$

is just the complex conjugate of (u, v).

28.2.4 Norm

What is the inner product of v with itself? This is a real nonnegative number:

$$(v, v) = \sum_{j=1}^{n} \bar{v}_j v_j = \sum_{j=1}^{n} |v_j|^2 \geq 0.$$

Could this vanish? Only if v was the zero vector:

$$(v, v) = 0 \quad \Leftrightarrow \quad v = \mathbf{0}.$$

Thus, (v, v) has a square root. This will be the norm (or length, or magnitude) of v:

$$\|v\| \equiv +\sqrt{(v, v)}.$$

This way,

$$\|v\| \geq 0,$$

and

$$\|v\| = 0 \quad \Leftrightarrow \quad v = \mathbf{0},$$

as a magnitude should be.

28.2.5 Normalization

Every complex number $c \in \mathbb{C}$ can now be pulled out of our new norm (in magnitude):

$$\|cv\| = \sqrt{(cv, cv)} = \sqrt{\bar{c}c(v, v)} = \sqrt{|c|^2(v, v)} = |c|\sqrt{(v, v)} = |c| \cdot \|v\|.$$

In particular, if $v \neq \mathbf{0}$ (or $\|v\| > 0$), then pick $c = 1/\|v\|$, to produce a new (normalized) unit vector: $v/\|v\|$. This is the unique vector of norm 1 that is proportional to v.

28.2.6 Other Norms

This is the l_2-norm (also denoted by $\|v\|_2$). But there are also other norms: the l_1-norm

$$\|v\|_1 \equiv \sum_{i=1}^{n} |v_i|,$$

and the l_∞-norm (the maximum norm)

$$\|v\|_\infty \equiv \max_{1 \leq i \leq n} |v_i|.$$

They have the same properties: they vanish if and only if $v = \mathbf{0}$. Furthermore, every complex number $c \in \mathbb{C}$ can be pulled out (in magnitude):

$$\|cv\|_1 = |c| \cdot \|v\|_1, \quad \text{and} \quad \|cv\|_\infty = |c| \cdot \|v\|_\infty.$$

Here, we mostly use the l_2-norm (denoted simply by $\|v\| \equiv \|v\|_2$.

28.2.7 Inner Product and The Hermitian Adjoint

Assume again that A is an $m \times n$ (rectangular) complex matrix. Let u and v be complex vectors of different dimensions: u is m-dimensional, and v is n-dimensional. This way, Av is m-dimensional again. So we can now take the inner product of two m-dimensional vectors:

$$(u, Av) \equiv u^* Av.$$

This is a well-defined complex number (Section 28.2.1). Next, look at A^*: an $n \times m$ (rectangular) complex matrix. Thus, A^*u is n-dimensional. So, the inner product (A^*u, v) is a well-defined complex number as well. Is it the same? To find out, recall that both u and v can also be viewed as narrow matrices. Thanks to associativity,

$$(u, Av) = u^*(Av) = (u^*A)v = (A^*u)^*v = (A^*u, v).$$

Let's use this in a special case: a square matrix.

28.2.8 Inner Product and Hermitian Matrix

Let's look at an interesting case: $m = n$, so A is now a square matrix. Assume that A is also Hermitian:

$$A = A^*$$

(Section 28.1.2). In this case, the above formula reduces to

$$(u, Av) = (Au, v)$$

(for every two n-dimensional complex vectors u and v). Next, let's work the other way around. Assume that we didn't know that A was Herminian. Instead we only knew that A was a square complex matrix of order $m = n$, satisfying

$$(u, Av) = (Au, v)$$

for every two n-dimensional complex vectors u and v. Could we then deduce that A was Hermitian?

To find out, let's pick u and v cleverly. For each pair of natural numbers $1 \le i, j \le n$, pick u and v as standard unit vectors, with one nonzero component only:

$$u_k = \begin{cases} 1 & \text{if } k = i \\ 0 & \text{if } k \ne i \end{cases} \quad (1 \le k \le n)$$

$$v_k = \begin{cases} 1 & \text{if } k = j \\ 0 & \text{if } k \ne j \end{cases} \quad (1 \le k \le n).$$

With this choice,

$$a_{i,j} = (u, Av) = (Au, v) = \bar{a}_{j,i},$$

implying that A is indeed Hermitian (Figure 28.1).

28.3 Orthogonal and Unitary Matrix

28.3.1 Inner Product of Column Vectors

Assume again that A is an $m \times n$ rectangular complex matrix. For $1 \le j \le n$, let $v^{(j)}$ be its jth column: an m-dimensional column vector. Now, look at two columns: the ith and jth columns $(1 \le i, j \le n)$. What is their inner product? Well, look at the (i, j)th element in A^*A:

$$
\begin{aligned}
(A^*A)_{i,j} &= \sum_{k=1}^{m} (A^*)_{i,k}\, a_{k,j} \\
&= \sum_{k=1}^{m} \bar{a}_{k,i} a_{k,j} \\
&= \sum_{k=1}^{m} \bar{v}_k^{(i)} v_k^{(j)} \\
&= \left(v^{(i)}, v^{(j)}\right).
\end{aligned}
$$

This will be useful below.

28.3.2 Orthogonal and Orthonormal Column Vectors

Consider two complex vectors: u and v (of the same dimension). Assume that their inner product vanishes:

$$(u, v) = 0.$$

Then they are *orthogonal.* Furthermore, assume that they also have norm 1:

$$(u, v) = 0, \quad \text{and} \quad \|u\| = \|v\| = 1.$$

Then they are *orthonormal.*

Consider again our $m \times n$ rectangular complex matrix. This time, write it column by column:

$$A \equiv \left(v^{(1)} \mid v^{(2)} \mid \cdots \mid v^{(n)}\right)$$

$(m \geq n)$. Assume that its columns are orthonormal:

$$\left(v^{(i)}, v^{(j)}\right) = 0, \quad 1 \leq i, j \leq n, \ i \neq j,$$

and

$$\left\|v^{(i)}\right\| = 1, \quad 1 \leq i \leq n.$$

In this case, what is the (i, j)th element in A^*A? Well, it is either zero or one:

$$(A^*A)_{i,j} = \left(v^{(i)}, v^{(j)}\right) = \begin{cases} 1 & \text{if } i = j \\ 0 & \text{if } i \neq j \end{cases}$$

(Section 28.3.1). In other words, A^*A is just the identity matrix of order n:

$$A^*A = I.$$

Thus, A preserves the inner product of any two n-dimensional vectors u and v:

$$(Au, Av) = (A^*Au, v = (Iu, v)) = (u, v)$$

(Section 28.2.7). In particular, by picking $u = v$, A preserves norm as well:

$$\|Av\|^2 = (Av, Av) = (v, v) = \|v\|^2.$$

Next, let's work the other way around. Assume that we didn't know that A had orthogonal columns. Instead, we only knew that A^*A was the identity matrix:

$$A^*A = I.$$

Could we then deduce that A must also have orthonormal columns? Yes! Indeed, thanks to Section 28.3.1,

$$\left(v^{(i)}, v^{(j)}\right) = (A^*A)_{i,j} = I_{i,j} = \begin{cases} 1 & \text{if } i = j \\ 0 & \text{if } i \neq j, \end{cases}$$

as asserted. In summary, A has orthonormal columns if and only if

$$A^*A = I$$

is the identity matrix of order $n \leq m$.

28.3.3 Projection Matrix and Its Null Space

So far, we looked at A^*A. But what about AA^*? This is different: after all, there is no commutative law here.

Recall our assumptions: $m \geq n$, and A has orthonormal columns. Look at the above equation, and multiply by A (from the left):

$$A\left(A^*A\right) = AI = A.$$

Thanks to associativity,

$$\left(AA^*\right)A = A\left(A^*A\right) = AI = A.$$

Thus, AA^* has no effect on the columns of A: it leaves them as is. In this sense, it is just like the identity matrix.

Still, this is not the whole story. Since $m \geq n$, there are more m-dimensional column vectors, orthogonal to all columns of A. Let v be such a column vector:

$$A^*v = \mathbf{0}.$$

Multiply this by A (from the left):

$$A\left(A^*v\right) = A\mathbf{0} = \mathbf{0}.$$

Thanks to associativity,

$$\left(AA^*\right)v = A\left(A^*v\right) = A\mathbf{0} = \mathbf{0}.$$

Thus, v is in the null space of AA^*. In summary, AA^* splits: on the columns of A, it acts like the identity matrix. On orthogonal vectors, on the other hand, it acts like the zero matrix.

What is the algebraic key to all this? Associativity:

$$\left(AA^*\right)^2 = \left(AA^*\right)\left(AA^*\right) = A\left(A^*A\right)A^* = AIA^* = AA^*.$$

This means that AA^* is a projection matrix: same as its own square.

28.3.4 Unitary and Orthogonal Matrix

Assume now that A is a square matrix of order $m = n$. If A has orthonormal columns, then A is called unitary. In this case,

$$A^*A = AA^* = I$$

(the identity matrix of order n). This could also be written in terms of the inverse matrix:

$$A^* = A^{-1}, \quad \text{and} \quad A = (A^*)^{-1}.$$

But the inverse of the transpose is just the transpose of the inverse. Thus, you can drop these parentheses, and simply write

$$A = A^{-*}.$$

If A is also real, then it is also called an orthogonal matrix. In this case,

$$A^tA = AA^t = I,$$

so

$$A^t = A^{-1}, \quad \text{and} \quad A = (A^t)^{-1} = A^{-t}.$$

28.4 Eigenvalues and Eigenvectors

28.4.1 Eigenvectors and Their Eigenvalues

Let A be a square complex matrix of order n. An eigenvector of A is a nonzero vector $v \in \mathbb{C}^n$, satisfying

$$Av = \lambda v,$$

for some number $\lambda \in \mathbb{C}$ (the eigenvalue). In other words, applying A to v has the same effect as multiplying v by λ.

Note that v is not unique: for every nonzero number $c \in \mathbb{C}$, cv is an eigenvector in its own right:

$$A(cv) = cAv = c\lambda v = \lambda(cv).$$

Thus, v is not defined uniquely, but only up to a (nonzero) scalar multiple. How to pick c best? Well, since $v \neq \mathbf{0}$, $\|v\| > 0$ (Section 28.2.4). Thus, best pick $c \equiv 1/\|v\|$. This would produce a unit eigenvector:

$$A\left(\frac{v}{\|v\|}\right) = \lambda\left(\frac{v}{\|v\|}\right), \quad \left\|\frac{v}{\|v\|}\right\| = 1.$$

This is the unique unit eigenvector proportional to v.

28.4.2 Singular Matrix and Its Null Space

What are the algebraic properties of the eigenvector v? First, it is nonzero:

$$v \neq \mathbf{0}$$

(the zero vector). Furthermore, v also satisfies

$$(A - \lambda I)\, v = \lambda v - \lambda v = \mathbf{0},$$

where I is the identity matrix of order n. Thus, the matrix $A - \lambda I$ maps v to $\mathbf{0}$. In other words, v is in the null space of $A - \lambda I$.

For this reason, $A - \lambda I$ must be singular (noninvertible). Indeed, if there were an inverse matrix $(A - \lambda I)^{-1}$, then we could apply it to $\mathbf{0}$, and obtain v back again:

$$v = (A - \lambda I)^{-1}\mathbf{0} = \mathbf{0},$$

in violation of the very definition of v as a (nonzero) eigenvector.

28.4.3 The Hermitian Adjoint and Its Eigenvalues

And what about A^*? It has its own eigenvalue: $\bar{\lambda}$. Indeed,

$$(A - \lambda I)^* = A^* - \bar{\lambda} I$$

is singular too: it has the same determinant: 0. Still, how to prove this without using determinants? By contradiction: if it were nonsingular (invertible), then there would be some nonzero vector u, mapped to v:

$$\left(A^* - \bar{\lambda} I\right) u = v,$$

leading to a contradiction:

$$0 = (\mathbf{0}, u) = ((A - \lambda I)\, v, u) = \left(v, \left(A^* - \bar{\lambda} I\right) u\right) = (v, v) > 0.$$

So, $A^* - \bar{\lambda} I$ must be singular. As such, it must map some nonzero vector to $\mathbf{0}$:

$$\left(A^* - \bar{\lambda} I\right) w = \mathbf{0},$$

or

$$A^* w = \bar{\lambda} w.$$

Thus, $\bar{\lambda}$ is indeed an eigenvalue of A^*, as asserted.

28.4.4 Hermitian Matrix and Its Eigenvalues

Assume now that A is Hermitian:

$$A = A^*.$$

Thanks to Section 28.2.8,

$$\lambda(v, v) = (v, \lambda v) = (v, Av) = (Av, v) = (\lambda v, v) = \bar{\lambda}(v, v).$$

Now, since v is a nonzero vector, $(v, v) > 0$ (Section 28.2.4). Thus, we can divide by (v, v):

$$\lambda = \bar{\lambda},$$

so λ must be real:

$$\lambda \in \mathbb{R}.$$

28.4.5 Hermitian Matrix and Its Eigenvectors

And what about eigenvectors? Let u and v be two eigenvectors of the Hermitian matrix A:

$$Av = \lambda v \quad \text{and} \quad Au = \mu u,$$

where $\lambda \neq \mu$ are two distinct eigenvalues. Thanks to Sections 28.2.8 and 28.4.4,

$$\mu(u, v) = \bar{\mu}(u, v) = (\mu u, v) = (Au, v) = (u, Av) = (u, \lambda v) = \lambda(u, v).$$

Now, since $\lambda \neq \mu$, we must have

$$(u, v) = 0,$$

leading to orthogonality, and even orthonormality: $u/\|u\|$ and $v/\|v\|$.

28.5 Positive (Semi)Definite Matrix

28.5.1 Positive Semidefinite Matrix

Consider a Hermitian matrix A. Assume that, for every complex vector v,

$$(v, Av) \equiv v^* Av \geq 0.$$

We then say that A is positive semidefinite. In particular, pick v to be an eigenvector of A, with eigenvalue λ. This way,

$$\|v\|^2 \lambda = (v, v)\lambda = (v, \lambda v) = (v, Av) \geq 0.$$

Divide by $\|v\|^2 > 0$:

$$\lambda \geq 0.$$

Thus, all eigenvalues of A are positive or zero.

And how about the other way around? If a Hermitian matrix has eigenvalues that are positive (or zero) only, must it be positive semidefinite? Yes! Indeed, given a complex vector, just expand it as a linear combination of (orthogonal) eigenvectors.

28.5.2 Positive Definite Matrix

Consider again a Hermitian matrix A. Assume that, for every nonzero vector $v \neq \mathbf{0}$,

$$(v, Av) \equiv v^* Av > 0.$$

We then say that A is not only positive semidefinite but also positive definite. In particular, pick v to be an eigenvector of A, with eigenvalue λ. This way,

$$\|v\|^2 \lambda = (v, v)\lambda = (v, \lambda v) = (v, Av) > 0.$$

Divide by $\|v\|^2 > 0$:

$$\lambda > 0.$$

thus, A has positive eigenvalues only. In particular, A is nonsingular: 0 can't be an eigenvalue any more.

And what about the other way around? If a Hermitian matrix has positive eigenvalues only (so it is positive semidefinite and nonsingular at the same time), must it be positive definite as well? Yes! Indeed, given a nonzero vector, just expand it as a linear combination of (orthogonal) eigenvectors.

28.6 Exercises: Generalized Eigenvalues

28.6.1 The Cauchy–Schwarz Inequality

1. Let u and v be n-dimensional (complex) vectors:

$$u \equiv (u_1, u_2, \ldots, u_n)^t \text{ and } v \equiv (v_1, v_2, \ldots, v_n)^t.$$

Prove the Cauchy–Schwarz inequality:

$$|(u, v)| \le \|u\| \cdot \|u\|,$$

or

$$\left|\sum_{i=1}^{n} \bar{u}_i v_i\right|^2 = |(u, v)|^2 \le (u, u)(v, v) - \left(\sum_{i=1}^{n} |u_i|^2\right)\left(\sum_{i=1}^{n} |v_i|^2\right).$$

Hint: pick two different indices between 1 and n: $1 \le i \ne j \le n$. Note that

$$\begin{aligned}
0 &\le |u_i v_j - u_j v_i|^2 \\
&= (\bar{u}_i \bar{v}_j - \bar{u}_j \bar{v}_i)(u_i v_j - u_j v_i) \\
&= \bar{u}_i \bar{v}_j u_i v_j - \bar{u}_i \bar{v}_j u_j v_i - \bar{u}_j \bar{v}_i u_i v_j + \bar{u}_j \bar{v}_i u_j v_i \\
&= |u_i|^2 |v_j|^2 - (\bar{u}_i v_i)(u_j \bar{v}_j)) - (\bar{u}_j v_j)(u_i \bar{v}_i) + |u_j|^2 |v_i|^2,
\end{aligned}$$

so

$$(\bar{u}_i v_i)(u_j \bar{v}_j)) + (\bar{u}_j v_j)(u_i \bar{v}_i) \le |u_i|^2 |v_j|^2 + |u_j|^2 |v_i|^2.$$

Do the same with the plus sign, to obtain

$$|(\bar{u}_i v_i)(u_j \bar{v}_j)) + (\bar{u}_j v_j)(u_i \bar{v}_i)| \le |u_i|^2 |v_j|^2 + |u_j|^2 |v_i|^2.$$

Finally, sum this over $1 \le i < j \le n$.

2. Could the Cauchy–Schwarz inequality be an exact equality as well? Hint: only if u is a scalar multiple of v.

28.6.2 The Triangle Inequality

1. Conclude the triangle inequality:

$$\|u + v\| \le \|u\| + \|v\|.$$

Hint:

$$\begin{aligned}
\|u + v\|^2 &= (u + v, u + v) \\
&= (u, u) + (u, v) + (v, u) + (v, v) \\
&= \|u\|^2 + (u, v) + (v, u) + \|v\|^2 \\
&\le \|u\|^2 + 2|(u, v)| + \|v\|^2 \\
&\le \|u\|^2 + 2\|u\| \cdot \|v\| + \|v\|^2 \\
&= (\|u\| + \|v\|)^2.
\end{aligned}$$

28.6.3 Generalized Eigenvalues

1. Let A be a square complex matrix of order n. Let B be a Hermitian matrix of order n. Let v be an n-dimensional vector satisfying

$$Av = \lambda Bv \quad \text{and} \quad (v, Bv) \neq 0.$$

   Then the (complex) scalar λ is called a generalized eigenvalue of A.
2. Could v be the zero vector? Hint: no! After all, v must satisfy $(v, Bv) \neq 0$.
3. Conclude that $A - \lambda B$ maps v to the zero vector.
4. Conclude that v is in the null space of $A - \lambda B$.
5. Is $A - \lambda B$ singular? Hint: otherwise,

$$v = (A - \lambda B)^{-1} \mathbf{0} = \mathbf{0},$$

   in violation of $(v, Bv) \neq 0$.
6. Is $A^* - \bar{\lambda} B$ singular as well? Hint: otherwise, there would be a nonzero vector u mapped to v:

$$\left(A^* - \bar{\lambda} B\right) u = v,$$

   leading to a contradiction:

$$0 = (\mathbf{0}, u) = ((A - \lambda B)\, v, u) = \left(v, \left(A^* - \bar{\lambda} B\right) u\right) = (v, v) > 0.$$

7. Prove this in yet another way. Hint:

$$\det\left(A^* - \bar{\lambda} B\right) = \det\left(A - \lambda B\right) = 0.$$

8. Conclude that $A^* - \bar{\lambda} B$ has a nontrivial null space.
9. Conclude that $\bar{\lambda}$ is a generalized eigenvalue of A^*.
10. Assume now that A is Hermitian as well. Must the generalized eigenvalue λ be real? Hint:

$$\lambda(v, Bv) = (v, \lambda Bv) = (v, Av) = (Av, v) = (\lambda Bv, v) = \bar{\lambda}(Bv, v) = \bar{\lambda}(v, Bv).$$

    Finally, divide this by $(v, Bv) \neq 0$.
11. Let u and v be two such vectors:

$$Av = \lambda Bv \quad \text{and} \quad Au = \mu Bu,$$

    where $\lambda \neq \mu$ are two distinct generalized eigenvalues. Show that

$$(u, Bv) = 0.$$

    Hint:

$$\begin{aligned}
\mu(u, Bv) &= \mu(Bu, v) \\
&= \bar{\mu}(Bu, v) \\
&= (\mu Bu, v) \\
&= (Au, v) \\
&= (u, Av) \\
&= (u, \lambda Bv) \\
&= \lambda(u, Bv).
\end{aligned}$$

12. Assume now that B is not only Hermitian but also positive definite. Define the new "inner product:"
$$(\cdot,\cdot)_B \equiv (\cdot, B\cdot)\,.$$
Show that this is indeed a legitimate inner product. Hint: show that it is indeed bilinear, skew-symmetric, and positive, and vanishes only for the zero vector.
13. Look at the new matrix $B^{-1}A$. Show that it is "Hermitian" with respect to the new "inner product:"
$$\left(B^{-1}A\cdot,\cdot\right)_B = (A\cdot,\cdot) = (\cdot, A\cdot) = \left(\cdot, B^{-1}A\cdot\right)_B\,.$$
14. Show that the above u and v are eigenvectors of $B^{-1}A$.
15. Show that they are indeed orthogonal with respect to the new "inner product."
16. Normalize them with respect to the new "inner product."
17. For example, set $B \equiv I$. Is this familiar?

References

1. Adachi, H., Mukoyama, T. and Kawai, J.: *Hartree-Fock-Slater Method for Materials Science: the DV-X Alpha Method for Design and Characterization of Materials.* Springer, 2006.
2. Alcubierre, M.: Introduction to 3+1 Numerical Relativity. Oxford Univ. Press (2012).
3. Anthony, M., and Harvey, M.: *Linear Algebra: Concepts and Methods.* Cambridge University Press, Cambridge, 2012.
4. Baumgarte, T.W., and Shapiro, S.L.: Numerical Relativity: Solving Einstein Equations on the Computer. Cambridge Univ. Press (2010).
5. Bransden, B.H., and Joachain, C.J.: *Quantum Mechanics.* Prentice-Hall, 2000.
6. Calkin, M.G.: *Lagrangian and Hamiltonian Mechanics.* WSPC, 1996.
7. Carlson, B.: *Lecture Notes on Quantum Mechanics,* 2019.
8. Carroll, S.: *Something Deeply Hidden: Quantum Worlds and the Emergence of Spacetime.* Oneworld Publications (2019).
9. Chirvasa, M.: Finite Difference Methods in Numerical Relativity. VDM Verlag (2010).
10. Courant, R. and John, F.: *Introduction to Calculus and Analysis* (vol. 1–2). Springer, N.Y., 1998–1999.
11. Einstein, A.: *Relativity: The Special and the General Theory.* Martino Fine Books, 2010.
12. Fischer, C.F.: *The Hartree-Fock Method for Atoms: a Numerical Approach.* Wiley, 1977.
13. Fowles, G.R.: *Introduction to Modern Optics.* Dover Publications, N.Y., 1989.
14. French, A.P.: *Newtonian Mechanics.* Norton, 1971.
15. Gibson, C.C.: *Elementary Euclidean Geometry: An Introduction.* Cambridge University Press, 2003.
16. Gilmore, R.: *Lie Groups, Lie Algebras, and Some of Their Applications.* Dover, UK, 2005.
17. Glendinning, P.: *Stability, Instability and Chaos: An Introduction to the Theory of Nonlinear Differential Equations.* Cambridge University Press (1994).
18. Goeke, K., and Reinhard, P.G.: *Time-Dependent Hartree-Fock and Beyond: Lecture Notes in Physics: 171.* Springer-Verlag, 1983.
19. Griffiths, D.J., and Schroeter, D.F.: *Introduction to Quantum Mechanics (Third Edition).* Cambridge University Press, 2018.
20. Hall, B.: *Lie Groups, Lie Algebras, and Representations: an Elementary Introduction (Second Edition).* Springer, NY, 2015.
21. Hecht, E.: *Optics (Fourth Edition).* Addison-Wesley, Boston, MA, 2001.
22. Henrici, P.: *Applied and Computational Complex Analysis,* Vol. 1–2. Wiley-Interscience, N.Y., 1977.

23. Henrici, P. and Kenan, W.R.: *Applied and Computational Complex Analysis: Power Series-Integration-Conformal Mapping-Location of Zeros.* Wiley-IEEE, Piscataway, NJ, 1988.
24. Hrabovsky, C.: em Classical Mechanics: The Theoretical Minimum. 2014.
25. Humphreys, J.E.: *Introduction to Lie Algebras and Representation Theory.* Springer, N.Y., 1972.
26. d'Inverno, R.: Approaches to Numerical Relativity. Cambridge Univ. Press (2008).
27. Jones, H.F.: *Groups, Representations and Physics (Second Edition).* CRC Press, N.Y., 1998.
28. Kac, V.G., Raina, A.K.:, and Rozhkovskaya, N.: *Bombay Lectures on Highest Weight Representations of Infinite Dimensional Lie Algebras (Second Edition).* World Scientific, Singapore, (2013).
29. Kadanoff, L.P.: *Quantum Statistical Mechanics.* Taylor and Francis, 2019.
30. Kannan, A.: *Lecture Notes on Lie Algebras,* 2019.
31. Karatzas, I. and Shreve, S.E.: *Brownian Motion and Stochastic Calculus (Second Edition).* Springer, New York, 1991.
32. Kibble, T.: *Classical Mechanics (fifth Edition).* Imperial College Press, London, 2004.
33. Knudsen, J.M., and Hjorth, P.G.: *Elements of Newtonian Mechanics: Including Nonlinear Dynamics.* Springer, NY, 2013.
34. Kuratowski, K., and Musielak, J.: *Introduction to Calculus.* Pergamon Press, 1961.
35. Maloney, A.: Lecture Notes on Classical Mechanics, 2010.
36. Neumaier, A., and Westra, D.: *Quantum Mechanics via Lie Algebras.* De Gruyter, 2024.
37. Robinett, R.: *Quantum Mechanics: Classical Results, Modern Systems, and Visualized Examples (Second Edition).* Oxford Univ. Press, 2006.
38. Romano, A., and Furnari, M.M.: *The Physical and Mathematical Foundations of the Theory of Relativity: A Critical Analysis.* Birkhouser, NY, 2019.
39. Rosenlicht, M.: *Introduction to Analysis.* Dover Publications, N.Y., 1985.
40. Schieve, W.C., and Horwitz, L.P.: *Quantum Statistical Mechanics.* Cambridge Univ. Press, 2009.
41. Schuller, M.F.: *Lectures on the Geometrical Anatomy of Theoretical Physics and Quantum Theory.* Lecture notes, independently published, 2019.
42. Schwerdtfeger, H.: *Geometry of Complex Numbers: Circle Geometry, Moebius Transformation, Non-Euclidean Geometry.* Dover Publications, N.Y., 1980.
43. Schwarz, A.: *Quantum Mechanics and Quantum Field Theory from Algebraic and Geometric viewpoints.* Springer, N.Y., 2024.
44. Shankar, R.: *Fundamentals of Physics I: Mechanics, Relativity, and Thermodynamics, Expanded Edition (open Yale courses).* Yale Univ. Press, 2019.
45. Shankar, R.: *Fundamentals of Physics II: Electromagnetism, Optics, and Quantum Mechanics, Expanded Edition (the open Yale courses).* Yale Univ. Press, 2020.
46. Shapira, Y.: *Mathematical Objects in C++: Computational Tools in a Unified Object-Oriented Approach.* CRC press, Boca Raton, FL, 2009.
47. Shapira, Y.: *Solving PDEs in C++: Numerical Methods in a Unified Object–Oriented Approach (Second Edition).* SIAM, Philadelphia, PA, 2012.
48. Shapira, Y.: *Set Theory and its Applications in Physics and Computing.* World Scientific, Singapore (2022).
49. Shapira, Y.: *Linear Algebra and Group Theory for Physicists and Engineers (second edition).* Birkhouser, Springer Nature, N.Y. (2023).
50. Shapira, Y.: *Quantum Mechanics and Algorithms: An Algemraic-Geometric Perspective.* World Scientific, Singapore (2026).
51. Shibata, M.: Numerical Relativity. World Scientific Publishing (2016).
52. Steinbauer, R.: *The Penrose and Hawking Singularity Theorems Revisited.* Math Faculty, Vienna Univ., 2006.

53. Strang, G.: *Introduction to Applied Mathematics.* Wellesley-Cambridge Press, 1986.
54. Strang, G.: *Introduction to Linear Algebra (Third Edition).* SIAM, Philadelphia, 2003
55. Strocchi, F.: *An Introduction to the Mathematical Structure of Quantum Mechanics: a Short Course for Mathematicians (Second Edition).* World Scientific, Singapore, (2008).
56. Strogatz, S.H.: *Nonlinear Dynamics and Chaos: With Applications to Physics, Biology, Chemistry, and Engineering (Second Edition).* Westview Press (2014).
57. Tauvel, P., and Yu, R.W.T.: *Lie Algebras and Algebraic Groups.* Springer, NY, 2005.
58. Taylor, J.R.: *Classical Mechanics.* University Science Books, Herndon, VA, 2005.
59. Torres del Castillo, G.F.: *An Introduction to Hamiltonian Mechanics.* Birkhouser, NY, 2018.
60. Vaisman, I.: *Analytical Geometry.* World Scientific, 1997.
61. Varchenko, A.: *Multidimensional Hypergeometric Functions and Representation Theory of Lie Algebras and Quantum Groups.* World Scientific, Singapore, (1995).
62. Vazirani, U.V.: *Lecture Notes on Quantum Mechanics and Quantum Computation.* UC, Berkeley.
63. Woodhouse, N.M.J.: *Special Relativity., (Second Edition).* Springer, London, 2007.

Index

www.ingramcontent.com/pod-product-compliance
Lightning Source LLC
LaVergne TN
LVHW051054210926
848486LV00002B/107

9789819831388